64097800

The International Handbook on Environmental Technology Management

Edited by

Dora Marinova

Associate Professor, Institute for Sustainability and Technology Policy, Murdoch University, Australia

David Annandale

Senior Lecturer, School of Environmental Science, Murdoch University, Australia

John Phillimore

Honorary Research Associate, Institute for Sustainability and Technology Policy, Murdoch University and Director – Intergovernmental Relations Policy Division, Department of the Premier and Cabinet, Government of Western Australia, Australia

Edward Elgar
Cheltenham, UK • Northampton, MA, USA

Published by
Edward Elgar Publishing Limited
Glensanda House
Montpellier Parade
Cheltenham
Glos GL50 1UA
UK

Edward Elgar Publishing, Inc.
William Pratt House
9 Dewey Court
Northampton
Massachusetts 01060
USA

A catalogue record for this book
is available from the British Library

Library of Congress Cataloguing in Publication Data
The international handbook on environmental technology management / edited by Dora Marinova, David Annandale, and John Phillimore.
    p. cm.—(Elgar original reference)
    Includes bibliographical references and index.
    1. Industrial management—Environmental aspects. 2. Environmental protection—Technological innovations—Management. 3. Environmental management. 4. Technology—Management. 5. Sustainable development. I. Title: Handbook on environmental technology management. II. Title: Environmental technology management. III. Marinova, Dora. IV. Annandale, David. V. Phillimore, John. VI. Series.
    HD30.255.I58 2006
    338.9′27—dc22
                                        2006005898

ISBN 978 1 84064 687 0 (cased)

Printed and bound in Great Britain by MPG Books Ltd, Bodmin, Cornwall

Contents

Contributors

Martin Anda, School of Environmental Science, Murdoch University, Australia

David Annandale, School of Environmental Science, Murdoch University, Australia

Diana Arbelaez-Ruiz, Monash University, Australia

Anthony Arundel, Maastricht Economic Research Institute on Innovation and Technology, University of Maastricht, the Netherlands

Suzanne Benn, University of Technology Sydney, Australia

Carmen Bizzarri, Università 'La Sapienza', Rome, Italy

Shanit Borsky, Monash University, Australia

Michael Braungart, McDonough Braungart Design Chemistry, Charlottesville, Virginia, USA

Roger Burritt, Faculty of Economics and Commerce, Australian National University, Australia

Felix Chan, School of Economics and Commerce, University of Western Australia, Australia

Chris Cocklin, Monash University, Australia

John Connolly, The Michael Smurfit Graduate School of Business, Department of Marketing, University College Dublin, Ireland

Nicole Cook, Australian Expert Group in Industry Studies, University of Western Sydney, Australia

Dexter Dunphy, University of Technology Sydney, Australia

David Evans, University of Melbourne, Australia

Tim Foxon, Centre for Energy Policy and Technology, Department of Environmental Science and Technology, Imperial College London, UK

Gerry Gillespie, Zero Waste, Australia

Paul J. Gollan, Department of Industrial Relations, London School of Economics, UK

Andrew Griffiths, University of Technology Sydney, Australia

Steve Halls, International Environmental Technology Centre, United Nations Environment Programme, Osaka, Japan

Stephan Harding, Schumacher College, UK

Karlson 'Charlie' Hargroves, The Natural Edge Project, Australia

Joe Herbertson, The Crucible Group Pty Ltd, Australia

Andrew Higham, Policy Division, Department of the Premier and Cabinet, Western Australia

Malcolm Hill, Russian and East European Industrial Studies, Loughborough University Business School, UK

Ruth Hillary, Corporate Environmental Strategy, International Journal of Corporate Sustainability, London, UK

Goen Ho, School of Environmental Science, Murdoch University, Australia

Doug Holmes, Monash University, Australia

Robert Hughes, School of Environmental Science, Murdoch University, Australia

Sharon Jones, Department of Civil and Environmental Engineering, Lafayette College, USA

René Kemp, Maastricht Economic Research Institute on Innovation and Technology, Maastricht University, the Netherlands

L. Hunter Lovins, Natural Capitalism Inc., USA

Ian Lowe, School of Science, Griffith University, Australia

Dorit Maor, School of Education, Murdoch University, Australia

Jane Marceau, University of Technology Sydney, Australia

Dora Marinova, Institute for Sustainability and Technology Policy, Murdoch University, Australia

Kuruvilla Mathew, School of Environmental Science, Murdoch University, Australia

Michael McAleer, School of Economics and Commerce, University of Western Australia, Australia

Pierre McDonagh, Faculty of Business, Dublin Institute of Technology, Ireland

William McDonough, William McDonough & Partners, Architecture and Community Design, and McDonough Braungart Design Chemistry, Charlottesville, Virginia, USA

Wilhelm Nell, Centre for Agriculture Management, University of Free State, South Africa

Alexis Nelson, The Natural Edge Project, Australia; currently pursuing an undergraduate degree in Environmental Studies from the University of Colorado

Peter Newman, Institute for Sustainability and Technology Policy, Murdoch University; Western Australian Sustainability Roundtable; Sustainability Commissioner New South Wales, Australia

Chris Page, Integrative Design Team, The Rocky Mountain Institute, USA

Saeed Parto, Maastricht Economic Research Institute on Innovation and Technology, Maastricht University, the Netherlands

John Phillimore, Institute for Sustainability and Technology Policy, Murdoch University, Australia

Caroline Plunkett, Supreme Court Judge Associate ACT, Australia; former researcher at the Australian National University

Michael Polonsky, School of Hospitality, Tourism and Marketing, Victoria University, Australia

Andrea Prothero, The Michael Smurfit Graduate School of Business, Department of Marketing, University College Dublin, Ireland

Giulio Querini, Università 'La Sapienza', Rome, Italy

Robert Repetto, Professor in Economics of Sustainable Development, Yale School of Forestry and Environmental Studies, USA

Harald Rohracher, Inter-University Research Centre for Technology, Work and Culture, Austria

Stuart Ross, University of Melbourne, Australia

Stefan Schaltegger, Centre for Sustainable Management, University of Lüneburg, Germany

Steven Schilizzi, School of Agricultural and Resource Economics, University of Western Australia, Australia

Michael H. Smith, The Natural Edge Project, Australia

Hardin Tibbs, Synthesys Strategic Consulting Ltd, London, UK

Christopher Tipler, The Natural Step, Australia

Tim Turpin, Australian Expert Group in Industry Studies, University of Western Sydney, Australia

René van Berkel, Centre for Cleaner Production, Curtin University of Technology, Australia

Piers Verstegen, Policy Division, Department of the Premier and Cabinet, Western Australia

Stanley Yokwe, Institute for Sustainability and Technology Policy, Murdoch University, Australia

Acronyms and abbreviations

& – and
£ – British pound
€ – euro
$ – dollar
3BL – triple bottom line
ABCD process – backcasting methodology
ABEC – Australian Building Energy Council
ABI – Association of British Insurers
ACF – Australian Conservation Foundation
ACIRRT – Australian Centre for Industrial Relations Research and Training
ACT – Australian Capital Territory
AD – Anno Domini (used before dates after the supposed year Christ was born)
ADB – Asian Development Bank
ADF – augmented Dicky-Fuller test
AEGIS – Australian Expert Group in Industry Studies
AGO – Australian Greenhouse Office
AMM – abandoned mine methane
ANC – African National Congress
ARC – Australian Research Council
ARCH – autoregressive conditional heteroskedasticity
B&C – building and construction
BASIX – Building Sustainability Index
BAT – best available technology
BC – before Christ
BC – British Columbia, Canada
3BL – triple bottom line
bcm – billion cubic metres
BEES – Building for Environmental and Economic Sustainability (software)
BEP – best environmental practice
bn – billion
BP – British Petroleum
BS – British standard
BSc – Bachelor of Science
BSR – Business for Social Responsibility
BTX – benzene/toluene/xylene
C – carbon
CAFE – Corporate Average Fuel Economy
CAS – Chemical Abstracts Service
CBA – cost–benefit analysis
CCA – chromated copper arsenate
CCGT – combined cycle gas turbine

CCX – Chicago Climate Exchange
Cd – cadmium
CD – construction and demolition
CECP – Centre of Excellence in Cleaner Production
CEO – chief executive officer
CFB – circulating fluidized beds
CFC – chlorofluorocarbon
CH_4 – methane
CHP – combined heat and power
CICA – Canadian Institute of Chartered Accountants
Cl – chlorine
CMA – Chemical Manufacturers Association
CMI – Cockerill Mechanical Industries
CMM – coal mine methane
CO_2 – carbon dioxide
CSD – Commission on Sustainable Development
CSIRO – Commonwealth Scientific and Industrial Research Organisation
CSR – corporate social responsibility
CTA – constructive technology assessment
CUB – Carlton & United Breweries
DALY – disability-adjusted life years
DBSA – Development Bank of Southern Africa
DEFRA – Department for Environment, Food and Rural Affairs, UK
DESA – Department of Economic and Social Affairs
DfE – Design for Environment
DMC – domestic material consumption
DMI – direct material input
DNR – Department of Natural Resources
DPO – domestic processed output
DRAFT – Direction Régionale de l'Agriculture et de la Forêt, France
DRIRE – Directions Régionales de l'Industrie, de la Recherche et de l'Environnement, France
DTI – Department of Trade and Industry, UK
DTIE – Division of Technology, Industry and Economics
DTO – Sustainable Technology Development Programme
EC – European Commission
ECCE – European Council of Civil Engineers
ECE – Economic Commission for Europe
EDF – Electricité de France
EDR – Earth Day Resources
EEA – external environmental accounting
EEVA – environmental economic value added
EGARCH – exponential generalized autoregressive conditional heteroskedasticity
EIA – environmental impact assessment
EMA – environmental management accounting
EMAS – Eco-Management and Audit Scheme

EMS – Environmental Management System
EPA – Environmental Protection Agency
EPR – Extended Producer Responsibility
ERFA – environmentally related financial accounting
ESD – ecologically sustainable development
ESM – environmental systems modelling
EST – environmentally sound technology
ETIS – Environmental Technology Innovation Scheme
ETM – environmental technology management
EU – European Union
EUETS – European Union's Emissions Trading Scheme
EVA – economic value added
FASB – Financial Accounting Standards Board
FICA – Federal Insurance Contributions Act
GAAP – generally accepted accounting principles
GARCH – generalized autoregressive conditional heteroskedasticity
GCC – gasification combined cycle
GDP – Gross Domestic Product
GE – General Electric
GEO – Global Environment Outlook
GERD – general expenditure on research and development
GHG – greenhouse gas
GIS – geographical information system
GJR – Glosten, Jagannathan and Runkle
GNP – Gross National Product
GPI – Genuine Progress Indicator
GRI – Global Reporting Initiative
GSP – Gross State Product
GST – Gross State Product
GTL – gas to liquid
GWP – global warming potential
ha – hectare
HIA – Housing Industry Association
HR – human resources
HRM – human resource management
HRSG – heat recovery steam generator
IAS – International Accounting Standards
IASB – International Accounting Standards Board
ICC – International Chamber of Commerce
IEA – International Energy Agency
IGCC – integrated gasification combined cycle
IGO – intergovernmental organization
IISD – International Institute for Sustainable Development
IMP – intelligent materials pooling
IPCC – Inter-governmental Panel on Climate Change
IPPC – Integrated Pollution Prevention and Control

IR – industrial relations
ISO – International Organization for Standardization
IT – information technology
IWM – integrated waste management
kg – kilogram
km – kilometre
kWh – kilowatt-hour
LBIO – literature-based innovation output
LCA – lifecycle assessment
LCD – lifecycle development
LCI – lifecycle inventory
LCIA – lifecycle impact assessment
LEED – Leadership in Energy and Environmental Design (rating system)
LIBS – laser-induced breakdown spectroscopy
LIFE – the Financial Instrument for the Environment programme
LM – Lagrange multiplier
LOHAS – lifestyles of health and sustainability
LPG – liquid petroleum gas
m^2 – square metre
MAC – marginal abatement cost
MAS – Mongolian Academy of Sciences
MBDC – McDonough Braungart Design Chemistry
MD – marginal damage
MEA – monoethanolomine
MEEA – monetary external environmental accounting and reporting
MEMA – monetary environmental management accounting
MFP – multifactor productivity
MJ – megajoule
MRF – material reclamation facility
Mt – million tonnes
MW – megawatt
MWhr – megawatt-hour
N/A – not applicable
NatHERS – Housing Industry Association's Partnership Advancing the Housing
    Environment
NBEET – National Board of Employment, Education and Training, Australia
NCAT – National Center for Appropriate Technology, Australia
NDA – National Department of Agriculture, South Africa
NEPA – National Environmental Policy Act, USA
NESEA – Northeast Sustainable Energy Association
NETTLAP – Network for Environmental Training at the Tertiary Level in Asia and the
            Pacific
NGO – non-governmental organization
NHS – National Health Service
NHTZ – New High Technology Zone
Ni – nickel

NIP – National Insulation Programme
NOx – nitrogen oxides
NPD – New Product Design
NRDC – National Research and Development Corporation, India
NSESD – National Strategy for Ecologically Sustainable Development
NSW – New South Wales, Australia
OECD – Organisation for Economic Co-operation and Development
OH&S – occupational health and safety
OPEC – Organization of the Petroleum Exporting Countries
OTA – Office of Technology Assessment, USA
P – phosphorous
PACE – pollution abatement costs and expenditures
PATHE – Partnership Advancing the Housing Environment
Pb – lead
PBEP – Performance Bretagne Environnement Plus
PCB – polychlorinated biphenyl
PDCA model – plan-do-check-act model
PEEA – physical external environmental accounting and reporting
PEMA – physical environmental management accounting
PET – polyethylene terephthalate
PFBC – pressurized fluidized bed combustion
PLA – polylactic acid
PP – Phillips-Perron
ppm – parts per million
PR – public relations
PSI – Product Stewardship Institute
PSRC – Photovoltaics Special Research Centre
PTO – Patent and Trademark Office
PV – photovoltaics
PVC – photovoltaic cells
PVC – polyvinyl chloride
QMLE – quasi-maximum likelihood estimator
R&D – research and development
REACH – registration, evaluation and authorization of chemicals
REPA – resource and environmental profile analysis
RFCD – research fields, courses and disciplines
RMI – Rocky Mountain Institute
RoHS – restrictions on hazardous substances
ROS – Raster Output Scanner
RSA – Royal Society for the Arts
RSMS – Restricted Substance Management Standard
S&T – science and technology
SABRC – State Agency Buy Recycled Campaign, California
SEAAR – social and ethical accounting, auditing and reporting
SEC – US Securities and Exchange Commission
SEDA – Sustainable Energy Development Authority

SEO – socio-economic objective
SME – small and medium-sized enterprise
SNM – strategic niche management
SO_2 – sulphur dioxide
SoEAC – State of the Environment Advisory Council
SOx – sulphur oxides
SRI – socially responsible investment
SSCD – Stationary Source Compliance Division
ST – STMicroelectronics
SUV – sport utility vehicle
t – tonne or ton
TBL – triple bottom line
TDO – total domestic output
TIFAC – Technology Information, Forecasting and Assessment Council, India
TMC – total material consumption
TMR – total material requirement
TNEP – The Natural Edge Project
TORs – terms of reference
TQM – total quality management
TsKTI – Central Boiler and Turbine Institute, Russia
TUD – Technical University of Delft
TWh – Tera Watts/hours. 1 TWh = 10^{12} Wh
UK – United Kingdom
UN – United Nations
UNCED – United Nations Conference on Environment and Development
UNDESA – United Nations Department of Economic and Social Affairs
UNDP – United Nations Development Programme
UNDTIE – United Nations Environment Programme, Division of Technology, Industry, and Economics
UNEP – United Nations Environment Programme
UNEP FI – United Nations Environment Programme Finance Initiative
UNESCO – United Nations Educational, Scientific and Cultural Organization
UNFCCC – United Nations Framework Convention on Climate Change
UNGA – United Nations General Assembly
UoB – University of Bonn
UPT – Technical Service Unit
US/USA – United States of America
US$ – United States dollar
USEPA – US Environmental Protection Agency
UTS – University of Technology Sydney
USSR – Union of Soviet Socialist Republics
VCBM – virgin coalbed methane
VISs – voluntary initiatives for sustainability
VISITs – voluntary initiatives for sustainability in tourism
VISTs – voluntary initiatives for sustainable tourism
VOC – volatile organic compound

VSM – voluntary simplicity movement
WA – Western Australia
WASIG – Western Australian Sustainable Industry Group
WB – World Bank
WBCSD – World Business Council for Sustainable Development
WEEE – waste electrical and electronic equipment
WHC – Westside Health Center, USA
WIR – Wet Investeringsregeling
WRI – World Resources Institute
WSSD – World Summit on Sustainable Development
WTO – World Tourism Organization
WW – wastewater
WWF – Worldwide Fund for Nature, formerly World Wildlife Fund
YLL – years of life lost
ZERI – Zero Emissions Research and Initiatives
ZEW – Zentrum für Europäische Wirtschaftsforschung/Center for European Economic
Research

Preface

Environmental technology management as a field of research and practice requires people who are committed to making the planet Earth a better place than the current prevailing unsustainable way of using natural resources, generating enormous amounts of often harmful and dangerous waste and showing little concern about local communities and individual people. This process of transformation is what the *International Handbook on Environmental Technology Management* is about. It contains an elaborate selection of writings by people most of whose names are easily recognizable not only in the academic world but also in the business realm because of their practically oriented work. They have been able to cross the boundaries of disciplines, organizations, countries and cultures to build the hope for a healthier, smarter and more pleasant way of bringing together the economic, social and environmental facets of our activities.

The power of this *Handbook* is in what it aspires to achieve and give to the readers. There are at least five ways in which it can be a valuable resource. First, it provides expert insights through the analyses, reflections and practical examples of the 58 contributors from ten countries and four continents. Second, it covers a wide range of aspects related to the management of technology and its environmental implications. Third, it is a rich source of reference in order to understand the changes that have been happening in the last decade and the way academic thinking has evolved. Fourth, the ideas and concepts presented here have a strong practical orientation and the readers should be able to take them to the real world. Fifth, it offers an enjoyable and thought-provoking read for people with a wide range of interests, including researchers, students, practitioners, policy-makers and members of the public.

Personally, it has been my pleasure and honour to work on the *Handbook* and I feel beholden to thank all the contributors for the excellent quality of their work, their passion for the outstanding job they are doing and their most obliging cooperation with the editors. The publishers at Edward Elgar have been extremely generous with their support and patience, for which they have my gratitude. My two co-editors also deserve a special mention, as they are both exceptional people who manage to break out of the ivory towers of academia and link research and policy development in the real world.

The whole area of sustainability is impossible to comprehend unless you think about the intergenerational links and this is where I am indebted to my late parents for giving me the love and appreciation of nature, respect and trust in other people and desire to pursue research. My two daughters have had the challenge of experiencing a number of transformations in their young lives and I wish to thank them for their endless love, inspiration and hope for the future. Finally, my thanks for their support and encouragement go to all my relatives in Bulgaria, my colleagues in Australia and my friends all around the world.

Dora Marinova
Perth, Western Australia

PART I

INTRODUCTION

1 Understanding environmental technology management as a move to sustainability

David Annandale*, Dora Marinova and John Phillimore

Perhaps surprisingly, it is relatively uncommon to find reference to 'environmental technology management' (or ETM) as a term of study or an academic discipline in its own right. Yet we can easily understand what it is about if we consider it as stemming from the intersection of three separate but inter-related fields of study. First, there is environmental technology. This is primarily an engineering-based discipline, concerned with the development of technology aimed at improving the environmental performance of existing products and processes, or at developing new products and processes. Environmental technology is particularly concerned with preserving the natural environment by recycling waste products produced by human activities, and is therefore concerned with technologies such as waste management, waste recycling, water purification, sewage treatment and so on. Second, there is environmental management. This combines environmental science and business studies and attempts to reduce environmental impact by applying particular management techniques within and between businesses. It often involves the advocacy, adoption and analysis of specific Environmental Management Systems (EMSs), a number of which are discussed in this book. Third, there is technology management, or the management of technology. This is primarily a business studies discipline, although it can also involve innovation studies and R&D and technology policy more generally. Technology management tends to focus on the firm (or less commonly, the industry sector) and aims to improve competitive performance by more effectively and efficiently managing technology within the enterprise. It occasionally incorporates insights from the sociology of technology, technology studies and science and technology policy, which have traditionally taken a more critical view of management and are concerned with the interaction between society and technology as much as with the way in which firms may improve the efficient application of technology.

In our view, 'environmental technology management' is a valuable term that captures the inescapable fact that businesses, engineers, managers and public policy actors face multifaceted challenges in attempting to steer companies, technology and society in a more environmentally sustainable direction. Analysis and change from any one direction will inevitably be partial and insufficient – but possibly indispensable nonetheless. The term is broad enough to allow a wide range of management techniques, technologies and policy options to be considered. It is an umbrella concept that enables different academic and professional communities to relate to one another while maintaining their own distinctive cohesiveness.

Our goal in this edited collection is to cover a wide range of approaches to environmental technology management. The need for such a publication, however, can easily be understood against the background of the current economic and technological revolution, which continues to fail to deliver a more sustainable society. While there are extraordinary levels of consumption in some parts of the world, in others there are still terrible problems with entrenched poverty. There is also burgeoning economic growth and consequent increases in disposable income in societies that have succeeded to give hundreds of millions of people a chance to experience life beyond basic survival.

Clearly the pressure on natural ecosystems continues to increase and the carrying capacity of the planet is constantly challenged. If everyone on the planet reached the current consumption level of affluent nations, the resource equivalent of five Earths would be needed to support it. Underlying these predictions are extraordinary current consumption statistics in the developed countries as well as in the Asian region as it rapidly industrializes. For example, in 2003 half of the world's cement output, a third of its steel, a quarter of its copper and a fifth of its aluminium went to the fast-developing People's Republic of China (ADB, 2005).

These statistics coupled with the incomparable levels of energy and materials consumption in the West paint a sobering picture. Clearly, it would be foolish not to think about the strain that these economic activities put on natural resources and the environment if they remain in their current shape. Careless economic actions that continue to erode the 'natural capital base' will eventually lead to a collapse of the economic system itself. There are already strong indications that we have 'overshot' the Earth's ecological limits in relation to some resources and waste sinks (Meadows et al., 2004). The challenge is how to make growth environmentally and socially sustainable, so that it can continue to serve people while conserving the natural assets and ecosystem services that make economic and social advancement possible.

Arguably, the move to sustainability requires attitudinal change, adoption of transformative technologies by industry and policy encouragement, and these are the issues that this book on environmental technology management deals with.

Much of the literature surrounding technological development assumes that innovation is – and should be – essentially market-driven. In this book we take a different view. We believe that the move towards a sustainable society requires conscious choice of environmentally transformative technologies. This belief is underpinned by strong evidence that there is a growing imbalance between the opportunistic directions of overall technological innovation, and the specific requirements of a long-term sustainable economy. In making this distinction, a number of commentators have suggested that to reach sustainability, societies need to choose 'real' technology. This is variously defined as technology that will 'do more with less'; reduce the ecological footprint; increase efficiency; enhance resources; improve signals; and end material deprivation (Meadows et al., 2004). This is why the book focuses so strongly on the concept of environmental technology *management*. In our view, technology is not an automatic panacea for the environmental problems associated with economic development. Technology needs to be 'chosen' actively and carefully so that, combined with an entrepreneurial attention to the market demand for 'green' products and services, it may offer a positive way forward.

All evidence suggests that the strategic introduction of technology cannot happen without the intimate involvement of governments and strategic partnerships with the corporate sector. The question remains as to what governments can do to 'nudge' corporations to adopt 'real' technology and modernize from an ecological perspective. One of the most important interventions is the adoption of overarching national innovation policies that make it clear how appropriate technologies will be defined. Too often, innovation policies have been designed around facilitating the rapid commercialization of new 'hard-ware' or 'software', without consideration of the socio-political and ethical/personal factors that are equally important determinants of whether technology will 'fit' in a given circumstance. Careful research over the last three decades has indicated that judgments about whether technologies are appropriate or not should only take place within a cultural framework that recognizes the equal importance of the technical-empirical, socio-political, economic and ethical/personal aspects of technology choice.

It is with the above ideas in mind that we start the book proper. In Part II we outline state-of-the-art thinking about innovation in environmental technology policy. *Ecomodernization and Technology Transformation* includes ten chapters that address a wide range of issues, helping the reader to understand the changes that are occurring with the pursuit of a more sustainable way of living. Chapter 2, *Industrial ecology* by van Berkel presents the concept of using nature's metaphor of balancing materials and energy flows through industrialized societies and mimicking nature in the operation of industry. Three cases of industrial ecology are presented because of their significance to the development and implementation of environmental technologies. These are: industrial symbiosis, biomimicry and greening of chemistry and engineering. This sets the scene for resolving tensions between industry and the natural environment but also calls for strong interdisciplinary and intersectoral collaboration. Chapter 3, *Remaking the way we make things* by McDonough and Braungart, argues that the way things are designed predetermines their environmental impact. By endorsing a cradle-to-cradle framework, designers, engineers and manufacturers can develop safe, cost-effective products and processes that generate a wide spectrum of economic, ecological and social value reducing unintended effects on humans and the environment. The transition to such a new concept of quality encourages creativity, generates innovation and provides competitive advantages. Including the users of technology in the development of technologies that have the potential to be more sustainable, is the focus of Chapter 4, *Users as a source of learning* by Rohracher. Such a change can speed up the adoption and diffusion of environmental technologies and the user–producer interaction can have beneficial effects in both directions. Chapter 5, *Cleaner production and eco-efficiency* by van Berkel, examines historical perspectives and current interpretations of environmental strategies to reduce human impact on the environment and achieve enhanced business performance. Thirteen examples from Australia demonstrate how this could be done and send the message that a change in philosophy can capitalize on common sense and creativity empowered by sound technical skills and management practices. The richness of examples from business and government is the focus of the inspirational Chapter 6 by Lovins, *Natural capitalism: path to sustainability*. It argues for a way of living within a capitalist society that does not destroy the value of nature and does not deplete natural resources and ecoservices on which we all depend.

The second half of Part II talks about the role of policies to achieve ecomodernization and technological transformation. Chapter 7, *Sustainable strategies towards human resources* by Gollan focuses on human resources and their role in any company's commitment to sustainable strategies. Developments in human resource management are exceptionally important to bridge the gap between human needs, imperatives from business and environmental concerns. The impact of environmental policies is another tool that can be used to change the course of technology development, but is this working? Chapter 8, *Innovation impacts of environmental policies* by Foxon and Kemp argues that these policies so far have contributed to the diffusion of existing technologies and have had limited impact on innovation, hence making the case that a specific technology focus is needed. The process of development of policies to encourage a more sustainable development is at the centre of Chapter 9, *Sustainable production and consumption policy development: a case study from Western Australia* by Higham and Verstegen. The authors use the example of Western Australia to outline the political framework for the sustainability debate and depict the use of deliberative democracy to shape the future of this state's economy.

The remaining two chapters in Part II discuss the link between the technology-rich countries and those that can potentially be affected from their technological trajectories. Chapter 10, *Managing research for sustainable development* by Turpin alerts us to the fact that the technologies developed in the North can exacerbate rather than ameliorate unfavourable environmental impacts in the South, and suggests how technology policies in the developing countries can capture benefits from the global research effort without adverse social and environmental impact. Finally, Chapter 11, *Technology transfer and uptake of environmentally sound technologies* by Halls covers a wide range of important issues for the success of environmental technologies, including capacity enhancement of stakeholders, needs assessment, creating an enabling political and regulatory environment.

Attempts to direct technology and have it serve the needs of people have traditionally been the subject of public policy decisions rather than private entrepreneurialism. However, conscious choice of technology does not fit only within the realm of public policy. The private sector has a role to play in leading the adoption of new business processes or practices and in the introduction of specific technology. The move toward an ideal ecologically sustainable economy will require innovation on the part of the business sector to help design a more 'eco-efficient' economy. Eco-efficiency innovations can be categorized as one or more of the following:

- production process, for example, 'good housekeeping' waste minimization or clean production;
- product innovation, for example, advanced materials, electric vehicles;
- innovation in producer–consumer relationships, for example, car sharing or recycling; and
- changes in consumption patterns, for example, reusable paper, recycling and composting (OECD, 1998).

All of these types of innovation involve increasing the efficiency of resource use. Institutional innovations associated with the planning and/or facilitation of eco-efficiency are also needed. In Part III and Part IV we focus on some of these facilitative changes taking place in business management practices. Our purpose in these sections of the book is to point to initiatives that companies are taking of their own accord. Variously defined as 'self-management' or 'voluntarism', these activities are rarely, if at all, driven by public policy. In recent years a number of techniques have been experimented with that allow firms to make unilateral performance improvements. Part III and Part IV introduce the practical tools of EMS; corporate environmental reporting; lifecycle analysis and environmental accounting.

Part III, Environmental Technology Management in Business Practices, highlights some of the most important aspects of changes being made by farsighted companies. Chapter 12, *Lifecycle assessment* by Evans and Ross provides an excellent overview of the method that requires all aggregate effects of all activities related to the manufacturing, use and eventual disposal of a product or service to be examined in order to assess its environmental burden.

The Natural Step framework presented and analysed in Chapter 13, *The Natural Step Framework* by Herbertson and Tipler, helps develop a common language and understanding about the underlying principles of a move towards sustainability. Chapter 14, *Integrating human and ecological factors* by Benn, Dunphy and Griffiths, argues for an integration between human and ecological considerations when corporations address the issues of sustainability. It also provides empirical examples. Although the big corporations have the potential and resources to make a shift towards more sustainable business practices, the small and medium-sized companies can often be innovative in their approach to turn in their favour what appears to be a disadvantage. Chapter 15, *Using network approaches to engage small and medium-sized enterprises in Environmental Management Systems* by Hillary provides useful conceptual ideas how this can be achieved.

Chapter 16, *Green marketing and green consumers* by Connolly, McDonagh, Polonsky and Prothero, reflects on the green marketing strategy adopted by business and the fact that if adopted properly, it requires more than cosmetic changes. Similarly, the triple bottom line accounting can remain a rhetoric only if environmental and social accounting is not integrated with finance. More discussion of these issues is presented in Chapter 17, *Business, environmental management and the triple bottom line* by Schilizzi. The changes required in tourism are discussed in Chapter 18, *Eco-management and Audit Scheme in sustainable tourism* by Querini and Bizzarri, where the authors argue that the main effect of the implementation of EMAS in tourist companies is to stimulate the introduction of technological innovation.

The list of chapters included in Part IV, Measuring Environmental Technology Management covers some challenging economic and assessment issues. Chapter 19, *Measuring the true productivity gains from environmental technology improvements* by Repetto stresses that most current methodologies to measure multifactor productivity are conceptually and empirically flawed. A better methodology, like the one suggested by the author, can provide a much better picture of the desired and undesired industrial outputs from the operation of an industry or firm as well as a better understanding of the relationship between environmental protection and economic growth.

The terms, scope and approaches to environmental accounting are outlined in Chapter 20, *Environmental accounting* by Schaltegger and Burritt. They discuss the variety of conventional and new tools that can inform internal and external stakeholders interested in the environmental aspects of sustainability. Chapter 21, *Indicators for environmental innovation* by Arundel, Kemp and Parto, takes the reader to the complicated world of developing indicators for describing environmental innovation. The authors call for the need of large-scale surveys that can generate reliable data on which to base these important indicators.

The international trends in socially responsible investments are a measure of the uptake of the sustainability culture by business and they are analysed in Chapter 22, *International trends in socially responsible investment* by Borsky, Arbelaez-Ruiz, Cocklin and Holmes. Positive shifts have been witnessed through direct links between good financial perform-ance and responsible social and environmental behaviour. Chapter 23, *Comparison of international strengths in sustainable technological solutions* by Marinova and McAleer, uses patent data as well as information from the *Green Pages* and ISO 14001 registrations to show that the process of adoption of sustainable technologies has just started with different countries exhibiting different national priorities. Although not specifically aimed at measuring, Chapter 24, *Sustainability assessment* by Newman, lays the foundations of sustainability assessment, a step further from environmental impact assessment, using examples from Western Australia in the case of policies, programmes, plans and with the assessment of buildings and developments in settlements. The final chapter in this part, namely Chapter 25, *Trends and volatility of ecological and anti-pollution technology patents in the USA* by Chan, Marinova and McAleer, is an empirical econometric study of the development of these crucial clusters of technologies.

Earlier, we suggested that the move towards sustainability requires the adoption of specific 'transformative' technologies by industry. In Part V we present some state-of-the-art case studies on new 'real' technologies that are both commercially successful, and environmentally positive.

Part V, Case Studies in New Technologies for the Environment, covers eight very inter-esting case studies, namely: integrated waste management (Chapter 26 by Hughes, Anda, Ho and Mathew); renewable energy (Chapter 27 by Lowe); the use of the Internet for online learning in relation to environmental technology management (Chapter 28 by Maor and Marinova); technologies for reduction of greenhouse gas emissions in Russia (Chapter 29 by Hill); the construction industry in Australia and the Sydney Olympic Village (Chapter 30 by Marceau and Cook); tools for environmental impact assessment and their application in the case of a health centre to be located on a Native American reservation in South-western USA (Chapter 31 by Jones); the use of technology by small-scale women farmers in South Africa (Chapter 32 by Yokwe, Nell and Marinova); and a chapter (Chapter 33) that summarizes examples of turning waste into value from The Natural Edge project based in Australia.

The last two parts of the book lead us naturally towards thinking about the future. Part VI, Environmental Technology Management and the Future includes two chapters that endorse an alternative futuristic approach to technology management. Chapter 34, *Environmental technology management* by Harding, conveys the disenchantment with the western analytical approach that looks at the whole as only a sum of its parts. It argues for a holistic approach and integration of thinking, ethics, sensing and intuition, which can be focused on participation rather than control of nature. The value loop paradigm as a business vision is the goal of Chapter 35, *The value loop – a new framework for busi-ness thinking* by Tibbs. The *Handbook* concludes in Part VII with final remarks about the lessons from today that can help for the transformation and adoption of a new sustain-ability culture (Chapter 36).

We hope that this compilation of essays from some of the most prominent writers in this field will make a contribution towards sustainable economic development. The main 'driver' for this book is a sense of urgency. Serious scholarship and thought leads to the conclusion that our planet is nearing its limits. We are of the view that the most hopeful way forward is a concerted effort on behalf of governments and business to alter the way markets operate, reconfigure regulation to be reflexive and supportive of voluntary environmental management tools, and establish policies that allow for the wise choice of technologies that support the environment rather than work against it.

Note

*   David Annandale is currently Senior Technical Advisor to the National Environment Commission of the Royal Government of Bhutan.

References

Asian Development Bank (ADB) (2005), 'Making Profits, Protecting Our Planet: Corporate Responsibility for Environmental Performance in Asia and the Pacific', *Asian Environment Outlook 2005*: Highlights, Manila: ADB.

Meadows, D., Randers, J. and Meadows, D. (2004), *Limits to Growth: The 30-Year Update*, Vermont: Chelsea Green Publishing Company.

Organization for Economic Cooperation and Development (OECD) (1998), *Eco-efficiency*, Paris: OECD.

PART II

ECOMODERNIZATION AND TECHNOLOGY TRANSFORMATION

2 Industrial ecology
René van Berkel

Introduction

Industrial ecology is the provocative and oxymoronic name of a rapidly evolving field of research, public policy and industrial practice that is primarily aimed at achieving symbiosis between industrial production and consumption and the natural ecosystem on which life on Earth ultimately depends. The field originated from taking a systems perspective of waste generation and resource consumption in the design, manufacturing, use and disposal of industrial processes, products and services. Although earlier references exist, industrial ecology was first put profoundly on the environmental technology map by Frosh and Gallopoulos' (1989) seminal paper 'Strategies for Manufacturing' published in the *Scientific American*. These authors reviewed the environmental trade-offs of selected 'environmental innovations', in particular of the extensive use of plastics for light-weighting, of the introduction of low emission iron and steel technology, and of the use of platinum in catalytic exhaust converters. They hypothesized the benefits of adopting a holistic, systems view of manufacturing, which dictates that 'wastes from one industrial process can serve as the raw materials for another, thereby reducing the impact of industry on the environment' (p. 94). Many different viewpoints have since surfaced that are both supportive and critical of the idea of using nature as a metaphor for industrial production and consumption. It is, however, uncontested that industrial ecology is 'characterised by the refreshingly unorthodox use of nature as a model appreciated as an expedient ideal in order to gain valuable insight for theory and to learn to deal with natural resources and services in practice' (Isenmann, 2003, p. 154).

Being a novel discipline there is no agreed taxonomy for industrial ecology and even its boundaries remain contested. In providing some insight into the theory and practice of industrial ecology, this chapter has been rather arbitrarily divided in three main sections. The next section (Perspectives) reviews some perspectives on the boundaries and goals of industrial ecology. The following sections cover systems applications of industrial ecology that are based on using natural ecosystems as metaphor (Systems applications), and product and process applications, which are based on exploring and mimicking natural processes (Product and process applications). The concluding section provides an outlook on the opportunities and constraints of using industrial ecology to drive efforts toward sustainable production and consumption.

Perspectives

Industrial ecology is both 'industrial' and 'ecological' (Lifset and Graedel, 2002, p. 3). It is industrial in that it focuses on product design and manufacturing processes. Industry is therefore viewed as the primary agent for environmental improvement as industry possesses the technological expertise, management capability and financial and other resources necessary for successful execution of environmentally informed design of products and processes. Industrial ecology is also ecological in at least two senses. First, it looks to non-human 'natural' systems as models for industrial activity. Mature ecosystems are extremely effective in recycling of resources and are therefore promoted as exemplary models for effective recycling in industry and society. Second, industrial ecology places industry – or technological activity – in the context of the larger ecosystems that support it. Thus industrial ecology examines the sources of resources used in industrial activity as well as the sinks that absorb and detoxify the wastes discharged by society.

Industrial ecology is inspired by biological ecology. A working definition of biological ecology is the study of the distribution of organisms and their interactions with the physical world (for example, Graedel and Allenby, 2003). All organisms must consume resources in order to live and go about their daily functions. The transference of these resources (nutrients and the embodied energy that they contain) from one organism to another, and then on to another, forms a 'food chain'. The principal steps in the chain are called the trophic levels, of which there are four: (1) extractors; (2) producers; (3) consumers (often at several levels); and (4) decomposers. The first levels consist of primary producers (for example, plants) that utilize energy and inorganic nutrients to produce biomass useable at higher trophic levels (seed, leaves and the like). The next levels consist of herbivores, carnivores and detritus decomposers (who receive residues from any other trophic level and from those residues regenerate materials that can again flow to primary producers).

Biological ecology can be studied from the perspective of food chains (that is, the nutrient and energy flows) as well as from the perspective of the organisms that transform the resource flows in the food chains (that is, the connecting nodes in the food web). These two perspectives are also possible in industrial ecology. In its study of industrial networks of material and energy transfer (that is, industrial food chains), industrial ecology employs the 'metaphor of natural ecosystems'. In its study of biological and other natural processes and efforts to mimic those in firms (that is, industrial organisms), industrial ecology employs the 'natural analogy'.

Industrial ecology has surfaced in a wide range of disciplines. The environmental sciences, environmental sociology and engineering being notable examples. Each discipline provides a distinctive perspective on its application and benefits.

The earth and environmental sciences have embraced industrial ecology for offering a fresh view on environmental management challenges and priorities. Socolow (1994) demonstrates this perspective by discussing six elements of industrial ecology:

- Long-term habitability: industrial ecology shifts the emphasis in environmental analysis from short-term abuse (pollution impacts) to the long-term habitability of planet Earth, and thereby puts the spotlight on persistence of toxic chemicals, depletion of natural resources and disruption of the grand life-supporting cycles (such as the global carbon, nitrogen and water cycles).
- Global scope: the spatial dimension expands from local abuse (localized negative impacts) to regional and global impact, where global includes both the global spatial scale (climate change, for example) as well as the global spread of universal local impacts (such as pesticide use, urban air quality and so on).
- The overwhelming of the natural system: nature is the measure of humans in the sense that manufactured emissions (carbon dioxide for example) and disturbances to the natural system (such as land clearing) are larger than natural cycles and variations.

- Vulnerability (or the overwhelming of the human system): humans are the measure of nature. Natural hazards, disease and ecosystem destruction can have disastrous impacts on communities, institutions and individuals.
- Mass flow analysis: application of engineering tools to track materials through time and space, and recognition of the environmental impacts associated with the size of the synthetic mass flows compared with the size of the natural mass flows.
- Centrality of the firm and the farm: recognition that firms and farms are primary agents of change for redirecting material flows through new products and services.

Environmental sociologists place industrial ecology in the context of ecological modernization. Huber (2000) discusses the limitations of the dominant concepts of 'sufficiency' (self-limitation and egalitarian distribution of scarce resources) and 'efficiency' (efficient use of materials and energy to achieve resource efficiency). He argues that both are valuable but are also limited by not addressing the compatibility of industry with the natural ecosystem. The notion of 'consistency' is therefore introduced, which achieves compatibility between industrial and natural metabolism through system innovations. Industrial metabolism and industrial ecology provide the direction for such 'consistency' efforts:

> Whereas the sufficiency version of sustainable development is a program for the conservation of nature and the efficiency version is a programme for improvement of existing technologies and infrastructures in order to economize on natural resources and sinks, the consistency version of sustainable development is a programme for innovation of new technologies, products and materials flows in order to change the qualities of the industrial metabolism, thus rendering possible a true industrial ecology. (Huber, 2000, p. 281)

The engineering fraternity looks upon industrial ecology as a source of normative ideas that can be incorporated to structure, enrich and solve the problems associated with process and product design activities. This is best reflected in the proliferation of 'Design for X' strategies in industrial ecology texts for engineers (for example, Behrendt et al., 1998; Graedel and Allenby, 2003). Design for X strategies inspired by industrial ecology, for example, include: design for recycling, design for dematerialization, design for toxics reduction, design for energy efficiency, design for refurbishment, design for maintenance and so on.

An often-quoted summary that incorporates these perspectives defines industrial ecology as 'the study of the flows of materials and energy in industrial and consumer activities, of the effects of these flows on the environment, and of the influences of economic, political, regulatory and social factors on the flow, use, and transformation of resources' (White, 1994, p. v). Isenmann (2003) suggests that industrial ecology is best described by its five distinctive characteristics:

1. Core idea (industrial symbiosis): thinking about economy–environment interactions, viewing economic systems and their surrounding natural systems in concert rather than in isolation, and looking for ways to make natural and economic systems compatible.
2. Fundamental perspective (nature as a model): using nature to learn from and discover new insights for dealing with industrial activity. It should not, however, be invoked as an uncritical justification for the replication of natural principles and concepts in the economic system.

3. Basic goal (balance industrial–ecological systems): to balance the development of industrial systems with the constraints of natural ecosystems.
4. Working definition (science of sustainability): key attributes are a systemic and integrated view of all material and energetic components of the industrial economy, including its relations with the biosphere, explicit emphasis on the biophysical substratum of industrial activities representing all materials and energy, and recognition that technological progress is a crucial but not exclusive element of the path to sustainability.
5. Main objects (products, processes, services and wastes): the focus is on products, processes, services and wastes at different aggregation levels of material and energy flows and including local, regional and global scales.

The range of perspectives on industrial ecology indicated above is illustrative of the diversity of efforts that define this emerging field of inquiry and practice. The industrial ecology practice is here codified in system applications as well as product and process applications.

System applications are primarily based on the metaphor of natural ecosystems and are principally concerned with the application of ecosystem principles – most importantly the circular flows of materials powered by solar energy – to industrial activity. Resource flows (materials, energy and so on) are the objects in the systems stream of industrial ecology, the strategic intent of which is to minimize the disturbance of natural resource flows by manufactured resource flows.

Product and process applications of industrial ecology are primarily based on the natural analogy. Biological and other natural processes are thus the objects in the products and processes stream of industrial ecology, which is principally concerned with mimicking natural, biological, chemical, physical and geological processes and materials in industrial applications. The strategic intent of these applications is to develop new processes and products that are more resource-efficient and compatible with nature than the current industrial processes and products they aim to replace.

System applications
In their use of a natural ecosystem's metaphor the system applications of industrial ecology are concerned with the assessment and improvement of industrial food chains, or the industrial metabolism; that is, the networks of material and energy transfer that take place between the producers, consumers and recyclers of industrial society. To provide insight into the theory and practice of industrial metabolism this section begins with principles and their implications as normative goals for industrial activity. Following is an introduction to materials flow analysis and the flow-based indicators that represent the analytical tools of industrial metabolism. Finally, the applications of industrial metabolism are described, the most profound of which is seen in industrial symbiosis or eco-industrial parks where industries are co-located to enable material and energy exchanges.

Industrial metabolism
'The word metabolism as used in its original biological context connotes the internal processes of a living organism. The organism ingests energy rich, low entropy materials (food), to provide for its own maintenance and functions, as well as a surplus to permit growth or reproduction' (Ayres, 1994, p. 23). Based on this metaphor industrial

metabolism is the whole integrated collection of physical processes that convert raw materials and energy, plus labour, into finished products and wastes in a (more or less) steady-state condition (Ayres, 1994). Figure 2.1 schematically displays the bio-geo-chemical cycles in natural ecosystems and their application to the industrial materials cycles.

The key implication of adopting industrial metabolism is that the long-run fate of materials in the industrial materials cycles is much more important than efficiency of their use in the materials cycles. If materials dissipate from the industrial materials cycles they will start to interfere with the bio-geo-chemical cycles, leading to disturbances in the natural ecosystem and the risk of impairing the natural ecosystem's ability to deliver vital resources (such as food and materials) and ecosystem services (the ability to provide fresh water and air for example). Using the language of environmental sociologists, the industrial materials cycles need to be 'consistent' (or compatible) with the natural metabolism through bio-geo-chemical cycles.

This notion of consistency in industrial and natural metabolism can be translated into normative aims for industrial ecology. The most profound example of such a normative goal has been developed using the first two of the four system conditions of The Natural Step project (see Chapter 13). These system conditions, developed by Robert (2003, p. 67) hold that in a sustainable society:

1. Nature is not subject to systematically increasing concentrations of substances extracted from the Earth crust.
2. Nature is not subject to systematically increasing concentrations of substances produced by society.
3. Nature is not subject to systematically increasing degradation by physical means.
4. The ability of humans to meet their needs worldwide is not systematically undermined.

System conditions 1 and 2 are most directly linked to industrial metabolism. In practice system condition 1 implies that dissipation of mined materials from the industrial system should be prevented unless these materials are already highly abundant in the natural ecosystem (so that any inflow of the substance will cause a negligible increase in concentrations in the environment). Likewise system condition 2 implies that dissipation of manufactured materials should be prevented unless these materials are easily and rapidly degraded into substances that are abundant in the natural ecosystem.

The natural ecosystem metaphor can be extended to encompass the principles that drive ecosystem development. The development of natural ecosystems is driven by four principles (Korhonen, 2001):

1. *Roundput.* Recycling of energy ('utilization of residual energy') occurs in food chains through a cascading effect, with the only driver of the system being the input of infinite solar energy.
2. *Diversity.* Diversity in species, organisms, interdependency and cooperation enhances ecosystem survival.
3. *Locality.* Actors in the ecosystem adapt to local environmental conditions and cooperate with their surroundings in diverse interdependent relationships.
4. *Gradual change.* Information transfer and change happen through reproduction, which is a slow process.

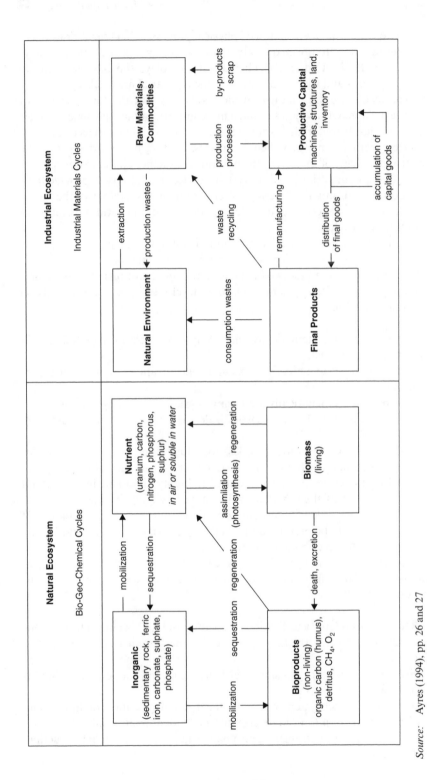

Source: Ayres (1994), pp. 26 and 27

Figure 2.1   Bio-geo-chemical cycles and their parallel in industrial materials cycles

Table 2.1 illustrates how these ecosystem principles can be applied to industrial systems.

Materials flow analysis and indicators
Understanding the structure and functioning of material and energy flows and exchanges in industry and society is at the core of industrial metabolism. Materials flow analysis enables such understanding by examining the throughput of process chains comprising extraction or harvest, chemicals transformation, manufacturing, consumption, recycling and disposal of materials (Bringezu and Moriguchi, 2002). To quantify the inputs and outputs of those processes materials flow analysis develops accounts in physical units (usually in terms of tonnes). The accounts can be generated for chemically defined substances (for example carbon or carbon dioxide) on the one hand and natural or technical compounds or 'bulk' materials (for example wood, coal) on the other hand.

Materials flow analysis studies are based on the industrial metabolism paradigm and use mass balancing as the methodological basis. There is, however, a considerable divergence in sustainability strategies pursued through industrial metabolism (Bringezu and Moriguchi, 2002). One basic strategy is *detoxification* of the industrial metabolism. It is concerned with mitigating the release of pollutants and other critical elements into the environment and thereby aims to minimize the negative impacts of materials flow on the functioning of the bio-geo-chemical cycles. Another complementary strategy is *dematerialization* of the

Table 2.1   Ecosystem principles applied to natural and industrial ecosystems

Ecosystem Principles	In Natural Ecosystem	In Industrial System
1. Roundput	• Recycling of matter • Cascading of energy	• Recycling of matter • Cascading of energy
2. Diversity	• Biodiversity • Diversity in species, organisms • Diversity in interdependency and cooperation • Diversity in information	• Diversity in actors, in interdependency and cooperation • Diversity in industrial input and output
3. Locality	• Utilizing local resources • Respecting the local natural limiting factors • Local interdependency and cooperation local actors	• Utilizing local resources, wastes • Respecting the local natural limiting factors • Cooperation between
4. Gradual change	• Evolution using solar energy • Evolution through reproduction • Cyclical time, seasons time • Slow time rates in the development of system diversity	• Using waste material and energy and renewable resources • Gradual development of system diversity

Source: Based on Korhonen (2001).

industrial metabolism. This is concerned with the scale and intensity of the materials flow in industrial societies and can be interpreted as aiming to reduce the scale and intensity of the material exchanges between the industrial and the bio-geo-chemical systems.

A first order typology of materials flow analysis studies is possible with regard to their focus on detoxification (studying specific environmental problems related to impacts per unit flow) or dematerialization (studying problems associated with the throughput of materials per unit of economic activity). Within each type of strategy a further subdivision is possible with regard to scale of the analysis. Table 2.2 provides an overview of the typology, where Type I studies address specific environmental problems and Type II studies focus on material throughput. Within this typology, Type Ia studies (also known as substance flow analysis) deal only with flows of chemicals of specific concern, to establish, for instance, the main entrance routes of pollutants into the environment (van der Voet, 2002). Type Ic studies can be regarded as subsets of lifecycle assessment, with total materials requirement being the sole category indicator used in the lifecycle impact assessment stage (Guinée, 2002). Type IIa studies deal with the material throughput of firms and thereby provide the physical basis (accounting in physical units) of environmental management accounting (Schaltegger and Burritt, 2000). Type IIc studies are concerned with the materials throughput of industrial parks, regions or nations (see also the section on industrial symbiosis below).

Ecological footprint analysis can be regarded as a specific application of Type II studies. An ecological footprint calculation starts with a physical account of the resource throughputs of a given economic or geographic entity. Each material input into the economy is converted into a specific land area, estimated to be the equivalent area required to grow and harvest that material. Each material output into the environment (waste, emissions) is also converted into a specific land area estimated to be the equivalent area disturbed by the pollutant outflow, or required to absorb and mitigate the

Table 2.2   Typology for materials flow analysis

Analytical concern	Analytical unit		
Specific environmental problems caused by impacts per unit of flow of	Substances e.g. Cd, Cl, Pb, Ni, P, C	Materials e.g. wood, energy carriers, biomass	Products e.g. batteries, cars
(Type I studies)	(Type Ia)	(Type Ib)	(Type Ic)
	Within certain firms, sectors, regions		
Problems of environmental concern related to the throughput of	Firms e.g. single plant, small to medium-sized enterprises	Sectors e.g. transport, chemical industry	Regions e.g. total material requirements
(Type II studies)	(SMEs) (Type IIa)	(Type IIb)	(Type IIc)
	Associated with substances, materials, products		

Source: Modified from Bringezu and Moriguchi (2002), p. 81.

pollutants. Summarizing the land areas associated with all materials and energy flows of a given economic activity leads to the total footprint; that is, the estimated total area of land required to support that economic activity that can be calculated, for example, for the average consumer of a given nation, for a city or for an island. This basic methodology is extended to consider five land categories: bioproductive land, bioproductive sea, energy land, built land and biodiversity land (Chambers et al., 2000).

Materials flow analysis has been applied to assess and compare the resource and pollution intensity of national economies. This has led to a number of aggregated indicators for economy-wide materials flow (including Bringezu and Moriguchi, 2002; Matthews, 2000):

- *Input indicators*. Measure the input of materials into the economy. Direct material input (DMI) only includes those materials that are of economic value and used in production and consumption. Total material requirement (TMR) also takes account of the upstream hidden materials flow associated with the import of materials into the economy.
- *Output indicators*. Measure the output of materials from the economy into the environment. Domestic processed output (DPO) is the total mass of materials that have been used in the domestic economy before being disposed into the environment. Total domestic output (TDO) adds the total domestic hidden flows to the DPO.
- *Consumption indicators*. Measure the materials used in an economy. Domestic material consumption (DMC) considers only materials directly used while total material consumption (TMC) also measures total primary material requirement from domestic consumption activities.
- *Balance indicators*. Deal with the accumulation of materials in either the industrial system or the ecosystem.
- *Efficiency indicators*. Link materials flow indicators to the size of the economy.

Economy-wide data to enable materials flow analysis are still scarce. A major comparative study coordinated by the World Resources Institute included five nations: Austria, Germany, the Netherlands, Japan and the USA. Table 2.3 provides the key efficiency indicators of DPO per unit of economic activity and per head of population. The data

Table 2.3 Material efficiency of nations, 1975–96

Country	Population (millions) %	DPO (million metric tons) %	GDP (own currency) %	DPO/ GDP* %	DPO/capita (metric tonnes per capita) %
Austria	+6	+18	+68	−29	+10
Germany	+32	+24	+93	−36	−6
Japan	+13	+20	+106	−42	+7
Netherlands	+14	+16	+62	−29	+2
USA	+23	+28	+74	−26	+5

Note: * Million metric tonnes per million constant monetary units, own currency.

Source: Modified from Matthews (2000), p. 20.

provide evidence for relative dematerialization of consumption (reflected in declining ratios of DPO/GDP), which is, however, not accompanied by dematerialization of society (as reflected in ongoing small increases in DPO/capita, with the exception of Germany, whose figures have, however, been influenced by the reunification after the fall of the Berlin Wall). Compounded by the absolute growth in population, this results in the materials throughput continuing to increase in absolute terms (reflected in ongoing increase of DPO).

Industrial symbiosis
Industrial symbiosis deals with the exchange of wastes, by-products and energy between closely situated firms. It is perhaps the best-known application of industrial ecology principles. Because of the many links among the firms, the industrial area is transformed into an 'industrial ecosystem' or 'industrial symbiosis'. Chertow (2000, p. 313) provides the following definition: 'Industrial symbiosis engages traditionally separate industries in a collective approach to competitive advantage involving physical exchange of materials, energy, water, and/or by-products. The keys to industrial symbiosis are collaboration and the synergistic possibilities offered by geographic proximity.'

A classical example of how an industrial ecosystem can develop is found at Kalundborg in Denmark (see, for example, Ehrenfeld and Gertler, 1997). At the core of this industrial system is the Asnaes Power Station (coal-fired), which provides steam to the Statoil Refinery and Novo Nordisk pharmaceutical plant. In return Statoil Refinery provides fuel gas and cooling and waste utility water to the Asnaes Power Station. The adjacent Gyproc Wallboard plant also utilizes fuel gas from the refinery while sourcing scrubber sludge from the power station. The fly ash from the power station is further processed as cement and road aggregate. Waste heat from both the refinery and the power station was initially provided to greenhouses but has recently been redirected to fish farms and district heating. Finally, sulphur from the refinery is shipped to the Kemira Acid Plant while Novo Nordisk's pharmaceutical plant provides treated sewage sludge as fertilizer to neighbouring farms.

The current system of exchanges has evolved dynamically over about four decades. The 'twinning' between the three key process industries (power station, refinery and pharmaceutical plant) is generally regarded as the critical success factor at Kalundborg since each provides high-volume waste streams with valuable by-products on a continuous basis (Ehrenfeld and Gertler, 1997). In the absence of two or more major process industries being co-located, it may not be possible to develop a significant industrial symbiosis. Manufacturing industries, often the largest segment at any industrial estate, typically produce comparatively small volumes of mono-material waste streams (resulting from cut-off from input materials) together with a mixed waste stream comprising production rejects, packaging and other production waste. The mono-material stream is most valuable for re-use or downcycling in other industries. It is only available in smaller quantities, however, often with large fluctuations in volume over time, and so is generally only valuable to industries in the same sector. These factors make it difficult to achieve positive economies for symbiosis on mono-material streams within any given industrial estate. The mixed waste stream from manufacturing industries is not fundamentally different from the general municipal waste stream, and likewise it has limited potential and economic value for direct re-use by industries. Together these factors mean that it may not be easy to design ex ante industrial symbiosis at the scale of the Kalundborg system. However, on a more modest scale it may well be possible to develop symbiotic relations between companies; for example the experience of the Business Council for Sustainable Development in the Gulf of Mexico (Mangan, 1998).

A useful framework for analysing and promoting eco-industrial parks – or industrial ecosystems – is provided by Lambert and Boons (2002). The primary distinction these authors draw is between industrial complexes (consisting of concentrations of materials and energy-intensive industries that are intrinsically interconnected) and mixed industrial parks (usually housing a variety of small to medium-sized enterprises). A third approach is to study the region that includes interactions between communities, industries and land users. Within each category, a distinction can be made between brownfield sites (redevelopment from existing land uses) and greenfield sites (first development of new industrial parks). The resulting typology is summarized in Table 2.4.

Table 2.4   Typology for eco-industrial park development

Industrial complexes (Geographically concentrated industrial activities, mainly process industries, with tight couplings of a relatively small number of materials and energy-intensive production processes)	*Greenfield* Development of a new complex of industrial activities with tight physical couplings, taking into account the ecological impact of the complex in the design phase
	Brownfield The revitalization of an existing industrial complex using reduction of the ecological impact as one of the goals
Mixed industrial parks (Industrial activities, mainly SMEs, which are concentrated in dedicated areas and are of a very diverse nature with little or no coupling of production processes)	*Greenfield* The development of a new industrial park, addressing ecological issues in the different stages of the development process
	Brownfield The revitalization of an existing industrial park taking the reduction of environmental impact as one of the goals
Eco-industrial regions (Industrial activities in a larger geographic or administrative area, usually referring to a diversity of industries but often with a definite specialization)	*Greenfield* Developing and industrializing a region according to a well-defined concept that includes the reduction of environmental impact
Also called virtual eco-industrial parks	*Brownfield* Restructuring an existing industrial region, often based on definite regional qualities and accounting for environmental performance

Source: Lambert and Boons (2002), p. 473.

Industrial complexes are usually spontaneously developing exchange networks for utilities, residual products and so on, as evidenced by the industrial complexes studied in Styria (Austria), Rijnmond (the Netherlands) and Kwinana (Australia). Most heavy industries attain economic benefits with materials exchange. However, a further extension of such networks to lower grade residual flows, such as residual heat, often appears to fail due to large preparation times in relation to internal system dynamics, high investment costs and uncertainties associated with resource pricing. Most promising are the opportunities for clustering in greenfield areas and for infill in brownfield areas. Whether and how further synergies and exchanges can be realized through concerted facilitating efforts in existing complexes remains largely unresolved.

Mixed industrial parks do not have an in-built incentive to facilitate industrial symbiosis. Material and energy costs represent a much lower share of total operating costs for the light industries typically located in these parks, and most manufacturing and service industries have significant variability in their resource consumption and waste and by-product generation over time. Government intervention for the operation of waste exchange services can be effective in establishing and improving recycling networks in mixed industrial parks (see, for example, Kincaid and Overcash, 2001; Kincaid, 2003).

Product and process applications

Product and process applications in industrial ecology use the natural analogy to develop products and processes that are more efficient and compatible with nature. The manner in which the natural analogy is incorporated structures the product and process applications. At one end of the spectrum, biomimicry seeks a natural process or material that fulfils a similar function to that required by industry and then mimics that natural process or material. At the other end of the spectrum, green chemistry and engineering apply more generic principles derived from natural processes to mainstream chemistry and engineering. Biomimicry and green chemistry are discussed in turn.

Biomimicry

In its use of the natural analogy, biomimicry looks to nature as a source of inspiration. Biomimicry is ultimately based on the recognition that nature has already solved many of the problems industrial societies are grappling with. Therefore, using principles and processes proven in nature will allow designers to create products, processes and policies that are well adapted to life on Earth over the long haul. The biomimicry challenge is to harness energy like a leaf, grow food like a prairie, spin fibre like a spider, grow ceramics like an oyster or run a business like a redwood forest (Benyus, 1997). This involves using nature as model, measure and mentor (see also Table 2.5) to inspire the development of cleaner products, materials and processes.

Biomimicry research to date has revealed a number of powerful cross-cutting similarities in biological processes, including (Benyus, 1997):

● Nature runs on sunlight.
● Nature only uses the energy it needs.
● Nature fits form with function.
● Nature recycles everything.
● Nature rewards cooperation.

Table 2.5 Biomimicry: using nature to guide innovation

Use Nature As	Biomimicry Leads To	Practical Examples
• Model	New science that studies nature's models and then imitates or takes inspiration from these designs and processes to solve human problems	*Artificial photosynthesis.* Various approaches are being investigated to mimic photosynthesis in industrial reactors to produce fuels or chemical feedstocks, such as use of chlorophyll-like organic pigments and photosensitive membranes as a photochemical cell to produce green electricity, the use of synthetic protein membrane systems to convert light into biochemical energy, which can sequester carbon dioxide, or photosynthetic bioreactors that use cyanobacteria to harvest light and produce isoprene (a liquid hydrocarbon that can form rubber and other compounds) (Davidson, 2003)

Blue mussel byssus and sealants. The byssus is a composite collagen/silk mix with a blended rather than abrupt interface between the two materials, which provides a design idea for composite materials that need both toughness and flexibility. The byssus sticks to a variety of underwater surfaces with glue-like adhesion polymers. Although the byssus formation is not yet fully understood, several mussel proteins have already been isolated for specific product applications such as a medical glue produced by Johnson and Johnson (Benyus, 1997; Skula, 2003; TUD, 2004) |
| • Measure | An ecological standard to judge the 'rightness' of innovations. Evolution has ensured that nature 'knows' what works, what is appropriate and what lasts | *Natural systems agriculture.* Mimicking prairie-type ecosystems for agricultural production to combat land degradation and to minimize dependency on non-renewable inputs of fertilizers and pesticides. The Land Institute's model for natural systems agriculture is based on prairie-like poly-cultivation of perennial crops specially bred for high yields of seeds to replace monoculture annual grain cultivation (Benyus, 1997; Land Institute, 2004)

Entropy carpet. A significant amount of waste is generated in the manufacture of carpet, often due to slight colour differences from batch to batch that also make the replacement of worn carpet tiles more difficult. Designers at Interface have explored natural models for design and colour, which have resulted in 'Entropy', a carpet that |

Table 2.5 (continued)

Use Nature As	Biomimicry Leads To	Practical Examples
		mimics the random colour palette of a grassland and forest floor, resulting in easier matching of replacement tiles, fewer discards and easier installation, all ultimately resulting in waste reduction (Interface, 2004)
• Mentor	A new way of viewing and valuing nature, changing the paradigm from what we can 'extract' from nature to what we can 'learn' from nature	*The lotus effect.* Plants like the lotus have self-cleaning surfaces that encouraged researchers at University of Bonn to investigate how nature cleans surfaces. It became apparent that nature does not use detergents but has a way of structuring surfaces with self-cleaning properties, essentially causing rain droplets to roll off the surface and thereby automatically removing dirt as they wash over. This insight has motivated material and product designers in sectors as diverse as coatings, textiles and plastics to design products with self-cleaning surfaces. A German company, ISPO, manufactures Lotusan, a building facade paint that dries with lotus-like bumps so that rainwater cleans the building (ISPO, 2004; UoB, 2004)
		Structural colour. Structural colour makes tropical butterflies and peacocks so gorgeous. Their 'colours' result from light scattering of regularly spaced melanin rods and interference effects through thin layers of keratin. Iridigm, based in San Francisco, is using structural colour ideas from tropical butterflies to create screens for Personal Digital Assistants that can be easily read in sunlight (Skula, 2003)

Source: Framework based on Benyus (1997).

- Nature banks on diversity.
- Nature demands local expertise.
- Nature curbs excesses from within.
- Nature taps the power of limits.

Green chemistry and green engineering principles reflect the first four similarities to a certain degree, even though neither green chemistry nor green engineering would normally be included as biomimicry initiatives.

Green chemistry
Green chemistry is commonly defined as the 'design, development, and implementation of chemical products or processes to reduce or eliminate the use and generation of hazardous and toxic substances' (Hjeresen et al., 2002, p. 260). Since the early 1990s green chemistry has gradually become recognized as both a culture and a methodology for achieving sustainability through chemical process and product development (Lancaster, 2002). At its core green chemistry is driven by concerns for the 'fate' and 'toxicity' of the raw materials, intermediaries and products involved in chemical synthesis:

- *Fate*. Green chemistry is aimed at reducing the dispersion and accumulation of synthetic chemical products in the environment. At the process level this translates into maximizing atom efficiency (which is a measure for how many atoms of the reactants end up in the final product and how many end up in by-products or waste). At the product level fate considerations translate into securing product degradation after its useful life to innocuous substances.
- *Toxicity*. Green chemistry is aimed at less hazardous materials, processes and products. In this context, 'hazardous' pertains to toxicity, flammability, explosion potential and environmental persistence. Toxicity is considered over the lifecycle of the chemical product, including the generation of toxic substances during raw material extraction and production, chemical synthesis, formulation, packaging, distribution and use of the chemical product and from its disposal after use.

The initial focus on fate and toxicity has gradually expanded to include a larger set of environmental considerations, particularly the resource efficiency of the process (its requirements for energy, water and process auxiliaries) and the fate of the raw materials (whether or not they are renewable as well as their relative abundance).

The 12 principles of green chemistry (Table 2.6) have structured the development of this research discipline and industrial practice. Initially formulated by Anastas and Warner (1998), these principles may be formulated as mutually reinforcing and partly overlapping design criteria for product and process chemistries.

The application of green chemistry design criteria generally improves not only the environmental performance of the product or process chemistry but also its financial performance through lower operating and/or lower capital expenses as illustrated by the following real-world examples:

Ibuprofen is a key active ingredient in much over-the-counter medication for pain relief. The traditional synthesis developed and patented by the Boots Company of England in the 1960s involved a six-step chemical synthesis with an atom efficiency of only 40 per cent

Table 2.6   Green chemistry design criteria

Green Chemistry Design Criteria	Principles of Green Chemistry (Anastas and Warner, 1998)
1. Design for waste prevention	It is better to prevent waste than to treat or clean up waste after it is formed
2. Design for atom economy	Synthetic methods should be designed to maximize the incorporation of all materials used in the process into the final product
3. Design for less hazardous synthesis	Wherever practical, synthetic methods should be designed to use and generate substances that pose little or no toxicity to people or the environment
4. Design of safer chemicals	Chemical products should be designed to effect their desired function while minimizing their toxicity
5. Design for safer solvents and auxiliaries	The use of auxiliary substances (e.g. solvents or separation agents) should be made unnecessary whenever possible and innocuous when used
6. Design for energy efficient synthesis	Energy requirements of chemical processes should be recognized for their environmental and economic impacts and should be minimized. If possible, synthetic methods should be conducted at ambient temperature and pressure
7. Design from renewable feedstocks	A raw material or feedstock should be renewable rather than depleting whenever technically and economically practicable
8. Design for minimization of derivatization	Unnecessary derivatization (use of blocking groups, protection/de-protection and temporary modification of physical/chemical processes) should be minimized or avoided if possible because such steps require additional reagents and can generate waste
9. Design for catalysis	Catalytic reagents (as selective as possible) are superior to stoichiometric reagents
10. Design for product degradation	Chemical products should be designed so that at the end of their functional use they can break down into innocuous degradation products and do not persist in the environment
11. Design for real-time analysis	Analytical methodologies need to be further developed to allow for real-time, in-process monitoring and control prior to the formation of hazardous substances
12. Design for accident prevention	Substances and the form of a substance used in a chemical process should be chosen to minimize the potential for chemical accidents, including releases, explosions and fires

(which means that for every tonne of product produced, at least 1.5 tonnes of process waste is generated from process reagents not incorporated into the final product – a figure that would be even larger if the process yield falls short of the theoretical yield). In the mid-1980s the Boots Company formed the BHC joint venture with the Hoechst Celanese Corporation to develop and implement a new synthesis route for ibuprofen. The green process is based on a three-step catalytic synthesis. The atom efficiency of the green synthesis is 77 per cent, which climbs to 99 per cent if one considers that the acetic acid generated in the first process step is regenerated (Cann and Connelly, 2000). This means that even in the absence of acetic acid recovery process waste generation from process reagents not incorporated into the final product drops to 0.3 of a tonne per tonne of product (an 80 per cent decrease compared with the traditional synthesis). The three catalysts used (hydrogen fluoride, Raney nickel and palladium) are all recovered and re-used. The simplification of the process chemistry has significant capital and operation cost advantages. The green synthesis was first implemented on an industrial scale at the world's largest ibuprofen manufacturing facility in Bishop (Texas) in 1992, which has since produced approximately 20–25 per cent of the world's annual supply of ibuprofen (Cann and Connelly, 2000).

Nature Works ™ *polylactic acid* (PLA) is an alternative for plastic consumer products using renewable feedstocks rather than petrochemicals (Hjeresen et al., 2002). Cargill Dow has begun operation of a full-scale production plant for PLA, initially using corn sugar to make polymers for fibres or plastic packaging. The intent is to replace corn sugar with agricultural by-products over time. The plant can produce 140 000 metric tonnes per year and will recycle by-products internally. By starting with a fermentation process, traditional solvent-based processes are avoided as are the disposal problems associated with them. Moreover, fossil fuel requirements are reduced by as much as 50 per cent. The wear properties of the product are excellent and the product can be recycled with ease, preferably through hydrolysis to return to lactic acid followed by reprocessing to make feedstock-grade polymer. Alternatively the product is compostable and will also degrade in landfills.

A green chemical process with many applications is the use of *supercritical CO_2*. This process is already established in the production of decaffeinated coffee and is often put forward as the ultimate alternative for dry-cleaning with (halogenated) solvents. More recently the efforts of Los Alamos National Laboratory in collaboration with the semiconductor industry have proven the ability of supercritical CO_2 to replace aggressive surfactants and cleaners in the photolithographic processes involved in wafer manufacturing (Hjeresen et al., 2002).

The application of green chemistry is closely intertwined with green engineering. Green engineering is also known as designing pollution prevention into chemical processes (CMA, 1993) or environmentally conscious design of chemical processes (Allen and Shonnard, 2002). It is based on the notion that pollution prevention (or eco-efficiency) options are most effective from the perspective of both cost and environment if they are considered early, throughout the process design and construction stages of chemical and other facilities.

In contrast to green chemistry, however, green engineering has not yet matured into an established research discipline and industrial practice. This might reflect the fact that greener engineering is not about doing engineering differently but rather about doing engineering better to address a broader set of design criteria. Essentially greener engineering adds environmental and resource efficiency considerations to the traditional process design criteria, which include capital and operating costs, process reliability, constructability, operation and maintenance. Green engineering provides analytical tools to evaluate systematically and improve the environmental performance of chemical processes. The key steps are (Allen and Shonnard, 2002):

- to specify the product to be manufactured and evaluate potential environmental fate, releases and exposures (the domain of environmental physics, chemistry and geology);
- to establish the input/output structure of the chemical process, including chemical synthesis pathways and potential by-products (the domain of green chemistry);
- to evaluate potential emission and environmental impacts associated with the conceptual process (a first-tier environmental release and risk assessment);
- to specify the unit operations and process flows and identify pollution prevention opportunities (the domain of cleaner production and eco-efficiency) (see also Chapter 5);
- to examine systematically the flowsheet to identify opportunities for environmental improvements and identify opportunities for energy and mass integration (the domain of process integration);
- to evaluate the environmental performance of the detailed process flowsheet (second-tier environmental release and risk assessments); and
- to evaluate the environmental costs associated with the process (application of environmental management accounting).

Outlook

Industrial ecology is best understood as using nature as a source of ideas to realize the environmentally sound provision of products and services with the ultimate aim of balancing industrial production and consumption with the sustainable use of natural resources, including materials, energy, water and the capacity of the environment to assimilate wastes and render valuable services. Industrial ecology has the grand ambition of reorganizing the industrial system so that it evolves towards a mode of operation that is compatible with the biosphere and is sustainable over the long term. Ultimately this revolves around the design of a quasi-cyclical economy, the minimization of dissipative losses, the increase of resource productivity and balancing the industrial diet (Erkman, 2003).

This chapter has discussed the practice of industrial ecology in terms of two major themes, dealing respectively with system applications and product and process applications. The systems level is driven by the natural system metaphor and its principal methodology is industrial metabolism, or the study of material, energy and other resource flows through the industrial system of production and consumption. The product and process level is more diverse but is governed in principle by the analogy with biological and other natural processes.

In sum, three subsets of activity in the industrial ecology field are already profoundly shaping the development and implementation of environmental technology:

1. *Industrial symbiosis.* The steep increase in activities on eco-industrial parks and networks globally is the best-known area of practical application of industrial ecology.

At its core, industrial symbiosis provides an extension and intensification of the well-known practices of process integration and manufacturing and recycling networks.

2. *Biomimicry*. By looking at natural processes and materials to solve industrial problems, biomimicry offers a refreshing perspective on industrial production and provides a launch pad for transformative technical change with the prospect of step changes in the environmental performance of products and processes.

3. *Green chemistry and engineering*. These processes are aimed at 'better AND greener' process and product chemistries and engineering designs and artefacts. The key challenge is to provide the tools and metrics to better incorporate environmental considerations in all stages of product and process chemistry development and engineering design. In many respects green chemistry and engineering can be seen as stretching the envelope of the technical dimensions in cleaner production and eco-efficiency.

References

Allen, D. and Shonnard, D. (2002), *Green Engineering: Environmentally Conscious Design of Chemical Processes*, Upper Saddle River, New Jersey: Prentice Hall Publishers.

Anastas, P. and Warner, J. (1998), *Green Chemistry: Theory and Practice*, Oxford, UK: Oxford University Press.

Ayres, R. (1994), 'Industrial Metabolism: Theory and Practice', in Allenby, B. and Richards, D. (eds), *The Greening of Industrial Ecosystems*, Washington: National Academy of Engineering, pp. 23–37.

Behrendt, S., Jash, C., Peneda, M. and van Weenen, H. (1998), *Life Cycle Design: A Manual for Small and Medium Sized Enterprises*, Berlin: Springer Verlag.

Benyus, J. (1997), *Biomimicry – Innovation Inspired by Nature*, New York: Harper Collins Publishers.

Bringezu, S. and Moriguchi, Y. (2002), 'Material Flow Analysis', in Ayres, R. and Ayres, L. (eds), *Handbook of Industrial Ecology*, Cheltenham, UK and Northampton, MA, USA: Edward Elgar, pp. 79–90.

Cann, M. and Connelly, M. (2000), *Real World Cases in Green Chemistry*, Washington, DC: American Chemical Society.

Chambers, N., Simmons, C. and Wackernagel, M. (2000), *Sharing Nature's Interest: Ecological Footprints as an Indicator of Sustainability*, London: Earthscan Publications.

Chemical Manufacturers Association (CMA) (1993), *Designing Pollution Prevention into the Process: Research, Development and Engineering*, Washington, DC: CMA.

Chertow, M. (2000), 'Industrial Symbiosis: Literature and Taxonomy', *Annual Review of Energy and Environment*, **25**, 313–37.

Davidson, S. (2003), 'Light Factories', *ECOS – Science for a Sustainable Future*, **117**, October–December, 10–13.

Ehrenfeld, J. and Gertler, N. (1997), 'Industrial Ecology in Practice: The Evolution of Interdependence at Kalundborg', *Journal of Industrial Ecology*, **1**(1), 67–79.

Erkman, S. (2003), 'Perspectives on Industrial Ecology', in Bourg, D. and Erkman, S. (eds), *Perspectives on Industrial Ecology*, Sheffield, UK: Greenleaf Publishing, pp. 338–42.

Frosh, R. and Gallopoulos, N. (1989), 'Strategies for Manufacturing', *Scientific American*, September, 94–102.

Graedel, T. and Allenby, B. (2003), *Industrial Ecology* (2nd edition), Upper Saddle River, New Jersey, USA: Pearson Education Inc.

Guinée, J.B. (2002), *Handbook on Life Cycle Assessment: Operational Guide to the ISO Standards*, Dordrecht, the Netherlands: Kluwer Academic Publishers.

Hjerisen, D., Kirchhoff, M. and Lankey, R. (2002), 'Green Chemistry: Environment, Economics and Competitiveness', *Corporate Environmental Strategy*, **9**(3), 259–66.

Huber, J. (2000), 'Industrial Ecology: Sustainable Development as a Concept for Ecological Modernization', *Journal of Environmental Policy and Planning*, **2**, 269–85.

Interface (2004), *Interface's Entropy*, www.interfaceinc.com/us/feature/entropy, accessed 5 February 2004.

Isenmann, R. (2003), 'Industrial Ecology: Shedding More Light on Its Perspective of Understanding Nature as a Model', *Sustainable Development*, **11**, 143–58.

ISPO (2004), *Lotusan Product Information* (in German), www.lotusan.de/lotusan/_02_wissen/index.jsp, accessed 5 February 2004.

Kincaid, J. (2003), 'Region-wide Eco-industrial Networking in North Carolina', in Cohen-Rosenthal, E. and Musnikow, J. (eds), *Eco-Industry Strategies: Unleashing Synergy Between Economic Development and the Environment*, Sheffield, UK: Greenleaf Publishing, pp. 317–21.

Kincaid, J. and Overcash, M. (2001), 'Industrial Ecosystem Development at the Metropolitan Level', *Journal of Industrial Ecology*, **5**(1), 117–26.
Korhonen, J. (2001), 'Four Ecosystem Principles for an Industrial Ecosystem', *Journal of Cleaner Production*, **9**, 253–59.
Lambert, A. and Boons, F. (2002), 'Eco-industrial Parks: Stimulating Sustainable Development in Mixed Industrial Areas', *Technovation*, **22**, 471–84.
Lancaster, M. (2002), *Green Chemistry: An Introductory Text*, Cambridge, UK: Royal Society of Chemistry.
Land Institute (2004), *Natural Systems Agriculture*, www.landinstitute.org/vnews/display.v/ART/1999/07/01/377bbbe53, accessed 5 February 2004.
Lifset, R. and Graedel, T.E. (2002), 'Industrial Ecology: Goals and Definitions', in Ayres, R. and Ayres, L. (eds), *Handbook of Industrial Ecology*, Cheltenham, UK and Northampton, MA, USA: Edward Elgar, pp. 3–15.
Mangan, A. (1998), *By-Product Synergy: A Strategy for Sustainable Development – A Primer*, Austin, Texas, USA: Business Council for Sustainable Development – Gulf of Mexico.
Matthews, E. (2000), *The Weight Of Nations: Material Outflows from Industrial Economies*, Washington, DC: World Resources Institute.
Robert, K. (2003), 'Integrating Sustainability into Business Strategy and Operations: Applying The Natural Step Approach and Framework and Backcasting from Principles of Sustainability', in Waage, S. (ed.), *Ants, Galileo and Gandi: Designing the Future of Business through Nature, Genius and Compassion*, Sheffield, UK: Greenleaf Publications, pp. 61–80.
Schaltegger, S. and Burritt, R.L. (2000), *Contemporary Environmental Accounting: Issues, Concepts and Practice*, Sheffield, UK: Greenleaf Publishing.
Skula, A. (2003), 'Back to Nature: Biomimicry Finds Engineering Solutions in the Natural World', *MIT Technology Review*, 28 March.
Socolow, R. (1994), 'Six Perspectives from Industrial Ecology', in Socolow, R., Thomas, V., Berkhout, F. and Andrews, C. (eds), *Industrial Ecology and Global Change*, New York: Cambridge University Press, pp. 3–16.
Technical University of Delft (TUD) (2004), *Marine Mussels Clue to Medical Adhesives*, http://www.delftoutlook.tudelft.nl/info/index2b34.html?hoofdstuk=Article&ArtID=3992, accessed April 2006.
University of Bonn (UoB) (2004), *The Lotus Effect®*, www.botanik.uni-bonn.de/system/bionik_flash_en.html, accessed 5 February 2004.
van der Voet, E. (2002), 'Substance Flow Analysis Methodology', in Ayres, R. and Ayres, L. (eds), *Handbook of Industrial Ecology*, Cheltenham, UK and Northampton, MA, USA: Edward Elgar, pp. 91–101.
White, R. (1994), 'Preface', in Allenby, B. and Richards, D. (eds), *The Greening of Industrial Ecosystems*, Washington, DC: National Academy Press, pp. v–vi.

3 Remaking the way we make things: creating a new definition of quality with cradle-to-cradle design
William McDonough and Michael Braungart

The culture of innovation within the field of environmental technology and management is bringing forth significant change in the world of industry. From the growing influence of green chemistry and engineering to the emergence of environmental concerns in corporate research and development, one can see promising new initiatives in nearly every sphere of industrial activity.

Change in fact is scaling up, with innovations in the design of single products paving the way for the development of ecologically intelligent design standards throughout entire industries, such as packaging and electronics. Indeed entire nations are changing course. In China, for example, President Hu Jintao called for the development of 'a circular economy' in which the clean production, recovery and re-use of intelligently designed materials will generate sustainable, long-term growth. Chinese officials meanwhile have titled the Chinese edition of our book, *Cradle to Cradle: Exploring Design for the Circular Economy* (McDonough and Braungart, 2004), suggesting a new awareness of the necessity and effectiveness of ecologically intelligent design.

While a significant number of industrial designers have begun to focus on cradle-to-cradle goals – which could be summarized as moving from the lean production of potentially degenerative technology to *the clean production of potentially regenerative technology* – many new developments in the field are limited by the 'eco-efficient' framework in which they are applied. A widely adopted business paradigm, eco-efficiency is often only a reductive agenda, its reforms rather narrowly aimed at minimizing the negative impacts of industry. New management tools based simply on efficiency for example, may allow industry to use fewer resources, produce less waste and minimize toxic emissions, but they tend not to proactively or positively change the fundamental design of products or industrial production. In other words, efficient may not be sufficient. As a result, even promising new technologies use energy and materials within a conventional cradle-to-grave system, diluting pollution and slowing the loss of natural resources without addressing the systemic design flaws that create waste and toxic products in the first place.

Global sourcing and lean production have standardized this state of affairs, with the result being a surfeit of products characterized by increasingly poor quality. Consider the off-gassing (release of gases) diagram of a name-brand children's toy marketed from the United States shown below in Figure 3.1, which identifies more than 30 chemicals known to be mutagenic, desensitizing, or even suspected or known carcinogens.

As the diagram illustrates, poor standards of quality result in everyday products that release hundreds of hazardous chemicals. These products – from electric shavers to carpets and upholsteries – are typically used indoors, where off-gassed chemicals accumulate. Energy-efficient buildings, which are designed to require less heating and cooling, and thus less air circulation, can make things worse. A recent study in Germany for

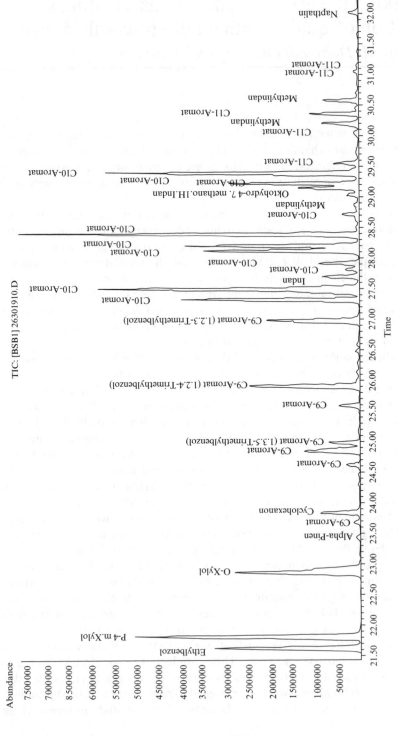

Source: © Copyright 2002 by Hamburger Umweltinstitut e.V.

Figure 3.1   Off-gassing diagram of a name brand United States children's toy

example, found that air quality inside several highly rated energy-efficient buildings in downtown Hamburg was nearly four times worse than on the dirty, car-clogged street.

The effects are hard to ignore. Where buildings with reduced air-exchange rates are common, so are health problems. In Germany where tax credits support the construction of energy-efficient buildings allergies affect 42 per cent of school-age children between six and seven years old, largely due to the poor quality of indoor air. This is what we would call chemical harassment. It is the result not of bad intentions but of poor design.

A new model for industry

Cradle-to-cradle design offers a clear alternative, a framework in which the safe regenerative productivity of nature provides models for wholly positive human designs. Working from this perspective, we do not aim to be less bad. Instead, our design assignment is to create a world of interdependent natural and human systems powered by the sun in which safe, healthful materials flow in regenerative cycles, elegantly and equitably deployed for the benefit of all.

Within this framework every material is designed to provide a wide spectrum of renewable assets. After a useful life as a healthful product, cradle-to-cradle materials are designed to replenish the earth with safe, fecund matter or to supply high-quality technical resources for the next generation of products. When materials and products are created specifically for use within these closed-loop cycles – the flow of biological materials through nature's nutrient cycles and the circulation of industrial materials from producer to customer to producer – businesses can realize both enormous short-term growth and enduring prosperity. As well, we can begin to redesign the very foundations of industry, creating systems that purify air, land and water; use current solar income and generate no toxic waste; use only safe, healthful, regenerative materials; and whose benefits enhance all life.

This positive industrial agenda identifies a new definition of quality in product, process and facility design. From the cradle-to-cradle perspective, quality is embodied in designs that allow industry to enhance the well-being of nature and culture while generating economic value. Pursuing these positive aspirations at every level of commerce adds ecological intelligence, social equity and cultural diversity to the conventional design criteria of cost, performance and aesthetics. When these diverse criteria define good design and when they are applied at every level of industry, productivity and profits are not at odds with environmental and social concerns. Indeed as cradle-to-cradle design matures we are increasingly able to design products and places that support life, that create ecological footprints to delight in rather than lament. This changes the entire context of the design process. Instead of asking: 'How do I reduce the impact of my work? How do I meet today's environmental standards?', we ask: 'How might I increase my ecological footprint and enhance its positive effects? How might I grow prosperity and celebrate my community? How might I create more habitat, more health, more clean water, more delight?'

The cradle-to-cradle paradigm

Cradle-to-cradle design refocuses product development from a process aimed at limiting end-of-pipe liabilities to one geared to creating safe, healthful, high-quality products right from the start. In the world of industry it is creating a new conception of materials and material flows. Rather than seeing materials as a waste management problem in which interventions here and there slow their trip from cradle to grave, cradle-to-cradle thinking sees materials as nutrients and recognizes two safe metabolisms in which they flow.

In the first, 'biological metabolism', the nutrients that support life on Earth – water, oxygen, nitrogen, carbon dioxide – flow perpetually through regenerative cycles of growth, decay and rebirth. Rather than generating material liabilities, the biological metabolism accrues natural fecundity. Waste equals food. The second, 'technical metabolism' can be designed to mirror natural nutrient cycles; it is a closed-loop system in which valuable, high-tech synthetics and mineral resources circulate in an endless cycle of production, recovery and remanufacture. Ideally the human systems that make up the technical metabolism are powered by the energy of the sun. See Figure 3.2.

By specifying safe, healthful ingredients, industry can create and use materials within these cradle-to-cradle cycles. Materials designed as 'biological nutrients', such as detergents, packaging or textiles for draperies, wall coverings and upholstery fabrics, can be designed to biodegrade safely and restore the soil after use, providing more positive effects, not fewer negative ones. Materials designed as 'technical nutrients', such as perpetually recyclable nylon fibre, can provide high-quality, high-tech ingredients for generation after generation of synthetic products – again a harvest of value.

Biological and technical nutrients have already entered the marketplace. The upholstery fabric Climatex Lifecycle® is a blend of pesticide-residue-free wool and organically grown ramie (vegetable fibre), dyed and processed entirely with non-toxic chemicals. All of its product and process inputs were defined and selected for their human and ecological safety within the biological metabolism. The result: the fabric trimmings are made into felt and used by garden clubs as mulch for growing fruits and vegetables, returning the textile's biological nutrients to the soil. The first product on the market designed as a biological nutrient, Climatex Lifecycle®, has been followed by many others since its introduction in the early 1990s.

Honeywell meanwhile is marketing a textile for the technical metabolism, a high-quality carpet yarn called Zeftron Savant®, which is made of perpetually recyclable nylon 6 fibre. Zeftron Savant® is designed to be reclaimed and repolymerized (taken back to its constituent resins) to become new material for new carpets. In fact Honeywell can retrieve old, conventional nylon 6 and transform it into Zeftron Savant®, upcycling rather than downcycling an industrial material. The nylon is rematerialized, not dematerialized – a true cradle-to-cradle product.

Ideally technical nutrients are designed as 'products of service', a key element of the cradle-to-cradle strategy. Products of service are goods designed by their manufacturer to be taken back and used again to make high-quality products. The product provides a service to the customer while the manufacturer, or the company that has sold the product, in effect maintains ownership of the product's material assets, taking them back directly from customers or through generalized recycling systems, and re-using them again and again through many product lifecycles.

The product of service concept can increase the effectiveness of companies in a wide variety of industries if one understands the subtleties of its application and deployment. First, an important distinction. The cradle-to-cradle product of service concept is quite different from product of service concepts founded on durability, which minimize a product's ecological footprint by amortizing its environmental costs over an extended

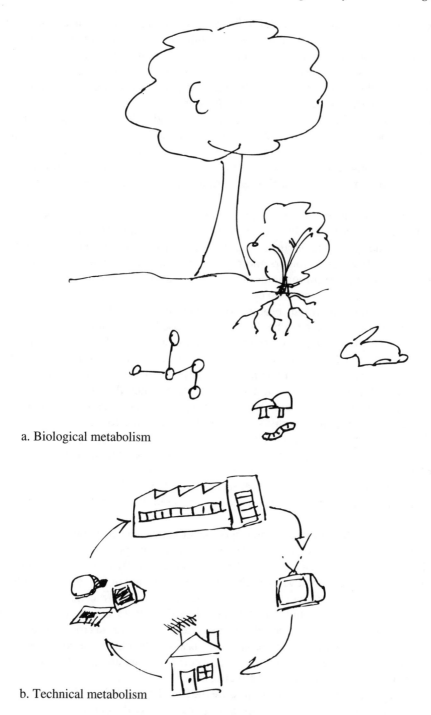

a. Biological metabolism

b. Technical metabolism

Figure 3.2   Biological metabolism/technical metabolism

period of time. According to the durability model, the manufacture of bulldozers is less harmful to the environment if bulldozers are built to last 20 years rather than five. To be sure, extending the life of bulldozers decreases current inputs, but strategies such as these typically do not take into account eventual end-of-life scenarios that ultimately lead to the loss of valuable materials. And, if the market demands more bulldozers, productivity increases eventually outweigh environmental gains.

The cradle-to-cradle product of service model includes durables but also many other consumer goods, including disposable products that may last a matter of months. The key is not durability over a long, linear life but design that creates a true, closed-loop lifecycle and a long-term relationship with customers. The flexibility and utility of the concept can be appreciated best through examples of four different product of service models:

Leading manufacturers in the carpet industry have engaged the product of service concept over the past decade and among them Shaw Industries has done so most effectively. By design, the company makes its carpets from safe, recyclable materials, takes back its carpets from its customers and recycles them into new, high-quality carpets. The flow of materials goes from manufacturer to customer to manufacturer, creating a safe, profitable and effective closed-loop cycle.

The second cradle-to-cradle product of service model applies to durable products such as automobiles, which can be designed to be used for a relatively short period, say three to five years, and then returned to their manufacturer for disassembly and re-manufacture. The relationship created by this model allows customers to frequently upgrade and enjoy the latest in technological advances – critical in the current shift to more energy-effective vehicles – and keeps valuable materials out of the landfill and flowing through the automakers' manufacturing processes. Already manufacturers in many sectors are designing for disassembly, both in response to new 'take-back' laws in the EU and in preparation for more fully realized material recovery systems.

A third model applies to generalized cradle-to-cradle recycling in which products and materials are not returned to a specific manufacturer but are sent to centralized recyclers. Products designed for this closed-loop cycle would include things like plastic beverage bottles, which might be returned to a 'polymer bank' that supplies an industry group.

The fourth model that includes products like athletic shoes distributed by design and marketing companies such as Nike, blends aspects of the previous three. In the Shaw model, carpet goes from the customer directly back to the carpet company and is made into new carpet; there is great value in the recovery and re-use of the material itself. In the design and marketing model, in which a company uses contract manufacturers, the product is returned to a retail outlet rather than directly to the manufacturer. While the shoe is ultimately recycled, the value of its materials in this case grows more out of the bond it creates between customer, retailer and brand; when the shoe comes back to the store, the retailer and the design and marketing company have an opportunity to continue to provide new shoe services to a valued customer.

Material recovery systems such as these are the foundation of the technical metabolism. Widely practised, the product of service concept can change the nature of production and consumption as companies deploy manufacturing and material recovery systems that allow them to re-use valuable materials through many product lifecycles.

The practice of cradle-to-cradle design

The cradle-to-cradle design framework incorporates nature's cyclical material model into all product and system design efforts through a process called lifecycle development (LCD). Product design within this framework is not the same as lifecycle assessment (LCA), which may simply catalogue an existing product's components at end-of-life. Lifecycle development (see Figure 3.3), on the other hand, uses 'lifecycle thinking' as a method of scientific inquiry and then applies it to the design of new, innovative products optimized for closed-loop systems.

Lifecycle development is a working, results-oriented method for evaluating products and processes as they are being re-designed. While observing how a material or final product flows through any of its various lifecycle stages (raw materials production, manufacturing, use and recovery/reutilization), and identifying human and environmental health impacts at each stage, LCD phases out undesirable substances and replaces them with preferable ones. The re-design process occurs during – not after – environmental and human health assessment. This simultaneous work saves costs for manufacturers and users and allows manufacturers to maintain market presence and continue to generate revenue as they improve their products.

The LCD process is made up of three phases, which follow an initial identification of a product's proper metabolism. Defining each product as either a potential biological nutrient or a potential technical nutrient sets up two different sets of design criteria and informs all phases of product development. Biological nutrients will need to be compostable for example, while the recovery of technical nutrients might require chemical recycling. All products, however, are assessed and developed through three phases:

- inventory of material flows;
- impact assessment according to the lifecycle of individual products; and
- optimization to produce a healthy, prosperous cradle-to-cradle lifecycle.

These phases represent an iterative process that can be engaged at many levels. The process can start with an idea as well as with a raw material (going into different products), a product (made of different raw materials) or a process.

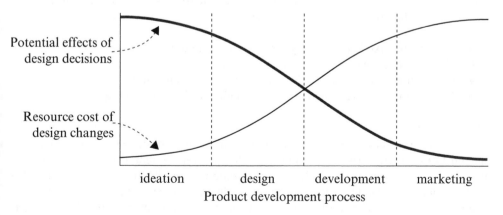

Figure 3.3   Lifecycle development process

Inventory
The first step in LCD is a material inventory designed to collect full information on every material used in the manufacture of a product. Each material is then inventoried for its chemical constituents. The inventory process results in a complete listing of components by CAS (Chemical Abstracts Service) number, name, function and percentage weight of the final material or product.

Impact assessment
Assessing materials encourages product transparency and the conscious selection of ingredients that will have the most positive impact on human and ecological health. A material's impact on human and environmental health is assessed in five basic categories:

1. Direct exposure covers the acute and chronic toxicological impacts on organisms that might be exposed to the materials, including carcinogenicity, endocrine disruption, irritation of skin and mucous membranes and sensitization.
2. The succession of generations includes potential impacts such as mutagenicity, reproductive and developmental toxicity, genetic engineering and persistence and biodegradation.
3. Food chains are evaluated according to their bioaccumulation potential.
4. Climatic relevance is evaluated based on global warming potential and ozone depletion potential.
5. Value recovery assesses a material's potential as a biological or technical nutrient. To recover value and maintain materials in closed-loop cycles, materials must either be returned safely to the soil or be perpetually recyclable. Evaluation of the value recovery potential of a material is based on the following considerations: (a) Is it technically feasible to compost or recycle the material? (b) Does a recycling or composting infrastructure exist for the material? (c) What is the resulting quality of the recycled material or compost?

In addition, products must have a defined end-of-use strategy and be designed for disassembly so that recovery of materials is possible. Evaluating the recoverability of materials in a product is based on the following questions:

•   What is the take-back strategy for the product and its materials?
•   Can dissimilar materials be easily separated?
•   Can common or readily available disassembly tools be used?
•   Can the material type be identified through markings, magnets and so on?

Optimization
Once all materials have been assessed, those with the most positive human and environmental health characteristics and highest value recovery potential may be selected for inclusion in a re-designed product. Optimization is an iterative process. Complete optimization of a product or material may initially be impossible due to time or financial constraints, or lack of materials that meet the criteria for environmental and human health and value recovery. When all problematic inputs cannot be substituted, they can

be prioritized for replacement and the manufacturing process can be re-designed to minimize exposure until a positive replacement is identified. Ultimately the optimization phase is designed to yield positively defined products that enhance commercial productivity, social health and ecological intelligence.

Getting results: cradle-to-cradle design at work

The LCD process is the foundation for designing biological and technical nutrients. Examining some of the details of the design of a particular cradle-to-cradle product illustrates the process at work and begins to suggest how it can yield extraordinary value.

Consider the technical nutrient carpet tile developed by Shaw Industries. Seeking a safe beneficial product for its commercial customers, Shaw undertook a thorough scientific assessment of the material chemistry of its carpet fibres and backing. Dyes, pigments, finishes, auxiliaries – everything that goes into carpet – were examined according to the cradle-to-cradle LCD and each ingredient was selected to meet its rigorous criteria. Out of this process has come the promise of a fully optimized carpet tile, a completely safe, perpetually recyclable, value-generating product. And a highly regarded product as well: Shaw's new design won the 2003 Presidential Green Chemistry Challenge Award.

Awards aside, Shaw's new product brings a much-needed alternative to the commercial carpet market. Typically carpet is made from two primary elements: a face fibre and a backing. Most face fibre today is nylon and most carpet backing is PVC (polyvinyl chloride). Commonly known as vinyl, PVC is a cheap, durable material widely used in building construction and a variety of consumer products, including toys, apparel and sporting goods. The vinyl chloride monomer used to make PVC is a human carcinogen and incineration of PVC can result in dioxin emissions. There are also concerns about the health effects of many additives commonly used in PVC, such as plasticizers, which off-gas chemicals known to be endocrine disrupters.

During conventional carpet recycling nylon face fibre and PVC backing are recycled together, which yields a hybrid material of lesser value. In effect the materials are not recycled at all but downcycled – and they're still on a one-way trip to the landfill or incinerator. There the PVC content of the material makes recycled carpet hazardous waste.

Responding to widespread scientific and consumer concern about PVC, Shaw developed an alternative, a safe polyolefin-based backing system with all the performance benefits of PVC, which it guarantees it will take back and recycle into safe polyolefin backing.

The face fibre of Shaw's technical nutrient carpet tile also changes the game. Shaw's face fibre is made from nylon 6, which can be easily depolymerized into its monomer (caprolactam) and repolymerized repeatedly to make high-quality nylon 6 carpet fibre. The main alternative face fibre – known as nylon 6,6 – while recyclable, is not as easily depolymerized. Following protocols for value recovery, Shaw is developing an effective take-back and recycling strategy for all of its nylon 6 fibre.

In effect, Shaw's new carpet tile does not just minimize waste, it eliminates the concept of waste, the fundamental goal of cradle-to-cradle design we identified in *The Hannover Principles* in 1992 (McDonough, 1992). The company now guarantees that all of its nylon 6 carpet fibre will be taken back and returned to nylon 6 carpet fibre, and its safe polyolefin backing taken back and returned to safe polyolefin backing. All the materials that go into the carpet will continually circulate in technical nutrient cycles. Raw material to raw material.

Waste equals food. This cradle-to-cradle cycle, which is altogether different from eco-efficient recycling, suggests the benefits of a positive approach to managing material flows.

Other industries are also achieving significant results. Working with McDonough Braungart Design Chemistry (MBDC), the footwear manufacturer Nike is employing the cradle-to-cradle framework to determine the chemical composition and environmental effects of the materials used to produce its line of athletic shoes. Focusing primarily on Nike's global footwear operations, the company's material assessment began with factory visits in China where teams collected samples of rubber, leather, nylon, polyester and foams, along with information on their chemical formulations.

In an ongoing partnership, when Nike and MBDC identify materials that meet or exceed the company's sustainable design criteria, those components are added to a growing palette of materials (a positive list) that Nike will use in its products. These ingredients are designed either to be safely metabolized by nature's biological systems at the end of the product's useful life or to be repeatedly recovered and reutilized in new products.

Nike's systematic effort to develop a positive materials palette has begun to produce tangible results, such as the phasing out of PVC. After two years of scientific review Nike set its sights on the elimination of PVC from footwear and non-screenprint apparel. In the spring of 2002, Nike highlighted two PVC-free products, Keystone Cleats and Swoosh Slides as a way to begin a dialogue with customers about its PVC-free commitment.

Integrating cradle-to-cradle design strategies

Many companies begin adopting cradle-to-cradle principles by applying them to a single product. Ultimately however the strategy's effectiveness depends on its deep integration into the product development process. The world-renowned modern furniture company, Herman Miller, has gone a long way toward that end, developing an interdisciplinary Design for Environment (DFE) team that implements cradle-to-cradle material assessments, translates design goals throughout the company, measures environmental performance and engages its supply chain in implementing design criteria.

Working closely with MBDC, the DFE team built a chemical and material assessment methodology that could be effectively used by the firm's designers and engineers. Throughout the design process the multifaceted assessment analyses materials for their human health and toxicological effects, recyclability, recycled content and/or use of renewable resources, and product design for disassembly.

The DFE team includes a chemical engineer who incorporates findings from assessments into an evolving materials database, and a purchasing agent who acts as a conduit and data source between the supply chain and Herman Miller's purchasing team. This strategy engages both groups as partners in implementing new design criteria, thereby ensuring the consistent procurement of safe materials. As one Herman Miller engineer has said, 'getting a handle on supply chain issues from an environmental standpoint has also helped us get a handle on the organization and prioritization of materials'. Now for example, Herman Miller can use the new database to record the volume and content of the raw materials it uses and distributes, measures the company had not previously tracked.

Herman Miller put the DFE team to work on the design of its new task chair, a complement to its popular Aeron office chair. After assessing production processes, as well as 500 chemicals in 850 materials, and integrating those findings into the overall design process, Herman Miller unveiled the Mirra. Noting its pioneering design, *Metropolis* magazine suggested that the environmentally sound, high-performance Mirra might be 'the next icon'. Perhaps. What is certain is that the Mirra's combination of ergonomic, aesthetic and environmental intelligence makes it not only extraordinarily comfortable and easy to adjust, but also a shining example of smart material and energy use.

Among other features, the Mirra is assembled using 100 per cent windpower. Recycled content comprises more than 40 per cent of its weight and nearly 100 per cent of its materials can be recycled, a strong step toward a cradle-to-cradle product. The elimination of PVC makes the chair environmentally safe and its overall design makes it easy to disassemble. It is a bold move into twenty-first-century product design.

Managing material flows with intelligent materials pooling

By defining product ingredients and engaging their respective supply chains, Shaw, Nike and Herman Miller are all taking steps toward developing a safe, profitable technical metabolism. This is a critical step in the cradle-to-cradle strategy. Ultimately the key to optimizing the assets of cradle-to-cradle materials lies in the intelligent management of regenerative material flows, just as in the world of energy the optimization of the strategy would lead us toward an effective use of renewable energy.

After eons of evolution, nature is well equipped to effectively manage the material flows of the biological metabolism. We need to be sure that the materials we design as biological nutrients can safely biodegrade and we need to set up recovery systems to be sure they are returned to the soil; but nature does not need our help to run its nutrient cycles. The technical metabolism, however, can only be managed by human design.

To manage the technical metabolism safely and effectively – the flows of polymers, rare metals and high-tech materials on which industry and the human future depend – we have developed a strategy for a 'nutrient management system', which we call intelligent materials pooling (IMP). A collaborative approach to materials flow management, IMP involves multiple companies working together to entirely eliminate hazardous materials. Partners in an IMP form a supportive business community, pooling information and purchasing power to generate material intelligence and profitable cradle-to-cradle materials flow.

The evolution of an intelligent materials pool unfolds in four phases. The first is a community-building phase in which companies committed to cradle-to-cradle design discover shared values and complementary needs. A business network of willing partners emerges as each agrees to work together to phase out a common list of toxic chemicals.

Out of this shared commitment comes a community of companies with the market strength to engineer the phase-out and develop innovative alternative materials. In this second phase the companies share the list of materials targeted for elimination and develop a positive purchasing and procurement list of preferred intelligent chemicals.

The third phase involves defining material flows within the partnership. The partners would specify for and design with preferred materials. They would also establish defined use periods for products and services and individually set up take-back programmes. This phase establishes the infrastructure that supports the 'product of service' concept, in which technical nutrients are designed to be returned to manufacturers for continual re-use. In effect this transforms the partners into a material bank with renewable assets. Their 'pool' of materials is not owned in common. Instead the partners' shared material specifications, effectively managed technical metabolism and combined purchasing power allow them to use positively defined, high-quality materials profitably.

The final phase of IMP is open-ended, as it involves the strengthening of the business partnership through ongoing support. This can involve such mutually beneficial activities as the creation of preferred business partner agreements, the sharing of information, the development of co-branding strategies and support for the mechanisms of the newly created technical metabolism.

Finding willing partners in the competitive world of business might be hard to imagine but it is hardly unprecedented. In the textile industry innovative mills like Victor Innovatex and Rohner Textil, along with MBDC and Designtex, have profitably collaborated on the design and production of ecologically intelligent fabrics. With the technology for recycling polyester under development, an industry-wide polyester coalition could begin to close the loop on the flow of this widely used industrial material. Or companies could participate in a de facto coalition, such as that that has kept in productive use 440 million tons of the 660 million tons of aluminium produced since 1880. So why not other industrial materials, and why not more than 66 per cent?

Design for the triple top line
The various aspects of the cradle-to-cradle strategy, from lifecycle development to intelligent materials pooling, together offer a framework for good design. While the protocols within the framework can be rigorous and exacting, they also create a space for enormous creativity. When a company decides to develop a biological or technical nutrient for example, the chemical assessment of materials is just one step toward a complete re-think of the design assignment. With a good scientific foundation and a positive rather than reductive agenda, one can begin to ask some very interesting design questions.

The conventional design questions revolve around cost, aesthetics and performance. Can we profit from it? Will the customer find it attractive? Will it work? Advocates of sustainable development have tried to expand those questions to include environmental and social concerns. While this 'triple bottom line' approach has given companies a useful tool for balancing economic goals with a desire to 'do better by the environment', the concept in practice often appears to centre only on economic considerations. For example, by using fewer materials or less energy to prevent environmental damage or waste, a company will likely leave a little bit more on the bottom line, but the social and environmental benefits gained from those economic considerations are simply the residues of less damaging activity, not revenue-generating innovations. Like eco-efficiency, this may not be sufficient. Simply using less fossil fuel does not lead to an optimized solution – clean renewable energy. And as a company grows and produces more goods, while it may be able to continue to record positive bottom line events on a per unit basis by using fossil fuels more efficiently, on a production basis, efficiency gains don't keep up with increased output. The bottom line hasn't changed.

But what if this triad of concerns – economic growth, environmental health and social equity – were addressed at the beginning of the design process as 'triple top line' questions rather than used as an accounting tool at the end? That's where the magic begins. Instead of meeting the bottom line through a series of compromises between economy, ecology and equity, designers can employ their dynamic interplay to generate revenue and value in all three sectors – triple top line growth. The goal is to create more positive effects rather than fewer negative ones. From this perspective, questions such as: 'How can I create more habitat? How can I create jobs?' become just as important as 'How much will it cost?' Often in fact a project that begins with pronounced ecological or social concerns can turn out to be tremendously productive financially in ways that would never have been imagined from a purely economic starting point, as we'll soon see in an astonishing example.

The fractal triangle

Working with our clients, we have found that a visual tool, the fractal triangle, helps us apply triple top line thinking throughout the design process. Representing the true ecology of human concerns, the fractal triangle (Figure 3.4) shows how ecology, economy and equity anchor a spectrum of value, and how, at any level of scrutiny, each design decision has an impact on all three. As we plan a product or system, we move around the fractal triangle inquiring how a new design can generate value in each category.

In the pure economy sector, we might ask: 'Can I make my product at a profit?' As we see it, the goal of an effective company is to stay in business as it transforms. The equity sector raises social questions: 'Are we finding ways to honour all stakeholders, regardless of race, sex, nationality or religion?' Moving to the ecology corner, the emphasis shifts to imagining ways in which humans can be tools for nature: 'Do our designs create habitat or nourish the landscape?'

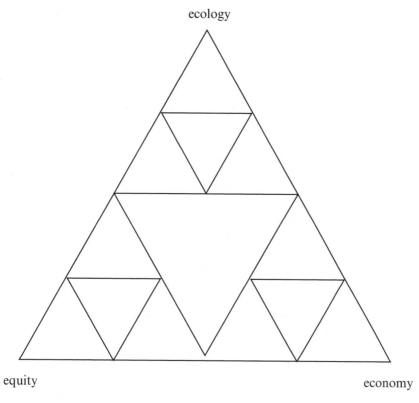

Figure 3.4   Fractal triangle

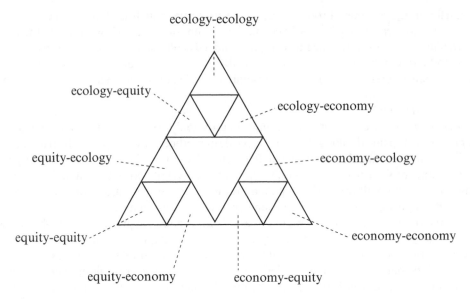

Figure 3.5   Fractal callouts

Fractal callout refers to the labelled triangles, each of which represents the larger-fractal at a finer level of detail. One 'calls out' or highlights a particular triangle to explore in finer detail the relationships at play between ecology, economy, and equity for a particular product or enterprise.

As we move around the triangle, questions expressing a complex interaction of concerns arise at the intersections of ecology, economy and equity. These are represented by the fractal callouts in Figure 3.5 which shows the relationships between the three sectors of the triangle. In the economy/equity sector for example, we consider questions of profitability and fairness: 'Are employees producing a promising product also earning a living wage?' As we continue on to equity/economy, our focus shifts more toward fairness. Here we might ask: 'Are men and women being paid the same amount for the same work?' 'Calling out' each sector of the fractal triangle is a formal way to explore in detail the many opportunities presented by new products, services and enterprises.

Often we discover our most fruitful insights where the design process creates a kind of friction in the zones where values overlap. An ecologist might call these areas 'ecotones', which are the merging, fluid boundaries between natural communities notable for their rich diversity of species. In the fractal triangle the ecotones are ripe with business opportunities.

Triple top line thinkers tap these opportunities not by trying to balance ecology, economy and equity but by honouring the needs of all three. In an infinitely interconnected world, they see rich relationships rather than inherent conflicts. Their goal: to maximize value in all areas of the triangle through intelligent design.

Getting results: generating value in the design process

In projects already underway – indeed, already completed – triple top line thinking has sparked an explosion of creativity in our clients' decision-making. Consider, for example, the restoration of Ford Motor Company's Rouge River plant in Dearborn, Michigan. In

May 1999 Ford decided to invest $2 billion over 20 years to transform the Rouge into an icon of twenty-first-century industry. As we approached the design process with Ford, many wondered if a blue chip company with a sharp focus on the economic bottom line could take a step toward something truly new and inspiring. Could inspiration and profits co-exist?

Well, yes. Using triple top line thinking and the fractal triangle, we explored with Ford's executives, engineers and designers innovative ways of creating shareholder value. Rather than using economic metrics to try to *reconcile* apparent conflicts between environmental concerns and the economic bottom line, the company began to ask triple top line questions. Innovations would still need to be good for profits but Ford's leaders began to examine how profits could be maximized by design decisions that also maximized social and ecological value.

Rather than trying to meet an environmental responsibility as efficiently as possible, Ford opted for a manufacturing facility that would create habitat, make oxygen, connect employees to their surroundings and invite the return of native species. The result: a day-lit factory with a 450 000 square foot roof covered with growing plants – a living roof. In concert with porous paving and a series of constructed wetlands and swales, the living roof will absorb and filter stormwater run-off, making expensive technical controls and even regulations, all but obsolete. All this with tremendous first cost savings, with the landscape thrown in for free. According to Ford, the natural stormwater system alone, compared with conventionally engineered water treatment systems, generated a first cost saving of $5 million, the equivalent of an order for $100 million of new cars at a 5 per cent margin. This is the economic power of triple top line design.

Now, when designing a manufacturing facility, one can imagine the value of asking: 'How can this project restore more landscape and purify more water? How much social interaction and joy can I create? How do I generate more safety and health? How much prosperity can I grow?' Questions such as these allow us to remake the way we make things. Today.

Toward a cradle-to-cradle world
Designs that celebrate this diverse range of concerns bring about a process of industrial re-evolution. Our products and processes can be most deeply effective when they resonate with the living world. Inventive machines that use the mechanisms of nature instead of harsh chemicals, concrete or steel are a step in the right direction but they are still machines – still a way of using technology to harness nature to human purpose. New technologies do not themselves create industrial revolutions. Unless we change their context, they are simply hyper-efficient engines driving the steamship of the first Industrial Revolution to new extremes.

Natural systems take from the environment but they also give something back. The cherry tree drops its blossoms and leaves while it cycles water and makes oxygen; the ant community redistributes nutrients through the soil. We can follow their cue to create a more inspiring engagement – a partnership – with nature.

Expressed in designs that resonate with and support natural systems, this new partnership can take us beyond sustainability – a minimum condition for survival – toward products and commercial enterprises that celebrate our relationship with the living earth.

We can create fabrics that feed the soil, giving us pleasure as garments and as sources of nourishment for our gardens.

We can build factories that inspire their inhabitants with sunlit spaces, fresh air, views of the outdoors and cultural delights; factories that also create habitat and produce goods and services that re-circulate technical materials instead of dumping, burning or burying them.

We can tap into natural flows of energy and nutrients, designing astonishingly productive systems that create oxygen, accrue energy, filter water and provide healthy habitats for people and other living things.

As we have seen, designs such as these are generators of economic value too. When the cradle-to-cradle principles that guide them are widely applied, at every level of industry, productivity and profits will no longer be at odds with the concerns of the commons. We will be celebrating the fecundity of the earth, instead of perpetuating a way of thinking and making that eliminates it. We will be creating a world of abundance, equity and health and well on our way to an era of sustaining prosperity.

This is not a path one must travel alone. GreenBlue, a new non-profit organization established to encourage the widespread adoption of cradle-to-cradle thinking, is now providing the theoretical, technical and information tools required to transform industry through intelligent design. Its mission is to make commercial activity an ecological and socially regenerative force, and its tools are designed to empower designers to participate in this transformation. Leading companies such as Estee Lauder/Aveda, MeadWestvaco, Starbucks, PepsiCo and Unilever, for example, worked with GreenBlue to create the Sustainable Packaging Coalition to pursue a 'positive, robust environmental vision for packaging' that includes developing cyclical material flows and increasing demand for 'environmentally intelligent, cradle-to-cradle materials'.

MBDC meanwhile announced in June 2005 that it will provide Silver, Gold and Platinum certification for cradle-to-cradle products that meet its criteria for material health and safety, material recovery and re-manufacturing systems, energy and water use and social ethics. Providing benchmarks that are both inspiring and scientifically sound, the certification programme will help a wide range of industries achieve a new level of quality in product and process design.

And so we invite you to join us in leaving behind design strategies that yield tragic consequences, and in taking up a strategy of hope, a strategy that allows us to create a world of interdependent natural and human systems powered by the sun in which safe, healthful materials flow in regenerative cycles, elegantly and equitably deployed for the benefit of all. Doing so is ultimately an act of love for the future, an act that allows us to take steps toward not simply loving our own children, but loving all children, of all species, for all time.

References

McDonough, W. (1992), *The Hannover Principles: Design for Sustainability*, http://www.mcdonough.com/principles.pdf, accessed 1 December 2005.

McDonough, W. and Braungart, M. (2004), *Cradle to Cradle: Exploring Design for the Circular Economy*, Shanghai, China: Tongji University Publishing House.

4 Users as a source of learning in environmental technology management
Harald Rohracher

Introduction

The role of users in technological innovation processes has often been neglected in practice, although empirical studies since the 1970s have revealed that users are an important factor of successful innovations. As Chris Freeman points out, a 'major determinant of innovative success lies in the nature and intensity of the interaction with contemporary and future users of an innovation. In the case of incremental innovations especially, but also for radical innovations, this has often been shown to be a decisive factor' (Freeman, 1994, p. 470).

Users are an important source of learning for firms – learning by interacting, as Lundvall (1988) calls it – not only in the classical sense employed by market research of segmenting markets into lifestyle groups and learning about the desires of consumers or the potential acceptance of product ideas, but also because various characteristics of products – which may require adaptation and product improvements – are only revealed in the course of actual product use in 'real life' contexts. In other words, technologies and use practices often co-evolve during the innovation and adoption phase, a process that will be more satisfactory for the various product stakeholders if it is organized in close interaction between product designers and producers, and product users. Beyond this role as a source of information for firms engaged in product development processes, many cases can be identified (even if they are not always evident at first sight) where users even act as innovators and developers of new technology and products or improve products in the diffusion phase – diffusion researcher Rogers (1995) calls these cases 're-invention'. Thus users are active agents of the social shaping of technologies predominantly during the implementation and early diffusion process (see Rohracher, 2003). Especially in the case of environmental technology, users may show a high level of engagement with the design and use of such products.

The core question of this chapter is: under which circumstances, in which cases, and with which strategies can companies and public policy better integrate users into the innovation process? More specifically, drivers and characteristics of environmental technologies and innovations will be identified with respect to the role of users and possible consequences for environmental technology management.

The first part of this chapter is dedicated to a better understanding of technological innovation and in particular the role users play in this process (for a comprehensive overview see also Oudshoorn and Pinch, 2003; Rohracher, 2005). Innovation and the role of users will be analysed in the context of concepts developed in evolutionary economics and social studies of technology. This analysis results in specific strategies, such as the 'lead user method', suggested by innovation researchers to establish closer links to users and to learn from the interaction with users.

The second part of the chapter will then specifically examine the relevance of these concepts and tools for environmental technology management. The chapter focuses on strategies and methods to improve user–producer relations: through marketing and communication, in supply chain management, in technology procurement or by setting up socio-technical experiments.

Before immersing in the discussion about innovation and the role of interactions between users, producers and other actor groups, let us give some thought to the two key concepts used in this chapter: 'environmental innovations' and 'users'.

What are environmental innovations?
Although primarily dealing with environmental technologies, the term 'environmental innovations' will be used in this chapter as it is broader and also comprises organizational innovations. These may, in conjunction with environmental technologies, be an important factor for the environmental performance of a company and may also be of special relevance for the organization of interactions with actors inside and outside the company, such as users.

However, even the narrower notion of environmental technologies is not easy to define. One possibility is to produce different categories of environmental technologies, such as (Kemp, 1998, cited in Markusson, 2001, p. 15):

- pollution control technologies;
- waste management;
- clean technology;
- recycling;
- clean products;
- clean-up technology; and
- monitoring and assessments technology.

A classic distinction has also been drawn between end-of-pipe technologies (such as pollution control technologies) and integrated environmental technologies (or cleaner technologies), which have increasingly moved to the centre of discussions about environmental innovations. However, as Markusson (2001, p. 15) points out, some innovations (such as information technologies) may have structural effects on companies and generate 'unintentional' environmental effects in companies or industries even if one would not categorize such innovations as 'environmental'. Virtually all technologies can have environmental effects and may contribute to environmental innovations even if better environmental performance was not the intention of the innovator.

Given the ambiguity of the term 'environmental innovation', Schaefer and others point out that: 'What is – and what is not – considered to be an environmentally important issue to be addressed within a company is often externally mediated, through regulation, market forces and, to some extent, other stakeholders' actions' (Schaefer et al., 2003, p. 212). However, it may well be the corporate actors themselves who are 'active agents in the taking and making of greening' and who may significantly influence how greening is defined – as 'a public relation issue, a marketing challenge, a legal problem, an engineering solution . . .' (Fineman, 2001, p. 24). Taking on such a discursive approach, users certainly also play a role in defining what is perceived as an environmental technology and what is not.

Applying a rather broad notion of environmental innovations classifies the majority of companies as environmental innovators. The Mannheim Innovation Panel in Germany asked 2264 German companies about innovations, based on potential areas of environmental innovations (including product/process integration, process external/internal recycling, end-of-pipe measures). Commenting on the results of this study, Cleff and Rennings (1999) note that: 'This produces 72% (929 companies in total) environmental innovators among the innovators as a whole. Such a high percentage is plausible, as virtually every large company today makes use of or innovates in environmental technology or organizational environmental protection in one form or another' (p. 194).

However, not all of these innovations are of direct relevance for users. In our case it may be useful to differentiate between process and product innovations, with process innovations relating to environmental regulation and the environmental image of a company, and product innovations being more strongly geared towards customers and users (although regulation and image may be important for product innovations too). Moreover, there will be product innovations that have little effect on use characteristics (using environmentally friendly materials, improving the energy efficiency of products), while other innovations will be more radical and more directly related to improving environmental performance (ultra-light electric vehicles for example).

Users – a heterogeneous category
And what about users? The most apparent aspect of the concept of users is the heterogeneity of the group of actors that is addressed by this term. Users are merely defined in relation to a product or technology, but they may be a firm, an organization or an individual. In the case of firms the company may be a user with respect to one technology and a producer with respect to another. Therefore von Hippel (1988) defines users and manufacturers through a functional relationship with the product, asking whether the firm benefits from using, producing or supplying it.[1] Depending on product fields the functional source of innovation may be predominantly users, or predominantly manufacturers or suppliers of products. In many instances the benefit of product innovations is different for different users.

Users may also be different in different stages of the 'career' of a product. Depending on which aspect of product use we look at, the installer of an energy management system in a building is a user of this technology during the implementation phase, whereas the inhabitant of a building (or in between: the building developer that owns the building and is responsible for its maintenance) is a user when the building is operational. For the purpose of this chapter, the term 'intermediary users' will be used to refer to users who are mediating between producers and end-users (architects, for example, who are users of building technologies but who also speak on behalf of the tenants as end-users). However, even the end-user of a product is not a homogeneous social category at all but brings 'to those processes a multiplicity of perspectives, ideas, social backgrounds, and life histories, not to mention wants and needs' (Rose, 2001, p. 68).

The final distinction to keep in mind will be that between customers and users. Customers of a company (be it individuals or organizations) are not necessarily the users of the technology they buy. Whereas a company may have a department or person responsible for procurement activities, we know from consumer studies that even individual consumers purchasing goods (in a family for example) are not necessarily those who actually use it. In this respect, customers may be one more intermediary in the chain between producers and end-users of a product.

Users and environmental innovations
Why should environmental technologies be different with respect to users than other technologies? Most general findings about the role of users in technological innovations are certainly equally valid for environmental technologies but there are a number of special characteristics of environmental technologies that make a specific analysis of the links between users and environmental technology development and management advisable. Environmental technologies often have characteristics that require special emphasis on the societal embedding of technologies – with users and product usage being an important dimension of this embedding process.

On the one hand, the specific use of products and technologies may have a substantial influence on the sustainability of such products and on their proper functioning, as the environmental impacts of a technology in many cases are influenced by the way such a technology is used (energy-efficient ventilation systems in buildings are a case in point). On the other hand, environmental technologies can be characterized as 'mixed goods' (Brown, 2001),[2] integrating characteristics of public goods and private goods, which may result in a specific relation of environmentally conscious users with environmental products; for example, an increased willingness to use such products and to contribute to product improvements. Thus there might be a higher chance of engaging environmentally aware customers and users into joint learning processes to improve such products. Finally, environmental technologies are subject to specific policy objectives and are strongly affected by environmental policy instruments such as regulations and norms.

These characteristics mean that environmental technologies play a special role with respect to broader policy aims (environmental policy, public good characteristics) and often depend on the development of specific use practices to increase their environmental effects. Improving learning processes between firms and users in the process of technology development and dissemination can certainly be an important means for a faster and better social embedding of these technologies.

In a literature survey of studies on drivers of environmental innovations in firms, Markusson (2001) came to a similar conclusion:

> Environmental innovations are driven by specific environmental demands and may contribute to decreased environmental impact, but they also seem to possess other properties that single them out in comparison to other innovations. Among other things environmental innovations involve social awareness, draw upon knowledge from a very wide range of fields and require complementary organisational innovations. (p. 46)

From this perspective closer and more conscious interaction between users and producers is in the interest not only of companies but also of public policy and other actors. As a consequence, strategies to improve user–producer relationships can take on a variety of different forms depending on the main actors employing these strategies (including producer companies, users and public policy-makers). Even if companies are in the centre of our discussions, the chapter will also present some approaches affecting user–producer relations depending on the initiative of other actors.

The role of users in innovation processes

What we have seen even from these introductory remarks is that technologies are closely interwoven with various actor relations and social contexts. Here we are already getting to the core of social studies of technology that analyses the creation, diffusion and use of technical artefacts as a social activity taking place in networks of actors and interests. Summing up the rationale of social studies of technology approaches, Russell and Williams (2002) highlight the social embeddedness of technologies:

- Technologies are produced and used in particular social contexts, and the processes of technological change are intrinsically social.
- Technologies function as such in an immediate setting of knowledge, use practices, skills, meanings and values, problems and purposes, and objects that they act on.
- Technologies in many applications are best considered to operate as socio-technical systems or configurations.
- Technological change is always part of a socio-technical transformation – technology and social arrangements are co-produced in the same process (p. 48).

The use of technologies and products – and this is an important point for our further analysis – is an integral part of the socio-technical configurations mentioned above, which finally constitute a technology in operation. Technology designers and producers as well as technology users are important poles within such configurations. Thus there are good reasons to approach the innovation process from both sides: product design and use. Let us briefly discuss two concepts approaching innovation either from the side of design or use.

Conceptualizing innovation and diffusion as a reconfiguration of actor networks, Weyer (1997) points out that bringing a product from its conception to a wide distribution on the market requires constant restructuring of the actor relations important for the product. It can require, for example, moving from a rather small and specialized network of actors in the development phase, who mainly focus on single artefacts and are often dominated by very specific interests, to a broader network of different and heterogeneous user groups with evolving practices of product use and different groups of professionals. As Vergragt and others conclude from several case studies in industry, 'the development of a new technology by a firm is accompanied by the formation of internal and external coalitions of actors; and the analysis of the social structure of these coalitions provides insights to the dynamics and direction of technology development' (Vergragt et al., 1992, p. 243).

Appropriation of products and technologies points to the active side of product use where users do not just submit themselves to the scripts or 'preferred readings' of a technology but actively define their way of usage: 'People may reject technologies, redefine their functional purpose, customize or even invest idiosyncratic symbolic meanings in them. Indeed they may redefine a technology in a way that defies its original, designed and intended purpose' (Mackay and Gillespie, 1992, pp. 698–9). An artefact thus only acquires its meaning in use. This is what is often called 'appropriation' or 'domestication' of a technological object, the way technology is 'incorporated into the routines and rituals of everyday life, the way it is used, and the ways it becomes functional' (Vestby, 1996, p. 68).

It should be kept in mind, however, that both processes – the work of designers, and the appropriation and evolving practices of users – remain situated in a broader context of social structures and technological regimes (see Rip and Kemp, 1998). As Shove (2003, p. 204) points out, 'it is clear that the use of any one device or appliance takes place in an already crowded social, symbolic and material world. Furthermore, use is connected intimately to the reproduction of what people take to be normal ways of life'. Users, whose expectations are formed by such visions of normal society and conventions of everyday life, obviously have only limited space for an active appropriation of these products, or a restricted 'room for manoeuvre'.

Learning from users
Such concepts of design and use emphasize that developing new technologies is not just a matter of technical expertise but is equally characterized by a social dynamics that involves designers and producers along with users and other actors. Managing technologies also means managing these kinds of social relations and sparking off processes of social learning between the actors involved.

Apart from the knowledge accumulation about expectations and the structure of demand from users, interaction between users and producers is often necessary to improve products, as many user requirements and product characteristics can only be discovered if the product is actually used. Communicating this information to the producer presupposes that there are mechanisms in place to pass on the economically and technically relevant information to producers (Habermeier, 1990). A special type of such user–producer interactions is market research, however, this is more concerned with learning about users than about learning from users. Nevertheless testing a product in trial markets, as market researchers also do, is already coupled to actual use of products. Basically the process of learning about product characteristics and user requirements during actual use and subsequent product improvements is an open-ended process that does not necessarily converge. Although there are other possibilities, such as laboratory testings (which usually cover only some aspects of the product), 'the natural and human environment in which the system operates is often not completely understood, at least in the beginning, when the technology is relatively new' (Habermeier, 1990, p. 276). This is also a reason why field testing may be difficult as good knowledge about the context of use is required. Thus 'it is often much more efficient to let users tell you in which way or how to improve a product once they have had some experience with it' (Habermeier, 1990, p. 280).

To sum up, interaction with users is an essential source of learning in innovation processes in general, and certainly for environmental innovations in particular. Empirical evidence, however, often shows that learning during technology introduction is often incomplete and insufficient and – as a typical case – product developers often refer to their own use experience without systematically taking a broader spectrum of user perceptions and experiences into account (see Akrich, 1995; Rohracher, 2003). A recent study analysing social learning processes in multimedia applications found that 'designers do not possess clear representations of the user and the user's setting and instead work with rather implicit and poorly specified models based on incomplete and often unreliable sources of information' (Williams et al., 2000, p. 17). From the perspective of a successful embedding of innovations and a better balance between the interests of a broad range of stakeholders, there is certainly a need for improved strategies to create conditions for a more conscious, reflexive and inclusive learning process between the actors involved.

User-initiated innovations

However, the role of users in innovation processes may even go further. As the evidence from an increasing number of empirical studies indicates, users are in many cases the source of innovation. Much of this research is connected with the name von Hippel who started out by analysing the development of scientific instruments, where more than three-quarters of all innovations turned out to originate from users (von Hippel, 1988). However, this diagnosis is not restricted to high-tech areas with highly specialized users. An investigation of the primary sources of innovation in the building industry (stressed-skin panels) found that more than 80 per cent of the innovations were initially created by users (often small residential builders) only for use in their own building projects (Slaughter, 1993). Many of these innovations were significant enough to be widely adopted in the manufacturing industry. Such user innovations mainly occurred in situations where the cost of user solutions was low but the cost of delay for users was high, or if it was easier for users to get regulatory approval for their solutions.

Moreover, there is a number of documented cases where the loosely coupled networks of early innovation include early users. Examples include the early user-clubs' involvement in the development of the personal computer (Weyer, 1997), and innovative milieus (constituted by outsiders and early users), which were the main source for the early development of lightweight vehicles in Switzerland (Truffer and Dürrenberger, 1997). An empirical study of innovations originating from informal sports communities also points to the importance of the innovative milieu – or the 'community-based innovation system' as Franke and Shah (2003) call it – provided by the sports community: 'The community-based system provides the user-innovator with information, assistance, and links to other individuals; simply put, it provides the innovator with access to resources' (Franke and Shah, 2003, p. 172). Central members of the community, as the study points out, 'are likely to both innovate and to have an exceptionally good knowledge of user-innovations developed by other community members' (p. 175).

Monitoring such innovative user-communities may be an efficient method for identifying commercially appealing innovations made by users and could be part of an extended portfolio of technology management. Similar community-based innovation systems certainly exist in environmental technologies as well. An example is the success of self-building groups for solar collectors and biomass boilers in Austria, which have also sparked off a number of product improvements by users (Ornetzeder and Rohracher, 2006). Acknowledging the potentially active roles of users in innovation processes may open new perspectives for policy-making: 'Important implications, particularly for policymakers, lay in discovering that the locus of product innovation can be shifted strategically by altering incentive structures and re-ordering relationships between users and producers' (Edquist and Hommen, 1999, p. 71).

Integrating users in the innovation process

Based on the role of users as a source of learning and innovation, strategies of consciously involving a limited number of users have been suggested and are often successfully practised. One of these strategies is focusing on the identification and involvement of a group of 'lead users' (von Hippel, 1986). 'Lead users' refers to a category of 'advanced' users of a technology who are expected to lead technical trends in terms of experience and intensity of needs that later will be general in a marketplace. They should be positioned to benefit significantly by obtaining a solution to those needs.

Meanwhile there is growing empirical evidence that lead users are of value as 'early adopters, sources of new product ideas, market research potential, and their role in fueling the diffusion process' (Morrison et al., 2004, p. 361). When such 'lead users' can be successfully identified they can serve as a 'need-forecasting laboratory' for marketing research. Again such experienced users are not only to be found with complex product systems, as has been demonstrated in applying the lead user method to some 'simple' products, for example in the construction-related industry (pipe hangers) (Herstatt and von Hippel, 1992). In a step-by-step process, an expert survey was conducted to identify important trends in the evolution of user needs with respect to the specific product. Then firms who used the product were screened and interviewed for lead users. (Were they interested in such innovations? Did they have ideas for improvements?) Finally, these lead users joined a three-day product concept development workshop. Finally the concepts were tested to establish whether they also appealed to ordinary customers. A comparison with ordinary market research methods showed that focusing on lead users in product generation was significantly faster and cheaper.

Nevertheless, apart from the difficulty of identifying lead users, one may ask whether lead users are indeed a 'good representation of the remaining 97.5% of the market' (Magnusson, 2003, p. 229). Despite such doubts, evidence is growing that a careful selection and involvement of users may speed up and improve the quality of innovations. An empirical study on new end-user services for mobile telephony, which compared groups of professional developers and user groups who were both given the task of creating new products in the respective area, found that 'the users produced more original ideas than the professional developers did' (Magnusson, 2003, p. 235). Magnusson (2003) thus concludes that involving users as innovators can result in more innovative services of greater use value.

As an additional strategy in the interaction of users and producers, von Hippel suggests that there may be products that may be adapted to user needs by outsourcing 'key need-related innovation tasks to their users, after equipping them with appropriate "user toolkits for innovation"' (von Hippel, 2001, p. 247). As Thomke and von Hippel (2002) explain, '[e]ssentially these companies have abandoned their efforts to understand exactly what products their customers want and have instead equipped them with tools to design and develop their own products, ranging from minor modifications to major new innovations' (p. 5). Examples for redesigned products where users can carry out design tasks and trial-and-error improvements on their own include software to develop integrated printed circuits, toolkits for companies to customize the composition of plastics to their specific needs, or computer monitors at hairdressing salons where customers can design and modify their hairstyle with their own image captured by a video camera. Nonetheless the number of products where iterative learning processes between design and use can be completely shifted to the user side is limited.

There is no doubt that the principles and strategies of user–producer interactions discussed so far equally apply to environmental technologies and would also help to better manage the processes of environmental innovation and technology dissemination. The following section looks more specifically at environmental technologies and possible extensions of the portfolio of technology management strategies in this area.

User–producer relationships in environmental technology management
Looking at the relationship between users and producers of environmental technologies, a key question is in which respect the exchange between both groups – the information flow between users and producers – is a driver of environmental innovations. Let us first review some empirical studies covering this question.

Drivers of environmental innovation
The perception of users and customers is certainly in the focus of companies when it comes to environmental products or technologies. This is not unexpected, as Schaefer et al. (2003) explain: 'Put crudely, if there is a perception that customers (industrial or end consumers) attach importance to environmental issues, companies are likely to take this seriously and concentrate their environmental efforts in such areas' (p. 215). The bottle-neck in this interaction, however, rather appears to be the interest of customers in environmental issues. This is what Foster and Green (1999) found when they investigated various companies interviewing environmental management, R&D management and other departments on the way the green agenda affected their organization and work: users, or more specifically customers, did play an important role in R&D and innovation management. R&D managers' perception was 'that their work programmes are closely specified by their "customers", both internal and external. Therefore, the green perfor-mance of products and services in the course of development work only receives significant attention if this is requested by the customer. Such requests are rare except under certain circumstances' (Foster and Green, 1999, p. 3). Similar responses came from company salespersons: 'Green issues are seldom on the agenda of the salesperson–user dialogue, unless a user has a particular concern relating to legislative compliance' (Foster and Green, 1999, p. 3).

In drawing up networks of information flow inside and outside the company, Foster and Green are especially interested in the exchange with users in both directions: as signals from users regarding green performance requirements (and efforts of businesses to improve this process by, for example, applying tools to better understand customer needs) and as information flow towards users, especially the educational role of sales and mar-keting groups to inform users about green aspects of product performance. In their con-clusion, Foster and Green identify three different situations where R&D departments considered green issues, including:

> Cases in which green performance is a key issue for the user, and one of the main driving forces behind the innovation project. Here green issues are treated in very much the same way as any other performance requirement in terms of evaluation and assessment . . . Cases in which there is a new market resulting from the green agenda. Again, green issues get serious consideration in innovations that seek to exploit these opportunities . . . The rest. In these cases, green issues receive only limited attention, most commonly in form of a compliance check'. (Foster and Green, 1999, p. 30)

However, looking on all the cases investigated, there appeared to be little communication between customers and suppliers about green issues and limited understanding by users of the environmental issues related to the acquisition of products and services. The main reason for customers making green product performance a key issue was when they felt the force of regulation (see also Foster and Green, 2002). Only in these cases users defined their environmental performance needs clearly. Moreover, businesses and public organizations with corporate environmental management systems often express strong commitments to encouraging environmental performance in their product chain. What is therefore needed, Foster and Green (2002) conclude, are efforts to create more environmentally driven users, with strong regulation as the key strategy.

However, these results are not undisputed in the light of other studies. An empirical study by the World Business Council for Sustainable Development about drivers of sustainable innovations in firms found that 'direct demands from regulators, customers and special interest groups were considered to be least important. Firm image and brand value were thus considered much more important than customer or regulator demands' (Dearing, 2000, cited in Markusson, 2001, p. 19). In the already mentioned Mannheim Innovation Panel (including more than 2000 firms), 'complying with existing legislation' comes on top as a driver of environmental innovation but at the same time 'environmental awareness, both internal – e.g. among staff – and external, in the associated improvements in public image, are seen by the companies as particularly important strategic goals in relation to environmental innovation' (Cleff and Rennings, 1999, p. 195).

Low customer awareness is not necessarily a barrier to environmental innovation, however. Noci and Verganti point out that:

> although a company's immediate customer seems to have low concern for environmental issues (or, at least, confers more value to product quality, cost and other parameters than eco-efficiency) this does not entail that environment is not a strategic issue. In fact changes in distribution, disposal and consumption regulation . . . will probably change the requirements of both dealers and customers. (Noci and Verganti, 1999, p. 11)

To some extent the importance of 'regulation' and 'public image' as drivers of environmental innovation also depends on the type of innovation. While environmental innovators at the product level predominantly 'pursue the goal of maintaining or increasing their market shares . . . it is compliance with existing legislation and anticipation of future laws' that is important for process innovations (Cleff and Rennings, 1999, p. 198). The authors assume that the reason for the higher importance of product innovations is the added benefit for the customers, which is in the strategic interest of companies.

Strategic interests of a company are certainly key to the importance of environmental innovations. 'Green' innovation will be managed indirectly when the environment is seen as a secondary factor; it will only be managed explicitly when the environment is considered as a major strategic challenge. Then the various implications of environmental innovation are systematically examined. Noci and Verganti (1999) call this type of firm behaviour an 'innovation-based "green" strategy', which usually 'involves the introduction of new technologies that radically improve the environmental performance of current technologies or the creation of new market needs as a consequence of environmentally friendly products' (p. 12). It is most likely that by adopting this kind of strategic orientation of a company environmental technology innovation particularly benefits from close interaction between the company and users, and the organization of joint learning processes.

As these investigations of the importance of users/customers for environmental innovation in companies show, user–producer relationships are important at several levels. At the highest level is the question of whether environmental issues are placed on the agenda at all – whether there are customer requirements for green performance or whether companies try to create awareness by green marketing efforts.

Even if there is a generally low awareness of environmental issues, niche markets do exist with a high demand for environmental innovations and technologies, either as specialized markets for environmental technologies (recycling, pollution control and so on) or where there are groups of customers with a high awareness of environmental problems who are motivated (or can be motivated) to use environmental technologies (such as green products, energy-efficient cars, sustainable buildings and so on). Firms focusing on such market segments will probably be those who see environmental innovations as a strategic issue and follow an 'innovation-based "green" strategy'. Developing and improving environmental innovations in close interaction with users will be of particular relevance in this sub-sector of environmental markets, which is characterized by environmentally aware customers and a demand for more radical environmental innovations. In the broad market, where there is no strategic focus on environmental issues, incremental improvements and compliance with existing regulation will prevail and diverse marketing instruments will be the main way of communicating with customers.

Managing user–producer relationships
As mentioned in the earlier section about the role of users in innovation processes, learning between users and producers may focus on communication processes about demand, user expectations and environmental performance requirements of products. Alternatively it may focus on the 'co-construction' of use practices and technologies, that is, iterative adaptations between the design of new products and the ways of usage that develop in phases of early product use. Beside strategies discussed earlier, such as the selection and integration of lead users or the systematic scanning of user communities for interesting innovations, there are a number of other approaches where the relation between users and producers plays an important role. Especially in the area of environmental technologies, where innovation is closely interlinked with a broader public interest, the initiative for the management of innovation processes does not necessarily come from producer companies but may emanate from public policy or stakeholder groups.

Marketing/communication
Communication with users is one of the strategic issues of importance to firms. Still, it is astonishing that there is often rather little attention paid to customer contacts about environmental issues. Prentis and Bird (1999) list several plausible reasons that may be responsible for this situation:

- Early environmental pressure was focused on manufacturing organizations from NGOs and not directly from customers.
- Long distribution chains for goods and services place many organizations at a distance from their end-user customers.
- Many direct customer contacts on environmental issues are perceived as negative, that is, customer complaints or queries.

- Environmental people tend to be technical, scientific by training rather than communication, marketing focused.
- Sales and marketing people rarely move into environmental roles.
- Failure of 'green' products on the marketplace (p. 53).

Prentis and Bird (1999) thus suggest a role for 'stakeholder dialogues' involving customer organizations and the team of environmental companies as a means for both parties to develop a better grasp of the environmental options available.

Information flow between users and producers is important in both directions, as has been pointed out earlier. Companies may learn about customers from various techniques of market research and they may communicate to customers with an 'educational' perspective through marketing and PR (public relations). 'Green marketing' is an important step in 'shifting from a supply- to demand-driven focus in which business should look at what sensible things can be done proactively in the environmental domain, rather than wait to be driven by external developments' (Stevels, 2001, p. 89).

'True green marketing', Polonsky emphasizes, 'involves environmental issues becoming as much a strategic organisational focus as are issues such as quality or customer service, and this will most probably require a change in the organisational mind-set as well as in organisational behaviour' (Polonsky, 2001, p. 286). We will not go into the details of green marketing strategies, but as Polonsky points out, it involves much more than promotion or product design activities and should include related topics such as targeting, green pricing, green design and product development, green positioning, the greening of logistics, the marketing of waste, the greening of promotion and green alliances. In any case, as a strategic aim of a company, green marketing has to be closely linked up with environmental technology and product development, and aims at an intense flow of information between (potential) customers or users and the producer firm.

Supply chain management
A supply chain is 'a metaphor for the linear sequence of interconnecting links between organisations . . .' (Green et al., 2000, p. 215). It maps the flow of materials and components for a certain product 'from raw material firms and through component firms, assembly firms, retail firms, transport firms, service firms, and finally, disposal and recycling firms' (Green et al., 2000, p. 215). On the one hand, these companies are linked in a way by user–producer relations themselves; on the other hand, end-user requirements cannot be met by the last chain link, which would be the retailer, or further upstream by the assembly firm but rely on the cooperation of all or many parts of the supply chain.

Finding strategies to manage such supply chains is thus an important element in organizing the relationship, and communication, between users and producers. In many cases external sources of information are central to meeting the information and knowledge needs faced by firms engaged in environmental innovations. For instance: 'Equipment manufacturers play an important role in the supply of information to the suppliers of environmental innovations. Suppliers of materials and components provide important information for developing environmentally friendly products and reducing consumption of materials' (Markusson, 2001, p. 19). However, as Kärnä and Heiskanen point out, 'product chain studies have, until now, usually not included consumers as stakeholders in the product chain. The focus is, however, changing. New initiatives are focusing on demand management and consumer involvement in the development of radically new product concepts' (Kärnä and Heiskanen, 1998, p. 28).

Supply chain management also points to the fact (as also emphasized in the first section of the chapter) that '[i]nnovations and continuous improvements occur often not due to one actor alone. On the contrary, actors along the chain should start networking. Therefore these approaches should be stimulated and financially supported with the intention of forming eco-innovation alliances' (Rubik, 2001, p. 228). There are certainly a number of conceivable strategies (and examples do exist) regarding how to network and manage supply chains. The strategy of setting up joint platforms for further product development and standards belongs to this group, as does setting up intermediary organizations to manage supply chains and communicate with customers or politics. Examples include the Californian Electric Vehicle initiatives CALSTART, or the Electricar-Synergy EV Group, acting as central nodes between industry, utilities, government or research in the emerging market of electric vehicles (see Schot et al., 1994).

Often supply chains may be managed by strong actors or 'system organizers' within the chain who define standards for suppliers and who have a whole portfolio of strategies to define these relationships:

> First, there is greater scope to play a significant role in the specification and design of the product and service on offer. Second, there may exist scope for making requirements of suppliers relating to the process and the product; an example would be using the adoption of environmental management systems (such as ISO 14000) as a prerequisite for approval to supply . . . Third, there is sometimes scope for projects that entail the direct transfer of knowledge and technology between buyers and sellers. (Green et al., 2000, p. 213)

This list can be complemented and specified by: 'Communicating environmental expectations through written policies and communication materials . . . providing questionnaires or the carrying out of audits; organising supplier meetings; offering supplier training and providing technical assistance' (Charter et al., 2001, p. 113).

Although not all of these strategies would be generally subsumed under the notion 'supply chain management', they address similar issues of organizing communication flows in innovation processes distributed over various supplier and producer firms and customers.

Technology procurement

A different category of environmental technology management along user–producer relationships is set up by green procurement strategies. In this case it is not the producer or supplier company that organizes the information flow, but customers or users who define the environmental requirements of products and send strong signals to producers.

Obviously customers being in the position and qualified enough to formulate requirements for procurement activities are large companies or public institutions, which, due to their huge demand volume, are strong enough to influence the market. Procurement by organizations should not be underestimated:

> First, in the term of the scale of economic activity, personal traditional (end) consumption very probably is dwarfed by the spending of public and private organizations. Sales between organizations – in which one organization will sell and the other will buy (and then consume in some way) – are of major importance in any economy. (Green et al., 2000, p. 208)

Public procurement in the European Union amounts to around 14 per cent of EU GDP (Lackner, 2003) and one can easily figure out what market pressure could be created by coordinating environmental performance criteria in procurement initiatives.

However, especially in large buyer organizations – where users are different from purchasers who are different from environmental management staff – coordinated green procurement is an organizational challenge and requires that 'various departments or divisions, in particular those charged with environmental care and purchasing, exchange their view and co-operate with respect to concrete purchasing orders' (Lackner, 2003, p. 2).

Of specific interest for inducing and managing environmental innovations is 'technology procurement', which is a special type of green procurement activity intended to stimulate innovations through invitations for tenders that demand eco-innovative solutions not yet available on the market. Again, usually public agencies or government institutions are the only users able to develop the competence to articulate such a specific demand for technologies or products that do not yet exist and that at the same time are socially desirable, technically possible and economically feasible. Such procurers must have 'a "vision of the future" (defining needs) and can then translate this into functional characteristics' (Edquist, 1996, p. 142). As Edquist goes on to point out, '[t]echnology procurement is an important force on the demand side, precisely because it involves interaction between users and producers. One of its principal advantages lies in the fact that it establishes a direct communication between innovation policy and usable outcomes' (Edquist, 1996, p. 146).

Especially in the area of energy-efficient technologies, technology procurement initiatives have been carried out by various public actors. Examples include the Swedish technology procurement programme on energy-efficient lighting (development of high-frequency electronic ballasts) in the early 1990s; a procurement programme of Nordic countries for more reliable, cheaper and improved heat pumps; a programme of the US Department of Energy on sub-compact fluorescent lamps; and the European Union's Energy+ Refrigerator/Freezer Procurement Programme in 1999. An evaluation of such programmes concluded that technology procurement 'is a useful approach when technologies are close to being ready for the market but require additional development with a particular eye on what potential buyers are looking for' (OECD/IEA, 2003, p. 28). Understanding the needs of potential buyers in-depth and exploring the scope for supplier response to these needs, have been identified as essential prerequisites for successful technology procurement strategies. What is effectively happening in such strategies is that environmental technology management functions, and particularly the interaction and information flow between users and producers, is looked after by public institutions using their purchasing power as a lever to stimulate environmental innovations.

Socio-technical experiments, strategic niche management
A further group of strategies for 'public technology management' includes firms as important actors but usually depends on a public agency as coordinator or change agent. These strategies, which come under different labels but are all based on a similar theoretical background and principles, include: 'constructive technology assessment' (see, for example, Schot and Rip, 1996), 'strategic niche management' (Hoogma et al., 2002; Kemp et al., 2001), and 'bounded socio-technical experiments' (Szejnwald Brown et al., 2003). One of the key issues of such approaches is to organize constellations that allow for social

learning about design and use, for example, by creating a nexus between designers and groups outside the design process, such as consumer associations or NGOs. An example for such a nexus in companies, which is nonetheless widely distributed, is the setting up of environmental departments uniting tasks of monitoring, external contacts (with government, the wider public and so on) and innovation and policy development (Schot, 1992, p. 49).

Broadening the design process in such a way increases the chances of developing widely accepted products that are better adapted to the needs of users. For instance, constructive technology assessment (CTA) 'proposes bringing together all interested parties early in the design process . . . Thus, in CTA, technology is assessed from many points of view throughout the entire process of design and redesign, and the interests of all parties can be incorporated from the beginning' (Schot, 2001, p. 40). Successful CTA processes should have certain characteristics: CTA should be anticipative as users participating in the design process are expected to be more likely to bring up social issues and acceptance problems very early; CTA should be reflexive in the sense that it encourages actors to recognize their own and others' perspectives and to consider technology design and social design as one integrated process; CTA should finally lead to social learning processes, including second 'order learning' (that is, not only articulating market demands but also questioning existing preferences and requirements in order to open up possibilities for more radical developments).

Strategic niche management (SNM) and bounded socio-technical experiments both specifically refer to the creation of protected spaces (market niches and controlled field experiments for example). Technological niches are 'a specific domain for application of a new technology functioning as a testbed where, under temporary protection from market and other institutional pressures, producers, users, and sometimes government develop it to maturity' (Weber and Hoogma, 1998, p. 548). Such niches do not necessarily have to be managed by public institutions. Private companies or NGOs may well be suitably qualified as niche managers. A central aim of the development of niches is to learn about needs, problems and possibilities connected with the environmental innovation experimented with, and to help articulate design specifications, user requirements or side-effects of the innovation. Managing the development of environmental technologies in niches (and finding the right timing to open these niches to the wider market and competition) is certainly one of the most advanced and reflexive forms of managing environmental innovations and technologies by organizing social learning processes involving producers, technology designers and users.

Experiences with SNM have been gained, for example, in the area of sustainable transport. An evaluation of a number of these examples discusses niche management processes at the level of transport technologies (such as various field experiments with electric cars and ultra-light electric vehicles) and on the level of experiments with the aim of reconfiguring mobility, such as car-sharing initiatives, bicycle-pool schemes or pilot projects on individualized public transport in France (Hoogma et al., 2002).

Conclusion

Let us finally sum up our argument. Although it is widely acknowledged that the interaction with customers – mainly through market research or marketing activities – is of strategic importance for successful technological innovation processes in companies, the practice of technology management often falls behind. Moreover, strategies to improve learning via the interactions between users and producers in the innovation and dissemination process are rarely systematically developed as an issue of technology management by companies or public institutions.

As our discussions in this chapter aimed to demonstrate, there are good reasons that strategies of learning between users and producers go beyond conventional market research and should consciously organize innovation processes as a co-construction of technical design and evolving social contexts, such as practices of use or changing regulatory requirements. This is especially true for environmental technologies, where company interests may align or conflict with the aims of public policy, environmentally aware customers and users and other stakeholders. Managing environmental technologies is thus closely aligned with managing social processes and the creation of stable actor networks.

Although the explicit environmental interests of users are presently restricted to a comparatively small segment of the public and companies, or to a number of niche markets, there are a number of strategies available (though often not trading under the name of environmental technology management) that would improve communication and cooperation processes with users and other stakeholders and thereby increase the chances for the success of innovative environmental technologies. What we have also seen is that environmental technology management is not only a task of producer companies but in many instances also of public bodies, user groups or others.

Making wider use of the suggested instruments of environmental technology management might help to improve conditions for a more conscious, reflexive and inclusive learning process between designers, intermediaries and users in product creation processes as well as to foster the development of environmentally friendly and politically desirable technologies and products.

Notes

1. An airline thus is a user of aircraft, even if it is the source of aircraft innovation. It is not uncommon that a user, who improves a product, even becomes a manufacturer of this product or at least markets it or licenses out the technology (Foxall, 1989).
2. Brown's example is the electric vehicle, which he calls a mixed good as its consumption is personal and exclusive and payment is individual but the effect of pollution reduction is a contribution to public welfare.

References

Akrich, M. (1995), 'User Representations: Practices, Methods and Sociology', in Rip, A., Misa, T.J. and Schot, J. (eds), *Managing Technology in Society: The Approach of Constructive Technology Assessment*, London: Pinter, pp. 167–84.
Brown, M.B. (2001), 'The Civic Shaping of Technology: California's Electric Vehicle Program', *Science, Technology, & Human Values*, **26** (1), 56–81.
Charter, M., Young, A., Kielkiewicz-Young, A. and Belmane, I. (2001), 'Integrated Product Policy and Eco-Product Development', in Charter, M. and Tischler, U. (eds), *Sustainable Solutions. Developing Products and Services for the Future*, Sheffield: Greenleaf Publishing, pp. 98–116.
Cleff, T. and Rennings, C. (1999), 'Determinants of Environmental Product and Process Innovation', *European Environment*, **9**, 191–201.
Dearing, A. (2000), *Sustainable Innovation: Drivers and Barriers*, World Business Council for Sustainable Development.
Edquist, C. (1996), 'Government Technology Procurement as an Instrument of Technology Policy', in Teubal, M., Foray, D., Justman, M. and Zuscovitch, E. (eds), *Technological Infrastructure Policy*, Dordrecht, the Netherlands: Kluwer Academic Publishers, pp. 141–70.
Edquist, C. and Hommen, L. (1999), 'Systems of Innovation: Theory and Policy for the Demand Side', *Technology In Society*, **21**, 63–79.
Fineman, S. (2001), 'Fashioning the Environment', *Organization*, **8**(1), 17–31.

Foster, C. and Green, K. (1999), 'Greening the Innovation Process', in *Proceedings of the Greening of Industry Network Conference*, Kathalys, Centre for Sustainable Product Innovation: TNO/DUT.

Foster, C. and Green, K. (2002), 'Environmental Innovation in Industry: The Importance of Environmentally-Driven Users', *International Journal of Environmental Technology and Management*, **2**(4), 303–14.

Foxall, G.R. (1989), 'User Initiated Product Innovations', *Industrial Marketing Management*, **18**, 95–104.

Franke, N. and Shah, S. (2003), 'How Communities Support Innovative Activities: An Exploration of Assistance and Sharing among End-users', *Research Policy*, **32**, 157–78.

Freeman, C. (1994), 'The Economics of Technical Change', *Cambridge Journal of Economics*, **18**, 463–514.

Green, K., Morton, B. and New, S. (2000), 'Greening Organizations. Purchasing, Consumption, and Innovation', *Organization & Environment*, **13**(2), 206–25.

Habermeier, K.F. (1990), 'Product Use and Product Improvement', *Research Policy*, **19**, 271–83.

Herstatt, C. and von Hippel, E. (1992), 'Developing New Product Concepts via the Lead User Method: A Case Study in a "Low-Tech" Field', *Journal of Product Innovation Management*, **9**, 213–21.

Hoogma, R., Kemp, R., Schot, J. and Truffer, B. (2002), *Experimenting for Sustainable Transport. The Approach of Strategic Niche Management*, London: Spon Press.

Kärnä, A. and Heiskanen, E. (1998), 'The Challenge of "Product Chain" Thinking for Product Development and Design – The Example of Electrical and Electronic Products', *The Journal of Sustainable Product Design*, January, 26–36.

Kemp, René (1998), 'Environmental Regulation and Innovation. Key Issues and Questions for Research', in *Proceedings of the Expert Meeting on 'Regulation and Innovation'*, Seville.

Kemp, R., Rip, A. and Schot, J. (2001), 'Constructing Transition Paths through the Management of Niches', in Garud, R. and Karnøe, P. (eds), *Path Dependence and Creation*, Mahwah, New Jersey/London: Lawrence Erlbaum Associates, pp. 269–99.

Lackner, B. (2003), 'Fostering Environmental Technologies by Public Procurement', paper presented at the conference *Towards Joint EU Action in Environmental Technologies*, Vienna, 26–27 May.

Lundvall, B.-Å. (1988), 'Innovation as an Interactive Process: From User–Producer Interaction to the National System of Innovation', in Dosi, G., Freeman, C., Nelson, R., Silverberg, G. and Soete, L. (eds), *Technical Change and Economic Theory*, London/New York: Pinter, pp. 349–69.

Mackay, H. and Gillespie, G. (1992), 'Extending the Social Shaping of Technology Approach: Ideology and Appropriation', *Social Studies of Science*, **22**, 685–716.

Magnusson, P.R. (2003), 'Benefits of Involving Users in Service Innovation', *European Journal of Innovation Management*, **6**(4), 228–38.

Markusson, N. (2001), *Drivers of Environmental Innovation*, Stockholm: VINNOVA – Verket för Innovationssystem.

Morrison, P.D., Roberts, J.H. and Midgley, D.F. (2004), 'The Nature of Lead Users and Measurement of Leading Edge Status', *Research Policy*, **33**, 351–62.

Noci, G. and Verganti, R. (1999), 'Managing "Green" Product Innovation in Small Firms', *R&D Management*, **29**(1), 3–15.

Organisation for Economic Co-operation and Development/International Energy Agency (OECD/IEA) (2003), *Creating Markets for Energy Technologies*, Paris: IEA.

Ornetzeder, M. and Rohracher, H. (2006), 'User-led Innovations and Participation Processes: Lessons from Sustainable Energy Technologies', *Energy Policy*, **34**(2), 138–50.

Oudshoorn, N. and Pinch, T. (eds) (2003), *How Users Matter. The Co-Construction of Users and Technologies*, Cambridge, MA: The MIT Press.

Polonsky, M.J. (2001), 'Green Marketing', in Charter, M. and Tischler, U. (eds), *Sustainable Solutions. Developing Products and Services for the Future*, Sheffield: Greenleaf Publishing, pp. 282–301.

Prentis, E. and Bird, H. (1999), 'Customers – the Forgotten Stakeholders', *The Journal of Sustainable Product Design*, January, 52–6.

Rip, A. and Kemp, R. (1998), 'Technological Change', in Rayner, S. and Malone, E.L. (eds), *Human Choice and Climate Change: Resources and Technology, Vol. 2*, Columbus, Ohio: Batelle Press, pp. 327–99.

Rogers, E.M. (1995), *Diffusion of Innovations*, New York: The Free Press.

Rohracher, H. (2003), 'The Role of Users in the Social Shaping of Environmental Technologies', *Innovation*, **16**(2), 177–92.

Rohracher, H. (ed.) (2005), *User Involvement in Innovation Processes. Strategies and Limitations from a Socio-Technical Perspective*, Munich: Profil Verlag.

Rose, D.A. (2001), 'Reconceptualizing the User(s) of – and in – Technological Innovation: The Case of Vaccines in the United States', in Coombs, R., Green, K., Richards, A. and Walsh, V. (eds), *Technology and the Market. Demand, Users and Innovation*, Cheltenham, UK and Northampton, MA, USA: Edward Elgar, pp. 68–88.

Rubik, F. (2001), 'Environmental Sound Product Innovation and Integrated Product Policy (IPP)', *The Journal of Sustainable Product Design*, 219–31.

Russell, S. and Williams, R. (2002), 'Social Shaping of Technology: Frameworks, Findings and Implications for Policy with Glossary of Social Shaping Concepts', in Sørensen, K.H. and Williams, R. (eds), *Shaping Technology, Guiding Policy: Concepts, Spaces and Tools*, Cheltenham, UK and Northampton, MA, USA: Edward Elgar, pp. 37–131.

Schaefer, A., Coulson, A., Green, K., New, S. and Skea, J. (2003), 'Sustainable Business Organisations?', in Berkhout, F., Melissa, L. and Scoones, I. (eds), *Negotiating Environmental Change. New Perspectives from Social Science*, Cheltenham, UK and Northampton, MA, USA: Edward Elgar, pp. 209–30.

Schot, J. (1992), 'Constructive Technology Assessment and Technology Dynamics: The Case of Clean Technologies', *Science, Technology, & Human Values*, **17**(1), 36–56.

Schot, J. (2001), 'Towards New Forms of Participatory Technology Development', *Technology Analysis & Strategic Management*, **13**(1), 39–52.

Schot, J. and Rip, A. (1996), 'The Past and Future of Constructive Technology Assessment', *Technological Forecasting and Social Change*, **54**, 251–68

Schot, J., Hoogma, R. and Elzen, B. (1994), 'Strategies for Shifting Technological Systems: The Case of Automobile System', *Futures*, **26**(10), 1060–76.

Shove, E. (2003), 'Users, Technologies and Expectations of Comfort, Cleanliness and Convenience', *Innovation*, **16**(2), 193–206.

Slaughter, S. (1993), 'Innovation and Learning during Implementation: A Comparison of User and Manufacturer Innovations', *Research Policy*, **22**, 81–95.

Stevels, A. (2001), 'Five Ways to be Green and Profitable', *The Journal of Sustainable Product Design*, 81–9.

Szejnwald Brown, H., Vergragt, P., Green, K. and Berchicci, L. (2003), 'Learning for Sustainability Transition through Bounded Socio-Technical Experiments in Personal Mobility', *Technology Analysis & Strategic Management*, **15**(3), 291–315.

Thomke, S. and von Hippel, E. (2002), 'Customers as Innovators. A New Way to Create Value', *Harvard Business Review*, April, 5–11.

Truffer, B. and Dürrenberger, G. (1997), 'Outsider Initiatives in the Reconstruction of the Car: The Case of Lightweight Vehicle Milieus in Switzerland', *Science, Technology, & Human Values*, **22**(2), 207–34.

Vergragt, P.J., Groenewegen, P. and Mulder, K.F. (1992), 'Industrial Technological Innovation: Interrelationships between Technological, Economic and Sociological Analyses', in Coombs, R., Saviotti, P. and Walsh, V. (eds), *Technological Change and Company Strategies: Economic and Sociological Perspectives*, London/New York: Harcourt Brace Jovanovich, pp. 226–47.

Vestby, G.M. (1996), 'Technologies of Autonomy? Parenthood in Contemporary "Modern Times"', in Lie, M. and Sørensen, K.H. (eds), *Making Technology our Own? Domesticating Technology into Everyday Life*, Oslo: Scandinavian University Press, pp. 65–90.

von Hippel, E. (1986), 'Lead Users: A Source of Novel Product Concepts', *Management Science*, **32**(7), 791–805.

von Hippel, E. (1988), *The Sources of Innovation*, Oxford/New York: Oxford University Press.

von Hippel, E. (2001), 'User Toolkits for Innovation', *Journal of Product Innovation Management*, **18**, 247–57.

Weber, M. and Hoogma, R. (1998), 'Beyond National and Technological Styles of Innovation Diffusion: A Dynamic Perspective on Cases from the Energy and Transport Sectors', *Technology Analysis & Strategic Management*, **10**(4), 545–66.

Weyer, J. (1997), 'Konturen einer netzwerktheoretischen Techniksoziologie', in Weyer, J., Kirchner, U., Riedl, L. and Schmidt, J.F.K. (eds), *Technik, die Gesellschaft schafft*, Berlin: Edition Sigma, pp. 23–52.

Williams, R., Slack, R. and Stewart, J. (2000), *Social Learning in Multimedia. Final Report*, Edinburgh: Research Centre for Social Sciences, The University of Edinburgh.

5 Cleaner production and eco-efficiency
René van Berkel

Introduction

The intent of cleaner production is essentially simple and straightforward: minimize wastes and emissions by eliminating their root sources and causes rather than handle or treat these wastes and emissions after they have been created. During the 1980s, progressive companies and governments identified that cleaner production was not only compellingly logical and conceptually simple but also made good business sense. By the end of the 1980s environmental managers, industrial planners and policy-makers started to look more favourably upon cleaner production as a real alternative to 'cleaning' and 'clean-up' forms of environmental technologies.

Simultaneously, proactive industrialists joined forces to consider and formulate a positive business response to the global environmental and social challenges being raised for discussion at the 1992 Rio United Nations Conference on Environment and Development. The term eco-efficiency was coined to capture the strategic intent of business to deliver more value with less environmental impact.

The post-Rio decade has seen a great deal of experimentation and consolidation of the cleaner production and eco-efficiency concepts and industrial practices. These evolved into complementary strategies for the effective and efficient use of materials, energy, water and other natural resources in business. They are now common improvement and innovation strategies with well-established track records in many industry sectors around the globe, in particular in the processing and manufacturing industries, and increasingly in the service, hospitality, transport, agribusiness and mining, minerals and energy sectors. In practice, cleaner production and eco-efficiency can be regarded as 'two sides of the same coin', with cleaner production highlighting the *means* of the change process and eco-efficiency the *objectives* (van Berkel, 2000). It is becoming common practice to use the terms interchangeably.

Cleaner production and – to a lesser extent – eco-efficiency have perhaps not yet received the strategic recognition they deserve, with the result being that they are often stigmatized as environmental technologies. It is illustrated here that cleaner production and eco-efficiency are strategic priorities for business that can be realized by combining common sense with sound business management, good engineering practice and innovation. The chapter is therefore structured in three main sections. The next section explains how the ideas of cleaner production and eco-efficiency grew from visionary thinking to industrial practice. A selection of practical examples from the food processing, metal fabrication and services sectors is then provided to illustrate how cleaner production and eco-efficiency deliver economic and financial returns. The final section discusses how to get started, with a particular focus on the diagnosis of products and processes for the development of improvement opportunities, and on the creation of a corporate culture conducive to eco-efficiency.

From ideas to actions
Cleaner production and eco-efficiency have strong ideological bases and are at their core better understood as pertinent ideas rather than well-defined sets of practical solutions. This section goes back to the roots of the ideas and follows their development into their current status.

Setting the scene
Society's response to industrial environmental problems has progressed over time, and has resulted in quite distinct classes of environmental technologies and practices (OTA, 1994; Reijnders, 1996; van Berkel, 1996; van Weenen, 1990). Four key stages in the evolution of environmental management strategies are (see also Figure 5.1):

1. *Dispersion.* Minimizing the impacts of wastes and emissions by enhancing the dispersion of waste and emissions so that exposure, in particular for humans, is being minimized. Even today dispersion remains an important component in the environmental management toolbox, in particular for risk management through proper stack heights for discharge of air emissions, and industrial zoning to separate industrial activity (including waste management) away from residential areas. At its base, however, dilution is no solution to pollution – the assimilative capacity of the environment for waste and emissions is limited.
2. *Control.* Collecting and treating waste and emissions from industrial processes to avoid or at least minimize their potential environmental impact. The control concept has

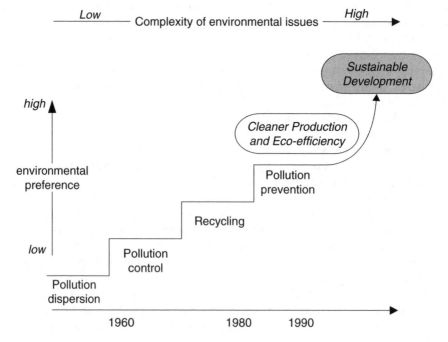

Figure 5.1   Evolution of environmental management paradigms

given rise to a broad spectrum of 'end-of-pipe' or 'clean-up' technologies such as biological, chemical and physical waste water treatment, air pollution control technology (for example, flue gas desulphurization, electrostatic precipitation and so on) and waste management (such as controlled landfill and incineration). The limitations of this control and treat strategy are also obvious. The technologies are expensive to install and run, and they produce their own waste products that have to be disposed of (such as waste water treatment sludge and contaminated air pollution control filters).

3. *Recycling*. Utilization of the waste (material, water, heat) for the same or another purpose through re-use, recycle and recovery. The recycling strategy has been successfully applied in industrial contexts (in particular for heat and process chemicals recovery and water re-use and recycling) as well as residential applications (composting of organic wastes, recycling of paper and plastics). Although much preferred over control and treatment, recycling also has its environmental downsides. Most often the recycled material is contaminated, rendering it unsuitable for use in the same application. Moreover, the recycling processes themselves generate waste and consume energy and other natural resources.

4. *Prevention*. Minimizing, or even eliminating, the generation of waste and emissions through modification of the products and processes. In many cases this is achieved through more efficient and innovative process technology and equipment.

At the conceptual level, prevention is by far the preferred strategy followed in turn by recycling, control and dispersion (van Weenen, 1990). In practice the net environmental benefit of practical solutions in a lower ranked strategy might be bigger than that achieved by solutions in a higher ranked strategy, depending on the efficiency and effectiveness of the specific solutions available within those strategies (Reijnders, 1996; van Berkel, 1996).

Preventative environmental management strategies have gained widespread recognition since their inception at pioneering American chemical process industries in the mid-1970s (Zosel, 1994). They are now widely acclaimed by both the business community and the public sector for their proven ability to deliver cost-effective or even cost-cutting reductions in the environmental impacts of organizations across a wide range of industry sectors (UNEP, 2002; WBCSD, 2000).

Both cleaner production and eco-efficiency are rooted in the prevention strategy. They go beyond pollution prevention by explicitly incorporating conservation of materials, energy and other natural resources, and by strengthening the value-adding aspect of goods and services. They are signposts on the business journey towards sustainability. Alone, however, they are insufficient as the business contribution to sustainable development because they do not explicitly aim for the integration of social advancement, environmental protection and economic development.

Turning waste into profit
Cleaner production is by origin an environmental management strategy and its principal thrust is best captured in the phrase 'turning waste into profit'. Cleaner production starts with recognizing that wastage of materials and energy into emissions and waste is an expensive affair and that reducing such wastage should therefore be cost-effective.

Cleaner production is about preventing waste and emissions, including the wastage of energy, materials, water and other natural resources, rather than dealing with them once they have been generated. More precisely, cleaner production is most often defined as 'the continuous application of an integrated preventative environmental strategy to processes, products and services to increase efficiency and reduce risks to humans and the environment' (ANZECC, 1998, p. 80; UNEP, 1994, p. 32). Cleaner production aims at progressive reductions of the environmental impacts of processes, products and services. It considers all relevant environmental aspects and impacts and is therefore not confined to one environmental impact category like most end-of-pipe technologies. Moreover, it serves economic and ecological efficiency ('eco-efficiency') and contributes to the realization of environmental risk reduction and management objectives for humans and the environment.

The cleaner production idea of reducing the generation of wastes and emissions at the source can be achieved in various ways. A division of five 'prevention practices' is most common (de Hoo et al., 1991; USEPA, 1988; USEPA, 1992; van Berkel, 1995; van Berkel, 2000):

1. Product modifications change the product characteristics such as shape and material composition. The lifetime of the new product is expanded, the product is easier to repair or the manufacturing of the product is less polluting. Changes in product packaging are generally also regarded as product modifications.
2. Input substitution refers to the use of less polluting raw and adjunct materials and the use of process auxiliaries (such as lubricants and coolants) that have a longer service lifetime.
3. Technology modifications include improved process automation, process optimization, equipment redesign and process substitution.
4. Good housekeeping refers to changes in operational procedures and management in order to eliminate waste and emission generation. Examples are spill prevention, improved instruction of workers and training.
5. On-site recycling refers to the useful application of waste materials or pollutants at the company where these have been generated. This could take place through re-use as raw material, recovery of materials or useful application.

These prevention practices are generally best thought of as starting points for developing site-, process- and product-specific cleaner production solutions (van Berkel, 1995). Such solutions would in many cases combine several practices. The establishment of a near closed-loop water system at a mine or power station, for example, is based on the recycling practice but will generally require supportive changes in technology (technology modification) and operating practices (good housekeeping). Table 5.1 contains both low and medium- to high-cost examples of each for the process industry.

Doing more with less
Eco-efficiency is by origin a business efficiency strategy that capitalizes on the deeply entrenched business imperative to continuously improve the efficiency of use of all business resources (staff, technology, skills and so on). The underlying imperative in eco-efficiency is 'to do more with less'; that is, to produce more value with lower ecological impact through the development and delivery of better products and services that are competitive and better meet customer needs. More precisely, it entails 'the delivery of

Table 5.1   Illustrative cleaner production options for the process industry

Prevention Practice	Typical Low-/No-cost Examples	Typical Medium-/High-cost Examples
1. Product modification	• High solids paint and inks to reduce solvent use in production and product application • Environmentally preferred packaging (e.g. less or re-usable packaging, recyclable and recycled materials)	• Develop premium products for longer service lifetime (e.g. coolants, lubricants) • Develop 'greener' products that are safer to the customer and pose lower risks to the environment
2. Input substitution	• Use biodegradable detergents and cleaners • Use higher purity materials by renewable material • Make use of renewable energy	• Replace toxic catalyst by less toxic catalyst • Replace non-renewable material
3. Technology modification	• Installation of appropriate process instrumentation to measure and optimize process conditions • Use mechanical tank wall wipers to scrape product from tank walls after a product batch has been emptied (e.g. paint, resins, etc.)	• Better mixing design of reactors to reduce by-product generation • Adopt alternative synthesis pathway to avoid toxic by-product or toxic process intermediary • Convert from batch to continuous processes • Develop more selective catalyst
4. Good housekeeping	• Training employees in proper material storage and handling procedures • Spill and leak detection and prevention programmes • Use spill and drip trays to recover losses from manual material transfer operations	• Maximize batch sizes, and follow with a similar product that may not require equipment cleaning between batches • Dedicated equipment for large volume products
5. On-site recycling	• Use counter-current washing, heating, etc. • Dry-clean equipment before hosing down and re-use recovered material	• Rinse with compatible solvent and store solvent for re-use in make-up of next batch of compatible product • Heat recovery from hot process streams

Source: van Berkel (2001).

competitively priced goods and services that satisfy human needs and bring quality of life, while progressively reducing ecological impacts and resource intensity throughout the lifecycle, to a level at least in line with the earth's estimated carrying capacity' (DeSimone and Popoff, 1997; WBCSD, 2000, p. 32). Eco-efficiency is therefore concerned with three broad objectives, namely (WBCSD, 2000):

1. Reducing the consumption of resources: this includes minimizing the use of energy, materials, water and land, enhancing recyclability and product durability, and closing material loops.

2. Reducing the impact on nature: this includes minimizing air emissions, water discharges, waste disposal and the dispersion of toxic substances as well as fostering the sustainable use of renewable resources.
3. Increasing product or service value: this means providing more benefits to customers through product functionality, flexibility and modularity; providing additional services and focusing on selling functional needs with the customers receiving the same functional need with fewer materials and less resources.

Practical experience in the implementation of these objectives revealed seven key dimensions that every business should be taking into account when developing products, introducing process changes or taking other actions with environmental implications. These are (DeSimone and Popoff, 1997; WBCSD, 2000):

1. reduce the material intensity of goods and services;
2. reduce the energy intensity of goods and services;
3. reduce toxic dispersion;
4. enhance material recyclability;
5. maximize the sustainable use of renewable resources;
6. extend product durability;
7. increase the service intensity of goods and services.

Box 5.1 provides a brief discussion of each. Although eco-efficiency has its roots in the manufacturing and processing industries, it has become evident that it is likewise applicable in the agribusiness, extractive and services sectors. Table 5.2 contains examples of existing and emerging agribusiness practices and technologies that deliver eco-efficiency.

BOX 5.1   SEVEN ECO-EFFICIENCY DIMENSIONS

1. *Materials intensity*. Materials require energy and create pollution and waste throughout their lifecycle, from mining or growing, via refining, manufacturing, transport and use, through to their final disposal. The total mass consumed in the lifecycle of delivering products or services can therefore be a useful measure of environmental impact in its own right and also a proxy for others. The material intensity of goods and services can be reduced through lean product designs; the use of less material-intensive raw materials; minimization of off-cuts and off-specification products in manufacturing; and effective re-use and recycling of products at the end of their useful life. Water conservation is also an integral element of reducing material intensity.

2. *Energy intensity*. All stages of the lifecycle of any product or service require energy. The production, distribution and use of this energy create pollution and waste materials, and generally deplete non-renewable fossil fuels. Reduction of the energy-intensity of goods and services therefore has great environmental benefits. This can be achieved in numerous ways, including use of energy-efficient production and distribution processes and

equipment; exchange of heat between processes; insulation of hot process equipment; and better utilization of daylight and natural ventilation.

3. *Toxic dispersion.* Toxic or otherwise harmful substances are of particular environmental concern, especially when such substances do not quickly break down in the environment in which they escape or are being emitted. Reduction of the dispersion of toxics from the lifecycle of goods and services is therefore essential. Use of safer substitutes that are less toxic or break down more easily is the preferred approach. Alternatively, technical and managerial changes may be required to better contain toxics throughout their lifecycle, and safely dispose of them after their use.

4. *Recyclability.* The utilization of material and energy can generally be improved by re-use in the same or another system. Eco-efficiency stresses the need for recycling to maximize the financial value of the recycled materials and minimize the environmental impacts of their processing. Recyclability requires products to be made of recyclable materials, in ways that facilitate their recovery, for instance through material marking, minimization of material diversity and design for disassembly. Vice versa, material specifications in product design should not unnecessarily disadvantage recycled materials.

5. *Renewable resources.* Utilization of renewable rather than non-renewable resources reduces the pressure on finite resources, and tends to create less environmental impact over the entire lifecycle. The most obvious renewable resource is solar power, whether captured directly through sunlight into heat or electricity or indirectly through biomass, waves and winds. Care should be taken in the use of the renewable resources so that the natural environment is not compromised through over-harvesting of biomass, degradation of soil quality, or high use of energy and materials in the extraction and refining of useful substances from biomass.

6. *Durability.* Extension of the useful life of products will result in less frequent replacements, which in turn can mean less waste and the use of fewer material and energy resources. Product durability can generally be improved by more robust product designs, use of appropriate materials and better maintenance.

7. *Service intensity.* Increasing service intensity means creating additional value for customers while reducing or holding constant environmental impacts. It can be achieved in a number of ways such as shared use, multifunctionality and easy upgrading.

Source: Based on DeSimone and Popoff (1997) and WBCSD (2000).

Convergence of concepts

Cleaner production and eco-efficiency tend to converge around the effective utilization of materials, energy, water and other natural resources. This takes place amongst the backdrop of a boom in other preventive environmental management strategies (van Berkel, 2000). Figure 5.2 shows the relative positions of key concepts regarding:

Table 5.2   Eco-efficiency elements illustrated for the agribusiness sector

Eco-efficiency elements	Example agribusiness applications	
	Farm level	Food and beverage processing
1. Reduce material intensity	• Drip irrigation • Partial root-zone drying (vineyards)	• Optimize process control and quality management • Use water-efficient nozzles and spray guns
2. Reduce energy intensity	• No till farming • 'Thermomass' cool storage rooms	• Variable speed drives • Co-generation
3. Reduce dispersion of toxic substances	• Integrated pest management • Organic farming • Hydroponics (closed-loop nutrient cycling)	• Use of biodegradable cleaning agents • Use of environmentally friendly packaging materials
4. Enhance recyclability	• Re-use of spent marc (grape seeds and skin residue) on vineyards • Bio-digesters	• Re-use of process water for irrigation purposes • Use of recyclable packaging materials and/or refillable containers
5. Maximize use of renewables	• Solar-powered pumps and fences • Natural-based insecticides	• Use of bagasse (sugar cane fibre) as alternative fuel in sugar mills
6. Extend product durability	• Low moisture harvesting	• Snap freezing of perishable produce
7. Increase service intensity	• Precision agriculture • Strip grazing • Integrated salinity and biodiversity management	• Prepared fresh foods (premixed salads, etc.)

Note: Many applications serve multiple eco-efficiency dimensions.

Source: van Berkel (2002b).

- discharge routes covered, in particular whether only one category of discharges is targeted (single media or compartment such as waste water discharges, air emissions) or several (multimedia);
- primary motivation driving the respective environmental management strategy: division is made between environmental regulation and corporate responsibility as the key drivers;
- reactive versus preventive approaches: whether the environmental strategy addresses waste and pollution once it has been generated or aims to avoid waste and emissions initially;

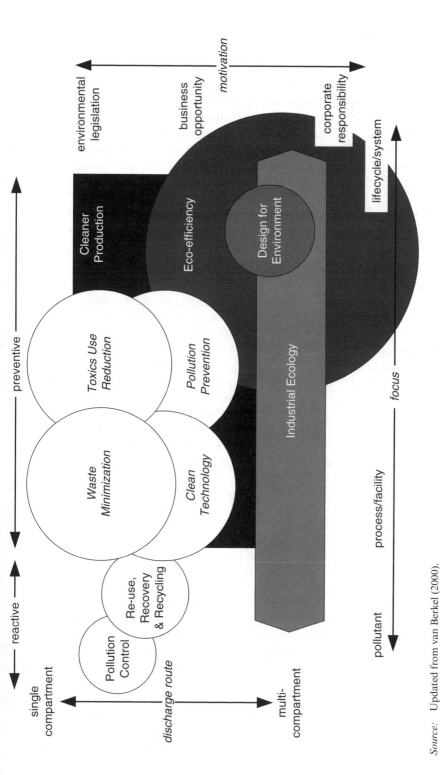

Source: Updated from van Berkel (2000).

Figure 5.2   Cleaner production and eco-efficiency in relation to other preventive environmental management strategies

- focus: whether the environmental management concept centres on waste streams, production facilities or product lifecycles.

Cleaner production and eco-efficiency can be regarded as common descriptors for the most frequently used preventive approaches. The older preventive approaches are waste minimization, clean technology, pollution prevention and toxics use reduction. Their use tends to be geographically bound and to a certain extent regulated (see Box 5.2). The newer preventive approaches explicitly target reduction of environmental impacts along the product's lifecycle, by focusing on product design (design for the environment) or on new approaches for value-adding activities (eco-efficiency).

BOX 5.2   CONCEPTS RELATED TO CLEANER PRODUCTION AND ECO-EFFICIENCY

1. *Waste minimization.* This term surfaced as the umbrella term for the preferred hazardous waste management strategies in the 1984 Hazardous and Solid Waste Amendments to the USA Resource Conservation and Recovery Act (RCRA). In the context of RCRA, waste minimization consisted of source reduction (reduction of the amount of waste at the source, through changes in industrial processes) and recycling (re-use and recycle waste for the original purpose, such as materials recovery or energy production) (USEPA, 1988). In the 1990s, the United Kingdom and other Commonwealth Nations began to use waste minimization in the context of reducing all non-product losses of natural resources (for example, Envirowise, 2002). In doing so, waste minimization has become almost synonymous with cleaner production and aligned with the concept of muda (waste)-elimination in the Japanese Kaizen (improvement) management philosophy (Imai, 1997).

2. *Pollution prevention.* This is the original term used by the pioneering manufacturing industries in the USA (Zosel, 1994) since the mid-1970s and was embedded in legislation in the USA in the 1990 Pollution Prevention Act. The Act specifies that facilities that are required to report releases for the Toxic Release Inventory provide documentation of their procedures for preventing the release of or for re-using these materials (USEPA, 1992) and encourages maximum possible elimination of wastes of all types. Although pollution prevention is commonly regarded as the North American equivalent of cleaner production, its legislative framework appears to have limited its application to prevention of regulated environmental impacts and not necessarily to the efficient use of materials, energy, water and other natural resources.

3. *Toxics use reduction.* Concerned with the reduction of input, throughput, and product and non-product output of toxic materials in industrial processes (Rossi et al., 1991). It is specifically defined in state legislation in a number of states in the USA (pioneered by Minnesota, Massachusetts, etc.) thereby effectively strengthening the requirements of the Pollution Prevention Act with regard to reportable emissions under the Toxic Release

Inventory. It can also be regarded as a synonym for the 'toxic dispersion reduction' dimension in eco-efficiency.

4. Clean(er) Technology: A term more commonly used in Europe to differentiate preventive environmental technologies from end-of-pipe clean-up technologies. Over the course of the 1990s, cleaner technology has evolved into the equivalent terminology for cleaner production in a number of countries (Scandinavia and UK for example).

5. Design for Environment: Also known as eco-design or life-cycle design, design for environment is about considering environmental factors in all stages of product design in order to reduce the net environmental impact of the product or service over its entire product lifecycle (Brezet and van Hemel, 1997). Design for the environment can be regarded as the application of cleaner production and eco-efficiency to product design.

6. Industrial Ecology: Is a field of academic study, industrial practice and public policy that is 'industrial' in that it focuses on product design and manufacturing processes, and 'ecological' in using natural models as inspiration for industrial models, and by placing human activity in the broader context of the natural systems that support human activity (Lifset and Graedel, 2002). (See also Chapter 2.)

Eco-efficiency and cleaner production are truly complementary concepts, with eco-efficiency focusing on the strategic dimension (*value creation*) and cleaner production on the operational dimension (*production*) (van Berkel, 2000). Accordingly, it is becoming common practice to substitute cleaner production for eco-efficiency and vice versa. These are the starting points for business to contribute to sustainable development (Holliday et al., 2002). They focus on persistent incremental improvements in existing products and operations that combine progressive reduction of environmental impacts of products and operations and of reduction of exposure of employees, neighbours and customers, with enhancement of business efficiencies.

Double dividends

Cleaner production and eco-efficiency have gained wide acceptance by industry and government alike for their proven ability to deliver benefits to business and the environment. Assessment of the size of such benefits is complicated by the diversity of practical solutions found by businesses when applying the cleaner production and eco-efficiency mindsets. As an illustration this section provides selected Australian case studies from food and beverages, metal products and service sectors.

Benefits to food industry

The environment and resource use profile of the food and beverage industry is quite diverse. Many operations have high energy and water usage and create significant volumes of waste water and organic waste. Moreover, waste associated with packaging is common in the food and beverage industry. Cleaner production and eco-efficiency have been successfully applied throughout the industry, regardless of the sub-sector and size of operations.

Tony's Tuna International (Port Lincoln, South Australia) processes locally farmed blue fin tuna (about 28 tonnes per day during season) into frozen and fresh sashimi (raw, sliced) tuna primarily for the Japanese markets (DEH, 2004). It is one of the most modern tuna processing plants in the area. The company is a significant water user in an area where water is a scarce resource, and is a large contributor to the sea discharges through the city waste water treatment plant. Detailed investigations of water usage and waste water generation during 1999–2000 found that pilchard thawing was the single largest consumer of mains water. The pilchard thawing operation was improved by changing the water inlet to the base of the thaw-out bins and by pulsing water exchange through solenoid valves. Quarantine concerns were addressed by using backflow prevention valves on the main water inlet pipe. The previous water consumption of at least 12m^3 water use per tonne of pilchards was reduced to 3.3–5.6m^3 per tonne, depending on mains water inlet temperature. The investment of approximately AUD1000 was quickly recovered as direct savings from avoided water consumption were at least AUD24000 per annum. Moreover, clean-up procedures were improved, introducing amongst others a dry clean-up before hosing down of equipment and work areas.

NSW Sugar Milling Cooperative Ltd (New South Wales, Australia) (DEH, 2004) grows sugar cane and mills and refines it into white sugar in three sugar mills and refineries. During 1997–2001 the Broadwater Mill and Refinery implemented various waste minimization and energy efficiency initiatives. Production waste has effectively been eliminated through use of bagasse (sugar cane fibre) as fuel, re-use of cleaned mill mud as fertilizer for sugar cane growing, use of molasses as stock feed and recirculation of mill washings. Meanwhile numerous improvements were made in the use of steam, virtually eliminating the use of coal for steam production. A co-generation plant was built to burn bagasse for the simultaneous generation of heat/steam and electricity. This converted the mill from a net importer of electricity to a net exporter of electricity (4MW continuously during the crushing season). The financial benefit from energy efficiency and sales of exported electricity mounts up to approximately AUD1 million annually and reduces greenhouse gas emissions by approximately 10000 tonnes CO_2-equivalent.

Cadbury Schweppes produces at its Ringwood Confectionary Plant (Victoria) more than 25000 tonnes per annum of chocolate products, including the brand names Cherry Ripe, Freddo Frog and Crunchie. In 1992 the plant participated in a cleaner production demonstration programme with the Victorian Environmental Protection Authority. A strong cleaner production ethos was created and utilized for a number of projects. Initiatives over the first five years included (DEH, 2004):

- *Improved segregation of solid waste.* The recycling rate for solid waste increased from 5 to 50 per cent after all operations were reviewed and worksite-specific provisions made for collection of recyclable materials. Waste food wrappers are collected and converted into plastic pallets used for in-house product storage.
- *Insulation.* The company insulated 2 km of pipe-work and 12 storage tanks used for chocolate to reduce heat loss, improve productivity and reduce internal air conditioning loads. The planned investment for a capacity increase of the hot water system was no longer necessary.
- Enclosing the boilers has reduced external noise and increased boiler efficiency by up to 5 per cent.

- *Process optimization.* Process parameters for solid chocolate moulding and caramel extrusion were reviewed and improved to reduce product wastage (and thereby waste water generation).
- *Dust collection.* The filter bags were replaced with innovative cyclones, enabling the re-use of fine sugar dust, and avoiding the need to wash filter bags (with associated waste water discharges).
- *Re-design of wash bays.* The wash bays for cleaning of equipment, storage containers and utensils were redesigned to cut waste water generation from washing operations.

The total investment in cleaner production initiatives was just over AUD1 million, between 1992 and 1997. The collective savings from these initiatives are AUD780 000 per annum.

Cascade Brewing Company holds a unique position in the Australian cultural history as the oldest operating brewery in the country. The Hobart (Tasmania) operations include a brewery, and soft drink and cider operations. Since 1996 the company has utilized cleaner production principles primarily for water conservation, reduction of waste water discharges, reduction of waste generation and energy conservation. Some of the more prominent initiatives have been (DEH, 2004):

- Water use optimization measures as part of the redevelopment of the cider-making plant, including: the discontinuation of once-through water coolers on the fermenters; collection, re-chilling and re-use of used chilling water from cooling jackets; collection of cooling water from the pastueriser for re-use elsewhere in the plant; replacement of heat exchangers with more efficient units; steam condensate collection and re-use in boiler; and re-assignment of existing internal storage tanks to collect clean waste water for redirection to a variety of less critical water use areas. The combined benefit of these measures is a reduction of water use in the cider plant by 200m³ per day.
- Reduction of waste water generation is being achieved by loss reduction initiatives and optimization of cleaning procedures and practices. Profound examples of loss reduction were: replacement of manual keg filling with robotic keg filling; installation of ultra filters (instead of centrifuges) in beverages and cider processing; and replacement of flexible transfer hoses with fixed pipes. A centralized bulk detergent handling and reticulation system in the brewery reduced cleaning agent loss to sewage by over 60 per cent. Moreover, the cleaning in place system was extended into the beverages section.
- The extensive recycling and re-use initiatives now include: use of malting dust as composting material; use of fruit pomace (fruit residue – stems, skin, pips etc.) as animal feed or compost; recycling of boiler ash for either land tailoring or composting; use of all excess yeast as either animal fodder or for compost; use of spent grain, hops and malt culms (stems) as animal fodder; and separate collection for off-site recycling of glass, metal, shrink-wrap, labels, crown seals and cartons.
- Energy conservation has been achieved through incremental improvements in operating and maintenance practices (driven largely by annual energy efficiency targets for each area) and by designing energy-efficiency into new capital expenditure projects. A good example of the latter was the redevelopment of the cider plant, which

included: energy efficient heat exchangers; steam condensate collection system; elimination of live steam use for cleaning; optimizing final product chilling; and better instrumentation and process controls overall.

Cascade Brewery realized tremendous environmental and financial benefit from its focus on cleaner production and eco-efficiency, as illustrated by the reduction of the specific water consumption per unit of product by 60 per cent in the brewery, by 55 per cent in the beverage and 30 per cent in malting between 1996 and 2002.

Australian Country Choice is one of Australia's leading integrated beef companies (QEPA, 2004). Its Cannon Hill (Queensland) slaughterhouse and meat facility has capacity to process 5000 cattle a week. The company developed a strategic cleaner production plan in collaboration with the Queensland Environmental Protection Authority. The first stage involves opportunities that require none or very minimal capital expense, such as fitting of efficient spray nozzles to hoses, maintaining minimum flow rates, reducing water pressure and using water efficient shower hoses. The forecast savings from these low-cost options will be approximately AUD1 million per annum. The second stage requires plant alterations involving an intermediate level of capital for replacement of sensors and timer controls on wash stations, insulation of knife sterilizers and installation of motor optimizers. Major innovations in stage three and four will require process redesign to allow for alternative sources of energy and water, for instance, rainwater harvesting for non-food processing wash down, greater internal water recycling and heat recovery from refrigeration compressors.

Benefits to metal products industry
Although there are only a limited number of key operations in the metal products industry (for example, machining, welding, forging, casting, cleaning and coating), the industry is quite diverse in terms of its final products and size of operations, ranging from engineering and fabrication contractors working from backyard premises to multinational car companies and their suppliers. The resource and environmental profile includes air emissions (from welding, casting and coating operations), waste water discharges (from cleaning operations), oily and solvent wastes (from machining and coating operations) and metal wastes (off-cuts and so on). Moreover, energy consumption is generally high, in particular in casting operations. Many metal fabricators have been able to benefit from the cleaner production approach.

Besides benefiting the environment by being the world leader in solar hot water systems, *Solahart* has actively pursued cleaner production initiatives at its Welshpool (Western Australia) manufacturing plant to reduce energy consumption and waste generation (DEH, 2004). The first round of energy-saving initiatives took place between 1994 and 1996 and included: review and, where feasible, replacement of compressed air tools; regrouping of factory lights by functional areas in the workshop; laser alignment of the 400 tonne press; and installation of timers and solenoid valves on boiler and chroming and degreasing baths. The total investment of approximately AUD17000 reduced the annual factory greenhouse gas emissions by 240 tonnes CO_2-equivalent. Between 1997 and 2000 some higher-cost modifications were made to increase the efficiency and throughput of the enamel line; equip the chroming line with hexifloats; optimize the chrome line rectifier; and streamline the phosphating line (including reduction of the number of cleaning steps). The total investment in this stage was AUD270000, and enabled further reduction of annual factory greenhouse gas emissions by 700 tonnes CO_2-equivalent. Meanwhile Solahart also started waste minimization initiatives through better-quality management (reduction of reject product), working with suppliers to reduce packaging waste from part supplies and establishment of a waste water treatment and recycling plant.

Bradken is Australia's largest combined foundry and engineering group with 11 operating sites around Australia. The company is committed to continual environmental improvement and is achieving such with significant involvement from its employees, using the Japanese Kaizen management philosophy (CECP, 2004; Chaplin, 2002). Major improvements have been achieved at its foundry in Welshpool (Western Australia) in regard to waste reduction and energy conservation. Bradken upgraded the sand binders to less hazardous alternatives causing less fumes and odour concerns in the workplace. In combination with temperature and humidity controls this has improved mould-making, leading to lower rejects of moulds and cast product. Moulding practices have been improved, including stack moulding as and when possible to increase the ratio of cast product to moulding sand use. Bradken also installed a dry mechanical recovery system allowing for up to 75 per cent recovery and re-use of the moulding sand. The use of refractory materials on the ladles was also reviewed and practices were improved to extend the interval between consecutive changes of the refractory material from 45 to 125 heats/uses, avoiding 5.5 tonnes of refractory waste per annum. Change of the melting practices provided the largest energy saving. Scrap metal acceptance criteria were therefore strengthened to improve melting. Melting was changed from day to night to take advantage of off-peak tariffs. Moreover, compressed air use was reviewed and where possible replaced by more efficient power. From 1999 to 2002 the improved energy practices saved the company AUD250 000 per annum with a further opportunity to save AUD320 000 with an investment of AUD560 000 in power factor improvement. Over the same period cost savings on waste sand disposal increased to AUD290 000 per year.

OneSteel Market Mills operates a steel tube manufacturing plant in Kwinana (Western Australia). The company introduced a gain sharing system to encourage its small core staff (17 employees) to come forward with suggestions to improve the operation (CECP, 2004). In this system the company shares gains with the participating employees in a variable pay plan with 20 per cent of the gains shared in the first year and 10 per cent of the gains shared in the second and third year. In the first year of establishment of the gain sharing system, the company implemented 18 cost-reduction projects to achieve annual savings worth AUD340 000, of which employees have received AUD40 000 in the first year. These included a paint cleaning and recycling booth that recovers excess powder paint from the paint booth and cleans it for re-use. The configuration of the packaging station was improved to reduce handling and minimize the need for strapping and clips. Customers have been provided with cages to collect the wood gluts for re-use, already leading to a 15 per cent reduction of the timber use for packaging.

ACL Bearing Company is Australia's sole manufacturer of precision engine bearings and a supplier for passenger vehicles for all four Australian car manufacturers. Operating from Launceston (Tasmania), the company embarked on a very innovative and challenging cleaner production initiative in 1997 to produce lead-free high-performance engine bearings (DEH, 2004). After initial investigations, ACL concluded that steel-backed aluminium-based lining material was the best way forward. This required a new casting process developed by ACL and extensively tested by the Commonwealth Scientific and

Industrial Research Organisation (CSIRO). The AUD1.7 million research and development (R&D) effort confirmed the suitability of the material, which was then offered to engine manufacturers for testing. Following their acceptance of the new material the manufacturing plant was built in Launceston costing AUD2 million. The first lead-free engine bearings were supplied to one Australian engine manufacturer in 2003 with expected scaling up to other Australian and American manufacturers in the near future.

Ford Australia has initiated a diversity of cleaner production and eco-efficiency initiatives. An early example of Ford's cleaner production efforts was at its Broadmeadows Assembly Plant (Victoria) (DEH, 2004). The skids used to carry car bodies through the painting line were historically cleaned with a dip in a hot caustic causing serious environmental and occupational health concerns. Upon extensive review it became possible to engineer a high-pressure water hose to clean old paint and other debris from the skids. The new process not only eliminated a hazardous operation and the use of caustic but also saved approximately AUD300 000 per year in operating costs for an initial investment of AUD120 000. As the skid cleaning operation is now less cumbersome, it is performed more regularly, thereby contributing to the 5 per cent increase in first-time right car body painting. More recently Ford has concentrated on working with its suppliers to reduce the environmental impacts of vehicles and the manufacturing operations (Caon, 2003). In September 1999 it announced to all its suppliers that first-tier suppliers should have a certified ISO 14001 environmental management system for at least one of their operations by December 2001 and for all their operating sites by July 2002. In return, Ford assisted its suppliers with a workbook, workshops and technical resources. Although the initial deadline was not achieved in full, progress has been remarkable with 63 per cent of the Australian suppliers certified by end of 2002. Complementary to the environmental management system requirements Ford has its Restricted Substance Management Standard (RSMS). The standard describes how toxic and hazardous materials need to be managed in production and supply of parts to Ford companies. Introduced in the 1980s with a view to occupational health and safety, the scope of the standard has extended to include other hazardous substances (for example, ozone-depleting substances, heavy metals and so on).

Ford Australia has also entered into performance-based contracts with some of its main suppliers to increase the eco-efficiency of selected inputs and processes. A total waste management contract was signed in 2000 with a single contractor to cover all waste management services with a payment schedule encouraging the contractor to seek ways to avoid waste and increase recycling (Ford, 2004). In the first year recycling increased by 168 per cent and general waste was reduced by 47 per cent. In late 1999 Ford entered into a performance-based contract with PPG Industries, where PPG would be paid on a 'per car body' rate for the supply of paint to the Broadmeadows Plant. PPG has six employees located at the Ford plant as chemical and process engineers overseeing and monitoring the painting operation. A series of improvements and innovations has been made since then, including continuous monitoring of phosphate and electrocoat baths; progressive introduction of a lead-free electrocoating system; installation of auto-sludge discharge in the phosphate bath; and application of new spray gun technology (DEH, 2004). As a result the emission of volatile organic compounds has dropped by 27 per cent per vehicle between 1997 and 2002, waste water generation has fallen by 15 per cent per vehicle between 2000 and 2003 and water consumption reduced by 32 per cent per vehicle between 2000 and 2003. Meanwhile, product quality has steadily improved, leading to a lower number of re-sprays and optimum paint thickness.

Benefits to service sector

The following examples demonstrate that cleaner production principles and practices can also deliver real-time benefits to service providers.

Atlas Drycleaners operates a comparatively large dry-cleaning business in East Perth (Western Australia). The facilities were upgraded in 1999 to improve plant layout and efficiency (DEH, 2004). Properly sized and lagged pipes replaced undersized, unlagged and often irrational steam and condensate pipes. The company was able to save 30 per cent on its gas consumption. Meanwhile the heat load in the facility was significantly reduced, resulting in lower air conditioning loads (enabling 10 per cent savings on electricity consumption) and a much better workplace. Since then Atlas has further optimized the operation of the dry-cleaning machines to prevent solvent emissions in the workplace, and most recently to recover up to 80 per cent of the solvents from its dry-cleaning waste, previously disposed of as controlled waste.

Perth Zoo (Western Australia) is one of the world's outstanding zoos, successfully combining conservation, recreation, education and research (DEH, 2004). When Brian Easton took over as CEO in 1999 he insisted on a more systematic approach to environmental improvements, in particular for water and energy use and waste, to underpin the zoo's educational mission. The purpose-built operations centre won an award for its innovative application of passive solar design principles and has since set a benchmark for other commercial buildings. More energy savings are being achieved by converting battery-powered electric fences to solar, and by employing alternative forms of transport, including gas- and electricity-powered vehicles and bicycles with cargo capacity. Rain meters have been installed on bore and reticulation controllers to minimize pumping energy consumption and bore water use. Energy use in absolute terms remained equal despite a significant increase in visitor numbers and operations. Water initiatives have focused on ways to slow down algae blooms in the main lake using barley straw, thereby extending the intervals between consecutive dumps of the 300 000 litre-capacity lake. Flow restrictors have been installed on all ponds (for example, the penguin exhibit) as well as in all showers and basins. Dual flush cisterns have been progressively introduced. In the three years to 2001 bore water consumption reduced from 433 000m³ per year to 244 000m³ per year. Waste management practices have been improved by investing in a chipper and chipping all green waste for mulch in the gardens, shredding office paper for bedding, composting all animal waste except where prohibited by health regulations, and establishing recycling stations for visitors' waste. Waste generation has been almost halved, saving about AUD3500 per year and avoiding the need to invest about AUD80 000 in a waste disposal and recycling area.

St John of God Health Care in Subiaco (Western Australia) is one of Australia's largest private hospitals. With a 481-bed capacity it is a comprehensive health care facility with a wide range of clinical specialities and easy access to a range of on-site diagnostic services. Although the hospital had implemented several environmental projects in the past, such as a 1.2 MWhr co-generation plant commissioned in 1998, a systematic approach was missing (CECP, 2004). The hospital instigated cleaner production initiatives through a major organizational change and the introduction of an environmental management system in May 2002, which, through caregiver participation, has succeeded in reducing energy, water and waste costs and as a result has helped preserve the environment. Numerous smaller and larger initiatives have been proposed by staff and since implemented. Examples include (CECP, 2004):

- *Materials efficiency and waste reduction.* Reusable cotton blankets replace the disposable blankets used in the operating suite. Likewise, laryngeal masks have been introduced to replace the alternative disposal product. Recyclables (glass, office paper, cardboard, newspaper, scrap metal, aluminium and steel cans, cartons and plastic film) are collected separately for external recycling. Office plastic bin liners are only replaced if soiled, some of the food scraps are composted on site with garden waste, and some 35 tonnes of grease, fat and oils are collected and re-used as stock feed. Staff have been encouraged to reconsider the need for paper use and instructed to use duplexers on copiers and printers enabling departments to save up to 50 per cent on paper consumption. Moreover, the hospital has changed over to 50 per cent recycled paper content office paper. The educational programme on clinical waste was strengthened and as a result clinical waste dropped by 8–18 per cent due to non-clinical waste items no longer unnecessarily being disposed of with the clinical waste.
- *Energy efficiency.* The hospital agreed to purchase 2.5 per cent green power from renewable energy sources. Lighting circuits are being progressively rewired to enable division of rooms into sections, controlled by separate switches. Corridor lights only have every second light in operation. Energy awareness has been encouraged through stickers and other information. As a result of all energy initiatives, total energy consumption remained equal in 2003 compared with 2002, despite a 3 per cent increase in used space in the hospital with gas consumption dropping by 18 per cent between 2001 and 2003.

The business case
The examples in the previous sections illustrate that investments in cleaner production and eco-efficiency generally have attractive economics. These are achieved through (van Berkel, 2002a):

- *Reduction of expenditures on input materials, energy and water.* Cleaner production options can reduce the material, energy, and/or water consumption per unit of product produced and hence savings are made on the costs of these natural resources.
- *Reduction of expenditures on waste (water) treatment.* The amount and pollutant load of the various process waste streams (including solid waste, waste water, air emissions) is reduced and hence expenditures for treatment and disposal of waste and emissions are reduced.
- *Increase of production revenues.* Cleaner production most often increases the efficiency of the production processes resulting in higher levels of production output and thereby an increase in revenues. And/or
- *Better product quality.* The application of cleaner production normally improves the level of control over the various production processes, which will normally increase the overall product quality level.

The business examples also show that companies benefit from the simultaneous application of the five mutually reinforcing prevention practices (see earlier section on 'Turning waste into profit') and pursuit of multiple eco-efficiency elements (see earlier section on 'Doing more with less'). Table 5.3 illustrates how multiple prevention practices and eco-efficiency elements were used in each of the company examples reviewed here.

Table 5.3   Classification of company examples

Company	Prevention practices (*)					Eco-efficiency elements (**)						
	GH	IS	TM	PM	R	MI	EI	TD	MR	SR	PD	SI
Food processing industry												
1. Tony's Tuna International	◆		◆◆			◆◆	◆					
2. NSW Sugar Mills	◆	◆	◆◆		◆	◆◆	◆◆			◆◆		
3. Cadbury Schweppes	◆		◆◆		◆◆	◆◆	◆◆					
4. Cascade Brewery			◆◆		◆◆	◆	◆◆	◆◆				
5. Australian Country Choice	◆◆	◆	◆◆		◆◆	◆◆	◆◆	◆				
Metal products industry												
6. Solahart Industries	◆		◆◆				◆	◆◆				
7. Bradken	◆	◆	◆◆		◆◆		◆	◆◆	◆			
8. OneSteel	◆◆			◆	◆		◆		◆◆	◆		
9. ACL Bearings		◆◆		◆◆				◆◆				
10. Ford	◆	◆◆		◆		◆	◆	◆	◆◆		◆	
Service sector												
11. Atlas Drycleaners	◆		◆◆			◆	◆	◆◆	◆◆			
12. Perth Zoo	◆			◆		◆	◆	◆◆			◆	◆
13. St John of God Health Care	◆	◆◆		◆		◆	◆	◆◆	◆◆			◆

Notes: ◆ ◆ = Primary practice or element; ◆ = Secondary practice or element.
(*) GH = Good Housekeeping; IS = Input Substitution; TM = Technology Modification; PM = Product Modification: R = Recycling.
(**) MI = Reduce Material Intensity; EI = Reduce Energy Intensity; TD = Reduce Toxics Dispersion; MR = Enhance Materials Recyclability; SR = Sustainable Use of Renewable Resources; PD = Extension of Product Durability; SI = Increase in Service Intensity.

Getting started

Cleaner production and eco-efficiency have developed significantly over the last two decades, as have the views on how to best enable their implementation in different industry settings (USEPA, 1992; USEPA, 2001; van Berkel, 1994; van Berkel and Lafleur, 1997; van Berkel et al., 1997). Cleaner production methods emphasize the effective identification, evaluation and implementation of practical solutions that avoid waste and emission generation. The eco-efficiency model stresses the strategic and organizational requirements for organizations to become eco-efficient (DeSimone and Popoff, 1997). This section provides a brief introduction to these complementary approaches.

Cleaner production methods

The traditional cleaner production method has a top-down focus and evolves around the application of sound engineering practice. The emergence of the ISO 14000 series of environmental management standards (Sheldon, 1997) catalyzed the development of the management systems method for cleaner production implementation. More recently there is growing interest in using quality models to establish cleaner production as a dimension of business excellence (Pojasek, 1997; USEPA, 2001).

The *engineering* method is based on engineering evaluations of production processes, referred to as cleaner production assessment. This is essentially a systematic procedure for the identification and evaluation of cleaner production options for a production facility (USEPA, 1988; de Hoo et al., 1991). It is nowadays generally divided into five phases, respectively: planning and organization; pre-assessment; assessment; feasibility studies; and implementation and continuation (van Berkel, 1996 and see Figure 5.3). As compared with earlier versions, the assessment procedure now places greater emphasis on sustaining cleaner production after the implementation of the first set of opportunities, through the creation of appropriate management and information systems, allocation of tasks and responsibilities and/or extension into an environmental management system. There is a great degree of flexibility in how the procedure is actually being implemented for any company taking into consideration the level of formalization of the environmental management function in the company as well as its management and information systems, process and technology development potential and corporate culture (van Berkel, 2002a).

The ISO 14001 Environmental Management System (EMS) standard places emphasis on improvements in *organizational* environmental performance by establishing and maintaining a systematic management plan designed to continually identify and reduce the environmental aspects of an organization's activities, products and services (Sheldon, 1997). The EMS standard is flexible with regard to the actual means of improvement of the *operational* environmental performance through preventive approaches like cleaner production, or pollution control and end-of-pipe treatment, as well as the desirable environmental performance level. The EMS standard has spurred a management systems method for cleaner production, involving five steps: getting started (recognizing the need); policy and commitment; planning; implementation; and evaluation and management review (USEPA, 2001 and see Figure 5.4). The prime difference with standard EMS practice is the focus and incentive for cleaner production, in particular through the management commitment to 'prevention of pollution' and the application of cleaner production principles and tools for the identification of aspects and significant impacts (step 3.2, Figure 5.4) and for the establishment of environmental management programmes (step 3.5, Figure 5.4). From a cleaner production perspective the main benefit is that the search for cleaner production opportunities is being embedded in the environmental management system, is being reiterated as part of the planning and review of the management system, and focused by the ongoing identification and update of environmental impacts resulting from the organization's activities, products and services (van Berkel, 2002a).

The engineering and management systems models have suffered from a reputation of being something that only environmental personnel do or direct others to do. Integration of cleaner production into core business practices will allow serious consideration by senior management and workers. The *quality-based method* therefore seeks to extend business excellence to deliver environmental performance and outcomes using proven quality and operations management approaches. This was developed into a quality-based method having five critical steps (USEPA, 2001 and see also Figure 5.5):

1. plan and develop cleaner production programme;
2. develop cleaner production opportunities;
3. implement cleaner production programme;
4. maintain cleaner production programme;
5. measure progress toward zero waste and emissions.

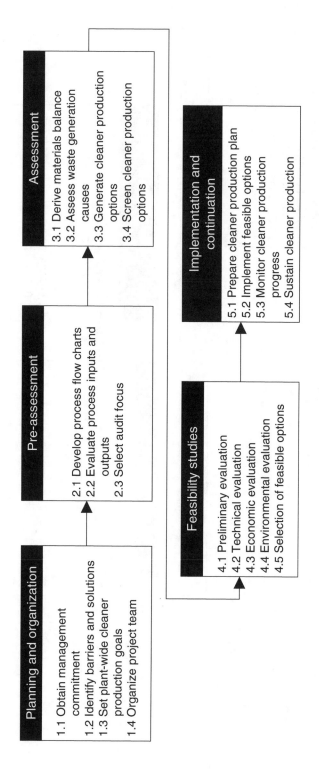

Source: van Berkel (1996).

Figure 5.3 Engineering method for cleaner production

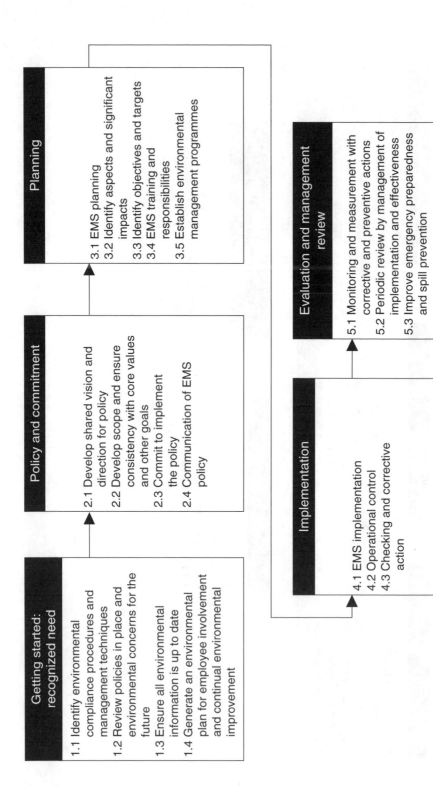

Getting started: recognized need

1.1 Identify environmental compliance procedures and management techniques
1.2 Review policies in place and environmental concerns for the future
1.3 Ensure all environmental information is up to date
1.4 Generate an environmental plan for employee involvement and continual environmental improvement

Policy and commitment

2.1 Develop shared vision and direction for policy
2.2 Develop scope and ensure consistency with core values and other goals
2.3 Commit to implement the policy
2.4 Communication of EMS policy

Planning

3.1 EMS planning
3.2 Identify aspects and significant impacts
3.3 Identify objectives and targets
3.4 EMS training and responsibilities
3.5 Establish environmental management programmes

Implementation

4.1 EMS implementation
4.2 Operational control
4.3 Checking and corrective action

Evaluation and management review

5.1 Monitoring and measurement with corrective and preventive actions
5.2 Periodic review by management of implementation and effectiveness
5.3 Improve emergency preparedness and spill prevention

Source: USEPA (2001).

Figure 5.4   Management systems method for cleaner production

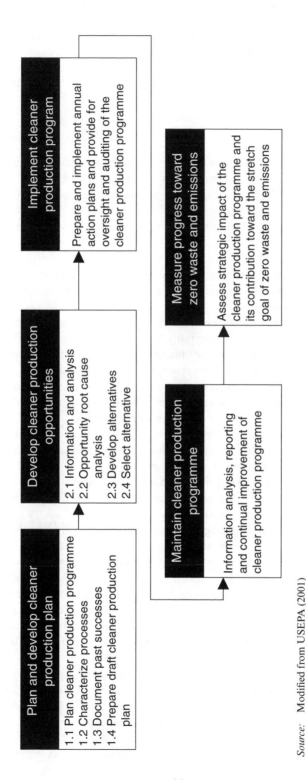

Source: Modified from USEPA (2001)

Figure 5.5   Quality method for cleaner production

The advantage of this quality-based method is its business orientation, which is particularly reflected in the planning and development of the cleaner production programme and the measurement of progress toward environmental excellence (van Berkel, 2002a). It can also be extended into a system that provides criteria that define 'best in class', offer a set of guiding principles (or core values), and enable rigorous rating of environmental excellence (USEPA, 2001).

Eco-efficiency elements

Eco-efficiency models have focused on the drivers and shapers that make an organization eco-efficient. The process of overcoming internal barriers is both radical and incremental; radical because it involves a complete change from business-as-usual thinking; incremental in that the operational transition to eco-efficiency is a step-by-step process. In either case there are nine key eco-efficiency elements that allow progress to be made (DeSimone and Popoff, 1997):

1. *Leadership*. Alerting staff, developing a shared sense of eco-efficiency, motivating and empowering staff, demonstrating implications for day-to-day actions and supporting the eco-efficiency champions.
2. *Foresight*. Looking into the future and establishing what competencies and technologies will be required for the company to succeed into the future.
3. *Culture*. Establishing a culture of continuous improvements in resource efficiency and environmental performance through personal commitment and empowerment of staff.
4. *Management tools*. Environmental management systems, environmental management accounting and environmental performance measurement help with the identification and selection of eco-efficiency opportunities.
5. *Lifecycle management*. This involves focused efforts to understand and manage the entire lifecycle of products and services, through the application of lifecycle assessment and design for environment approaches.
6. *Research and development*. Adopting a longer-term view on the eco-efficiency challenges faced, R&D provides the ingredients for design and lifecycle management of future eco-efficient products and services.
7. *Production and operations*. Focusing on operational means to improve the eco-efficiency of today's products and services through changes in process or manufacturing techniques, replacing input materials, modifying product features and re-using materials.
8. *Marketing and procurement*. Lifecycle management requires an equal emphasis on reduction of environmental impacts caused by suppliers and consumers, which can be addressed through new partnerships and collaboration in the value chain.
9. *After sales service and disposal*. Companies have a responsibility and a business opportunity to consider the environmental impacts created by use and disposal.

As with any organizational change process there is no single formula to transform an organization into an eco-efficient organization. The path toward eco-efficiency has to be designed with due regard to key business competencies, principal products and services along with their markets and environmental and social profiles, and the organization's history and corporate culture. Some businesses might benefit from bold visions with stretch goals and targets set by their boards or chief executives. Other organizations might fit the quiet achiever category and should be encouraged to continue to do the right thing but enriched with a broader eco-efficiency vision. Achieving buy-in across the organization in many cases also involves recognizing and rewarding those past initiatives that had eco-efficiency benefits, although they were not undertaken specifically for eco-efficiency reasons in the first place. The art of becoming eco-efficient lies perhaps in striking a balance between encouraging the view that eco-efficiency is achievable whilst simultaneously developing an awareness that the business-as-usual approach is no longer sufficient to deal with the current sustainable development challenges.

Closing remarks

Cleaner production and eco-efficiency have evolved into complementary approaches for driving improvements in resource efficiency and environmental performance of organizations. They are both concerned with a philosophy – 'turning waste into profit' through cleaner production and 'doing more with less' through eco-efficiency – not with a well-defined set of unique technological solutions or pieces of hardware. Both capitalize on common sense and creativity and are empowered by sound engineering and management tools, practices and skills. The business imperative and the environmental benefits are intuitively accepted and underpinned by ongoing streams of real-life examples from all quarters of business, and indeed all corners of the world.

Regardless of their compelling logic, cleaner production and eco-efficiency remain unnoticed to many in business, government and the community alike. When confronted with resource and/or environmental constraints it often appears easier to fix the current problem than to eliminate its source and cause (e.g. Hirschhorn, 1997; Hirschhorn and Oldenburg, 1991; UNEP, 1994). Despite their good intent, environmental regulations may stand in the way of experimentation with new products and services that could potentially eliminate major environmental hazards. Consumers (private and institutional as well as businesses) are often more focused on initial capital outlay than on lifecycle costs, thereby forgoing real opportunities to increase their own quality of life while preserving the environment. Barriers like these can and should be challenged – what were the right choices in the past might no longer be the valid business propositions in the future because of changed environmental expectations and realities.

The important message from the examples discussed in this chapter is that cleaner production and eco-efficiency can be done and should be done to achieve quality of life and superior business performance while preserving the environment and our natural resource base on which life on Earth depends.

References

Australia and New Zealand Environment and Conservation Council (ANZECC) (1998), *Towards Sustainability: Achieving Cleaner Production in Australia*, Canberra: ANZECC.

Caon, M. (2003), 'How Ford Australia Greens its Supply Chain', in *Best Practice Seminar on Greening the Supply Chain*, Perth, Australia: WA Sustainable Industry Group.

Centre of Excellence in Cleaner Production (CECP) (2004), *Cleaner Production Case Studies (Western Australia)*, Perth, Australia: CECP, Curtin University of Technology.

Chaplin, G. (2002), 'Bradken: Progressing Towards Continual Environmental Improvement', Waste and Recycle 2002 Conference, Perth, Australia: Keynote Conferences.

de Hoo, S., Brezet, H., Crul, M. and Dieleman, H. (1991), 'Manual for the Prevention of Waste and Emissions', in Crul, M. (ed.) *PREPARE Manual and Experiences*, The Hague: Ministry of Economic Affairs (for EuroEnviron).

Department of Environment and Heritage (DEH) (2004), *Eco-efficiency and Cleaner Production Case Studies*, Canberra: DEH.

DeSimone, L. and Popoff, F. (eds) (1997), *Eco-efficiency: The Business Link to Sustainable Development*, Cambridge, MA: The MIT Press (for WBCSD).

Envirowise (2002) *Good Practice Guides (GG367)*, Oxford, UK: Envirowise.

Ford (2004), *Total Waste Management*, Broadmeadows (Vic), Australia, Ford Motor Company.

Hirschhorn, J. (1997), 'Why the Pollution Prevention Revolution Failed and Why it Ultimately Will Succeed', *Pollution Prevention Review*, **1997** (Winter), 11–31.

Hirschhorn, J. and Oldenburg, K. (1991), *Prosperity without Pollution: The Prevention Strategy for Industry and Consumers*, New York: Van Nostrand Reinhold.

Holliday, C., Schmidheiny, S. and Watts, P. (2002), *Walking the Talk: The Business Case for Sustainable Development*, Sheffield, UK: Greenleaf Publishing.

Imai, M. (1997) *Gemba Kaizen: A Commonsense, Low-Cost Approach to Management*, New York: McGraw Hill.

Office of Technology Assessment (OTA) (1994), *Industry, Technology and the Environment: Competitive Challenges and Business Opportunities*, Washington, DC: US Congress, OTA.

Pojasek, R. (1997), 'Improving Environmental Performance with a Systems Approach', *Environmental Manager*, **9**(1), 1–5.

Queensland Environmental Protection Agency (QEPA) (2004), *Meat Processing Industry – The Way Forward through Eco-efficiency*, Brisbane, Australia: QEPA.

Reijnders, L. (1996), *Environmentally Improved Production Processes and Products: An Introduction*, Dordrecht, the Netherlands: Kluwer Academic Publishers.

Rossi, M., Ellenbecker, M. and Geiser, K. (2001), 'Techniques in Toxics Use Reduction', *New Solutions*, **2**(2), 25–32.

Sheldon, C. (ed.) (1997), *ISO 14001 and Beyond: Environmental Management Systems in the Real World*, Sheffield, UK: Greenleaf Publications.

United Nations Environment Programme (UNEP) (1994), *Government Strategies and Policies for Cleaner Production*, Paris: UNEP.

United Nations Environment Programme (UNEP) (2002), *Industry as a Partner for Sustainable Development*, Paris: UNEP.

United States Environmental Protection Agency (USEPA) (1988), *Waste Minimization Opportunity Assessment Manual*, Cincinnatti, OH: USEPA.

United States Environmental Protection Agency (USEPA) (1992), *Facility Pollution Prevention Guide*, Washington, DC: USEPA.

United States Environmental Protection Agency (USEPA) (2001), *An Organizational Guide to Pollution Prevention*, Cincinnati, OH: USEPA.

van Berkel, R. (1994), 'Comparative Evaluation of Cleaner Production Working Methods', *Journal of Cleaner Production*, **2** (3–4), 139–52.

van Berkel, R. (1995), 'Introduction to Cleaner Production Assessments with Applications in the Food Processing Industry', *UNEP Industry and Environment Review*, **18**(1), 8–15.

van Berkel, R. (1996), 'Cleaner Production in Practice: Methodology Development for Environmental Improvement of Industrial Production and Evaluation of Practical Experiences', PhD thesis, IVAM Environmental Research, Amsterdam: University of Amsterdam.

van Berkel, R. (2000), 'Cleaner Production in Australia: Revolutionary Strategy or Incremental Tool?', *Australian Journal of Environmental Management*, **7**(3), 132–46.

van Berkel, R. (2001), 'Cleaner Production – Designing Systems to Prevent Waste', *Chemical Engineering in Australia*, **26**, 22–7.

van Berkel, R. (2002a), 'Application of Cleaner Production Principles and Tools for Eco-efficient Minerals Processing', *Proceedings Green Processing 2002: International Conference on the Sustainable Processing of Minerals*, Cairns, Australia: Australian Institute of Mining and Metallurgy.

van Berkel, R. (2002b). 'The Application of Life Cycle Assessment for Improving the Eco-efficiency of Supply Chains', in Batt, P. (ed.), *Proceedings of the MURESK 75th Anniversary Conference, From Farm to Fork – Linking Producers to Consumers through Value Chains*, Perth, Australia: Curtin University of Technology, pp. 1–16.

van Berkel, R. and Lafleur, M. (1997), 'Application of an Industrial Ecology Toolbox for the Introduction of Industrial Ecology in Enterprises', *Journal of Cleaner Production*, **5** (1–2), 27–38.

van Berkel, R., Willems, E. and Lafleur, M. (1997), 'Development of an Industrial Ecology Toolbox for the Introduction of Industrial Ecology in Enterprises', *Journal of Cleaner Production*, **5** (1–2), 11–26.

van Weenen, J. (1990), 'Waste Prevention: Theory and Practice', PhD thesis, Industrial Design, Delft, the Netherlands: Technical University of Delft.

World Business Council for Sustainable Development (WBCSD) (2000), *Eco-efficiency: Creating More Value with Less Impact*, Geneva, Switzerland: WBCSD.

Zosel, T. (1994), 'Pollution Prevention in the Chemical Industry', in Edgerly, D. (ed.), *Opportunities for Innovation: Pollution Prevention*, Gaithersburg, USA: National Institute of Standards and Technology, pp. 13–25.

6 Natural capitalism: path to sustainability?
L. Hunter Lovins

Business as usual is not a safe place to stay

Corporate executives worked for years comfortable in the conventional wisdom that their role was to look after their enterprise. Issues of the larger world belonged to the realm of policy. Executives might pay lobbyists to promote policies friendlier to their activities, but felt little professional responsibility for a greater good. Those interested in philanthropy either belonged to service clubs or created a corporate foundation.

Milton Friedman praised this approach, stating that for 'businessmen to preach the pure and unadulterated socialism' of corporate social responsibility would be a 'fundamentally subversive doctrine in a free society . . . in such a society, there is one and only one social responsibility of business – to use its resources and engage in activities designed to increase its profits.'[1] Former CEO of General Electric (GE), Jack Welch, echoed that business is socially responsible only when it is creating jobs and shareholder value.[2]

Margaret Mead said that the only person who likes change is a wet baby. The growing numbers of executives who believe that good management demands that they take responsibility for more than next quarter's share price, or their own well-padded future, thus appear uniquely courageous. These leaders, the vanguard of the corporate social and environmental responsibility movement, are demonstrating that this approach is, contrary to Friedman's outmoded views, also the best way to achieve their traditional goal: enhanced shareholder value. Far from managing a 'Triple Bottom Line', in which attention to the environment and social causes is a cost centre bolted on to the traditional financial metric, leading executives are practising an 'Integrated Bottom Line'[3], in which the best way to enhance shareholder value is to behave in ways that are responsive to the larger world.

Companies and communities,[4] and their people – us – stand on the edge of a crumbling cliff. Between pervasive loss of the health of ecosystems, increasingly acute challenges such as global warming, and the desperation of two-thirds of the world's population living in developing countries, the ground we have taken for granted is collapsing. Some deny the need for change and seek to shore up the precipice even as it erodes. Most people sense that a migration strategy is necessary, but lack the visionary leadership to begin it. Some far-sighted executives, however, are now pointing to a realm of stability across the chasm that they call sustainability. They are finding that it is simply better business to protect their people, and the world at large.

What's wrong with where we are?

In April 2005, the United Nations released the Millennium Ecological Assessment.[5] This study by 1360 experts in 95 nations from 22 national science academies, reported that over the past 50 years a rising human population has polluted or over-exploited two-thirds of the ecological systems on which life depends.

Healthy ecosystems provide clean water, productive soils, beautiful tourist destinations, the ability to detoxify society's wastes and dozens of other services that we take for granted, but which we would sorely miss if they were to cease to function. Such scientists as Gretchen Daly and economists like Robert Costanza estimate that the economic value these services provide to our economy is at least $30 trillion a year (Costanza and Folke, 1997), or the same as the entire value of the economy that *is* counted.[6] None of this 'capital' appears on conventional balance sheets, so 'business as usual' treats these 'eco-system services' as having a value of zero. Because the way in which people around the world meet their needs does not take protection of the environment as a priority, every major ecosystem on the planet is in decline. The loss of the services that these ecosystems used to provide to us for free will force businesses and communities to pay for replacements – assuming that humans are capable of creating substitutes.

Lester Brown of the Earth Policy Institute points out that:

> Accounting systems that do not tell the truth can be costly. Faulty corporate accounting systems that leave costs off the books have driven some of the world's largest corporations into bankruptcy. The risk with our faulty global economic accounting system is that it so distorts the economy that it could one day lead to economic decline and collapse.

In the same article, Brown also quotes Øystein Dahle, former Vice President of Exxon for Norway and the North Sea, who said: 'Socialism collapsed because it did not allow the market to tell the economic truth. Capitalism may collapse because it does not allow the market to tell the ecological truth.'[7]

In March 2004, Reuters reported:

> The world's second largest re-insurer, Swiss Re, warned on Wednesday that the costs of natural disasters, aggravated by global warming, threatened to spiral out of control, forcing the human race into a catastrophe of its own making.
> In a report revealing how climate change is rising on the corporate agenda, Swiss Re said the economic costs of such disasters threatened to double to $150 billion (£82 billion) a year in 10 years, hitting insurers with $30–40 billion in claims, or the equivalent of one World Trade Centre attack annually.[8]

Swiss Re subsequently opined that if a company does not take its carbon footprint seriously, perhaps it does not wish to insure that business – or its officers or directors.[9] Coming a year before the Gulf hurricanes' cost to insurance agencies of $70 billion,[10] this warning made Swiss Re appear prescient.

In January 2005, Dr Rajendra Pachauri, the chairman of the official Inter-governmental Panel on Climate Change (IPCC), told an international conference attended by 114 governments: 'Climate change is for real. We have just a small window of opportunity and it is closing rather rapidly. There is not a moment to lose . . . We are risking the ability of the human race to survive'[11]. Pachauri also agreed that the impacts of climate change will fall disproportionately on the poor.

Both the Millennium Assessment and the threat of climate change represent a loss of natural capital on a global scale, caused by conducting business as usual. Unlike many of the crises that gave rise to the environmental movement, the problem is not so much local losses of environmental amenities or that we are running out of a particular resource.[12] Rather the integrity of the earth's life support systems is now threatened.

The business case for change

Business leaders have an important role to play in addressing these challenges. Most of the threats result from the conduct of business or the consumption of its products. However, business may be the only institution on the planet big enough, resourceful enough and well enough managed to be able to respond effectively. Business leaders seem now to be realizing that they can no longer afford to ignore these issues.

In 2005 a group from the UK called the Carbon Disclosure Project[13] sent a questionnaire to the Financial Times 500, the largest companies in the world. The document asked about the carbon footprint of the companies: for what emissions of climate changing gases were they responsible. They had sent this survey out for a number of years but this time over 60 per cent of the companies surveyed answered in extensive detail. Ford Motor Company replied with an accelerated timetable for the release of its hybrid cars.

Why the difference? The Carbon Disclosure Project now represents institutional investors managing $31 trillion in assets. If, as a corporate manager, you wish to go to the capital markets for financing you might wish to answer their questions. However, a bigger reason may be that under Sarbannes Oxley, the new US business ethics law, corporate executives who fail to release to shareholders information that might materially affect the value of share-price can personally be held criminally liable.

Global warming is driving significant changes in the way that businesses respond to the environment. The advent of the Chicago Climate Exchange (CCX) carbon trading mechanism has proven that businesses can reduce emissions and remain profitable. The CCX, North America's only and the world's first, greenhouse gas (GHG) emission registry, created a market under which companies and other organizations emitting GHGs can cut emissions, and sell these reductions.[14] It is a self-regulatory, rules-based exchange designed and governed by its members.

Opening with 16 members in December 2003, CCX now has over 200 members (including such businesses as DuPont, American Electric Power, IBM, Ford Motor Co, Motorola, Dow Corning, Waste Management and Baxter Health Care), representing over 8 per cent of all direct US greenhouse gas emissions. The state of New Mexico, cities such as Chicago and Boulder, universities including Presidio School of Management, Tufts and University of Oklahoma and a wide array of smaller businesses and non-profit groups are also members.

The CCX is only the first of a growing number of efforts to create carbon markets in the United States. The seven Northeastern states have approved the Regional Greenhouse Gas Initiative, a mandatory regulatory scheme. Over 20 states have already either passed or proposed legislation on CO_2 emissions, or have developed carbon registries.

In August 2006, California became the first state in the US to impose mandatory limits on greenhouse gas emissions, requiring a 25 per cent cut in greenhouse gases by 2020 that would affect companies from automakers to manufacturers. The state is the twelfth largest carbon emitter in the world despite leading the nation in energy efficiency standards and its lead role in protecting its environment.[15] The California Chamber of Commerce opposed the bill, but such business groups as A New Voice for Business supported the measure, stating that it would create jobs and help to launch a whole new industry in California. Many believe the legislation will be the turning point in the country's global warming policy.

There is now such a proliferation of inconsistent carbon reduction regimes that in early 2006, a group of major businesses called on the US Congress to pass National legislation capping carbon emissions to relieve them of having to navigate the competing schemes. The Senate Energy and Natural Resources Committee heard statements from leaders representing major energy companies, including GE, Shell and the two largest owners of utilities in the United States, Exelon and Duke Energy. Six of the eight said they would welcome or accept mandatory caps on their greenhouse gas emissions. Wal-Mart executives also supported restrictions. The companies stated that federal regulations would bring stability and sureness to the market.[16] At subsequent Senate hearings on global warming, Senator Bingaman asked representatives of CCX whether there were any reasons that the US should not simply implement CCX as the basis for a regulated US carbon market.

Building a bridge to tomorrow
The world's leading companies are showing how to make the transition from business-as-usual to the greater profitability, lowered risk, enhanced brand equity and stronger shareholder value that more sustainable behaviour can confer. Many are implementing elements of what has been called Natural Capitalism (Hawken et al., 1999). The books *Factor Four* (von Weizsäcker et al., 1997), and its successor, *Natural Capitalism* (Hawken et al., 1999), detail the massive savings achieved by companies that increase the productivity by which they use resources. Natural Capitalism's basic principles are:

- Radical resource efficiency: Radically increase the productivity with which all resources are used, including energy, water, materials and people. The growing number of profitable solutions to environmental challenges can buy the time needed to implement measures that will attain true sustainability.
- Design for sustainability: Such innovative design processes as Biomimicry, Cradle to Cradle and other forms of green design enable businesses to harness nature's wisdom to drive innovation. These approaches allow us to create systems that eliminate waste and toxics while delivering superior products and services.
- Manage for prosperity and sustainability: Employ the emerging practice of sustainable management to restore, enhance and sustain the natural and human capital needed for continuing prosperity.

Eco-efficiency, the first step, can result in enormous cost reductions, while improving the company's reputation, brand equity and reducing its environmental footprint. For example over a 12-year period, Dow's Louisiana plant was able to save enough energy implementing worker-suggested savings measures to add $110 million each year to the bottom line (Hawken et al., 1999). Each measure also reduced Dow's carbon footprint. In April 2000, British Petroleum (BP) announced a commitment to reduce its carbon emission to 10 per cent below its 1990 levels by 2010. It only took two years to achieve this. Doing it is now saving the company $750 million. The results, and the thinking that led to the commitment also convinced BP to announce a re-branding to 'Beyond Petroleum', and to publicize its efforts to become a more sustainable company. BP sees that it cannot remain on the cliff's edge, and has begun to build a bridge to the other side.

Financial savings are not the only reason that companies engage in such behaviour. Rodney Chase, a senior executive at BP, subsequently reflected that even if the programme had cost BP money, it would have been worth doing because it made them the kind of company that the best talent wants to work for.[17] It is reducing costs, gaining market share, and attracting and retaining the best talent.[18]

The World Business Council for Sustainable Development (WBCSD) helps members, including 160 major corporations, capture such opportunities. In the WBCSD book, *Walking the Talk*, the CEOs of DuPont, Anova and Royal Dutch Shell state that sustainability's business case is strengthened in the way it encourages innovation (Holiday et al., 2002). By capitalizing on this, a company can do very well in gaining customer success, brand strength, first mover advantage, motivated employees and potentially more profits.

For example, STMicroelectronics (ST), a Swiss-based $8.7 billion semiconductor company, set a goal of zero net GHG emissions by 2010 while increasing production forty-fold.[19] Its strategy is to reduce on-site emissions by investing in co-generation (efficient combined heat and electricity production)[20] and fuel cells (efficient electricity production). By 2010 co-generation sources should supply 55 per cent of ST's electricity with another 15 per cent coming from fuel switching to renewable energy sources. The rest of the reductions will be achieved through improved energy efficiency (hence reducing the need for energy supply) and various projects to sequester carbon. ST's commitment has driven corporate innovation and improved profitability. During the 1990s its energy efficiency projects averaged a two-year payback (a nearly 71 per cent after-tax rate of return).[21] Making and delivering on this promise has also driven ST's corporate innovation and increased its market share, taking the company from the number 12 micro-chip maker to the number six in 2004.[22] By the time ST meets its commitment it will have saved almost a billion dollars.

Other companies are implementing practices that mimic the way nature does business, running on sunlight, using waste as a valued input rather than a cost, designing products to be inherently non-toxic and re-manufacturable. Far from imposing a penalty on business, such innovative designs frequently work better and cost less. The book *Biomimicry* (Benyus, 1997) details this approach that represents the future of industry.

Commitment to sustainability: the hallmark of an investment worthy company
Some financial advisors already state that a commitment to sustainability is the hallmark of good corporate governance and the best indicator of management capacity to protect shareholder value.[23] The Dow Jones Sustainability Index outperforms the general market, and the Domini Index of Socially Responsible companies has outperformed the Standard and Poors for over a decade. In spring 2005, the socially responsible investment research firm, Innovest[24] released a report showing that in whole industry sectors, from forest products and paper, to oil, gas and electric utilities, environmental leaders in each sector are outperforming the environmental laggards.

Adopting an aggressive sustainability programme can be highly profitable for companies and cost-effective for non-profit (including government) organizations (Lovins and Lovins, 1999). Companies that commit to behave in responsible ways, especially in the context of a broader whole-system corporate sustainability strategy, will achieve multiple benefits for shareholders beyond reducing their contribution to global climate change. Governments that take a similar course will accrue similar benefits to their citizen stakeholders.[25]

A corporate commitment to sustainability enhances every aspect of shareholder value. Implementing more responsible behaviour allows a company to strengthen what is called 'The Integrated Bottom Line'. As the examples above show, using resources more efficiently, redesigning products in ways that mimic nature and managing to restore and enhance human and natural capital offer ways to:

1. Enhance financial performance from energy and materials cost savings in:
   - industrial processes;
   - facilities design and management;
   - fleet management; and
   - government operations.
2. Enhance core business value:
   - sector performance leadership;
   - greater access to capital;
   - first mover advantage;
   - improved corporate governance;
   - the ability to drive innovation and retain competitive advantage;
   - enhanced reputation and brand development;
   - market share capture and product differentiation;
   - ability to attract and retain the best talent;
   - increased employee productivity and health;
   - improved communication, creativity, and morale in the workplace;
   - improved value chain management; and
   - better stakeholder relations.
3. Reduce risk:
   - insurance access and cost containment;
   - legal compliance;
   - ability to manage exposure to increased carbon regulations;
   - reduced shareholder activism; and
   - reduced risks of exposure to higher carbon prices.

Taken together, the elements of sustainability confer the ability to be 'First to the future'. The companies that practise this approach will be the billionaires of tomorrow.

Ability to capture opportunities
Business success in a time of technological transformation demands innovation. Since the Industrial Revolution, there have been at least six waves of innovation in which the technologies that underpinned economic prosperity have shifted (see Figure 6.1). In the late 1700s textiles, iron mongering, water-power and mechanization enabled modern commerce to develop. The second wave saw the introduction of steam power, trains and steel. In the 1900s electricity, chemicals and cars began to dominate. By the middle of the century it was petrochemicals, and the space race, along with electronics. The most recent wave of innovation has been the introduction of computers, the digital or information age. As economies move on, older industries suffer dislocations, unless they join the increasing number of companies implementing the array of sustainable technologies that will make up the next wave of innovation.

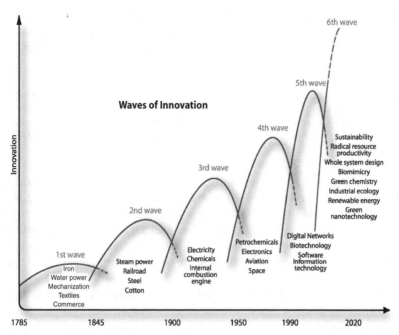

Source: This figure used courtesy of The Natural Edge Project, Australia, www.naturaledgeproject.net/

Figure 6.1 Waves of innovation

Aidan Murphy, vice president at Shell International, stated in 2000:

> The Kyoto treaty has prompted us to shift some of its [Shell's] focus away from petroleum toward alternative fuel sources. While the move has helped the company make early strides toward its goal of surpassing treaty requirements and reducing emissions to 10% less than 1990 levels, *Shell is being driven largely by the lure of future profits* . . . We are now involved in major energy projects involving wind and biomass, but I can assure you this has nothing to do with altruism . . . *We see this as a whole new field* in which to develop a thriving business for many years to come. Capital is not the problem, it's the lack of ideas and imagination.[26]

In May 2005, a tipping point occurred after which major companies will find it hard to continue business as usual. Jeffrey Immelt, the man who replaced Jack Welch at the helm of GE, stood with Jonathan Lash, the President of World Resources Institute, a leading environmental organization, to announce the creation of GE Ecomagination. The two co-authored an article in the *Washington Post* entitled, 'The courage to develop clean energy'.[27] Immelt committed GE to implement aggressive plans to reduce emission of greenhouse gases, spending $1.5 billion a year on research in cleaner technologies. As part of the initiative, Immelt promised to double GE's investment in environmental technologies to $1.5 billion by 2010, and reduce the company's GHG emissions by 1 per cent by 2012; without any action, emissions would have gone up 40 per cent.[28]

GE is the sixth largest company in the world, and the only company, which, had there been a Fortune 500 around in 1900, would have been on it and is still on it today. So its announcement is significant but it was followed in the spring of 2006 by a far larger green commitment from Wal-Mart, one of the largest companies in the world. If Wal-Mart were a country it would be the twentieth largest in the world, or the fifth largest US city. The company's CEO Lee Scott stated in the pages of *Fortune* Magazine:

> There can't be anything good about putting all these chemicals in the air. There can't be anything good about the smog you see in cities. There can't be anything good about putting chemicals in these rivers in Third World countries so that somebody can buy an item for less money in a developed country. Those things are just inherently wrong, whether you are an environmentalist or not.[29]

He set a goal of supplying his stores with 100 per cent renewable energy. Wal-Mart is experimenting with green roofs, and green energy (which is now used to power four Canadian stores, for a total of 39 000 megawatts – the single biggest purchase of renewable energy in Canadian history). The company has pledged to become the largest organic retailer and to increase the efficiency of its vehicle fleet by 25 per cent over the next three years. It will eliminate 30 per cent of the energy used in store and invest $500 million in sustainability projects.[30]

An unabashedly astonished article in the San Francisco Bay *Guardian* reflected:

> Wal-Mart's rationale for all of this, of course, has absolutely zero to do with any sort of deep concern for the planet (though it does make for good PR), nothing at all about actual humanitarian beliefs or honest emotion or spiritual reverence, and has absolutely everything to do with the corporation's rabid manifesto: cost-cutting and profit.
>
> The reason Scott promised that Wal-Mart will double the fuel efficiency of their huge truck fleet within a decade? Not to save the air, but to save $300 million in fuel costs per year. The reason they aim to increase store efficiency and reduce greenhouse gasses by 20% across all stores worldwide? To save money in heating and electrical bills, and also to help lessen the impact of global warming, which is indirectly causing more violent weather, which in turn endangers production and delivery and Wal-Mart's ability to, well, sell more crap. Ah, capitalism.[31]

Scott is reading the market. There exists a large and growing market in the US and Europe for goods produced in ways that do not harm the environment or people. Approximately 30 per cent of the adults in the US, or 63 million people, fit a demographic that Conscious Medium named, 'Lifestyles of Health and Sustainability' or LOHAS.[32] Such consumers are driving market changes by demanding goods and services that meet their desires to enhance health, environment, social justice, personal development and sustainable living. They are part of a tectonic shift in consumer awareness and behaviour, a worldwide market conservatively estimated at $228.9 billion, and growing. In the US, ecological lifestyles, including purchases of organic products constitute a market worth at least $81 billion a year.[33] Sociological research found that this population comprises a growing market segment of educated consumers who make conscientious purchasing and investing decisions based on social and cultural values (Ray and Anderson, 2000).

The growing sustainability movement, combined with the instant access to information that the Internet provides, has resulted in a more educated and discerning consumer than may have been apparent in past surveys of the general market. As a driver, it is unparalleled in US history, to the point where these consumers will determine the future of many businesses. Some analysts are now calling the sustainability movement the largest phenomenon in human history. Hundreds of thousands of organizations are working to achieve social justice, alleviate poverty, enhance standards of living for all, and achieve environmental protection, elements of a more sustainable world.[34] Hundreds of communities, states and even whole nations are setting sustainability indicators and promulgating implementation plans to enhance their human and natural capital. In mid-2006 such magazines as *Vanity Fair* and *Newsweek* devoted cover stories to the explosion in green lifestyles.

In the face of such trends, companies are increasingly realizing that next quarter's share price is not the sole component of shareholder value. The real value of a company is determined by an array of assets including the quality and utility of the products it makes, its intellectual capital, the reputation of the company, its brand equity, the ability to innovate and especially the ability to attract and retain the best talent.

Leadership, the ability to inspire and elicit the best from its people, is perhaps the most important asset that a company can have. Peter Drucker writing in the *Economist* stated, 'In the next society, the biggest challenge for the large company – especially for the multinational – may be its social legitimacy: its values, its missions, its visions.'[35]

The 2004 Gallup International and Environics International 'Voice of the People' Survey found that well under 50 per cent of people polled trusted large national or global companies.[36] Companies like Enron and WorldCom that demonstrated lack of integrity not only destroyed their own shareholder value, but eroded many companies' 'franchise to operate'. This puts any unpopular company at risk: in an Internet empowered world, a small group of determined citizens can de-legitimize any company, or country.

Nike, attacked by social activists[37] because it had erroneously thought that it could draw the boundary within which it had to concern itself with social responsibility at its American plant gates, considered eliminating the 'Swoosh', its multi-billion dollar brand symbol, when it was attacked by citizens' groups protesting the labour practices in the companies overseas from which Nike purchased its products. In response, Nike recommitted itself to sustainability, implementing third party verification of its human rights policies, and undertook a major effort to increase its environmental performance. In April 2005, Nike made its supply chain transparent, discovering that this made management of it much easier.

The unwelcome attention Nike attracted is part of a growing phenomenon. A 1999 survey on corporate social responsibility by Environics International, found 67 per cent of North Americans willing to make decisions to buy or boycott products on ethical grounds. Companies wanting to build lasting relationships with these customers must be willing to subject their values and practices to scrutiny.[38] Environics' 2001 survey found that social responsibility was the factor that most influenced a customer's impression of companies in almost half of respondents. The 2001 Cone Corporate Citizenship Study indicated that 81 per cent of American consumers said they would consider switching brands – price and quality being equal – to support a cause. Eighty per cent of American consumers said that a company's commitment to causes was important when deciding what to buy or where to shop.[39]

A 2004 survey of the world's leading CEOs, by the World Economic Forum at Davos, found that the leaders believe corporate reputation is now a more important measure of success than stock market performance, profitability and return on investment.[40] Only the quality of products and services edged out reputation as the leading measure of corporate success – 59 per cent of the respondents estimated that corporate brand or reputation represents more than 40 per cent of a company's market capitalization.

Needed: leaders

No business, even an imperilled one, will embark on a course of action that would compromise profitability. As companies realize that shifting their behaviour, taking the first steps across the bridge to greater sustainability, can make them more profitable, they will increasingly drive this process. Today business as usual is not a safe bet. Only a sustainability strategy can protect shareholder value in the long run. Companies that realize the seriousness of the challenges, and commit to a transition to more sustainable behaviour *and* that deliver on that commitment, will be the leaders in the coming decades. Corporate commitment to and follow-through on sustainability will come to be the hallmark of corporate integrity and management capacity. The ability to make such transitions will be the distinguishing characteristic of the effective executive.

In *The Lord of the Rings*, Gandalf said

> The rule of no realm is mine. But all worthy things that are in peril as the world now stands, these are my care. And for my part I shall not wholly fail if anything passes through this night that can still grow fair and bear fruit and flower again in the days to come. For I too am a steward, did you not know?[41]

This is not a bad approximation of the sort of leadership that is needed in today's business world. However, it is worth remembering that in the story it was two rather ordinary, playful and unassuming hobbits who undertook to save the world. Despite their fears, they took the future of the world on their shoulders, and in the end, all the wizards, kings and warriors could only play a supportive role.

Dana Meadows, perhaps the greatest systems thinker of our times, once said:

> It seems to me a powerful message worth repeating and repeating, that people want peace, simplicity, beauty, nature, respect, the ability to contribute and create. These things are much cheaper and easier to achieve than war, luxury, ugliness, waste, hate, oppression, manipulation. Some day, when everyone understands that nearly all of us truly want the same kind of world, it will take surprisingly little time or effort to have it.[42]

Notes

1. *New York Times* Magazine, 13 September, 1970. The New York Times Company. www.colorado.edu/ studentgroups/libertarians/issues/friedman-soc-resp-business.html.
2. At a Stanford Business School address in 2005, Welch stated: 'Winning companies are the only things that matter – without them, nothing else would work'. www.gsb.stanford.edu/news/headlines/vftt_welch.shtml.
3. A phrase developed by the socially responsible investment expert, Theo Ferguson.
4. These arguments apply equally to executives of communities and countries, but space constraints limit this discussion to the corporate world.
5. www.millenniumassessment.org/en/index.aspx.
6. W.K. Stevens, 'How much is nature worth? For you, $33 trillion', *New York Times*, 20 May, 1997.
7. www.earth-policy.org/Books/Seg/PB2ch12_ss 2.htm.
8. T. Atkins, 'Insurer warns of global warming catastrophe', Reuters, 3 March, 2004.
9. J. Ball, *Wall Street Journal*, 7 May, 2003.
10. J. Stempel, 'Berkshire profit falls 48 pct', Reuters, 4 November, 2005.
11. A. Kirby, 'Aid Agencies' warning on climate', BBC News website, 20 October, 2004, http://news.bbc.co.uk/1/hi/sci/tech/3756642.stm.
12. Although this may come to be an issue as well. Such resources as fisheries *are* depleted around the world, and if such theories as peak oil prove correct, absolute resource shortages may come to be a fixture of the policy world.
13. www.cdproject.net/.
14. www.chicagoclimatex.com/.

15. http://sfgate.com/cgi-bin/article.cgi?f=/c/a/2006/08/30/MNGBMKS7733.DTL.
16. http://www.salon.com/opinion/feature/2006/04/10/muckraker/index_np.html.
17. Personal communication with Hunter Lovins at the 2002 Fortune Magazine Annual Meeting, Aspen, Colorado.
18. BP now states this on its website and in its advertisements.
19. STMicroelectronics 2003 Sustainable Development Report, www.bl.uk/pdf/eis/stmicroelectronics2003is. pdf.
20. Conventional power stations that burn fossil fuels give off a lot of heat, wasting as much as 70 per cent of the energy they consume.
21. STMicroelectronics 2001 Environmental Report. It further reported that no energy efficiency project undertaken incurred more than a three-year payback. The source of the correlation of years payback to real after-tax rate of return is Hawken et al. (1999, p. 267).
22. IC Insights, 'IC insights announces 1Q 05 top ten semiconductor supplier ranking' from www.icinsights. com.
23. Personal Communication: Sasha Millstone, Senior Vice President of the Millstone Evans Group of Raymond James and Associates, Inc. Boulder, Colorado.
24. www.innovestgroup.com/.
25. For a detailed synthesis of this thesis refer to Hawken et al. (1999) and Hargroves and Smith (2005), including The Natural Edge Project (www.naturaledgeproject.net).
26. W. Drozdiak, 'Big corporations alter view of global warming', *Washington Post* Service, Friday, 24 November, 2000.
27. J. Immelt and J. Lash, *Washington Post*, 21 May, 2005, Page A19.
28. Ibid.
29. M. Gunther, 'The green machine', *Fortune* Magazine, 27 July, 2006, http://money.cnn.com/magazines/fortune/fortune_archive/2006/08/07/8382593/index.htm.
30. Ibid.
31. M. Morford, SF Gate, 'Can you still hate Wal-Mart? It's a shockingly eco-friendly plan from the world's most toxic retailer. Did hell just freeze over?', August 2006, http://www.sfgate.com/cgibin/article.cgi?file=/gate/archive/2006/05/24/notes 052406.DTL.
32. www.lohas.com/about.htm.
33. Ibid.
34. Paul Hawken, Natural Capital Institute, www.naturalcapital.org/Projects.html.
35. Peter Drucker, 'Will the corporation survive?' in 'A survey of the near future', *The Economist*, 3 November, 2001.
36. www.voice-of-the-people.net/.
37. G. Owen, Corporate Social Responsibility: rethinking the role of corporations in a globalizing world, www.21stcenturytrust.org/owen.htm.
38. www.corporateknights.ca/stories/kris_kringle.asp.
39. trevorcook.typepad.com/weblog/community_involvement/.
40. www.ethicalcorp.com/content.asp?ContentID=1608.
41. J.R.R. Tolkien (1992), *The Lord of the Rings*, London: HarperCollins, Book V, Chapter 1.
42. Personal communication. Dana was also the first person to use the term 'Sustainability'. It was in her landmark book, *Limits to Growth* in 1972.

References

Benyus, J. (1997), *Biomimicry: Innovation Inspired by Nature*, New York: HarperCollins Publishing.
Costanza, R. and C. Folke (1997), 'Valuing ecosystem services with efficiency, fairness, and sustainability as goals', in G. Daily (ed.), *Natures Services: Societal Dependence on Natural Ecosystems*, Washington, DC: Island Press, pp. 49–68.
Hargroves, K. and M. Smith (2005), *The Natural Advantage of Nations: Business Opportunities, Innovation and Governance in the 21st Century*, London: Earthscan.
Hawken, P., A. Lovins and L.H. Lovins (1999), *Natural Capitalism. Creating the Next Industrial Revolution*, London: Earthscan.
Holiday, C., S. Schmidheiny and P. Watts (2002), *Walking the Talk: the Business Case for Sustainability*, San Francisco: BK Publishers, and Sheffield: Greenleaf.
Lovins, A. and L.H. Lovins (1999), 'Climate: making sense and making money', in P. Hawken, A. Lovins and L.H. Lovins, *Natural Capitalism. Creating the Next Industrial Revolution*, London: Earthscan, pp. 234–59.
Ray, P. and S. Anderson (2000), *The Cultural Creatives*, New York, NY: Three Rivers Press.
von Weizsäcker, E., A.B. Lovins and L.H. Lovins (1997), *Factor Four: Doubling Wealth – Halving Resource Use. A New Report to the Club of Rome*, London: Earthscan.

7 Sustainable strategies towards human resources – a way forward

Paul J. Gollan

Introduction

With a growing emphasis on customized quality consciousness in world business and increased use of new technologies, a new form of worker has emerged. These new workers are often labelled as 'knowledge workers' and they possess certain key characteristics: they are highly skilled, qualified, trained and experienced in new and growing areas of business. In essence they can be defined as workers who deal with a high degree of complexity and uncertainty that requires a high degree of judgment (De Lacy, 1999). In the new knowledge-based society the notion of commitment has also been redefined: the maintenance of intellectual capital or the 'corporate memory' is now seen as dependent on employee commitment and satisfaction. The intellectual capital of organizations is the knowledge, experience and ideas of employees that management attempts to codify and formalize to produce greater organizational value. Kennedy (2000) has argued that corporations are 'at the crossroads' and the future lies in corporations having sustainable strategies towards human resources. Managers must consequently confront the challenge of aligning the interests and needs of their business with those of their most important resource, people, to achieve and maintain productivity and competitive advantage (Wilkinson et al., 2001).

This chapter will outline a number of issues for organizations to consider when pursuing sustainable human resource outcomes in the workplace: those outcomes that reinforce corporate profitability and corporate survival, and those that satisfy employee aspirations and needs in the workplace. It is suggested that developments in human resource management (HRM) in recent years have shifted the emphasis away from human management to a new focus on resource management. It is argued that this focus is misdirected; rather, the needs, potential and aspirations of individuals must take centre stage in the workplace. If employers do not bridge the current gap between their rhetoric and workplace reality, then the likely outcome will be an exodus of bright and enthusiastic people to organizations that do (Gollan, 2000). For true corporate sustainability an organization must recognize, value and promote the capability of its people. The chapter also examines and provides evidence of the link between workplace effectiveness and humanistic work structures in organizations and shows how reinforcing this link increases productivity and profits. Importantly, it is suggested that for human resource sustainability to be achieved, the HR policies and practices need to be integrated for sustained business performance and positive employee outcomes of equity, development and well-being.

The chapter will address the following key questions. How do organizations currently utilize and apply human resources? How do organizations depreciate or renew these resources, and what are the implications of such approaches for employers and their employees? How can we redefine the ways organizations use their human resources in order to ensure human sustainability? To what extent do corporations need to exercise social responsibility as well as economic responsibility? How can employers balance the interests of different stakeholders in organizations while maintaining a sustainable work environment for employees?

Organizational sustainability and human sustainability

Organizational sustainability is more of a 'symbol' rather than a scientific concept. In general terms, organizational sustainability is a focus for a new value debate about the shape of the future. It is a signpost and a roadmap pointing and showing the general direction we should take (Dunphy and Benveniste, 2000). However, some have suggested that most business leaders still use the term organizational sustainability when it suits them, and is defined in ways that meet their immediate needs (Elkington, 2001). Many people still doubt the sincerity of pronouncements on sustainability, arguing that initiatives of this type are often sponsored by marketing or corporate affairs departments and designed to put a good 'spin' on an organization's activities (Trapp, 2001). Organizational sustainability requires organizations to maintain, protect and renew the viability of the environment; to enhance society's ability to maintain itself and enhance its capacity to solve major problems; to maintain a decent level of support and welfare for present and future generations; and extend the productive life of organizations and to maintain high levels of efficiency and performance to add value to society (Figure 7.1).

Environmental, social and economic sustainability encompass the main tenets of organizational sustainability, including what is commonly termed the triple bottom line – that is, economic, social and environmental outcomes that are of true value to society. Importantly, organizations must recognize that to be even more successful in business terms, they must make substantial contributions to the quality of life of their employees and the community as well as sustaining and renewing the natural environment (Dunphy and Benveniste, 2000).

However, a sustainable organization is one that is built to last, which continually adds value. For organizations to be successful it is not enough having a business strategy and then tacking a human resource onto it (Delany, 2001). Organizations need a people vision that supports the vision of the business.

While many organizations may see sustainability as 'soft' and 'optional' rather than 'grounded in commercial strategic need', there is an opportunity for the HR function and HR professionals to take the lead on the grounds that truly effective strategies in this area need to involve all staff (Trapp, 2001). Benefits such as improved people skills and deeper loyalty both from employees and from customers can be achieved by greater focus on the HR arena. Human resource sustainability requires the organization to recognize and place value on human capabilities, and this entails taking a more holistic and integrated approach to people management (Gollan, 2000). As can be seen in Figure 7.2, this chapter will focus on the relationship between the social and economic added value to human resource sustainability. As Peter Davies, Deputy Chief Executive of Business in the Community, recently stated:

Increasingly, people are making decisions about which organisation to work for on the basis of how a company demonstrates its social responsibility. As a starting point, being an employer of choice requires a respect for the talents of all individuals, regardless of their sex, race, age, disability

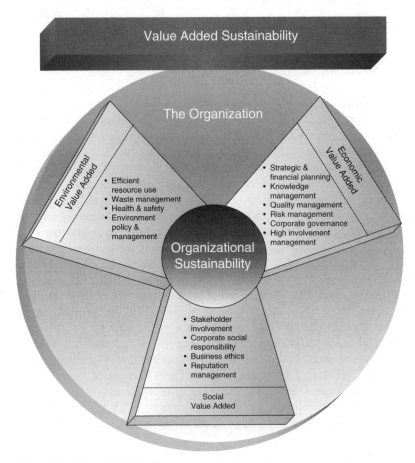

Value Added Sustainability

The Organization

Environmental Value Added

- Efficient resource use
- Waste management
- Health & safety
- Environment policy & management

Economic Value Added

- Strategic & financial planning
- Knowledge management
- Quality management
- Risk management
- Corporate governance
- High involvement management

Organizational Sustainability

- Stakeholder involvement
- Corporate social responsibility
- Business ethics
- Reputation management

Social Value Added

Source: Adapted from Delany (2001, p. 8).

Figure 7.1   The sustainability model

or sexual orientation. It involves investing in the people in the community in which the company operates, and building the social capital and health of that community. Such investment not only develops the skill base on which the company can draw; it also builds the morale and motivation of the existing workforce. The support for employee involvement in community projects develops skills, encourages innovation and strengthens staff loyalty. (As reported in Trapp, 2001)

Human resource sustainability
The debate of human resource sustainability has been advocated by a number of commentators (Dunphy and Griffiths, 1998; Gollan, 2000; Pears, 1998). They argue that there is a crisis facing the management of human resources with staff turnover increasing, loyalty declining, stress levels rising and productivity growth diminishing. As Pears (1998, p. 1) states: 'Managers confront the challenge of aligning the interests and needs of their business with those of their most valuable resource, their staff, so that business success can be achieved and maintained'. They see this challenge as human sustainability. In particular they see human sustainability as a shift in focus from short-term corporate survival

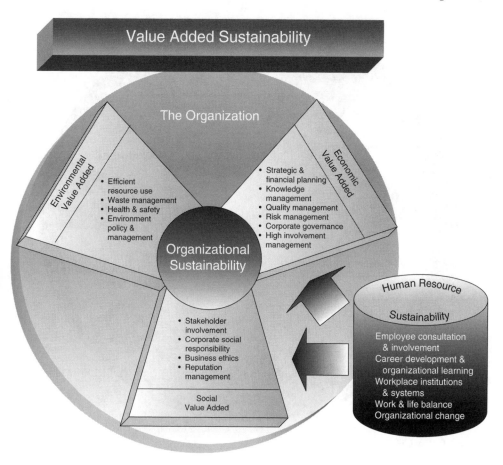

Source: Adapted from Delany (2001, p. 8).

Figure 7.2 The sustainability model and human resource sustainability

to long-term business success. This requires a shift in criteria from short-term financial profit to more broadly defined long-term returns. This involves changes in structures, operation, priorities and values that firms promote, with a clear focus on the context within which a business operates recognizing a fundamental requirement for long-term business success (Pears, 1998).

Dunphy and Griffiths (1998) have developed this thesis further by suggesting that this involves continuous ongoing investment in acquiring knowledge and skills. This they say requires a fundamentally different approach in leading and managing firms. In particular organizational competencies are needed for sustained corporate success that can be systematically developed. These include: strategically building a corporate knowledge and skills base; fostering productive diversity and building human potential; and creating communities of knowledge in organizations and developing competencies for continuous corporate renewal (Dunphy and Griffiths, 1998). They argue that this focus on investing in people rather than divesting them and actively promoting strategies of knowledge

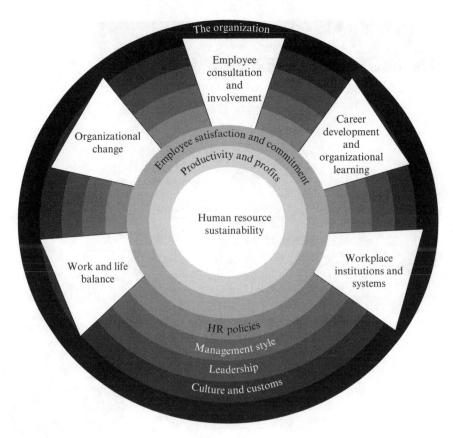

Source: Gollan (2000).

Figure 7.3   Factors and influences in human resource sustainability – an integrated approach

creation is the most viable path to sustaining societies' corporate and economic life (Dunphy and Griffiths, 1998, pp. 168–70).

Figure 7.3 represents the major factors, influences and outcomes of human resource sustainability in organizations. While not intending to be exhaustive, it identifies five major factors in the debate about human resource sustainability. Essentially the model defines human resources sustainability in terms of the capacity of organizations to create value within their organizations, thereby having the ability and capacity to regenerate value and renew wealth through the application of human resource policies and practices. This will entail investment in human knowledge through continuous learning, and the application and development of such knowledge through employee participation and involvement. In addition, the model identifies four main drivers for organizations trying to achieve corporate human resource sustainability and examines their impact on employee satisfaction and commitment and on the traditional organizational objectives of increased productivity and profits. Importantly, the model suggests that for human resource sustainability to be achieved, the HR policies and practices need to be integrated for sustained business performance and positive employee outcomes of equity, development and well-being.

The state of play

For many employees, careers as we have traditionally known them no longer exist. They were predicated on progression through an organizational hierarchy, by seniority and/or merit and on security of employment. Instead, careers are now becoming self-managed. Individuals assess their own worth and think strategically about placing themselves in the best possible employment position, acting as their own agent in a competitive, market-driven environment. This being so, companies are suddenly finding themselves in a position where much of the management skeleton that held the edifice together and the cultural 'oil' of loyalty that enabled this machine to function smoothly have largely disappeared. At the same time new management philosophies stress that workers should be seen as empowered members of firms rather than antagonists in a confrontation between capital and labour.

Some employees, and indeed what remains of middle management, may counter such schizoid demands with indifference or cynicism. Nevertheless the effect of such demands is likely to be the entrenchment of 'low-trust' employment relations where employees feel they are only a cog in a machine – simply a necessary component in the return on investment equation. With their expectations lowered, a more unforgiving mentality begins to emerge.

The danger for organizations is that power and knowledge are now invested in more people outside the traditional confines of the management hierarchy. Corporate knowledge and experience, the strategic competitive advantage in most organizations, is now located at the lower levels of the organization in a great many more people. It is vital for organizations to understand that employees have the knowledge that is essential for the organization's success, and so have the potential to shift the balance of power away from management and back to themselves.

Today many workers who have jobs are working long hours, with an increasing number of employees working more than 48 hours a week. However, as the trend to work longer and harder continues among the generation that has discovered the meaning of downsizing, organizational change and restructuring, and job insecurity, it should not be surprising that a new generation is developing different attitudes to work. Attracting and retaining talented staff presents a challenge for the HR function in understanding the wants, needs and values of the different generations who make the workforce (Rance, 2002). Importantly, selecting and retaining good staff is increasingly regarded as critical to how well organizations can adapt to change and build for a sustainable future.

Dr Robin Kramar, Director of the Centre for Australasian Human Resource Management at the Macquarie Graduate School of Management, has suggested that fostering high organizational performance requires an understanding of what motivates the new generation of the workforce:

> Generation X saw their parents working really hard, being loyal to organisations and then being retrenched – social research shows that has shaped their attitudes to work. They don't believe in loyalty – they will stay with an organisation only as long as it suits them. Unlike their parents, younger workers are more interested in experience and what they perceive as the quality of their lives rather than material possessions or financial security. This challenges HR professionals to find ways of increasing staff satisfaction at a time when technology is making many jobs routine and mundane. HR might need to redesign work to make it more challenging and also find ways of managing expectations. This may mean being a good corporate citizen with sound environmental and social policies. It is also likely to mean offering flexible work practices that help people find balance between their work and home lives. Younger

workers want learning opportunities because they know they will be moving on. They want good social relationships in their workplaces. These are expectations the baby boomers did not have. (As reported in Rance, 2002)

Interestingly these attitudes may have had an affect on the perceptions of work itself. Christopher Shen, head of the recruitment division of Hamilton James and Bruce in Melbourne argues that recent developments have created an atmosphere that people feel penalized for loyalty and commitment: 'People feel that unless they are constantly moving or receiving rapid promotion, something is wrong. In some warped way, tenure is seen as a sign of poor performance' (as reported in Rance, 2002).

There is evidence of the emergence of a new 'get-a-life' generation. A survey of 1000 young professionals in 73 countries found that 40 per cent planned to leave their existing employer within two years. The survey uncovered widespread disappointment at the dearth of company investment in their personal development, with many considering starting their own business (Donkin, 1999). This finding was further reinforced when the survey revealed that 41 per cent of respondents would value more choice over working hours, with many wanting to spend more time with friends and family and to slow down enough to keep healthy and balanced (Donkin, 1999). Moreover, a study conducted by Gemini Consulting of 10 300 employees in ten countries found that only 27 per cent of those surveyed felt their organizations were preparing them for successful futures. As a consequence, 44 per cent of respondents said they would leave their jobs tomorrow for a job that would provide them more opportunity for advancement (Gemini Consulting, 1998, p. 13). In Australia it is estimated that the cost of replacing an employee is put between AUD20 000 to AUD50 000, suggesting that the ability to establish loyalty and trust is a high financial priority as well as an operational one (Dabkowski, 2003).

To further reinforce the importance of achieving a work and life balance, a study on the quality of working life (Worrell and Cooper, 1998) of over 1300 managers found that nearly 60 per cent indicated that long hours were having a negative impact on their health. Just over 70 per cent said it affected their relationship with their partner and encroached on the time spent with their children, and some 55 per cent of managers suggested that it made them less productive at work. In addition, over half of the respondents felt a sense of 'powerlessness' and lack of control in their work environment, with less than one in five respondents agreeing that they had enough time to fulfil the requirements of their job. When respondents to the survey were asked what one piece of advice they would give to top management of their organization, 40 per cent of comments referred to the poverty of organizational communication and consultation strategies. The second most important piece of advice identified was for the organization to value and appreciate employees as the organization's most important assets (Worrell and Cooper, 1998).

The institutions for a sustainable future
The real question arising from this evidence is whether there can be any real progress towards corporate sustainability without a fundamental overhaul of the structures that underpin existing institutions and systems of employment. A report by Valerie Bayliss (1996) of the Royal Society for the Arts (RSA) may hold some clues. The RSA's report challenges traditional assumptions and calls for the dismantling of employment structures and education systems that have supported the traditional working model of a

'40/40 job'. This model assumes that people (mostly men) work 40 hours a week for 40 years. The report suggests that this outdated model was based on the growth of public administration in the 1930s and mass production in factories, supported by the concept of the job for life in the same organization. The problem, as identified in the RSA report, is that if people work less than 40 hours (or in some cases 35 hours) they are described as 'part-timers', peripheral and in fundamentally insecure employment relative to the main 'core' full-time workforce.

As the RSA report states, there are unlikely to be enough paid permanent full-time jobs to go around (Bayliss, 1996). The paradigms have changed, from where a majority of people will be employed in 40/40 jobs to where a majority of people will be working flexible hours in the temporary labour market. People themselves will be responsible for their working patterns, their education and training and redefining their own work. The report addresses five important issues for HR sustainability. Can the economy generate a consistently high supply of employment opportunities, however they are structured? Can the intermediate labour market develop and maintain opportunities for all those who need it? Do the nation's education and training systems meet the needs for a sophisticated and competent society? Does the country's infrastructure support people in and out of work, and match the needs of the new world rather than the old? Do individuals recognize their responsibility to develop their own employability, and is the right support available to assist and encourage them? (Bayliss, 1996).

An Australian report by Gollan et al. (1996) highlights the increasing acute socio-economic polarization between the highly skilled and the poorly skilled. Those that are highly skilled will be increasingly overworked and working long and anti-social hours, while the poorly skilled will be either unemployable or have their wages driven down to a point that renders them the 'working poor' or a 'ghettoized' worker. There will be an industrial relations system that offers few substantive protections for those who are in work, and a social security system that does not adjust to the change in working time arrangements and so creates inequitable outcomes for citizens, thus eroding the social fabric and well-being of society as a whole (also see ACIRRT, 1999). This will produce a society where the social dislocation and hostility we are now seeing will intensify, with the prospect of 'two nations' emerging within society. A book by Watson et al. (2003) highlight this state in working life as 'fragmented futures' arguing that increasing fragmentation in the labour market has limited choices and opportunities for many and produced an unsatisfactory balance between work and life outside work.

The danger has been highlighted in an editorial in the *Harvard Business Review* (1999, p. 8) where it is even feared that knowledge workers will walk away from deep involvement with society as a whole. After all, many of these workers already live in privately policed communities and send their children to private schools. This divide has been further highlighted in the US. American economist Paul Krugman highlighted this dilemma in a speech at the London School of Economics, where he outlined the fact that in 1970 the ratio between the average wage in the US and the average wage of the top 100 chief executives was 1 to 40. In the late 1990s this figure had risen to 1 to 1000. To further highlight this polarization in wealth, in the 1970s 15 per cent of the world population had 10 per cent of the wealth. By the year 2000 only 0.001 per cent of the world population had around 30 per cent of the world's wealth. In addition, the top 10 per cent of the world's population had 90 per cent of the world's wealth (Krugman, 2003).

Richard Sennett highlights the dangers for a civil society when he argues that this short-term capitalism 'threatens to corrode . . . character, particularly those qualities of character which bind human beings to one another and furnishes each with a sense of sustainable self' (Sennett, 1998, p. 27). Sennett suggests that the flexible, constantly changing, high-risk, bitty and short-term patterns of employment undermine any sense of meaning and identity in our work and life, producing a new flexible worker with no long-term vision, no coherent 'story' to tell of his or her life. He encapsulates this by stating: 'To imagine a life of momentary impulses, of short-term action, devoid of sustainable routines, a life without habits, is to imagine a mindless existence' (Sennett, 1998, p. 44). He also states: 'transposed to the family realm, no long-term means keep moving, don't commit yourself, and don't sacrifice' (Sennett, 1998, p. 25). It could be argued that this puts the moral character and self-discipline at the centre of the argument (Ackers, 2002).

Some organizations have acknowledged these concerns and have introduced volunteer programmes in which employees help the less fortunate in society. Such initiatives reflect a shift in corporate attitudes towards the community. For example, Royal Mail and Whitbread Brewing in the UK have introduced community action programmes. Royal Mail suggests that it is good for business because volunteering helps personnel development, especially in morale, motivation and working with other people, and in the introduction and deployment of new skills (Slavin, 1999, p. 16). Such actions, however, may ameliorate the social distress caused by current trends in employment but will not eliminate it.

Human resource sustainability and the bottom line

We often hear from people at the top echelons of business that 'people are our greatest asset'. In reality, however, despite this rhetoric many organizations seek to cut costs through downsizing and job enlargement. The problem for many organizations is that human resource managers often find it difficult to convince senior management executives that progressive HR strategies make a difference due to the intangible nature of the benefits of many HR programmes and the long timeframe for any efficiency gains to become apparent (Dunphy et al., 2003). As Dunphy et al. (p. 157) suggest, 'the contrast is often expressed in terms of short-term cost efficiencies but medium to long-term negative impact (as in the case of downsizing) versus short-term increased costs (such as in training and development) and medium to long-term benefits'. However, evidence now demonstrates that companies that do invest in people, rather than just talk about it, show positive financial outcomes.

For example, a comprehensive study, 'Impact of People Management Practices on Business Performance' (Patterson et al., 1997) in the UK has reconfirmed the importance and linkage of good people management practices to productivity and profits. The report showed that the management of people had a greater effect on a business's performance than the combined effect of its strategy, product or service quality, manufacturing technology and expenditure on R&D. The researchers also indicated that satisfaction and organizational commitment explain up to 5 per cent of the difference between the profitability of companies. These results demonstrate the importance of the relationship between employee attitudes and company performance and as such indicate the more satisfied workers are with their jobs, the better the company is likely to perform in terms of profitability and particularly productivity.

However, the study's most important finding was the linkage between 'good' human resource management practices and profitability and productivity. The report suggested that overall HRM practices (that is, appraisals, training and development, employee involvement and team working and so on) explain 19 per cent of the variations in profitability and 18 per cent of differences in productivity between companies and within organizations. As the researchers state, this is the most convincing demonstration in the research literature of the 'link between the management of people and the performance of companies' (Patterson et al., 1997).

A follow-up study of 5000 employees in 42 UK manufacturing companies conducted by Patterson and West (1998, p. 2) also suggests a strong link between employee satisfaction and productivity and profitability. In fact they indicate that aggregated job satisfaction within a company predicted up to 25 per cent of the variation between companies in performance and 12 per cent of profitability one year later. They argue that no other factor (such as competitive strategy, technology, market share, total quality management or R&D strategy) could so effectively predict company performance (Patterson and West, 1998, p. 3).

Similar results have been found by the Institute for Employment Studies in the UK (Bevan and Barber, 1999; Barber et al., 1999). Their survey of 65 000 employees and 25 000 customers in almost 100 retail stores showed a strong link between employee commitment and customer satisfaction and increased sales. The survey data indicated that a one-point increase in employee commitment led to a monthly increase of £200 000 per store.

Perhaps the most comprehensive reports of positive statistical relationships between the adoption of developed HR practices and business performance has been developed by Huselid (1995), Huselid and Becker (1996) and Pfeffer (1994, 1998). These studies all suggest that the way in which organizations treat their employees is at the heart of their success (Pfeffer, 1998). Pfeffer (1998) cites evidence that HR practices can raise shareholder value (that is, stock market value) by between $20 000 and $40 000 (US) per employee. In addition, a study by Huselid and Becker (1996) emphasizes the link between market value and sophistication of HR practices. In particular they suggest that business can achieve above-average returns when a wider range of practices are adopted for more employees and when these practices are integrated within the business strategy (Huselid and Becker, 1996).

Huselid (1995) also suggests that the utilization of a number of HR practices was associated with a level of sales revenue that was on average $27 000 (US) per year per employee, and the corresponding increase in shareholder value was $18 500 (US), and profits increased by nearly $4000 (US) per year per employee (Huselid, 1995). As Pfeffer (1994, 1998) claims, there is powerful evidence to suggest that deploying certain high-performance management practices makes bottom line sense (see also Richardson and Thompson, 1999). Importantly, the studies suggest that the personnel or HR function should move away from being seen as a peripheral administrative activity and be put at the centre of the organization's activities (Richardson and Thompson, 1999).

However, this is best illustrated in the case of hospitals and the National Health Service (NHS) in the UK, highlighting this most vividly in terms of life and death, or what can be termed as the sustainability of human life. Recent groundbreaking research has linked HR practices and mortality rates (West, 2002). In a study of 61 hospitals in England, there was a strong link between the extent and sophistication of appraisal systems in hospitals and lower mortality rates. In addition, there were also links with the quality and sophistication of training and the number of staff trained to work in teams. Other research into so-called 'progressive' HRM practices may give a clue to this linkage. This research suggests that these measures enhance productivity by improving the knowledge, skill, motivation and performance of employees (see evidence above, especially Huselid, 1995; Patterson et al., 1997). Importantly, HR policies like selection and training are associated with improved job performance by changing and enhancing employee behaviour, encouraging a more cooperative and helpful approach to colleagues and making better team workers (West, 2002).

Interestingly, the research also revealed that the introduction of self-managing teams created more cooperation and better inter-departmental communication, and most significantly created a better quality of working life among employees, with individuals reporting that they felt involved in the decision-making process, and benefited from the professional and emotional support they gained as part of a team and experienced greater job satisfaction, producing lower sickness absence rates (West, 2002). The research found that the appraisal system had the strongest relationship with patient mortality, with the extent of team working and the sophistication of training policies associated with lower patient mortality. These findings allowed for differences in size, regions and socioeconomic status.

For example, the strongest association was between deaths following admissions for hip fractures and appraisals – for hospitals of equal size and local population health needs, a significant improvement in the appraisal system would lead to the equivalent of 1090 fewer deaths per 100 000 admissions, equal to 1 per cent of all admissions, or in other words it could save 12.3 per cent of hospital deaths. A weaker association was found between team working and deaths following emergency surgery, with 25 per cent more staff working in teams could produce 275 fewer deaths per 100 000 admissions or 7.1 per cent of the total (West, 2002). The researchers conclude: 'If you have HR practices that focus on effort and skill; develop people skills; encourage co-operation, collaboration, innovation and synergy in teams for most, if not all, employees, the whole system functions and performs better. Another striking finding from the research is that when HR directors are members of the board of the hospital, the association between HR practices and patient mortality are even stronger . . . The resource implications are considerable. The magnitude of returns for investments in HR practice can be substantial' (West, 2002).

Sustainable strategies towards human resources – a way forward
The future capabilities in the organization, and as a consequence the improvement in performance, is premised on the belief that it is necessary to develop a new workplace culture that emphasizes the role of employees as assets rather than merely as a costly factor of production. Importantly, human sustainability is based on organizations pursuing an integrated strategy in which employee relations policies are integrated into all aspects of the organization's planning and implementation process (Gollan and Davis, 1997). It is important that an organization 'continuously develops the organisational capabilities to identify and integrate within its vision and strategy the conceptual shifts needed to respond to dynamic markets and the operational capability to deliver' and enhance performance (Ford, 1999, p. 3). Increasingly, the financial investors for superannuation and pension funds will realize that it is better to invest in companies where employers add value to their organization by adding value to employees.

Figure 7.4 presents a model of organizations pursuing a human resource sustainability approach. It emphasizes the influence of organizational culture and the impact of the

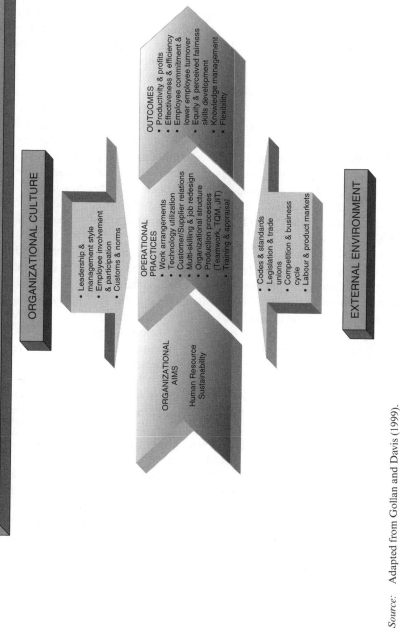

Source: Adapted from Gollan and Davis (1999).

Figure 7.4 Sustainable strategies towards human resources – a way forward

external environment on sustaining outcomes. The model shows that improved performance and productivity are predicated on the need for managers to use situationally attuned approaches to ensure that the potential of the organization is contributing to desired organizational aims. Thus there is no one best way for all organizations, only organic processes based on situational characteristics that satisfy the aims and objectives for the organization and its employees in a sustainable way.

The lessons for organizations are that for true corporate sustainability, recognition and value must be placed on the internal human capability of the organization. In addition, this chapter suggests that for human resource sustainability to be achieved, the HR policies and practices need to be integrated for sustained business performance and positive employee outcomes of equity, development and well-being.

In this process, the role of the human resource function is pivotal to the success of corporate sustainability policies. Dave Ulrich (1998) has suggested four ways that HR can deliver organizational excellence. First, by the HR function becoming a partner with senior and line managers in strategy implementation by forming the link between boardroom decisions and the marketplace. Second, by providing relevant expertise in the way work is organized and executed, thus ensuring administrative efficiency. Third, representing employees' concerns to management while simultaneously giving employees the capacity to increase their contribution to organizational decision-making. Fourth, the HR function should be an agent of change, shaping processes and culture that together improve and enhance an organization's capacity for change (Ulrich, 1998, pp. 124–25).

Importantly, this requires organizations to take a more holistic and integrated approach to people management. In particular, managers need to reassess the role and level of the HR function, specifically its role in persuading organizations to adopt practices that support a sustainable approach (Hunt, 1999). This requires devolved decision-making emphasizing medium- to long-term sustainability rather than the short-term horizons characteristic of more traditional, centralized corporate HR or personnel management approaches. The advocates of corporate-level human sustainability issues and practices need to be political movers and leaders putting forward an HR agenda to support sustainability. At lower levels, HR advocates need to be coordinators, mentors and integrators, linking and integrating human capabilities into organizational structures, technologies and practices of organizations (Hunt, 1999, p. 21).

The central challenge for the HR function will be to move organizations to adopt sustainable practices and structures towards different organizational strategies and create a climate whereby employees' potential can be released. Distinguished Professor Dexter Dunphy has added to the debate by suggesting that the HR traditional role of building human capital in organizations will intensify as the economy becomes more knowledge-based and intellectual capital is increasingly the key to an organization's success. Importantly, the HR function needs to be a strategic partner as part of business strategy (Dunphy, 2003).

Overall, the evidence presented in this chapter is that corporate sustainability is predicated on organizations recognizing the needs of employees and implementing sustainable policies and practices to reinforce its values and principles. Only by acknowledging the importance of employee satisfaction and commitment through the development of integrated employee consultation, organizational change, work and life policies, workplace institutions and comprehensive career development programmes will the organization achieve greater efficiency, productivity and increased profits.

References

Ackers, P. (2002), 'Reframing Employment Relations: The Case for Neo-Pluralism', *Industrial Relations Journal*, **33**, 1.
Australian Centre for Industrial Relations Research and Training (ACIRRT) (1999), *Australia at Work: Just Managing?* Sydney: Prentice Hall.
Barber, L., Hayday, S. and Bevan, S. (1999), *From People to Profits*, London: The Institute for Employment Studies.
Bayliss, V. (1996), *Redefining Work*, London: Royal Society for the Arts.
Bevan, S. and Barber, L. (1999), 'The Benefits of Service with a Smile', *Financial Times*, 26 June.
Dabkowski, S. (2003), 'The Cost of Job Churn', *Management Today*, August, p. 18.
De Lacy, A. (1999), 'Understanding the Knowledge Worker', *HR Monthly*, February.
Delany, K. (2001), 'Futurising Human Resources – How to Achieve Sustainability', *IBIS Best Practice Business Improvements*, London, pp. 8–9.
Donkin, R. (1999), 'Rewrite the Rule Book', *Financial Times*, 16 June, p. 19.
Dunphy, D. (2003), 'Tomorrow's People', *HR Monthly*, June, p. 10.
Dunphy, D. and Benveniste, J. (2000), 'An Introduction to the Sustainable Corporation', in Dunphy, D., Benveniste, J., Griffiths, A. and Sutton, P. (eds) *Sustainability – The Corporate Challenge of the 21st Century*, Sydney: Allen and Unwin, pp. 55–77.
Dunphy, D. and Griffiths, A. (1998), *The Sustainable Corporation: Organisational Renewal in Australia*, Sydney: Allen and Unwin.
Dunphy, D., Griffiths, A. and Benn, S. (2003), *Organisational Change for Corporate Sustainability*, London: Routledge.
Elkington, J. (2001), *The Chrysalis Economy: How Citizen CEO's and Corporations Can Fuse Values and Value Creation*, Oxford: Capstone/AJ Wiley.
Ford, J. (1999), 'The Learning Enterprise: Market Driven Organisational Transformation in Lead Enterprises in Sweden', Unpublished working paper, Bronte.
Gemini Consulting (1998), *Capitalising on the International Workplace Revolution*, London: Gemini Consulting Limited.
Gollan, P. (2000), 'Human Resources, Capabilities and Sustainability', in Dunphy, D., Benveniste, J., Griffiths, A. and Sutton, P. (eds) *Sustainability – The Corporate Challenge of the 21st Century*, Sydney: Allen and Unwin, pp. 55–77.
Gollan, P. and Davis, E. (1997), *The Implementation of HRM Best Practice: Beyond Rhetoric*, Unpublished, Sydney: Labour Management Foundation, Macquarie Graduate School of Management.
Gollan, P. and Davis, E. (1999), *High Involvement Management and Organisational Change: Beyond Rhetoric*, Unpublished, Sydney: Labour Management Foundation, Macquarie Graduate School of Management.
Gollan, P., Pickersgill, R. and Sullivan, G. (1996), 'Future of Work: Likely Long-Term Developments in the Restructuring of Australian Industrial Relations', ACIRRT Working Paper No. 43.
Harvard Business Review (1999), 'Editorial', *Harvard Business Review*, May–June, p. 8.
Hunt, J.W. (1999), 'Working in a Rich Human Vein', *Financial Times*, February 10, p. 21.
Huselid, M.A. (1995), 'The Impact of Human Resource Management Practices on Turnover, Productivity and Corporate Financial Performance', *Academy of Management Journal*, **38** (3), 635–72.
Huselid, M.A. and Becker, B.E. (1996), 'Methodological Issues in Cross-sectional and Panel Estimates of the Human Resource–Firm Performance Link', *Industrial Relations*, **35** (3), 400–22.
Kennedy, A. (2000), *The End of Shareholder Value: Corporations at the Crossroads*, Cambridge, MA: Perseus.
Krugman, P. (2003), 'America's New Fiscal Mess', Presentation to the London School of Economics, 24 June.
Patterson, M. and West, M. (1998), 'People Power: The Link between Job Satisfaction and Productivity', *Centre Piece*, **3** (3), autumn, London: Centre for Economic Performance, London School of Economics.
Patterson, M., West, M., Lawthom, R. and Nickell, S. (1997), 'Impact of People Management Practices on Business Performance', *Issues in People Management* No. 22, London: Institute of Personnel and Development.
Pears, A. (1998), 'Towards Sustainability: A Corporate Focus', Preamble, Unpublished, Sydney: Centre for Corporate Change, Australian Graduate School of Management, The University of New South Wales.
Pfeffer, J. (1994), *Competitive Advantage through People*, Boston: HBS Press.
Pfeffer, J. (1998), *The Human Equation*, Boston: HBS Press.
Rance, C. (2002), 'Stability in a Changing World', mycareer.com.au, 15 August.

Richardson, R. and Thompson, M. (1999), *The Impact of People Management Practices on Business Performance: A Literature Review, Issues in People Management*, London: Institute of Personnel and Development.

Sennett, R. (1998), *The Corrosion of Character: The Personal Consequences of Work in the New Capitalism*, New York and London: W.W. Norton and Company.

Slavin, T. (1999), 'Make a Profit – Do it for Free', *The Observer*, 30 May.

Trapp, R. (2001), 'Virtue and Reality – Corporate Social Responsibility', *People Management*, 11 October, 28–34.

Ulrich, D. (1998), 'A New Mandate for Human Resources', *Harvard Business Review*, January–February, pp. 124–34.

Watson, I., Buchanan, J., Campbell, I. and Briggs, C. (2003), *Fragmented Futures: New Challenges in Working Life*, The Federation Press, Sydney.

West, M. (2002), 'A Matter of Life and Death', *People Management Online*, http://www.peoplemanagement. co.uk/pm/sections/researcharchive/, accessed 1 December 2005.

Wilkinson, A., Hill, M. and Gollan, P. (2001), 'The Sustainability Debate', *International Journal of Operations and Production Management*, **21**(12), 1492–502.

Worrell, L. and Cooper, C. (1998), *The Quality of Working Life – The 1998 Survey of Managers' Changing Experiences*, London: Institute of Management.

8 Innovation impacts of environmental policies
Tim Foxon and René Kemp[1]

Introduction

This chapter offers an overview of the innovation effects of environmental policies. We begin in the next section by addressing the rationale for environmental innovation policies. The third section discusses the notion of innovation, saying that innovation implies novelty and uncertainty. Typically, innovation is the indeterminate outcome of a heterogeneous process with systemic, dynamic properties and inherent uncertainties – something that has implications for policy. The fourth section concerns itself with the seemingly simple question: what instrument promotes environmental innovation best? This question is addressed in the context of theoretical models and we examine the results from economic models of innovation in pollution control. The fifth section looks at evidence of actual technology responses to environmental policies and discusses the link between policy features and environmental technology responses. In the sixth section we compare policy instruments and examine how policy may be used for promoting 'system innovation' – the most far-reaching type of innovation with potentially the greatest environmental and social benefits.

Rationale for environmental innovation policies

The traditional economic argument for policy measures relating to environmental innovation has been based on correcting for two principal market failures:

1. First, knowledge can easily be copied once it has been created, so that innovators cannot appropriate the full benefits of their investment in the creation of that knowledge; that is, social returns to innovation exceed private returns, giving rise to a disincentive for private firms to undertake innovation (Arrow, 1962). This is the underlying rationale for public support for research and development (R&D).
2. Second, negative externalities exist; for example, unpriced environmental impacts (Pigou, 1932). This provides the rationale for economic instruments such as environmental taxes or emissions trading schemes to internalize those externalities.

Until very recently measures to correct these market failures have largely been applied and assessed in separate regimes, with *innovation policy* dealing with the first, and *environmental policy* dealing with the second. We argue that appropriate integration of environmental and innovation policies is both necessary and likely to be beneficial.

What is this thing called innovation?

To make progress it is helpful to consider what is meant by 'innovation'. Unfortunately innovation is used in both a 'broad' sense and a 'narrow' sense, often interchangeably. For example, the concept of an 'innovation system' is defined in the broad sense as 'the elements and relationships which interact in the production, diffusion and use of new, and economically, useful knowledge' (Lundvall, 1992, p. 2). The narrower sense dates back to Schumpeter's (1934) distinction between 'invention' – the first practical demonstration of an idea; 'innovation' – the first commercial application of an invention in the market; and 'diffusion' – the spreading of the technology or process throughout the market. As we discuss below, there are distinct stages in the whole innovation process with different drivers and barriers but modern innovation theory has moved beyond a simple 'linear' model.

Innovation implies novelty. The nature and degree of novelty will vary greatly: an innovation can be new for the company, new for the sector or new for the world. In this chapter the focus is on innovations new to the world. Innovations may be classified according to their degree of novelty. Most innovations are an incremental improvement of something that exists. Such *incremental innovation* should be distinguished from *radical innovation*, which consists of new processes and entirely new products such as solar cells. A third type of innovation is *system innovation*: changes to the whole system relating to provision of a particular function, involving change at multiple levels, including both organizations and institutions; for example, a move to decentralized rather than centralized electricity generation systems.

Most innovations require some amount of R&D but innovation is not a simple product of research. Rather than being categorized as a one-way, linear flow from R&D to new products, innovation is seen as a process of matching technical possibilities to market opportunities, involving multiple interactions and types of learning (Foxon, 2003; Freeman and Soete, 1997; Rip and Kemp, 1998). There are a number of clearly identifiable stages (shown in Figure 8.1) in the development of new technology: from R&D, to demonstration and commercialization, to diffusion. However, knowledge flows in both directions, for example, as information from early market applications feeds back into further product research. This means that the conventional drivers of *technology push* (from R&D) and *market pull* (from customer demand) can be reinforced or inhibited by *feedbacks* between different stages and by the influence of *framework conditions* such as government policy and availability of risk capital.

Thus the dynamic nature of the innovation process is emphasized, with feedbacks between different stages, which can either amplify or inhibit the basic technology push and market pull drivers.[2] Innovation is really a journey: a generative process, of which the final result is indeterminate, with parallel paths and many actors in changing networks who converge and diverge on ideas, in which there are many in-process assessments and spin-offs (van de Ven et al., 1999, p. 8). It may be set into motion by one particular impulse but is not governed by it.

The complexity of the innovation process and the extent to which this is recognized is a key factor in determining the effectiveness of innovation policies. A study of the lessons from US technology and innovation policies for climate change initiatives concluded that (Alic et al., 2003):

- Federal investments contribute to innovation not only through R&D, but also through 'downstream' adoption and learning.
- Public–private R&D partnerships may have particular advantages in fostering vertical collaborations.
- Smaller firms may be less able to absorb innovations without government assistance.

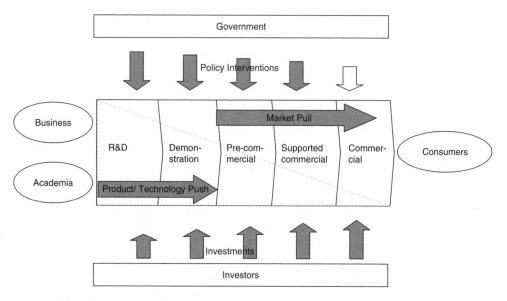

Source: Adapted from Carbon Trust (2002).

Figure 8.1. Stages of the innovation chain

- Diversity and pluralism of support mechanisms can foster innovation by encouraging the exploitation of many technological alternatives.
- Effective policies and programmes require insulation from short-term political pressures.
- Creating the space and opportunities for technological learning (learning-by-doing and learning-by-using) is crucial for promoting long-run innovation.

These factors demonstrate that feedbacks and frame conditions are as important as purely economic incentives for the creation and diffusion of environmental innovation.

Environmental innovations
Environmental innovations are new and modified processes, equipment, techniques, management systems and products to avoid or reduce harmful environmental impacts (Kemp, 1997; Kemp et al., 2000; Rennings, 2000). Environmental innovations may be technical, organizational or technical-organizational. The last category denotes those innovations that have both a 'hard' and a 'soft' aspect.

Innovation purpose
The positive environmental impact and the associated 'value added' may be the sole purpose of an environmental innovation or a by-product of innovation concerned with non-environmental aspects of the process. However, the greater part of environmental gains may very well be achieved through solutions that are not environmentally motivated. This suggests that it may be better to talk about 'innovation for the environment', which includes innovations that are not environmentally motivated. There is also a

second reason for not privileging environmental technologies, which is that they may lead to a transfer of environmental problems or add to the costs of doing business. For example, so-called end-of-pipe technologies still cause a waste problem unless the captured pollution and treated emissions are re-used, which depends on the existence of markets for waste and the economics of re-use with regard to the costs of waste disposal.

Sometimes the term 'sustainable innovation' is used though this is not a straightforward concept. Drawing on the three-pillar concept of sustainable development, sustainable innovations could be defined as new or modified processes, techniques, practices, systems and products with positive environmental, economic and social effects (for society at large and for workers). Unfortunately there are few innovations that will immediately have these triple benefits. In the short term there is often a conflict between the three goals and facilitating measures will be needed to overcome this conflict. Sustainable innovation would require a long-term commitment to both technological and institutional changes that would enable the multiple benefits to be realized.

Here we focus on environmental benefits that may be secured through 'system innovation'. System innovations consist of a set of innovations combined to provide a service in a novel way or to offer new services. System innovations involve a new logic (guiding principle) and new types of practices. Examples of green system innovation are high precision farming, integrated mobility and industrial ecology (the re-use of waste streams, process water and heat by other companies).

From the above it is clear that (environmental) innovation is not a simple thing. It requires learning-by-doing beyond R&D and the management of dynamic processes in a context of uncertainty. For an innovator there is the danger of imitation: the knowledge that is contained in an innovation or even a patent can be used by imitators, which undermines the incentive to innovate. Uncertainty is inherent to innovation and is especially endemic to radical and system innovation. This is reflected in approaches to understanding innovation based on evolutionary theory (Kemp and Soete, 1992; Lipsey and Carlaw, 1998; Metcalfe, 1998; Nelson and Winter, 1977, 1982).

Theoretical arguments on the best type of policy instrument to promote environmental innovation

This section discusses briefly the theoretical literature on the innovation impacts of different environmental policy instruments. Until recently this has consisted of neoclassical-type economic models, dealing with one type of environmental innovation only: innovation in pollution control, and usually focusing on the impact of a single instrument. These fall into two categories: models focusing on firm incentives for innovation in pollution control (Downing and White, 1986; Milliman and Prince, 1989) and models in which the behaviour of innovating firms is modelled (Fischer et al., 2003; Jung et al., 1996; Magat, 1978, 1979; Mendelsohn, 1984).

In the first type of approach researchers set out to assess firm incentives for innovation in pollution control under different regulatory regimes. These incentives are measured as cost savings in firm abatement costs in three forms: (1) direct costs of abatement (equipment expenses, operating costs); (2) associated transfer losses (payments made by the firm such as emission taxes); and (3) associated transfer gains (payments made to the firm, such as emission subsidies or patent royalties) (Milliman and Prince, 1989, p. 251). The central idea behind such analyses is that the cost savings under the different regulatory regimes are indicative for the probability that innovation in pollution control will occur.

Innovation in pollution control is modelled as a downward shift in the marginal cost curve of emission reduction – not just for some infra-marginal units of control. Polluters are modelled as profit maximizers, which means that the polluting firm will pursue innovations at the margin up to the point at which the marginal gains are equal to the marginal costs of innovation. In the models the regulator is assumed to possess perfect knowledge about the marginal conditions (the marginal cost curves and the damage function), which allows it to realize the socially optimal amount of emission reduction (before the innovation is available, and after).

Milliman and Prince (1989) assess and rank firm incentives to promote technological change in pollution control for polluting innovators, non-innovators and outside suppliers under two appropriability regimes (with and without patent protection), before and after optimal agency control, for all five regulatory regimes. They find that incentives under emission taxes and auctioned permits are equal to or higher than incentives under direct controls, free permits and emission subsidies in all cases, except for control adjustment with a non-industry innovator. Moreover, only emission taxes and auctioned permits clearly reward positive gains to an industry innovator from the entire process of technological change by providing economic incentives for continuous innovation. Thus the analysis finds that emission taxes and auctioned permits are better facilitators of technological change than regulation and free permits.

The superiority of incentive-based instruments is called into question by later studies using the second type of model. For instance the study by Fischer et al. (2003) found that no unequivocal ranking was possible between pollution taxes, auctioned permits and free (grandfathered) permits. The relative welfare ranking of instruments depends on the costs of innovation, the extent to which innovations can be imitated, the slope and level of the marginal environmental benefit function, and the number of pollution firms.

This research leads to the conclusion that there is not one instrument that is best in all circumstances – the choice depends on circumstances. Moreover, these types of theoretical model are not well-suited to analyse non-cost drivers of innovation, including expectations and political factors. For example, a technology-forcing standard may provide a greater spur to innovation than a low pollution tax because it creates expectations of future profits for companies that can successfully innovate to meet the standard. The innovation context has to be taken into account together with the political context (Bressers and Huitema, 2000; Majone, 1976). We have already seen that innovation is a complex process. Policy choices may affect technology choices (Kemp, 1997; Yarime, 2003). In general, environmental ends are best met by prevention at source rather than pollution control, and system innovation may be needed to achieve improvements in environmental efficiencies of the order of Factor 4 to 10 (von Weizsacker et al., 1998). Not every innovation is desirable. The costs may outweigh the benefits and the innovation may have some other undesirable effects in the form of hazards. For achieving efficient responses, economic instruments have a lot to recommend them, by giving industry greater flexibility with regard to the means of compliance and the time at which to comply. However, for successful innovation more is likely to be needed than purely economic incentives. For example, a knowledge-based approach aimed at learning and cooperation and the removal of institutional or regulatory barriers to innovation in order to enable the orientation of actors towards (from an industry point of view) disruptive solutions.

Empirical evidence of the impacts of environmental policy instruments on innovation
This section discusses empirical evidence of the impacts of environmental policy instruments, both regulatory and market-based, on innovation. In recent years there has been a significant expansion of the use of market-based instruments in environmental management in OECD countries but too little evaluation evidence to arrive at valid conclusions with respect to their effectiveness (Sprenger, 2000). There has also been an unresolved discussion of the effectiveness of regulations in promoting environmental innovation, with both sides able to point to evidence to back their case. Ashford (1993, 2000) and Porter and van der Linde (1995a, 1995b) have argued that as firms have bounded rationality (and so are profit-seeking rather than profit-maximizing), innovation-forcing regulations may lead to cost-reducing 'innovation offsets' that 'can not only lower the cost of meeting environmental regulations, but can even lead to absolute advantages over firms in foreign countries not subject to similar regulations' (Porter and van der Linde, 1995b, p. 98). Porter and van der Linde (1995a, 1995b) offer case studies of firms that have benefited from such innovation offsets – through first-mover advantages for example – whilst economists have typically been sceptical of the evidence of such benefits (Palmer et al., 1995).

Standards and regulations
We begin by looking at experiences with regulation in the form of emission limit values and product bans. There is a great deal of evidence of regulations that spurred innovation, but also a great deal of evidence to the contrary. Examples of regulations that promoted innovative responses are described in Ashford et al. (1985) and Strasser (1997) for the US. In Europe one finds fewer examples, since European regulation has been less hard-nosed and typically based on existing technologies. The studies available also show that the technology responses differed among firms; that is, firms chose different solutions for dealing with environmental problems with differing environmental gains and economic costs. Sometimes the solution is a technical failure as in the case of the fuel additive MMM, a lead replacement that plugged catalytic converters and was subsequently banned by the US EPA. In other cases the technology is environmentally benign but fails to make an impact in the market. Which solution will be chosen is sometimes hard to predict, especially when there are competing technologies and when there is uncertainty about future regulations.

What the studies find is that typically the search for technical solutions starts (long) before regulations actually came into effect. This happened in the case of PCBs and CFCs where firms both in and outside the chemical industry were looking for substitutes ten years before the use of PCBs and CFCs was banned. The certainty that their product or activity would be subject to regulations was an important factor.

The case studies also show that environmental policy may cause a modernization of the industry. This happened in the US textile industry where cotton dust standards caused firms to invest in advanced textile technology with higher productivity (Ashford et al., 1985). There were also some cost gains connected with cleaner production (Dieleman and de Hoo, 1993; Schmidheiny, 1992); for example, firms in the computer industry discovered that soap and water was a more cost-efficient way of cleaning and degreasing electronic parts than CFC-based cleaning agents. These findings are at odds with the widespread economic presumption (based as we saw in the previous section, on neoclassical economic models) that regulation adds to firms' costs and stifles normal business innovation. That regulation need not be at the expense of normal business innovation was demonstrated in a series of OECD studies. These studies found that, although there were examples where an R&D project in pollution control caused other innovation projects to be cancelled or delayed, environmental pressures often modified or accelerated a more general programme for improving production processes, as pollution abatement joined the initial objectives of energy and raw material conservation (OECD, 1985, p. 88).[3]

On the basis of his evidence, Ashford (1993) identifies seven conditions for regulatory initiatives to promote environmental innovation successfully:

1. shift attention from assessing risk to identifying technologies for risk reduction;
2. focus on appropriate targets – hazards, sectors or processes;
3. promote source reduction, through input substitution, product reformulation or process redesign;
4. maintain a multimedia focus, including worker health and product safety;
5. coordinate environmental, energy and industrial policies;
6. design regulation to promote improvements in technology, not just diffusion of existing technologies; and
7. set strict standards to promote innovation, but with flexibility as to the means.

Similarly, Porter and van der Linde (1995a) identify 11 design factors for innovation-friendly regulation:

1. focus on outcomes, not technologies;
2. enact strict rather than lax regulation;
3. regulate as close to the end-user as practical, while encouraging upstream solutions;
4. employ phase-in periods;
5. use market incentives;
6. harmonize or converge regulations in associated fields;
7. develop regulation in sync with other countries or slightly ahead of them;
8. make regulatory processes more stable and predictable;
9. require industry participation in setting standards from the beginning;
10. develop strong technical capabilities among regulators;
11. minimize the time and resources consumed in the regulatory process itself.

Needless to say, these factors are easier to state in theory than to apply simultaneously to real-world policy-making where other factors, such as political priorities, may be given higher priority.

Pollution taxes and emission trading

There is a view, widespread among economists, that economic instruments such as taxes or tradable permits provide a greater spur to innovation. This is because these instruments provide an incentive for the innovator to exploit an extra gain (in the form of a reduction of tax payments, subsidies for emission reduction or revenues from selling pollution rights) by reducing emissions beyond a particular emission reduction standard. This is shown in an indirect way by the studies of Newell et al. (1999) and Popp (2003) in which the direction of innovation was found to be responsive to energy price changes and by implication will also be responsive to energy taxes. The overview article of Jaffe et al. (2003) also provides evidence that economic incentives for pollution reduction and energy reduction induce innovation. The studies reported in the survey do not say much about the *nature* of the innovation, but this appears to be incremental innovation. The highly acclaimed system of sulphur trading in the US, heralded as a model for environmental protection in the world, was successful in stimulating innovation to reduce the cost of meeting an emissions target. However, this was not mostly due to R&D leading to new technological innovations, as predicted by neo-classical economic models, but to the stimulation of organizational innovation at the firm, market and regulatory levels, and of process innovation by electricity generators and upstream fuel suppliers (Burtraw, 2000) – which is in line with the more complex picture of innovation painted earlier in this chapter in the discussion on the rationale behind environmental innovation. This implies that the superiority of economic incentives to regulatory standards in these models may not always hold. As a case in point, in analysing the influence of energy price changes and fuel economy requirements on the fuel economy of new cars, Greene (1990) found that 'the standards were at least twice as important as market trends in prices, and may have completely replaced fuel price increases as a basis for long-range planning about miles per gallon [fuel economy]' (p. 55).

This result should not be interpreted as evidence that in general regulation is a stronger stimulus than economic incentives, because it is not. Whether an innovative response will occur depends on the level of the stimulus and the responsiveness of the problem sector to that stimulus: actors may decide not to comply with a legal standard (as happened in the case of the 1970 US federal Clean Air amendments that required 90 per cent reductions in carbon monoxide and hydrocarbons by 1975) or may pass on the bill from pollution taxes to consumers. In most cases a high pollution tax or energy tax will encourage innovation for the environment.

Experiences with innovation waivers

Another way to promote technological innovation is the use of innovation waivers. These are incentive devices built into environmental regulation. Generally they extend the deadline by which industry must install pollution control equipment to meet emission limitation requirements. They exempt industry from penalties during trial periods and offer the prospect of cost savings derived from superior technology (Ashford et al., 1985, p. 444). In theory, innovation waivers seem very attractive for both potential innovators and the regulating agency. In practice they have not achieved their intended effect. Ashford et al. (1985) give several reasons for their failure.

First, the acts were often ambiguous as to their requirements. For instance, the Clean Air Act required that the proposed technology had not yet been adequately demonstrated *and* proof that the technology would operate effectively. It is hard to see how it will ever be possible to provide proof that the technology operates effectively without proper demonstration. Second, the short and inflexible deadlines acted as a disincentive for innovation, especially for radical innovation with long development times. And third, there were shortfalls in the way in which the programme was administered. Under the Clean Air Act, the responsibility of issuing innovation waivers was given to the Stationary Source Compliance Division (SSCD) of the Environmental Protection Agency, a division with limited technical expertise, whose primary task was enforcement. As it turned out, the SSCD narrowly interpreted the waiver provisions, provided little guidance and the agency took a long time before it arrived at a decision.

In retrospect it is easy to comprehend why innovation waivers were unsuccessful in this case. This does not disqualify innovation waivers per se. There are several remedies to the problems encountered, many of which were given by the authors, including administration of the programme by people trained to interact with industry, the establishment of a technology review panel, delineation of eligibility criteria and longer time allowances. However, it does illustrate the practical difficulties associated with designing regulations that encourage technological change towards efficient conservation of environmental qualities.

Investment subsidies

Subsidies for green investments are a much-used instrument. In the Netherlands several studies have been undertaken into the technological impact of investment subsidies. This section describes the outcomes of these studies, showing that the behavioural effect was very low. The first study is a parliamentary study into the effectiveness of the investment subsidies under the 'Wet Investeringsregeling' (WIR). The study, based on the opinions of environmental authorities, found that investment subsidies for environmental technologies (being 15 per cent of total investment costs) induced only 8 per cent of the firms to undertake investments that they would not have undertaken otherwise (Tweede Kamer, 1987, p. 39). Curiously this led the Dutch government to conclude that the subsidy scheme was 'reasonably effective'. The effectiveness of three other types of investment subsidies was analysed by Vermeulen (1992) who arrived at a similar result. He found that the investment subsidies for the replacement of PCB equipment, quieter new trucks and the storage of manure were effective in only a small number of cases. In all three cases other factors were more important than the subsidies, such as: fuel economy, road performance and comfort in the case of silent trucks; health and safety considerations in the case of PCB replacement; and environmental regulations in the case of the manure storage. According to Vermeulen (1992), under the three programmes, 200 million guilders (about US$125 million) were spent without having any effect on polluters' purchasing decisions.

The effectiveness of investment subsidies for thermal insulation under the one billion National Insulation Programme (NIP) in the Netherlands is studied in Kemp (1997). The statistical analysis established that there was only a weak positive relationship between the subsidies for thermal home improvements and the diffusion of thermal insulation technologies. This result is confirmed by another Dutch study in which only 11 per cent of the respondents said that a subsidy was the primary motivation for investing in thermal insulation (Beumer et al., 1993, p. 42). Other evaluation studies of investment subsidies for environmentally beneficial technologies (including energy conservation, solar boilers and co-generation of heat and power) arrive at similar conclusions. With the exception of the investment subsidy for CHP (combined heat and power – co-generation), and possibly the subsidy for wind turbines, the effectiveness of the investment subsidies in the Netherlands was small (Evaluatiecommissie WABM, 1992).

The combination of investment subsidies with environmental taxes may, however, be an effective policy for accelerating the diffusion of clean technology, at least when a substitute technology is available with comparable service characteristics. As an example, in 1986 the Dutch government instituted a subsidy for clean automobiles and a tax for cars with high emission levels. The subsidies for clean cars were paid out of the extra tax revenues from the sales of highly polluting cars. This policy proved to be very effective: the share of clean cars in new car sales increased from 15 per cent in 1986 to 90 per cent in 1990. The same kind of policy was used to encourage the supply and distribution of unleaded gasoline in cars. Due to a differentiation in excise taxes, unleaded gasoline (first only 'regular' but later also 'super' gasoline) became cheaper than leaded gasoline. Oil companies quickly responded to these changes in the tax regime by offering unleaded gasoline for sale.

The experience with the Danish Clean Technology Development Programme, described in Georg et al. (1992), is more positive. Under this programme, firms and private and semi-governmental research institutions could apply for financial aid for developing and implementing clean technology. The programme was oriented at stimulating preventive process solutions and cooperation among technology suppliers, research institutes, consultancy firms and users. The Danish Environmental Protection Agency played an active role in selecting environmentally beneficial projects and in finding the right partner with whom to cooperate. That is, the agency acted as a match-maker to elicit environmentally innovative solutions, something that previous subsidy programmes had failed to do.[4] According to the authors, the Danish programme was a success. In almost all cases, appropriate technical solutions were found for the environment problems at hand. In more than half of the projects, substantial environmental improvements were achieved at low costs. Some projects led to net economic gains for the polluting firms.

R&D programmes and subsidies

What about the effectiveness of R&D programmes and subsidies for the *development* of environmentally preferable technologies? Environmental technology is often supported through special innovation programmes. Examples are the Dutch 'Milieu en Technologie' programme and the German BMBF research programme for environmental research and environmental technology, under which 1402 projects were supported in the 1980–92 period (receiving €646 million). Most of the projects consisted of treatment technology or cleaning technology, in keeping with the past orientation of environmental policy towards end-of-pipe solutions. An evaluation study by ISI (Angerer and Holland, 1997) found that the programme did encourage companies to do research they would not have done otherwise (36 per cent of the projects would not otherwise have been done and 38 per cent of the projects were done in a more elaborate way; only 5 per cent of the projects would have been done in the same way – free-rider effect). The high behavioural additionality of the German programme is probably caused by the fact that the subsidy was quite high (51 per cent on average). A second explanation might be that many of the projects are end-of-pipe solutions for which the incentives are less favourable than for normal innovations.[5]

'Additionality' is one criterion to evaluate R&D programmes and subsidy schemes. Another criterion is whether the supported innovations are used, together with the environmental benefit connected with their use. There is no information on the environmental impact and actual use of the innovations supported by the German programme. However, the evaluation study by ISI found that 16 per cent of the projects had a decisive influence on the design or implementation of environmental regulation and a further 20 per cent were supportive but not decisive to policy changes.

This compares favourably to the experiences with the Sustainable Technology Development Programme (DTO) in the Netherlands, an interdepartmental research programme for sustainable technologies, which ran from 1993 to 1997 (Weaver et al., 2000). The additionality of DTO was very high but the results were not widely utilized for policy, nor were they utilized by industry. The goal of the programme was to identify and work towards technology options offering a Factor 20 improvement in environmental efficiency while satisfying human needs in terms of nutrition, transport, housing, and water supply and protection. Industry was an important actor in the programme. Industrial opinion leaders, such as company research directors, were asked to think about long-term technological solutions offering high-magnitude environmental benefits. In total 25 million guilders (€11.3 million) was spent under the programme by the Dutch government. The financial contribution from industry was low, about 10 per cent of the costs of the illustration projects, in the form of money and time. The DTO programme led to the development and articulation of 14 illustration processes for sustainability.

The project was successful in tapping people's minds and imagination and led to ideas for system innovation and networks of collaboration but it failed to influence industries' research agenda in an important way – for the simple reason that the technologies were not economical. Their use would require a change in the frame conditions, giving the sustainable technologies a competitive edge. A five million guilder (€2.3m) programme of knowledge transfer called DTO-KOV followed the programme, but like the first programme it did not address the root problem of unfavourable frame conditions. The absence of a pull mechanism frustrated the further development of these technologies, preventing processes of co-evolution, which could have resulted in transformations and the creation of new systems. This attests to the need for policy coordination, an issue to which we will return in a later section (Implications for policy).

Covenants/voluntary agreements

Many developed countries engage in the use of covenants, negotiated agreements between industry and the government in which the industry promises to reduce the environmental burden of their products and activities. The effectiveness of eight product-related covenants in the Netherlands is analysed in Klok (1989), mostly relating to the substitution of an environmentally hazardous substance, such as mercury-oxide batteries, alkaline batteries, beverage packages, heavy trucks, PET bottles and the use of cadmium in beer cases, CFCs in aerosols and phosphates in detergents. According to Klok, the effectiveness of covenants was typically small: when environmental improvements were achieved, this was due more to autonomous technological change, external regulations (such as European Commission guidelines) and the evolution of market demand than to covenants. Furthermore there is little evidence that the covenants fostered technological innovation. An exception is the KWS-2000 programme in the Netherlands to reduce VOC (volatile organic compound) emissions, which stimulated research into low-solvent paints, especially for the housing market. Covenants are now used for achieving reductions in energy use in the Netherlands (but also Germany) where sectors promised to reduce their energy use by 20 per cent in 2000 compared with 1989 levels. Again the impact on innovation is likely to be limited as such reductions can be met with existing technology. In our view, this demonstrates a disadvantage of covenants and, perhaps of public–private partnerships. If policy-makers want to make further use of covenants, these should be more oriented towards innovation. An example of this is the voluntary agreement between the EU and European car manufacturers to improve the average fuel efficiency of new cars by 2008, though again the standard set is only likely to promote incremental innovation. Experience from the US shows that the Energy Star and Green Lights voluntary programmes have resulted in significant energy savings and less technology-specific programmes were more likely to result in the development of information directed towards innovation (Norberg-Bohm, 1999a).

Conclusions: the impacts of environmental policy instruments on innovation
From the studies examined above it is possible to draw three conclusions: (1) the effects of environmental policy instruments in the real world are governed by the complexity of the innovation processes and political considerations rather than the responses that would be predicted by simple theoretical models. As a consequence of this: (2) there is no single best (magical) instrument for promoting environmental innovation. Hence: (3) combinations of instruments will be needed: control policies combined with information instruments and innovation assistance instruments.

The evidence summarized above suggests that economic instruments heralded for promoting technological innovation have failed to achieve this in most cases. The same is true, with some exceptions in the US, for regulation.[6] Overall, we see that existing environmental policy mostly led to the diffusion of existing technologies; it did not spark many innovations new to the world. We propose that five important reasons for this are:

1.   the absence of an economic incentive to go beyond standards required by law;
2.   low level of taxes when used;
3.   regulatory preference for quick results, which leads regulators and permit writers to set unchallenging standards from an innovation point of view;
4.   poor innovation capabilities of the problem sector; and
5.   the problem sector's preference for non-disruptive technology responses.

The first two points relate to the fact that instrument choices are connected with politics: any attempt to introduce an environmental standard or economic instrument to promote the innovation or diffusion of more environmentally friendly technologies will be the subject of fierce political negotiation. In many cases, this will result in the level of the standard, tax or emission limit being set by what is judged to be politically acceptable, rather than the level needed to internalize the relevant externality, even if this could be calculated. This in turn nullifies one of the chief arguments for the use only of economic instruments – that they create an additional incentive to reduce pollution levels compared with a fixed standard. At best, the result will be to stimulate only incremental, rather than radical, improvements. This was the case with the UK Landfill Tax on disposal of waste to landfill. This was originally introduced at a level of £5 per tonne, which created little incentive for waste reduction. In 2002/03 regular annual increases in the levy were announced that will take it up to an 'economic level' of £30 per tonne by the end of the decade, with the aim of increasing the incentive for waste reduction (HM Treasury, 2002).

Political considerations also influence the regulatory preference for quick results, which leads regulators and permit writers to set unchallenging standards from an innovation point of view. Standards are likely to require long time periods and persistence on the part of policy-makers if they are to be set at a level that stimulates innovation (Anderson et al., 2001; Ashford et al., 1985; Gross and Foxon, 2003; Kemp, 1997). How such persistence may be encouraged will be considered in the following section.

The innovation capabilities of the problem sector will also influence the likely success of innovation-forcing measures. As we have argued, innovation is a complex process and factors such as the willingness and opportunity to share knowledge and information in the early stages of a technology's development can be as important as the level of R&D in the sector (see, for example, Alic et al., 2003).

The problem sector's preference for non-disruptive technology responses relates to the fact that the actors in a sector who are successful with current technologies have an incentive to discourage innovation in new technologies that may displace their market share. This is referred to as technological and institutional 'lock-in' (Arthur, 1989; Unruh, 2000, 2002). Christensen (1997) argues that most radical, disruptive innovation comes from outside the current mainstream in any sector.

So, amongst possible policies innovation policy appears to be the most important policy for environmental innovation.

Implications for policy

What does this tell us for policy? As we already argued, there is no single best instrument. Generally policy instruments should be combined with one another to benefit from synergistic effects. A combination of standards with economic instruments may be particularly useful since it combines effectiveness with efficiency. A good example of an effective and economically efficient environmental policy is the US corporate automobile fuel economy (CAFE) standards, which set progressive fuel economy targets for automobile manufacturers in the 1979–85 period under penalty of a fine of US$50 per car sold for each mile per gallon of shortfall. Tradable pollution permits also deserve to be used more as they too combine effectiveness with efficiency. In 1990 a nationwide tradable permits market for SO_2 emissions was set up in the US, with utilities trading SO_2 rights at the Chicago Board of Trade. This succeeded in stimulating organizational and process innovations, with the result that the cost of meeting the SO_2 emission cap was 40–140 per cent lower than projections (Burtraw, 2000). Apart from control policies, there is a need for policies that address and develop competence, networks and strategic orientation (Aggeri, 1999; see also Clayton et al., 1999; Hansen et al., 2002).

A second conclusion is that the focus of policy for the environment should be on *all* technologies. Any technology that uses less materials and energy is de facto an 'environmental technology', although some may object to the use of this word. Such technologies should be an important target point of a policy that tries to reconcile economic goals with environmental protection goals. It also suggests that apart from changing frame conditions for technical change in an environment-friendly way, there is a need for environmental policy to be explicitly – rather than implicitly – concerned with technical change and innovation.

This may be done through the creation of innovation capabilities through the fostering of linkages with knowledge holders and the use of goals. Goal setting should be done in consultation with industry and wider stakeholders. Here one should take stock of technical developments and utilize these for achieving environmental improvements. One way to do this is through foresight exercises involving industry and science. Foresight exercises can help to set challenging goals.

Niche policies
A focal point for transition policy could be the creation of niches for promising technologies. Historical analysis of the innovation process across a large number of industries shows that new technologies typically commercialize initially through small niche markets, in which experience is gained and cost reductions through learning can be made (see Foxon, 2003; Kemp et al., 1998; Utterback, 1994). Market development is driven not just by price signals and expectation of profits, but also by the development of appropriate knowledge and skills bases, and the formation of institutional structures that support the emerging new technologies (see Hoogma et al., 2002; Norberg-Bohm, 1999a, 1999b). These issues have been recognized in recent work by the International Energy Agency (IEA, 2003), which argues that policy initiatives designed to facilitate the adoption of cleaner energy technologies should combine three basic priorities:

- invest in niche markets and learning, in order to improve technology cost and performance;
- remove or reduce barriers to market development that are based on instances of market failure;
- use market transformation techniques that address stakeholders' concerns in adopting new technologies and help to overcome market inertia that can inhibit the take-up of new technologies.

System innovation
Greater attention should also be given to system innovation. In general, environmental policy has merely stimulated pollution control and pollution prevention through process changes, product changes and waste minimization; it has not stimulated system innovation. In our view it is important to have programmes for system innovation. The rationale for this is that system innovation may provide Factor 10 improvements in environmental impact compared with the Factor 2 improvements associated with incremental changes or Factor 5 improvements connected with partial system design (Factor 10 Club, 1994). System innovation is not only about environmental innovation but about system changes offering environmental benefits alongside other types of benefits – economic ones and social ones, although there may be trade-offs, especially in the early phase. However, this is not an easy thing for policy and there are many dangers. By engaging in system innovation policies, the following things require special attention:

- The need to be careful not to get locked into sub-optimal solutions. One way of circumventing lock-in is by exploring different configurations through the use of diverse portfolios of options.

- The need to embed transition policy into existing decision-making frameworks and legitimize transition policies.
- The need to ascertain a dynamic mechanism of change, making sure that the process does not come to a halt when positive results do not immediately materialize, due to setbacks.
- The need to engage in multi-level coordination: to coordinate top-down policies with bottom-up initiatives (to engage in vertical coordination as well as horizontal coordination). Local experiments should inform national policies and there should be strategic experimentation for system innovation.

Transition management
The above considerations influenced the model of transition management (Kemp and Loorbach, 2003; Kemp and Rotmans, 2001; Rotmans et al., 2000) that is used in the Netherlands as the model for managing the 'sustainability transitions' in energy, transport, water management and agriculture. Transition management for sustainable development consists of deliberate attempts to work toward social, economic and ecological objectives in a gradual, forward-looking, adaptive manner. Transition management seeks to overcome the conflict between long-term imperatives and short-term concerns. Because of its focus on the evolutionary dynamics of socio-technological innovation processes, transition management pays particular attention to learning, maintaining variety of options (through portfolio management) and institutional change – to avoid becoming locked into 'evolutionary traps' and to escape existing ones. Key elements of transition management are:

1. the reliance on markets for coordination and use of long-term goals chosen by society;
2. political commitment to transitions and the creation of a transition arena in which visions for the future are discussed;
3. programmes for system innovation offering societal benefits alongside user benefits;
4. the use of strategic experiments;
5. the exploration of different technology options and management of a portfolio of options;
6. the coordination of different policy areas: science policy, technology policy, environment policy and sector policies;
7. the involvement of the whole of society in the transition endeavour.

Transition management is not based on a simple blueprint. It relies on 'learning by doing', guided by long-term goals instead of detailed long-term planning. Long-term transitions are worked towards in a gradual, reflexive manner: policies and goals are regularly evaluated and adapted.

Three-layer policy approach for system innovation
A related approach to system innovation has been put forward by Butter (2002). He argues for a three-layer policy approach for sustainable development, incorporating ecological, economic and social concerns (Figure 8.2):

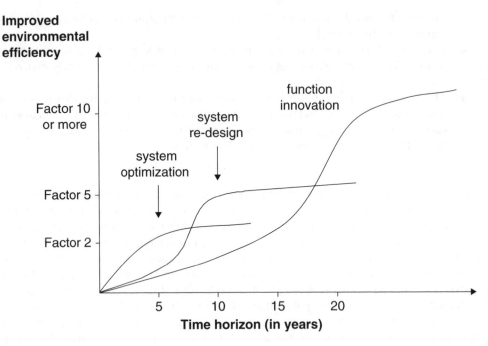

Source: Butter, 2002.

Figure 8.2   Benefits from system improvement, redesign and innovation

1. *Optimization of existing products.* This type of innovation modifies existing systems by improving existing products, processes or infrastructure, resulting in only incremental increases in system efficiency.
2. *System redesign.* The second kind of innovation involves redesign of system elements, such as products, processes or infrastructure, but with the overall system concept remaining largely unchanged.
3. *Functional innovation.* The third kind of innovation involves developing new systems that perform the same function in a better way, leading to large improvements in system efficiency.

The magnitude of environmental benefits and time periods are believed to differ for each trajectory: function innovation (called system innovation in this chapter) is believed to yield the greatest benefit in the long run; in the short term, however, the optimization of the existing system may bring greater benefits.

Butter argues that a policy approach to system innovations should have three layers:

- layer 1 – system innovations: the stimulation and alignment of innovations that will contribute to the system innovation;
- layer 2 – singular innovations: the development, dissemination and adoption of singular innovations in individual organisations;

- layer 3 – innovation climate: the creation of a supporting generic climate for sustainable innovations.

The basic assumption behind the three-layered approach is that, because of specific problems of uncertainty, complexity and specialized interests (resistance from incumbents), system innovation will need a different approach than singular innovation. The general innovation climate is seen as an important background factor that may also need to change.

Practical details of this scheme for policy need to be further worked out. However, the underlying message is that incentives for innovation, especially sustainable system innovation, require a multi-pronged, conceptualized strategy (see, for example, Foxon et al., 2005; Kemp, 1997, 2000; Norberg-Bohm, 1999a, 1999b; Rennings et al., 2003; Smith, 2000).

Policy styles
A final issue concerns policy styles. Comparing strategies for environmental policy in relation to industrial innovation in a number of countries and industries, Wallace (1995) has argued that policy styles are important for environmental innovation. He identified two important elements of environmental policy styles as: *consultation* and *long-term targets*. He concluded that 'firms are more comfortable innovating when risks are reduced; and risks are lower when environmental policy is stable and credible over the long term, and when regulatory processes are based on open, informed dialogue and executed by competent, knowledgeable regulators' (Wallace, 1995, p. *xx*). Blazejczak et al. (1999) arrive at similar findings. Consensus-based approaches may fail to pull radical innovations (Ashford, 2005). So far little systematic research has been done on the issue of policy styles and further research is needed.

Conclusion
In this chapter, we have reviewed the innovation effects of environmental policies, and highlighted recent attempts to promote systems innovations – combining environmental, social and economic benefits – that may be needed to achieve the grand vision of sustainable development. On the basis of the evidence accumulated so far, as well as a richer theoretical picture of innovation, we have argued that factors such as the longevity and consistency of policy measures and frameworks, and support for improving the innovation capabilities of industry sectors such as knowledge-sharing, may be more important to the promotion of environmental innovation than the type of instrument used (Foxon et al., 2005). Combinations of different types of policy instrument are likely to be needed in practice. Attempts to promote systems innovations for a transition to sustainability are likely to prove difficult and controversial. Their implementation will require further theoretical developments, rigorous empirical testing and a willingness by policy-makers to combine persistence in pursuing long-term frameworks with learning about the effectiveness of different policy measures in practice.

Notes
1. The chapter draws on Kemp (1997) and Kemp (2000).
2. An early attempt to represent the system feedbacks within the innovation process was made by Kline and Rosenberg (1986) in their 'chain linked' model. Foster and Green (2000) developed a flow scheme for environmental innovation.

3. Rothwell (1992) in weighing the evidence from studies into the innovation effects of regulation, arrived at a similar conclusion. He writes that the direct impact of regulation on the success or failure of individual innovation projects was generally weak. Regulation was one of many factors, and rarely the most important influencing innovatory outcome (Rothwell, 1992, p. 454).
4. The project was more than a subsidy programme. The programme brought together firms with an environmental problem and firms and institutes that could provide solutions to these problems.
5. The weak incentive for developing technologies with greater environmental efficiencies has to do with the fact that most regulations are based on existing technologies. Such regulations provide no incentive to go beyond existing control efficiencies; they only provide an incentive to develop solutions that are less expensive than those in use.
6. Although there exist examples of technology-forcing regulations, the regulatory process is also said to erect barriers for innovation (Heaton and Banks, 1999).

References

Aggeri, F. (1999), 'Environmental Policy and Innovation. A Knowledge-based Perspective on Cooperative Approaches', *Research Policy*, **28**, 699–717.
Alic, J.A., Mowery, D.C. and Rubin, E.S. (2003), *US Technology and Innovation Policies: Lessons for Climate Change*, Pew Centre on Global Climate Change.
Anderson, D., Clark, C., Foxon, T.J., Gross, R. and Jacobs, M. (2001), *Innovation and the Environment: Challenges and Policy Options for the UK*, London: Imperial College Centre for Energy Policy and Technology & the Fabian Society.
Angerer, H. and Holland, K. (1997), *Umwelttechnologie am Standort Deutschland. Der ökologische und ökonomische Nutzen der Projektförderung des BMBF. Evaluation of Fraunhofer ISI*, Heidelberg, New York: Physica-Verlag.
Arrow, K. (1962), 'Economic Welfare and the Allocation of Resources for Invention', in Nelson, R. (ed.), *The Rate and Direction of Inventive Activity*, Princeton, NJ: Princeton University Press, pp. 609–25.
Arthur, W.B. (1989), 'Competing Technologies, Increasing Returns, and Lock-In By Historical Events', *The Economic Journal*, **99**, 116–31.
Ashford, N.A. (1993), 'Understanding Technological Responses of Industrial Firms to Environmental Problems: Implications for Government Policy', in Fischer, K. and Schot, J. (eds), *Environmental Strategies for Industry: International Perspectives on Research Needs and Policy Implications*, Washington, DC: Island Press, pp. 277–307.
Ashford, N.A. (2000), 'An Innovation-based Strategy for a Sustainable Environment', in Hemmelskamp, J., Rennings, K. and Leone, F. (eds), *Innovation-oriented Environmental Regulation: Theoretical Approach and Empirical Analysis*, ZEW Economic Studies, Heidelberg, New York: Springer Verlag, pp. 67–107.
Ashford, N.A. (2005), 'Government and Environmental Innovation in Europe and North America', in Weber, M. and Hemmelskamp, J. (eds), *Towards Environmental Innovation Systems*, Heidelberg: Springer, pp. 159–74.
Ashford, N.A., Ayers, C. and Stone, R.F. (1985), 'Using Regulation to Change the Market for Innovation', *Harvard Environmental Law Review*, **9**, 419–66.
Beumer, L., van der Giessen, E.C., Olieman, R. and Otten, G.R. (1993), 'Evaluatie van de isolatieregeling (SES 1991) en de ketelregeling (SNEV)', Rotterdam: NEI.
Blazejczak, J., Edler, D., Hemmelskamp, J. and Jänicke, M. (1999),'Environmental Policy and Innovation – An International Comparison of Policy Frameworks and Innovation Effects', in Klemmer, P. (ed.), *Innovation Effects of Environmental Policy Instruments*, Berlin: Analytica, pp. 9–30.
Bressers, H.T.A. and Huitema, D. (2000), 'What the Doctor Should Know: Politicians are Special Patients. The Impact of the Policy-making Process on the Design of Economic Instruments', in Andersen, M.S. and Sprenger, R.U. (eds) *Market-based Instruments for Environmental Management. Politics and Institutions*, Cheltenham, UK and Northampton, MA, USA: Edward Elgar, pp. 67–88.
Burtraw, D. (2000), 'Innovation under the Tradeable Sulphur Dioxide Emission Permits Programme in the US Electricity Sector', in *OECD, Innovation and the Environment*, Paris: OECD, pp. 63–84.
Butter, M. (2002), 'A Three-layer Policy Approach for System Innovations', Paper presented on the 1st Blueprint Workshop 'Environmental Innovation Systems', Brussels, January, see Rennings et al. (2003).
Carbon Trust (2002), *Submission to DTI Energy White Paper Consultation*, September 2002.
Christensen, C. (1997), *The Innovator's Dilemma: When New Technologies Cause Great Firms to Fail*, Boston, MA: Harvard Business School Press.
Clayton, A., Spinardi, G. and Williams, R. (1999), *Policies for Cleaner Technology: A New Agenda for Government and Industry*, London: Earthscan.
Dieleman, H. and de Hoo, S. (1993), 'PRISMA: The Development of a Preventative, Multi Media Strategy for Government and Industry', in Fischer, K. and Schot, J. (eds), *Environmental Strategies for Industry:*

International Perspectives on Research Needs and Policy Implications, Washington DC: Island Press, pp. 245–75.

Downing, P.B. and White, L.J. (1986), 'Innovation in Pollution Control', *Journal of Environmental Economics and Management*, **13**, 18–29.

Evaluatiecommissie Wet Algemene Bepalingen Milieuhygiëne (1992), 'Financiële instrumenten in het beleidsproces', advies nr. 5, Den Haag, VROM.

Factor 10 Club (1994), *Carnoules Declaration of the International Factor 10 Club*, Carnoules, France.

Fischer, C., Parry, I.W.H. and Pizer, W.A. (2003), 'Instrument Choice for Environmental Protection when Technological Innovation is Endogenous', *Journal of Environmental Economics and Management*, **45**, 523–45.

Foster, C. and Green, K. (2000), 'Greening the Innovation Process', *Business Strategy and the Environment*, **9**, 287–303.

Foxon, T.J. (2003), *Inducing Innovation for a Low-Carbon Future: Drivers, Barriers and Policies*, London: The Carbon Trust, also available at http://www.thecarbontrust.co.uk/carbontrust/about/publications/ FoxtonReportJuly03.pdf, accessed 3 December 2005.

Foxon, T.J., Pearson, P., Makuch, Z. and Mata, M. (2005), 'Transforming Policy Processes to Promote Sustainable Innovation: Some Guiding Principles', Report for policy-makers, ESRC Sustainable Technologies Programme, Imperial College London, http://www.sustainabletechnologies.ac.uk/PDF/project%20reports/SI_policy_ guidance_final_version.pdf, accessed 2 December 2005.

Freeman, C. and Soete, L. (1997), *The Economics of Industrial Innovation*, (3rd Edition), London: Pinter.

Georg, S., Røpke, I. and Jørgensen, U. (1992), 'Clean Technology – Innovation and Environmental Regulation', *Environmental and Resource Economics*, **2** (6), 533–50.

Greene, D.L. (1990), 'CAFE or PRICE. An Analysis of Effects of Federal Fuel Economy Regulations and Gasoline Price on New Car MPG, 1978–89', *The Energy Journal*, **11**(3), 37–57.

Gross, R. and Foxon, T.J. (2003), 'Policy Support for Innovation to Secure Improvements in Resource Productivity', *International Journal of Environmental Technology and Management*, **3**(2), 118–30.

Hansen, O.E., Søndergård, B. and Meredith, S. (2002), 'Environmental Innovations in Small and Medium-sized Enterprises', *Technology Analysis & Strategic Management*, **14**(1), 37–56.

Heaton, G.R. and Banks, R.D. (1999), 'Toward a New Generation of Environmental Technology', in Branscomb, L.W. and Keller, J.H. (eds), *Investing in Innovation. Creating a Research and Innovation Policy that Works*, Cambridge, MA: MIT Press, pp. 276–98.

HM Treasury (2002), *Tax and the Environment: Using Economic Instruments*, London: Her Majesty's Stationery Office.

Hoogma, R., Kemp, R., Schot, J. and Truffler, B. (2002), *Experimenting for Sustainable Transport: The Approach of Strategic Niche Management*, London: SPON Press.

International Energy Agency (IEA) (2003), *Creating Markets for Energy Technologies*, Paris: OECD/IEA.

Jaffe, A., Newell, R. and Stavins, B. (2003), 'Technological Change and the Environment', in Mäler, K.-G. and Vincent, J. (eds), *Handbook of Environmental Economics*, Amsterdam: North-Holland/Elsevier Science, pp. 461–516.

Jung, C., Krutilla, K. and Boyd, R. (1996), 'Incentives for Advanced Pollution Abatement Technology at the Industry Level: An Evaluation of Policy Alternatives', *Journal of Environmental Economics and Management*, **30**, 95–111.

Kemp, R. (1997), *Environmental Policy and Technical Change. A Comparison of the Technological Impact of Policy Instruments*, Cheltenham, UK and Lyme, USA: Edward Elgar.

Kemp, R. (2000), 'Technology and Environmental Policy – Innovation Effects of Past Policies and Suggestions for Improvement', *OECD Proceedings Innovation and the Environment*, OECD, Paris, pp. 35–61.

Kemp, R. and Loorbach, D. (2003), 'Governance for Sustainability through Transition Management', Paper EAEPE conference, November, Maastricht, the Netherlands.

Kemp, R. and Rotmans, J. (2001), 'The Management of the Co-Evolution of Technical, Environmental and Social System', Paper for the international conference Towards Environmental Innovation Systems, September, Garmisch-Partenkirchen (forthcoming in Weber, M. and Hemmelskamp, J. (eds), *Towards Environmental Innovation Systems*, Springer Verlag).

Kemp, R. and Soete, L. (1992), 'The Greening of Technological Progress: An Evolutionary Perspective', *Futures*, **24**(5), 437–57.

Kemp, R., Schot, J. and Hoogma, R. (1998), 'Regime Shifts to Sustainability through Processes of Niche Formation. The Approach of Strategic Niche Management', *Technology Analysis and Strategic Management*, **10**(2), 175–95.

Kemp, R., Smith, K. and Becher, G. (2000), 'How Should We Study the Relationship between Environmental Regulation and Innovation?', in Hemmelskamp, J., Rennings, K. and Leone, F. (eds) *Innovation-oriented Environmental Regulation: Theoretical Approaches and Empirical Analysis*, Heidelberg, New York: Physica Verlag, pp. 43–66.

Kline, S. and Rosenberg, N. (1986), 'An Overview of Innovation', in Landau, R. (ed.), *The Positive Sum Strategy: Harnessing Technology for Economic Growth*, Washington, DC: National Academy Press, pp. 275–306.

Klok, P.J. (1989), 'Convenanten als instrument van milieubeleid. De totstandkoming en effectiviteit van acht produkt-gerichte milieu-convenanten en hierop gebaseerde verwachtingen omtrent de effectiviteit van convenanten', Enschede, Unversiteit Twente.

Lipsey, R.G. and Carlaw, K. (1998), 'Technology Policies in Neo-classical and Structuralist-Evolutionary Models', *STI Review 22*, Paris: OECD, pp. 31–73.

Lundvall, B.-A. (ed.) (1992), *National Systems of Innovation: Towards a Theory of Innovation and Interactive Learning*, London: Pinter Publishers.

Magat, W.A. (1978), 'Pollution Control and Technological Advance: A Dynamic Model of the Firm', *Journal of Environmental Economics and Management*, **5**, 1–25.

Magat, W.A. (1979), 'The Effects of Environmental Regulation on Innovation', *Law and Contemporary Problems*, **43**, 3–25.

Majone, G. (1976), 'Choice Among Policy Instruments for Pollution Control', *Policy Analysis*, **2**, 589–613.

Mendelsohn, R. (1984), 'Endogenous Technical Change and Environmental Regulation', *Journal of Environmental Economics and Management*, **11**, 202–7.

Metcalfe, J.S. (1998), *Evolutionary Economics and Creative Destruction*, London: Routledge.

Milliman, S. and Prince, R. (1989), 'Firm Incentives to Promote Technological Change in Pollution Control', *Journal of Environmental Economics and Management*, **17**, 247–65.

Nelson, R.R. and Winter, S.G. (1977), 'In Search of Useful Theory of Innovation', *Research Policy*, **6**, 36–76.

Nelson, R.R. and Winter, S.G. (1982), *An Evolutionary Theory of Economic Change*, Cambridge, MA: Belknap Press.

Newell, R., Jaffe, A. and Stavins, B. (1999), 'The Induced-innovation Hypothesis and Energy-saving Technological Change', *The Quarterly Journal of Economics*, **114**, 941–75.

Norberg-Bohm, V. (1999a), 'Stimulating "Green" Technological Innovation: An Analysis of Alternative Policy Mechanisms', *Policy Sciences*, **32**, 13–38.

Norberg-Bohm, V. (1999b), 'Creating Incentives for Environmentally Enhancing Technological Change: Lessons From 30 Years of US Energy Technology Policy', *Technological Forecasting and Social Change*, **65**, 125–48.

Organisation for Economic Co-operation and Development (OECD) (1985), *Environmental Policy and Technical Change*, Paris: OECD.

Palmer, K., Oates, W.E. and Portney, P.R. (1995), 'Tightening Environmental Standards: The Benefit–Cost or the No-cost Paradigm?', *Journal of Economic Perspectives*, **9**(4), 119–32.

Pigou, A. (1932), *The Economics of Welfare*, (4th Edition), London: Macmillan.

Popp, D. (2003), 'Pollution Control Innovations and the Clean Air Act of 1990', *Journal of Policy Analysis and Management*, **93**, 390–409.

Porter, M. and van der Linde, C. (1995a), 'Green and Competitive: Ending the Stalemate', *Harvard Business Review*, **73**(5), 120–34.

Porter, M. and van der Linde, C. (1995b), 'Towards a New Conceptualization of the Environment–Competitiveness Relationship', *Journal of Economic Perspectives*, **9**(4), 97–118.

Rennings, K. (2000), 'Redefining Innovation – Eco-innovation Research and the Contribution from Ecological Economics', *Ecological Economics*, **32**, 319–22.

Rennings, K., Kemp, R., Bartolomeo, M., Hemmelskamp, J. and Hitchens, D. (2003), 'Blueprints for an Integration of Science, Technology and Environmental Policy (BLUEPRINT)', Final Report of 5th Framework Strata project, available at http://www.insme.info/documenti/blueprint.pdf, accessed 2 December 2005.

Rip, A. and Kemp, R. (1998), 'Technological Change', in Rayner, S. and Malone, E.L. (eds), *Human Choices and Climate Change*, Vol. 2, Columbus, Ohio: Battelle Press.

Rothwell, R. (1992), 'Industrial Innovation and Government Environmental Regulation: Some Lessons from the Past', *Technovation*, **12**(7), 447–58.

Rotmans, J., Kemp, R., van Asselt, M., Geels, F., Verbong, G. and Molendijk, K. (2000), *Transities & Transitiemanagement. De casus van een emissiearme energievoorziening*. Final report of study 'Transities en Transitiemanagement' for the 4th National Environmental Policy Plan of the Netherlands (NMP-4), Oktober 2000, ICIS & MERIT, Maastricht.

Schmidheiny, S. (1992), *Changing Course: A Global Business Perspective on Development and the Environment*, Cambridge, MA: MIT Press.

Schumpeter, J.A. (1934), *The Theory of Economic Development*, Cambridge, MA: Harvard University Press.

Smith, K. (2000), 'Innovation as a Systemic Phenomenon: Rethinking the Role of Policy', *Enterprise & Innovation Management Studies*, **1**(1), 73–102.

Sprenger, R.-U. (2000), 'Market-based Instruments in Environmental Policies: The Lessons of Experience', in Andersen, M.K. and Sprenger, R.-U. (eds) *Market-based Instruments in Environmental Policies: Politics and Institutions*, Cheltenham, UK, Northampton, MA, USA: Edward Elgar, pp. 15–17.

Strasser, K. (1997), 'Cleaner Technology, Pollution Prevention, and Environmental Regulation', *Fordham Environmental Law Journal*, **9**(1), 1–106.

Tweede Kamer (1987), 'Evaluatie WIR-milieutoeslag', Dutch Parliamentary study, Tweede Kamer, vergaderjaar 1986–1987, 19858, No. 4.

Unruh, G.C. (2000), 'Understanding Carbon Lock In', *Energy Policy*, **28**, 817–30.

Unruh, G.C. (2002), 'Escaping Carbon Lock In', *Energy Policy*, **30**, 317–25.

Utterback, J.M. (1994), *Mastering the Dynamics of Innovation: How Companies Can Seize Opportunities in the Face of Technological Change*, Boston, MA: Harvard Business School Press.

van de Ven, A.H., Polley, D.E., Garud, R. and Venkataraman, S. (1999), *The Innovation Journey*, Oxford, New York: Oxford University Press.

Vermeulen, W. (1992), *De vervuiler betaald. Onderzoek naar de werking van subsidies op vier deelterreinen van het milieubeleid*, Utrecht, Holland: Jan van Arkel.

Von Weizsacker, E., Lovins, A.B. and Lovins, L.H. (1998), *Factor Four: Doubling Wealth, Halving Resource Use*, London: Earthscan.

Wallace, D. (1995), *Environmental Policy and Industrial Innovation: Strategies in Europe, the US and Japan*, London: Royal Institute of International Affairs.

Weaver, P., Jansen, L., Grootveld, G. and Vergragt, P. (2000), *Sustainable Technology Development*, Sheffield: Greenleaf Publishing.

Yarime, M. (2003), 'From End-of-pipe Technology to Clean Technology. Effects of Environmental Regulation on Technological Change in the Chore-Alkali Industry in Japan and Western Europe', PhD thesis, Maastricht University.

9 Sustainable production and consumption policy development: a case study from Western Australia

Andrew Higham and Piers Verstegen

Introduction

There are many policy dimensions and approaches that necessarily must be pursued simultaneously if human civilization is to achieve sustainability. There are no silver bullets – no one approach, it seems, can offer a complete and plausible solution to the unsustainable situation that industrialized societies find themselves in.

This chapter deals with one, albeit powerful and ambitious policy approach – known generally as *sustainable production and consumption* (see Ryan, 2002; UNDESA, 2003; UNGA, 1992). In particular it examines how this approach has been derived from the international policy framework for sustainability and how it is being applied at the scale of a large sub-national government in the western third of Australia.

If defining the normative characteristics of sustainability was a key feature of sustainability policy debate in the 1980s (see Davison, 2004) then the 1990s was the era where society moved forward to better understand the scale of the sustainability challenge in more quantitative and qualitative terms. This challenging task has occurred during a time of rapid advancements in theoretical and scientific understanding of how global ecosystems function, the capacity of global ecosystems to support human life and the extent to which human beings have appropriated the resources of the Earth and disrupted ecological processes. We now have a better understanding of where the boundaries to Earth's capacity to support life exist and we have learnt that these boundaries have been or are close to being transgressed by human activities (Millennium Assessment Board, 2005; Rees and Wackernagal, 1994; UNEP, 2000; WWF, 2004).

Records on the global production of materials have been collected since 1963. Since this time there has been a 2.4-fold increase in materials production globally, comprised of a 2.5-fold increase in minerals production; a 2.1-fold increase in metals; a 2.3-fold increase in wood products; and a 5.6-fold increase in fossil fuel-based synthetics (Gardner and Sampat, 1998). Best available data show that in 2001 humanity's ecological footprint exceeded the global biocapacity by 21 per cent, and that 'overshoot' of the Earth's capacity to support life occurred during the mid-1980s. Humanity is now spending the ecological capital of future generations (WWF, 2004). Research has also shown that the scale of the human economy now overwhelms many of the Earth's material cycles, such as nitrogen (Vitousek, 1994), sulphur (MacKenzie, 1997), carbon (Houghton et al., 1996), water (Postel et al., 1996) and trace metals (Nriagu, 1990).

And all the while, the world's attention has been drawn to the stark and growing contrast between rich and poor (UNDP, 1998). The Millennium Development Summit and World Summit on Sustainable Development highlighted the policy dilemmas of achieving equality in the allocation of resources within a finite and globalizing world. Furthermore research has highlighted the implications of a global aspiration system (for example, Gardner and Sampat, 1998; Ryan, 2002; UNDP, 1998; UNEP, 2001), which is driving large emerging economies to seek to achieve the consumption patterns and lifestyles of the industrialized world. This combined with evidence of an emerging trend of increased wasteful consumption (that is, consumption of products that are never consumed) particularly within younger people as income increases is of significant concern for sustainable resource management (Hamilton et al., 2005).

It has also been widely accepted that increases in resource efficiency alone will not lead to sustainability. In fact, this conclusion had already been reached in the international policy consensus by 1992 (UNGA, 1992, Chapter 4), although it has since been heavily contested particularly by business interests and some neo-classical economists who see such a conclusion as an affront to business 'welfare' and economic theory. Rather, there is recognition that radical system-wide transformations of the global economy are required over time spans that are in most cases unprecedented in recent human history (Jacob et al., 2004).

Moreover, in the absence of some other limiting factor, resource use will often increase as efficiency increases in response to pricing and behavioural effects (Gardner and Sampat, 1998; Ryan, 2000, 2002). Technical efficiencies can allow for reduced input costs leading to increased sales and increased total resource use. Furthermore they may cause indirect effects by increasing the incomes of consumers (and driving further consumption) or by causing structural changes through substitution of products or services that result in greater consumption. They may also ultimately result in economy-wide effects as economies are transformed by changing cultural norms, lifestyles and quality of life. In this way increased resource efficiency drives economic growth, which simply allows greater consumption of resources.

These feedback loops are referred to as rebound effects (Jalas and Plepys, 2001), and in the literature on microeconomics, as Jevons' Paradox after William Stanley Jevons who investigated in 1865 the potential to accelerate coal depletion as a result of increased production efficiencies (Jevons, 1865; Jevons and Foxwell, 1884). At the macroeconomic scale, rebound effects are clearly visible in the USA where substantial efficiency improvements in a number of different resource sectors including steel production, electricity production, transport and building have been outstripped by increased consumption leading to net increases in the use of materials in all of the above sectors (Ayres et al., 2005).

With this emerging knowledge, researchers have been developing a more sophisticated understanding of how patterns of production and consumption must change to achieve sustainability (for example, see Adriaanse et al., 1997; Bartelmus, 1999; Bringezu, 2002; Matthews and Hammond, 1999; Matthews et al., 2000; Michaelis and Lorek, 2004; Schutz and Welfens, 2000; Spangenberg and Lorek, 2002). Research at the global, national, regional and sectoral scales has found that increases in resource efficiency and reductions in resource consumption in industrialized countries of two- to 50-fold are required within one or two generations (Bringezu, 1997, 2002; IPCC, 1990; Meadows et al., 1992; von Weizsacker et al., 1997; Weterings and Opschoor, 1992).

This understanding has led the United Nations Environment Programme (2000) and the General Assembly (UNGA, 1997, paragraph 28), the OECD (1996, 1997), the World Business Council for Sustainable Development (Willums, 1998), the Nordic Council of Ministers (1999) and several leading governments (European Union, Norway, Japan, Western Australia) to adopt targets for sustainable production and consumption. These targets approximate the scale of change that is required in the use and impact of resources within the economy and are designed to stimulate innovation in society, business and government policy for sustainability.

International policy history

Whilst sustainable production and consumption were a feature of the policy debate leading up to and during the first global conference on the human environment in Stockholm in 1972, it was not until 1992 at the Rio Earth Summit (UNGA, 1992) where serious attention and commitment was made to recognize and address the issue. Chapter 4 of Agenda 21 'Changing Consumption Patterns' provides the foundation for international policy on the topic. Agenda 21 recognized unsustainable production and consumption, along with poverty, as the fundamental and overriding drivers of unsustainability. At the Earth Summit nations agreed to a dual strategy of: (1) increasing resource efficiency; and (2) changing consumption patterns to reduce the total global environmental burden.

Following on from the Earth Summit, the General Assembly of the United Nations reiterated the fundamental and critical nature of sustainable production and consumption patterns and since this time the issue has been a focus of each session of the UN Commission on Sustainable Development. At its Third Session in 1995, the Commission endorsed an international work programme to support the implementation of Chapter 4 of Agenda 21. This work plan included:

- the monitoring of trends in production and consumption;
- assessment of the impacts of unsustainable production and consumption patterns on developing countries;
- the development and assessment of specific policy measures that could be adopted by national governments;
- the establishment and adoption of voluntary commitments and indicators of sustainable production and consumption at the national level; and
- a review of the UN Guidelines for Consumer Protection to embed the principles of sustainable production and consumption.

In 1997 the UN General Assembly undertook a comprehensive five-year review of progress toward the implementation of Agenda 21. Concerned at the lack of progress, this led to the acceleration of international work programmes on sustainable production and consumption. The General Assembly also endorsed a recommendation for national governments to consider adopting targets of achieving a ten-fold increase in resource productivity over the long term and a four-fold increase in 20–30 years. In 1996, the 29 environment ministers of member nations to the OECD agreed to aim 'to achieve substantial improvements in resource productivity, for example by a factor of 4 and eventually of 10' (OECD, 1996).

The work of the UN Commission on Sustainable Development on sustainable production and consumption patterns continued throughout the decade following Rio, and its work programme expanded to examine the issues of effectiveness of policy development and implementation, natural resources management, globalization and urbanization. At the end of the decade and in the lead up to the 2002 Johannesburg World Summit on Sustainable Development (WSSD), various reviews concluded that while the policy framework for sustainable production and consumption had been advanced significantly, there was little effective uptake of policy by governments, and that unsustainable patterns of production and consumption worsened dramatically (Ryan, 2002).

Despite acute and alarming trends in consumption, little progress was made at the WSSD. Many progressive governments, business organizations and non-government organizations were hoping to see a strengthening of the General Assembly's 1997 resolution with stronger commitments to targets and concrete actions. The United Nations Secretary General's Report to the Social and Economic Council, which concluded a comprehensive review of implementation of Agenda 21, reported poor performance in sustainable production and consumption policy and encouraged the adoption of time-bound, Factor 4 and Factor 10 targets. However, the Plan of Implementation that was agreed at the Summit was disappointing – yet another vague commitment to a programme of work. However, it did move the policy agenda for sustainable consumption and production forward by strengthening the recognition that social and economic development should be kept within the carrying capacity of ecosystems.

The Marrakech Process is the key international process to drive further adoption of sustainable production and consumption policy. It began at the International Expert Meeting on Sustainable Consumption and Production held in Marrakech, Morocco in June 2003. The Marrakech Process currently consists of international and regional expert meetings on sustainable consumption and production being held throughout the world. The aim of these meetings is to create regional strategies and provide expertise that countries can draw upon when formulating national plans on sustainable consumption and production. It is through the Marrakech Process that the further ten-year framework of programmes for sustainable production and consumption is being developed, which will be presented for adoption at the Commission on Sustainable Development in 2010–11.

Sustainable consumption and production policy in Western Australia

At the World Summit on Sustainable Development the Australian Government maintained that it was inappropriate to take a target-based approach to sustainable production and consumption policy and like most other non-EU governments, actively lobbied against the setting of such targets. However, this is not the position endorsed by Australian State Governments, and Western Australia (WA) in particular, which acknowledge the importance of sustainability. The Western Australian State Sustainability Strategy *Hope for the Future* (Government of WA, 2003) was developed between 2001 and 2003. The Strategy is comprehensive, addressing 42 themes of sustainability, and has triggered the implementation of over 336 strategies and actions, and mandated the development of sustainability action plans by all government agencies.

The Strategy concluded that radical changes to present production and consumption systems are required in Western Australia to achieve sustainability. Consistent with the international policy framework for sustainable production and consumption, the Strategy established two ambitious stretch targets for the Western Australian economy:

- increase total resource efficiency of the state economy by a factor of 4 by 2020;
- halve the ecological footprint of the state economy by 2020.

These targets are complementary with the vision the State has of becoming a waste-free (or zero waste) economy within the same timeframe.

In response to the WSSD Plan of Implementation, the Strategy also recognized the need to develop a long-term programme of action to achieve the State's sustainable production and consumption policy targets. The State has established a high-level Sustainability Roundtable and will create sustainability legislation to give statutory effect to the State Sustainability Strategy.

The approach Western Australia has taken to the development of a programme of action for sustainable production and consumption is not dissimilar to approaches that have been taken in other jurisdictions. However, what is significant about this case study is that Western Australia has a very different economy from those that have approached the issue to date. Western Australia is known as 'the resource state' with 64 per cent of its material throughput exported and only 3 per cent of its material throughput imported. Moreover, 18 per cent of employment and 25 per cent of Gross State Product (GSP) can be attributed to the mining sector within the State's economy. The economy has been referred to as a hot, heavy and wet (ACF, 2002), and on any standard measure, it is a long way from sustainability.

According to our assessment, Western Australia is one of, and perhaps the most resource- and energy-dependent economies in the world. For example, based on 1998–99 figures Western Australia has an estimated total material requirement (TMR, for further clarification see Adriaanse et al., 1997) of 153 tonnes per capita per year, excluding hidden flows (which have been estimated to be an additional 150–500 tonnes per capita per year). In comparison, the European Union has a TMR of 50 tonnes per capita per year (based on 1997 figures) including hidden flows (Bringezu and Schutz, 2001). Based on our survey of the literature, Western Australia's TMR per capita exceeds all other nations where TMR has been calculated. Since 1999 when these figures were calculated, Western Australia has undergone a major resources boom that has dramatically increased TMR.

In regard to the ecological footprint (Rees and Wackernagel, 1994) of the Western Australian economy, comparative assessments indicate that, on average, Western Australians are very large consumers of natural resources by international standards. The ecological footprint for Western Australia has been calculated to be 27 299 454 hectares (ha) for 2000–01 or approximately 14.7 ha per person. Although the methodologies differ, this figure is double most other OECD countries and is much greater than the average Australian footprint of 9.4 hectares per person. In addition, an estimation of Western Australia's ecological balance of trade reveals that Western Australia is a net exporter of 'land' (− 32 520 438 ha) reflecting the heavy focus on export-based primary industries of the Western Australian economy.

As these statistics illustrate, Western Australia is a very unique economy, which will be affected like none other with the implementation of strict sustainable production and consumption targets and policies. This creates severe political impediments to the development and implementation of sustainable production and consumption policy. Despite the physical scale of the sustainability challenge, Western Australians are well placed to design solutions to sustainability with an innovative culture, strong and diverse knowledge and skills base, the natural advantage of relatively healthy ecosystems and an enormous natural resource base that can support a wide range of human endeavours.

The State Sustainability Strategy aims to make Western Australian businesses hyperefficient to increase economic performance and to bring the physical scale of the Western Australian economy within the productive capacity of the ecosystems in which it exists.

Once this is achieved, it is hoped that the State will be able to rely upon its natural advantages, make further global contributions to sustainability and substantially increase the competitiveness of its businesses. It is recognized that this will not only require greater efficiency, but depends to a large degree on the transformation to a new economy where patterns of production and consumption lead to reductions in total resource use and environmental impact.

As mentioned previously, the State Sustainability Strategy, having established the strategic framework for sustainable production and consumption policy, established a high-level Sustainability Roundtable to drive the development of a programme of action for further policy development and implementation. Three priority initiatives were identified to provide the foundation for a programme of action:

1. a multi-stakeholder and expert dialogue applying backcasting techniques to *design development pathways* that could achieve the targets outlined in the State Sustainability Strategy;
2. a feasibility study to develop and implement a *physical economy model* into the policy-making activities within the State Government, and which would be able to refine the development pathways identified by stakeholders and experts; and
3. the development and implementation of a *framework of indicators* of sustainability that could monitor and evaluate progress toward the achievement of targets.

Apart from the objective of deriving a concrete plan of action, the intention of this set of initiatives was also to build greater consensus about how sustainable production and consumption could be achieved in Western Australia and to increase the capacity for change within the public and private sectors.

Design of development pathways for sustainable production and consumption
Changes on a system level are often described as industrial transformations. They require combinations of technological, cultural, social, institutional and organizational changes, they affect many stakeholders when diffusing into society, and they involve complex processes of social change over long timeframes. Important questions arise about how to identify attractive and desirable system changes (system innovations, industrial transformations or transitions), how to get them started and implemented in practice, what are the roles of different stakeholder groups and stakeholder cooperation, and how do they relate to the role of government.

The Western Australian Government held a two-day workshop in April 2004 to explore these questions and to identify appropriate development pathways for sustainable production and consumption in Western Australia. Dr Hunter Lovins, co-author of *Factor Four* (von Weizsacker et al., 1997) and *Natural Capitalism* (Hawken et al., 1999), and international expert on sustainability and resource efficiency hosted the workshop and assisted participants to identify development pathways to sustainable production and consumption within the framework of natural capitalism. Expert advice was also provided by members of The Natural Edge Project, Karlson 'Charlie' Hargroves and Michael Smith (Hargroves and Smith, 2005), who assisted in design and facilitation of the workshop.

The concept of the workshop was inspired by the approach taken by the Netherlands Government's Sustainable Technology Development Programme (Weaver et al., 2000) with modifications that responded to the particular local circumstances and from what had been learned through conducted evaluations of similar sustainability research programmes in six other European countries and Japan (Hinterberger et al., 2003; Jansen et al., 2003; Whitelegg and Weber, 2001). Modifications included:

- broadening the development of pathways from a focus on technology development to include political and policy, economic, and social, cultural and behavioural innovations for sustainable production and consumption;
- participants being asked to identify personal and professional contributions that they would make to implement the development pathways in order to inspire immediate and personal commitments to act from the participants;
- enhancement of the methodology using deliberative democracy techniques aimed at improving government, stakeholder and expert interactions; and
- condensing deliberations into a shorter interactive format, consistent with the limited budget available for the initiative.

The workshop brought together 70 researchers and policy professionals, representatives of diverse stakeholder organizations and decision-makers from around Australia with expertise or an interest in the multidisciplinary area of sustainable production and consumption policy. Participants used a technique known as 'backcasting' (Quist and Vergragt, 2003; Weaver et al., 2000) to design development pathways for sustainable production and consumption. It is a scenario planning methodology that is ideally suited to complex problems and situations where it is necessary to plan in circumstances of extreme uncertainty and where there is a deliberative attempt to achieve a well-defined direction of social change. In the context of sustainability, it involves creating a normative description of society where sustainability has been achieved, or where clearly defined milestones along the pathway to sustainability have been reached. In contrast to forecasting, which tends to use existing trends to generate future states, the backcasting process works from the future back to the present. This avoids for analytical purposes, futures being 'generated' based on existing trends that are at the heart of the problem. It is a method where future desired conditions are envisioned and steps are then defined to attain those conditions.

In our case, the backcasting process involved the following steps (the first three were the subject of the workshop process):

1. Strategic problem orientation – defining the scale and nature of the sustainability challenge.
2. Construction of sustainable future visions or scenarios – development of a normative description of a desired future state, in this case a detailed vision of the world in 2020 when we have achieved our sustainable production and consumption targets.
3. Backcasting stage – participants projected back from 2020 toward the present and 'mapped' the economic, technical, social, cultural and behavioural pathways that would enable society to make the vision a reality.
4. Elaboration and defining follow-up activities and action agendas – having developed a detailed strategy for sustainable production and consumption, many participants were able to take the ideas generated and pursue them in their own business, or in

their research and policy work. However, in reality the workshop process only provided a patchy outline of the pathways to sustainable production and consumption rather than a blueprint for the future. Further development and refinement of this initial framework is planned through the Sustainability Roundtable's programme of action. One major mechanism is through the development of a physical economy model for Western Australia.

5. Embedding of action agenda, activities and generating follow-up and implementation – the outcomes of the workshop and subsequent modelling and assessment work will be used by the Sustainability Roundtable to implement a programme of action for sustainable production and consumption in Western Australia.

To maximize the benefits of the backcasting methodology, the workshop process also employed a deliberative democracy strategy adapted from America Speaks' 21st Century Town Meeting methodology. The diverse selection of stakeholders and experts engaged through in-depth roundtable discussions that were focused on the critical policy issues and design parameters for sustainable production and consumption. All participants were provided with extensive background material prior to the workshop to help create a wider common knowledge of sustainable production and consumption policy and practice, as well as information on the history and current status of our production and consumption patterns. Deliberations were assisted by independent facilitators, expert commentators, and with the use of computer software and a system of networked computers that linked all small groups together. The results of each phase of the backcasting process were displayed back to the workshop plenary. This enabled real-time synthesis of input from all groups so that participants were able to make common decisions on the development pathways that would lead to a sustainable future in Western Australia. Decision-makers actively participated in the process and were required to respond through a panel discussion on the consensus developed over the two days on how the development pathways for sustainable production and consumption might be achieved.

The workshop was structured into three main themes: nutrition, energy and materials. Using their own knowledge and information that was made available to them through workshop presenters and technical experts who were available to answer questions, participants provided the responses as summarized below. A full report on the outcomes of the workshop is available at www.sustainability.dpc.wa.gov.au.

Hopes and visions   In broad terms, participants described a world where resource use patterns were significantly different from the trends of the last 50 years and today. They envisioned technologies mimicking natural systems (biomimicry), and a re-conception of waste and resource use as the basis of a new sustainability revolution. There was optimism that Western Australia could achieve the Factor 4 target for increased materials efficiency and halve its ecological footprint by 2020. There was also optimism that Western Australia could become a future 'global sustainability superpower' by building on the foundations of the Western Australian State Sustainability Strategy.

Strategies and solutions   Workshop participants were placed in groups to focus on one of three main themes: food, energy, or metals and minerals. Technological, economic and socio-cultural solutions were identified in each of the three themes to achieve the sustainable world that participants had described previously. Participants also identified actions that should be undertaken immediately to shift current trends towards those solutions. This provided a detailed action plan for sustainable production and consumption. Workshop participants were also asked to identify personal and professional contributions that they intended on making immediately, so they were more likely to leave with a lasting impression and commitment to work toward the vision.

Taking workshop findings into the policy development process   To actually transform the Western Australian economy will require radical shifts in policy and practice. It will also require the collection and analysis of information on the physical basis of the economy. Without information there can be no policy or political understanding of the physicality of the economy and it will not be possible to develop rigorous policy or to evaluate progress toward political targets (Jänicke, 1997).

Furthermore, sustainable production and consumption requires a systems approach to policy analysis. Understanding the dynamics of complex systems can be assisted with the use of computer-based models that are capable of synthesizing for policy analysts the complexity of interactions within the physical economy (Anarow et al., 2003). From an economic perspective, unsustainable production and consumption is not well understood, information is poor and impacts are widely dispersed geographically and temporally. Unless these barriers are addressed, an institutional response to correct market failure is unlikely (Arrow et al., 1995) and the public is unlikely to 'demand (or accept) political action involving short term costs for the sake of uncertain long term gains' (Hatfield-Dodds, 2004).

This has led the State to create a consistent information system on the physical scale and intensity of the Western Australian economy. Through this information system, it will be possible to build an information and knowledge base and establish the capacity within government and society to develop policies and programmes that support sustainable production and consumption. Western Australia, like most other jurisdictions, does not possess the tools or policy practices to achieve this. Before we can generate policy change, the very art of thinking about the physical dimensions of sustainability must be institutionalized within the systems of governance and policy-making. This requires information systems that drive policy-making and new policy-making processes that are geared toward sustainable production and consumption.

Therefore, the State has established a collaborative arrangement with Australia's leading science and technology organization (the Commonwealth Science and Industry Research Organization or CSIRO) to develop and implement a model of the physical economy to underpin policy development processes across government and within key sectors of the economy.

Physical economy modelling

The proposed Western Australian Physical Economy Model is an integrated account of the human, natural and industrial sectors of the economy, which enables the development of policy within the physical, thermodynamic and environmental constraints imposed by natural systems within which the economy operates. The model will be an open and transparent framework that provides policy analysts across government with direct and live access to a common analytical framework, thus encouraging consistent policy-making activities across the various silos within government.

The model applies the Australian Stocks and Flows Framework (Turner and Poldy, 2001), which was developed by the CSIRO Resource Futures Group. This framework keeps track of all significant stocks (such as people, cars and oil reserves) and flows (such as energy and materials that have been removed from the stock and are transformed through industrial processes within the various parts of the economy, or wastes that result from the use of those resources within the economy) of all components of the socio-economic system. Each element of the framework is linked by a set of calculators that have been specifically designed and tested to represent real world transformations and relationships all grounded in 50–60 years of historical data. Impacts on stocks of land, air and water resources are also calculated, and the framework is also being enhanced to account for stocks of and impacts on biological diversity. Population numbers and demography drive the model, and the life histories of all stocks within the economy are represented, which ensures that inertia within the economy in regard to the feasibility of technology change can be characterized.

However, unlike traditional models of the monetary economy, behaviours such as people's responses to prices are not modelled, and feedback loops, such as rebound effects, are excluded. Rather, these are incorporated as assumptions or inputs to the scenarios that are created by policy analysts during modelling exercises. This ensures that assumptions are consciously chosen and are entirely transparent within the process of using the model.

The model enables the testing of policy scenarios across the major dimensions of the physical economy. The policy analyst is able to use the model to test the effects of any policy proposal, and can specify physical outcomes (such as a four-fold increase in resource efficiency or the halving of the ecological footprint of a population). It is intended that the outcomes of the workshop can therefore be tested within the model to determine whether the development pathways that had been envisaged will in fact achieve the State's sustainable production and consumption targets. We also seek to gain a much deeper appreciation and more detailed understanding of what changes will be necessary and effective. It will also enable the development of policy measures that will allow the State to meet its targets.

The model is also capable of generating a wide range of indicators, particularly indicators of environmental pressure, such as environmental intensity, total material requirement, direct material input, total factor resource efficiency, ecological footprint, Gross State Product, water and energy consumption (see Hatfield-Dodds, 2004). Therefore, it provides essential information for the State's sustainability monitoring, evaluation and reporting framework.

The political challenge

The processes outlined above represent an ambitious and progressive first step toward the design and development of a sustainable production and consumption policy for Western Australia. It is important though that they are understood within the broader context of Western Australia's development culture and its policy-making actors and environment. As we have noted, major themes of sustainable consumption and production policy include radically increased resource efficiency, and ultimately the use of instruments that cap and force down resource use, promoting a significant net dematerialization of the economy. These policy goals, however, present a direct affront to the primary modes of economic development in the Western Australian economy, and as such, they are counterintuitive to the State's dominant policy actors. In 2001, the State Government commitment to the development and implementation of a whole of government sustainability strategy for Western Australia resulted in one of the most ambitious and respected sustainability policy agendas in the world.

While the Western Australian sustainability agenda has been highly progressive, and in many policy areas it has affected significant positive change, the reality is that by 2005 this agenda has had very little measurable effect on planning and policy determining the resource efficiency and the sustainability of production and consumption underpinning the Western Australian economy. Despite the requirement in the 2003 State Sustainability Strategy for all agencies to incorporate sustainability principles into their strategic and operational plans, the 2004 Industry Policy Statement produced by the Department of Industry and Resources *Building Future Prosperity* (Government of Western Australia, 2004) and agreed by the State Cabinet, includes only the briefest mention of sustainability and completely ignores the targets for sustainable production and consumption in the Sustainability Strategy. Despite the resonant support for a strong sustainability agenda from the Western Australian community, the political mandate for implementation has been co-opted by strong industry agendas with manifest political relevance both within and outside government.

As the sustainable production and consumption aspects of the Western Australian sustainability agenda have been among the most difficult and controversial to implement, most of those responsible for implementation of the State Sustainability Strategy have chosen areas of more fertile ground to focus their efforts.

It is often suggested that private sector industries are in fact more responsive to the adoption of sustainability than governments. This is not the case when it comes to the sustainable consumption and production in Western Australia. The Western Australian Sustainable Industry Group (WASIG), a government- and industry-funded association, advocates a sustainable industry approach of increasing profitability for existing industries through improvements in efficiency. However, while there is a significant increase in government and business knowledge of eco-efficiency through the programmes initiated by this group, efficiency gains or reductions in the environmental burden of industry have not been demonstrated yet.

The social sustainability principles of intra- and intergenerational equity, which underpin sustainable production and consumption policy are also under challenge in Western Australia. The dominant view is that it is reasonable for Western Australia to claim a greater global proportion of the available global environmental space than an equitable and fair distribution would otherwise allow. For example, in relation to establishing mandatory global targets for sustainability, Western Australia often puts forward the argument, as Australia did while participating in the development of the Kyoto Protocol, that it is a special case and that it should be granted a greater quota of global environmental space (to emit greenhouse gas emissions, for example), so that it can do what it does best – extract and sell bulk commodities in the global economy. According to this argument, if Western Australia can be the most efficient and 'environmentally and socially responsible' producer of minerals and energy, then the international community should recognize that it would be far better to have global mineral and energy production focused in Western Australia, as this will result in a net environmental benefit. After all, the State reasons, during the great transition to sustainable production and consumption, raw materials will still be required in vast quantities. Just as this line of reasoning has been used to seek greater permitted greenhouse emissions, the government similarly might argue for greater allocations of total material requirement or global environmental burden. Unfortunately, serious attention to achieving the required transition is conveniently misplaced and neglected under this familiar scenario.

The government might similarly argue that Western Australia has a limited social and ecological responsibility, and that consumptive behaviours beyond our border are not within our immediate responsibility or sphere of influence. This view is consistent with industry responses to sustainable consumption and production, claiming that there is a market for virgin resources and if we do not fill that market, someone else will – probably in a developing nation and with less regard for direct social and environmental impacts of their activity. This view characterizes global industry as locked into a lose–lose race towards escalating and unsustainable resource throughput. A more sophisticated view, and one that is gaining ground under the name of extended producer responsibility, is that producers have as much of a role to play in driving demand as consumers, and should maintain responsibility for the impacts of the products they sell long after the transaction has been finalized.

Another prevalent view in Western Australia is that because mining has a relatively small land area impact relative to the total wealth generated by its activities, it is environmentally benign. This view is embedded in the Western Australian development psyche, which holds that the immense Australian outback represents a limitless frontier for human exploitation, and that the area is so large that the direct environmental impacts of extractive activities are insignificant. The ultimate cumulative impact of our exported natural resources as they move through the human economy is routinely neglected by this frontier ethos. In reality though the cumulative impacts of the various material transformations behind the profusion of products and services underpinning western consumption patterns represent a very significant burden on the global ecosystem.

The situations outlined above demonstrate that the sustainable production and consumption agenda in Western Australia are in fact acting against a tide of powerful countervailing trends. On one hand this provides an increased need and mandate for a far-reaching and ambitious sustainable production and consumption policy in Western Australia. On the other hand it makes the realization and implementation of that policy agenda fraught with political and practical difficulties.

The evidence presented above illustrates the extent to which Western Australia's present sustainability agenda is eclipsed and in some cases co-opted by the policy goals of maintaining a growing economy based on increased exploitation and international export of virgin resources. The few attempts that have so far been made at reducing the environmental impacts of these activities have been dominated by incremental and ecomodernist approaches towards ameliorating the environmental impacts of resource use, rather than seeking genuine dematerialization or improved total factor efficiency of the state economy.

By contrast, the programme outlined in this case study represents an ambitious policy agenda ultimately aimed at far-reaching and transformative change. Intuitively, the idea of a transformative sustainable production and consumption agenda being developed and progressed within a culture of entrenched support for unsustainable economic growth and increased material throughput seems to be an incongruous combination. How can this combination be possible?

So far the policy programmes outlined in this case study remain in development, and as such, unimplemented and unrealized. It may be that the most effective, or indeed only, way of achieving a transformative sustainability programme from within the policy environment described above lies in choosing those few policy avenues available that are acceptable to the politics of growth and the present ecomodernization agenda, but do not act to further lock in or embed unsustainable development patterns and therefore foreclose future sustainability options. Policies that meet these criteria may be few and far between, but must be the subject of increased emphasis by policy-makers. Examples may be investment in integrated urban planning and public transport infrastructure rather than subsidizing more efficient private motor vehicles, or establishing extended producer responsibility schemes for packaging products rather than passing the costs of kerbside recycling on to the householder. In this way, foundations can be laid for a time in the future when there may be stronger political support for transformative sustainable production and consumption policies.

Sustainability targets, growth and welfare and framework of indicators
It is important that any sustainable production and consumption policy programme has clearly defined targets, and a framework of indicators for measuring progress towards these targets. But how do the contestable and sometimes contradictory concepts of sustainability translate into concrete and measurable policy targets?

The two related targets provided in the Western Australian State Sustainability Strategy (increased resource productivity by a factor of 4 and halving the ecological footprint) arise from the combination of two established sustainability principles: (1) development not exceeding the carrying capacity of the Earth, and (2) intragenerational and intergenerational equity. These principles are combined to provide a measure of the fair share of the Earth's 'ecological space' that may be taken up by the population of Western Australia.

Measurement of environmental carrying capacity and the drivers of environmental impact is not an exact science and in many areas there is a complete lack of data. For this reason, the targets offered in the Sustainability Strategy provide an initial measure of the magnitude of change required, and not necessarily the exact targets to aim for. As the policy framework for sustainable production and consumption is developed, implemented and monitored, it will be necessary to further refine these targets to ensure they are consistent with internationally agreed principles of sustainability and new developments in scientific knowledge.

The Western Australian Sustainability Strategy adopts the Factor 4 target as a stretch goal that is easily understood by the industries and institutions that are the subject of sustainable production and consumption policy. It is acknowledged that Factor 4 improvements in resource productivity can actually be achieved by increasing the value derived from each unit of environmental impact, while the total amount of environmental pressure is held constant or is even allowed to increase. This is demonstrated when we consider the total factor efficiency ratio derived from the familiar I = PAT equation (Ehrlich and Ehrlich, 1990):

$$\text{Impact/Value} = \text{Total Factor Efficiency}$$

Or in its more commonly used form:

$$\text{Impact/GDP} = \text{Total Factor Efficiency}$$

To overcome this issue with the use of improved total factor efficiency as a policy target, the Sustainability Strategy combines this measure with the goal of halving the ecological footprint, which by definition requires a net decrease in environmental pressure, irrespective of the ratio of value derived by the activity causing that pressure.

The ecological footprint target, however, is largely intangible to the industries and institutions that are the subject of sustainable production and consumption policy. It is therefore becoming increasingly necessary to improve and refine the Factor X target (e.g. Factor 10) as a goal on which policy institutions and target industries can focus their efforts. We have already outlined one important issue with focusing on the level of impact per unit of value – that is, increasing production efficiency has a rebound effect of increasing consumption and therefore total resource throughput. A related issue is the situation described above where it is possible to make total factor efficiency gains without reducing levels of environmental impact simply by driving up the value ratio. As such, the Factor X target bears no direct relationship to the main sustainability principle informing it – the need for human activity to remain within the carrying capacity of the Earth.

Together these issues tend to combine and inform a technocentric and to some extent oxymoronic vision for resource productivity that does not consider the role of the market in determining the extent of environmental pressure associated with the provision of human welfare. The incrementalist approaches toward sustainability that are informed by this vision are at least partially consistent with present economic development trajectories and those supported by mainstream business organizations.

A related and equally fundamental issue with the use of the Factor X target is that economic value (measured at the state level as GSP) is becoming increasingly irrelevant in defining genuine progress or even the provision of human welfare. This has prompted the development of alternative indicators, including the Genuine Progress Indicator or GPI (Hamilton and Saddler, 1997; Lawn, 2003). The Genuine Progress Indicator calculated for Western Australia (see Figure 9.1) demonstrates the increasing disparity between the level of economic wealth in Western Australia and true well-being of its citizens. When we consider the widening gap between economic prosperity and Genuine Progress or well-being, it becomes evident that there must exist an amount of environmental impact associated with increasing GSP that is superfluous in the sense that it does not act to improve welfare or standards of living. Therefore, the use of GSP as a measure of value in the total factor efficiency equation may be providing the wrong policy signals. Driving up value as measured by GSP might improve total factor *economic* efficiency, but not necessarily with any real or measurable gain in welfare or benefit for society.

To resolve the above issues partially, we propose that a more useful policy target for sustainable production and consumption policy may be a modified version of Factor X, using GPI instead of GDP/GSP as the value measure in the total factor efficiency equation:

Environmental Impact/GPI = Total Factor Welfare Efficiency

Using the equation above, we have calculated that Western Australia's Total Factor Welfare Efficiency must improve by at least a factor of 17 in order to achieve a sustainable and equitable level of resource use by 2050 (Verstegen, 2005).

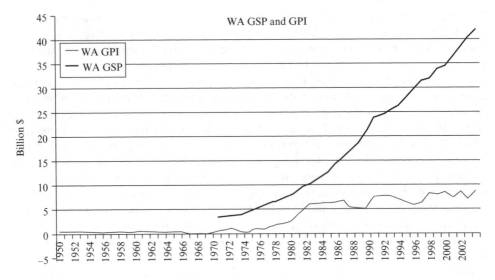

Source: Centre for Water Research (2004).

Figure 9.1   Western Australian Gross State Product and the Genuine Progress Indicator

It is hoped that the Total Factor *Welfare* Efficiency challenge will more clearly define the courses of action required in order to reduce substantially the environmental intensity associated with the provision of welfare for our growing population. A healthy economy will be required to keep welfare levels stable in a period of increasing population, but economic growth per se will not necessarily help reach the Factor X target. As a result, the new target is a closer approximation of reality, and rebound effects are visible as they increase the level of impact per unit of welfare delivered.

If the business community finds the measure of welfare or genuine progress less tangible, it will focus on the critical reductions in environmental impact required to achieve the new sustainable production and consumption challenge. Economic contributions are recognized by the environmental impact/GPI ratio, however, the business community will no longer be able to claim that they are contributing to sustainability simply by providing economic wealth. In this way, a Factor 20 target for Total Factor Welfare Efficiency provides an improved means of translating the integration and systems thinking required for sustainability into a measurable indicator.

Chapter 4 of Agenda 21 noted that consideration should be given to new concepts of wealth and prosperity that are more in harmony with the Earth's carrying capacity (UNGA, 1992). The Total Factor Welfare Efficiency challenge provides a means of operationalizing this notion. The redefined policy goal is therefore an example of a sustainability approach that can be pursued within the existing programme of economic growth and ecological modernization, but does not lock in this present agenda and thereby act to foreclose a more radical sustainability programme in the future. This is made possible by retaining the possibility of abandoning economic growth as the primary means of providing and measuring welfare.

Conclusion

In Western Australia, just like in many communities, there is a public discourse concerning the future of the global environment and how communities might respond to the dilemmas that this discussion discovers as our learning and understanding unfold. For those that are committed to a systematic and effective betterment of human society and its relationship with the global environment, our case study presents some lessons and intractable problems worth solving.

The Western Australian situation also draws attention to difficulties in the emerging approaches to transformational sustainability strategies, including those related to sustainable production and sustainable consumption, which must be resolved if we are to present a credible alternative to the failing and dominant neo-classical and liberal approaches of our time.

References

Adriaanse, A., Bringezu, S., Hammond, A., Moriguchi, Y., Rodenburg, E., Rogich, D. and Schütz, H. (1997), *Resource Flows: The Material Base of Industrial Economies*, Washington, DC: World Resources Institute.

Anarow, B., Greener, C., Gupta, V., Kinsley, M., Henderson, J., Page, C. and Parrot, K. (2003), *Whole-systems Framework for Sustainable Consumption and Production*, Environmental Project 807, Copenhagen: Danish Ministry of the Environment.

Arrow, K., Bolin, B., Costanza, R., Dasgupta, P., Folke, C., Holling, C.S., Jansson, B.-O., Levin, S., Mäler, K.G., Perrings, C. and Pimental, D. (1995), 'Economic Growth, Carrying Capacity, and the Environment', *Science*, **268**, 520–21, reprinted in *Ecological Economics*, **15** (1995), 91–5.

Australian Conservation Foundation (ACF) (2002), *Natural Advantage: Blueprint for a Sustainable Australia*, Melbourne: ACF.

Ayres, R., Ayres, L. and Warr, B. (2005), 'Is the U.S. Economy Dematerialising? Main Indicators and Drivers', in Van den Bergh, J. and Janssen, M. (eds), *Economics of Industrial Ecology – Materials, Structural Change and Spatial Scales*, Cambridge, MA: MIT Press, pp. 57–94.

Bartelmus, P. (1999), 'Economic Growth and Patterns of Sustainability', Wuppertal Paper Number 98, Wuppertal, Germany.

Bringezu, S. (ed.) (1997), *Analysis for Action: Support for Policy towards Sustainability by Material Flow Analysis*, Proceedings of the ConAccount Conference, November, Wuppertal, Germany.

Bringezu, S. (2002), 'Towards Sustainable Resource Management in the European Union', Wuppertal Paper Number 121, Wuppertal, Germany.

Bringezu, S. and Schutz, H. (2001), *Total Material Requirement of the European Union*, Technical Report 55, Copenhagen: European Environment Agency.

Centre for Water Research (2004), *The Index of Sustainable Functionality*, Perth: Centre for Water Research, University of Western Australia.

Davison, A. (2004), *Technology and the Contested Meanings of Sustainability*, New York: State University of New York Press.

Ehrlich, P.R. and Ehrlich, A.H. (1990), *The Population Explosion*, New York: Simon & Schuster.

Gardner, G. and Sampat, P. (1998), 'Mind over Matter: Recasting the Role of Materials in Our Lives', *Worldwatch Paper 144*, Washington, DC: Worldwatch Institute.

Government of Western Australia (2003), *Hope for the Future: The Western Australian State Sustainability Strategy*, Perth: Department of the Premier and Cabinet, www.sustainability.dpc.wa.gov.au, accessed 3 December 2005.

Government of Western Australia (2004), *Building Future Prosperity: Creating Jobs and Wealth Through Industry Development*, Industry Policy Statement, Perth: Department of Industry and Resources.

Hamilton, C. and Saddler, H. (1997), *The Genuine Progress Indicator: A New Index of Changes in Well-being in Australia*, November, Canberra: The Australia Institute.

Hamilton, C., Denniss, R. and Baker, D. (2005), 'Wasteful Consumption in Australia', Discussion paper 77, Canberra: The Australia Institute.

Hargroves, K. and Smith, M. (eds) (2005), *The Natural Advantage of Nations*, London: Earthscan.

Hatfield-Dodds, S. (2004), 'Economic Growth, Employment and Environmental Pressure: Insights from Australian experience 1951–2001', Working Paper to the Australian National University Environmental Economics Network, 18 November 2004.

Hawken, P., Lovins, A. and Lovins, L.H. (1999), *Natural Capitalism: Creating the Next Industrial Revolution*, London: Earthscan.

Hinterberger, F., Bosch, G. and Giljum, S. (2003), *Final Report of the Adaptive Integration of Research and Policy for Sustainable Development*, Vienna: Sustainable Europe Research Institute.

Houghton, J.T., Filho, L.G.M., Callander, B.A., Harris, N., Kattenberg, A. and Maskell, K. (eds) (1996), *Climate Change 1995*, New York: Cambridge University Press.

Inter-governmental Panel on Climate Change (IPCC) (1990), *Climate Change: the Intergovernmental Panel on Climate Change Scientific Assessment*, Geneva: World Meteorological Organization and United Nations Environment Programme.

Jacob, K., Binder, M. and Wieczorek, A. (eds) (2004), *Governance for Industrial Transformation*, Proceedings of the 2003 Berlin Conference on the Human Dimensions of Global Environmental Change, Berlin.

Jalas, M. and Plepys, A. (2001), 'Summary of Workshop 10: Sustainable Consumption and Rebound Effects', in *Proceedings of the 7th Conference of the European Roundtable on Cleaner Production*, Lund: International Institute for Industrial Environmental Economics.

Jänicke, M. (1997), 'The Role of MFA and Resource Management in National Environmental Policies', in Bringezu, S. (ed.), *Analysis for Action: Support for Policy Towards Sustainability by Material Flow Analysis*, Proceedings of the ConAccount Conference, Wuppertal, Germany.

Jansen, L., Bosch, G. and Weaver, P.M. (2003), 'Research and Technology Development Programmes: From the Very Start to the Very Finish', in Hintebuger et al., *Final Report of the Adaptive Integration of Research and Policy for Sustainable Development Project*, Vienna: Sustainable Europe Research Institute.

Jevons, W.S. (1865), *The Coal Question. An Inquiry Concerning the Progress of the Nation, and the Probable Exhaustion of Our Coal Mines*, London: Macmillan and Co.

Jevons, W.S. and Foxwell, H.S. (1884), *Investigations in Currency and Finance*, London: Macmillan and Co.

Lawn, P.A. (2003), 'A Theoretical Foundation to Support the Index of Sustainable Economic Welfare (ISEW), Genuine Progress Indicator (GPI), and Other Related Indexes', *Ecological Economics*, **44**, 105–18.

MacKenzie, J.J. (1997), *Oil as a Finite Resource: When is Global Production Likely to Peak?*, Washington, DC: World Resources Institute.

Matthews, E. and Hammond, A. (1999), *Critical Consumption Trends and Implications: Degrading Earth's Ecosystems*, Washington, DC: World Resources Institute.

Matthews, E., Bringezu, S., Fischer-Kowalski, M., Huetller, W., Kleijn, R., Moriguchi, Y., Ottke, C., Rodenburg, E., Rogich, D., Schandl, H., Schuetz, H., van der Voet, E. and Weisz, H. (2000), *The Weight of Nations. Material Outflows From Industrial Economies*, Washington, DC: World Resources Institute.

Meadows, D.H., Meadows, D.L. and Randers, J. (1992), *Beyond the Limits: Confronting Global Collapse, Envisioning a Sustainable Future*, Post Mills, VT: Chelsea Green Publishing Co.

Michaelis, L. and Lorek, S. (2004), *Consumption and the Environment In Europe: Trends and Futures*, Environmental Project No. 904, Danish Ministry of the Environment, Copenhagen, Denmark.

Millennium Assessment Board (2005), *Living Beyond Our Means: Natural Assets and Human Well-Being*, The Board of the Millennium Ecosystem Assessment, www.millenniumassessment.org, accessed 4 December 2005.

Nordic Council of Ministers (1999) *Factors 4 and 10 in the Nordic Countries*, TemaNord, 528.

Nriagu, J.O. (1990), 'Global Metal Pollution: Poisoning the Environment?', *Environment*, **32**, 7–11.

Organisation for Economic Co-operation and Development (OECD) (1996), *Meeting of OECD Environmental Policy Committee at Ministerial Level, Paris 19–20 February*, Paris: OECD Communications Division.

Organisation for Economic Co-operation and Development (OECD) (1997), *Sustainable Consumption and Production*, Paris: OECD.

Postel, S., Daily, G.C. and Erlich, P.R. (1996), 'Human Appropriation of Renewable Fresh Water', *Science*, 271, 785–8.

Quist, J. and Vergragt, P. (2003), 'Backcasting for Industrial Transformations and System Innovations toward Sustainability: Is it Useful for Governance', in *Proceedings of the 2003 Berlin Conference on the Human Dimensions of Global Environmental Change, Berlin*.

Rees, W. and Wackernagel, M. (1994), 'Ecological Footprints and Appropriated Carrying Capacity: Measuring the Natural Capital Requirements of the Human Economy', in Jansson, A.-M., Hammer, M., Folke, C. and Costanza, R. (eds), *Investing in Natural Capital: The Ecological Economics Approach to Sustainability*, Washington, DC: Island Press, pp. 362–90.

Ryan, C. (2000), 'No Sustainable Production without Sustainable Consumption: Thinking about Sustainable Systems', Theme Paper – Sustainable Consumption, UNEP CP6, Montreal, October.

Ryan, C. (2002), *Global Assessment Report: Sustainable Consumption*, Paris: UNEP UNDTIE.

Schutz, H. and Welfens, M. (2000), 'Sustainable Development by Dematerialisation in Production and Consumption: Strategy for the New Environmental policy in Poland', Wuppertal Paper 103, Wuppertal, Germany.

Spangenberg, J.H. and Lorek, S. (2002), 'Environmentally Sustainable Household Consumption: From Aggregate Environmental Pressures to Priority Fields of Action', *Ecological Economics*, **43**, 127–40.

Turner, G. and Poldy, F. (2001), 'Let's Get Physical: Creating a Stocks and Flows View of the Australian Economy', *Proceedings of the International Congress on Modelling and Simulation MODSIM 2001*, Canberra: Australian National University.

United Nations Department of Economic and Social Affairs (UNDESA) (2003), *International Expert Meeting on the 10-Year Framework of Programmes for Sustainable Consumption and Production*, http://www.unep.fr/pc/sustain/reports/events/MarrakechReport.pdf, accessed 4 December 2005.

United Nations Development Programme (UNDP) (1998), *Human Development Report*, New York: Oxford University Press.

United Nations Environment Programme (UNEP) (2000), *Global Environmental Outlook*, London: Earthscan.

United Nations Environment Programme (UNEP) (2001), *Consumption Opportunities – Strategies for Change*, Geneva: UNEP.

United Nations General Assembly (UNGA) (1992), *Report of the United Nations Conference on Environment and Development*, Rio de Janeiro, 3–14 June 1992, Paris: UNCED.

United Nations General Assembly (UNGA) (1997), *Special Session for the Purpose of an Overall Review and Appraisal of the Implementation of Agenda 21*, Fifty-first session.

Verstegen, P. (2005), 'Industrial Transformation – A Theoretical Overview and Policy Implications For Western Australia', Honours thesis, Perth: Murdoch University.

Vitousek, P.M. (1994), 'Beyond Global Warming: Ecology and Global Change', *Ecology*, **75**, 1861–76.

von Weizsacker, E., Lovins, A. and Lovins, L.H. (1997), *Factor Four: Doubling Wealth, Halving Resource Use*, London: Earthscan.

Weaver, P., Jansen, L., van Grootveld, G., van Spiegel, E. and Vergragt, P. (2000), *Sustainable Technology Development*, Sheffield, UK: Greenleaf Publishing.

Weterings, R. and Opschoor, J. (1992), *The Eco-capacity as a Challenge to Technology Development*, Advisory Council on Nature and the Environment (RIVM), Report No 74 A, Rijswijk, Netherlands.

Whitelegg, K. and Weber, K.M. (2001), 'National Research Activities and Sustainable Development. An ESTO Survey and Assessment of National Research Initiatives in Support of Sustainable Development', Workshop: *Setting Concepts into Motion – Sustainable Development and R&D Policies*, 28–29 November, Brussels.

Willums, J.O. (ed.) (1998) *The Sustainable Business Challenge. The Foundation for Business and Sustainable Development*, World Business Council for Sustainable Development & United Nations Environmental Programme, Sheffield: Greenleaf.

Worldwide Fund for Nature (WWF) (2004), *Living Planet Report 2004*, Gland, Switzerland: WWF.

10 Managing research for sustainable development: different countries, different contexts
Tim Turpin

Introduction

At least 500 billion US dollars are spent globally each year on research and development. Around 85 per cent of this research is devised, funded and carried out in the countries of the 'North' (UNESCO, 2000). Scientific, economic or political forces within these industrially developed economies generally drive research funding decisions, largely directing them toward the creation and development of new technologies. An underlying assumption behind these funding strategies is that advances in science will increase international competitiveness.[1]

Research funding policies are usually tied to a defined set of research priorities for national strategies for economic development. These research priorities set the goals for ex ante research evaluation and are formative in steering the ways researchers design and justify their research proposals (Cozzens and Turpin, 2000). This chapter is concerned with the question: to what extent, and how, can public research investments contribute globally to sustainable development? While much is learned of environmental consequences from research and new technologies (either positive or negative) it is usually well after the original research assessment and investment (ex post evaluation). Is it possible to assess the *potential* impact of research investments (ex ante) for their contribution to strategies for sustainable development?

Research evaluation issues

The term sustainable development has generally been used to convey a strategy for achieving a just, comfortable and secure future. It refers to the 'complex of activities that can be expected to improve the human condition in such a manner that the improvement can be maintained' (Munro, 1995, p. 53). This implies an interlinking of ecological, social and economic factors through time. Given the level of funding for research and development globally, it is somewhat surprising that national research evaluation assessments are not presented in terms of their overall contribution to sustainable development – the interlinking of these factors. Rather, they are presented in terms of contributions to specific socio-economic objectives, such as waste management, reduction in greenhouse gas emissions, alternative energy sources or environmental management generally.

Research investments and the technology trajectories set in play through national priority setting processes in the North can carry quite different implications for countries of the South. For these latter countries research is an expensive and consequently limited business. The research that is conducted in developing countries is steered by international science and technology trends and trajectories, rather than by local policies concerned with the development of an endogenous science and technology base. There are a number of obvious reasons for this. First, international research output such as publications and patents are dominated by the research agendas of the North. Over 75 per cent of international journal articles recorded in the ISI database globally are written by authors from the 'triad' – USA, European Union and Japan. Scientists in developing economies are consequently steered toward contributing to these debates in order to justify their international scientific credentials. Second, research funding in developing economies is largely dependent on international sources. For example, recent evidence from countries such as Mozambique and Mongolia show that more than 50 per cent of their national research effort is funded externally to national priority setting goals (see Turpin and Bulgaa, 2004; Turpin and Martinez, 2003). While most countries identify sustainable development as an overarching policy objective, few have managed to incorporate the potential impact of research investments on achievements toward this objective (Harger et al., 1996). How might this be achieved? What tools can be used to monitor and account for research funding contributions to sustainable development? This is the first group of questions addressed in this chapter.

Science and technology policy issues

A second set of questions concerns how science and technology policies in developing countries can draw appropriately from research investments in the North in order to promote local strategies for sustainable development. In short, how can the research outcomes of the North, with their very different industrial capacities and strategic priorities, serve to promote policies for sustainable development in the South? The idea of 'appropriate technologies' for development has been around since at least the 1960s (Clarke, 1985, pp. 56–7). For developing countries, this concept led to government efforts to align the acquisition, adaptation or development of technologies according to local economic priorities. More recently there has been a growing recognition that the emphasis on appropriate technologies has led to the emergence of two separate and somewhat disconnected streams of technology development: one focused on global scientific advances, and one focused on the adaptation and application of technologies for local development (see Jain, 2002; Krishna, 2001).

There is now a growing movement in some developing countries to reframe science and technology policies in order to reconnect these two streams of technology development. A key feature of this 'reframing' is to sharpen the policy focus so that it is less concerned with building solely on international and institutional links for science, and more concerned with the integration of 'knowledge systems'. At the national level this radically shifts the focus of technology management away from technologies themselves and toward the systems through which they are devised, diffused, adapted and used. Just as the idea of sustainable development seeks to interlink the social, ecological and economic world, science and technology (S&T) policies in some developing countries are now seeking to interlink scientific outcomes with economic and cultural objectives within local sub-sets of their economies.

From this perspective the management of research funding for sustainable development takes into account two streams of action. The first stream is driven primarily by research investments for the production of new knowledge. This stream is predominantly devised, funded and carried out in the industrially developed economies of the North. Research outcomes generally benefit those with the financial and industrial capacity to elaborately transform raw materials through manufacturing or to produce knowledge-intensive services for global markets. The second stream is driven by a local capacity to capture, absorb, adapt and build on technology for local development. This stream, in contrast to the first, concerns the life-world of small production units often in rural villages. Research outcomes in this stream are directed toward the development or improvement of local technologies in the South. Beneficiaries are local producers and the users of services such as energy, transport, health and education.

The central argument presented in this chapter is that it is the integration between these two streams of action that provides the challenge for achieving sustainable development in the South. Two key policy objectives follow. The first is to establish a system of indicators for monitoring the extent to which public research investments carry the potential to bridge the two streams. The second is to develop policies conducive to capturing the benefit of research outcomes driven from the North for appropriate application in the South.

Outline of the chapter
The next section in this chapter provides a discussion of the concept of sustainable development and introduces the problem of accounting for public research and development (R&D) in strategies for achieving sustainable development. The concept of sustainable development is a comparatively recent addition to policy imperatives in most areas of government activity. It is even more recent in national research policies. Although the objective of sustainable development is clearly interconnected with the creation and application of knowledge through national research efforts, measures of sustainable development have not been explicitly or systematically incorporated into research evaluation. An Australian case study is presented to provide an example of how such research accounting might be achieved.

The following section deals with S&T policy issues for sustainable development in developing countries. It is argued here that new policy approaches are emerging and that these are quite different from those that prevailed among developing countries through the 1980s and 1990s. Rather than focusing on building S&T infrastructure or links between institutions, there is a growing emphasis on building 'innovation networks'. The policy shift reflects efforts to avoid some of the negative consequences of globalization and to reconnect S&T developments in the North with the development imperatives of the South. This reconnection concerns the interlinking of ecological, economic and social factors referred to above, but in a local rather than a global context. Examples drawn from Mozambique, Kenya, Thailand, China, Mongolia and India are presented to illustrate this trend.

Research indicators for ecologically sustainable development
General and specific ESD issues: the Australian case
In 1992 following the Rio 'Down to Earth Summit', the Australian government endorsed a National Strategy for Ecologically Sustainable Development (NSESD). The strategy developed out of an increasing government concern in the 1980s about the state of the national and global environment and the connections between this and the path of economic growth and international trade. The Australian NSESD reinforced earlier observations that the world's current pattern of economic growth was not sustainable on ecological grounds (Commonwealth of Australia, 1992, p. 3). Subsequent to Australia's endorsement of 'Agenda 21' (which flowed from the Rio Summit) and the launch of the NSESD, a national commitment was also made to account for the extent to which public research investments contributed to the strategy for achieving ecologically sustainable

development (ESD). As a result some effort was directed toward developing national indicators of research for ESD.[2]

Events and ESD debates have since moved on and now focus more on separate pressing environmental issues such as controlling greenhouse gas emissions and soil salinity rather than the broader ESD goal. The Australian attempt to embed the ESD objective into an overarching research strategy has now drawn back to an emphasis on the separate priorities. While there is clearly a need to monitor and evaluate research outcomes in terms of their specific objectives, there remains a strong case to continue to monitor research more generally, and on the basis of the objective of interlinking ecological, social and economic advances in science and technology.

This section outlines the Australian effort to develop indicators to monitor research for ESD. It concludes with the observation that the basic concepts used in the Australian case could be adapted and applied in developing countries to monitor the role of research for national strategies for socio-economic development.

Most countries have systematically collected R&D statistics and used them as tools for monitoring and steering national R&D investments for some time now. In general attempts to collect this information have been based on the concepts, methodologies and standards laid down in OECD's *Frascati Manual* (OECD, 1991) and UNESCO's *Manual for Statistics on Science and Technology Activities* (UNESCO, 1984). The Australian research classification system provides three dimensions: the type of research to be carried out – basic, applied or experimental development; the research discipline in which the work is to be primarily carried out – research fields, courses and disciplines (RFCDs); and the objective to which the research is directed – socio-economic objective (SEO).

The problem with current indicators of 'research objective'

Although these manuals serve as templates for the development of national indicator systems, indicators for assessing the extent to which the research effort is directed toward *sustainable* socio-economic development are absent. The SEO classification describes the socio-economic objective toward which the research was intentionally directed and is the only indicator that goes close to monitoring the technological direction embedded in research.

Given the complex ways in which the social, physical and life sciences interact and contribute to (or detract from) sustainable development, research can potentially contribute to more than one socio-economic objective at the same time. However, the logic of the OECD and UNESCO classification systems tend to steer research reporting to one or other of the various SEO categories. Research therefore tends to be reported as *either* contributing to the development of knowledge about the natural *or* the social environment (Deville and Turpin, 1996). In the context of sustainable development, research would be better accounted for if it could be assessed according to its capacity to contribute to 'complementary' rather than just 'substitution' technologies and techniques.

Clearly some research that is carried out contributes toward the development of knowledge, processes and products that have the potential to harmonize the natural and social worlds. The problem is that the S&T classification systems (such as those developed by the OECD) used to monitor investments in R&D are rooted in a classification system that separates specific socio-economic objectives without regard to their collective contribution to aggregate objectives such as sustainable development or the reduction of poverty.

On the basis of this dominant research classification system, research projects or aggregates are measured according to research disciplines and their achieved contribution to one, or at best two, quite specific socio-economic objectives. The OECD, UNESCO and Australian classification systems include SEO categories relating to environmental protection, environmental impacts of economic development and the general classification 'environmental knowledge'. National data collected for R&D management and evaluation using these systems can be seen to classify research as *either* contributing toward natural *or* manufactured 'capital' by the logic of the SEO system. This classificatory schema eliminates the possibility of identifying research that is contributing toward developing technologies or techniques that are potentially complementary to, or which integrate, both natural and manufactured capital (Victor, 1991).

Toward a definition of ESD-related research
The Australian exercise of capturing the relatedness of research to ESD required a working definition of ESD-relatedness. The starting point was the National Strategy definition of ESD as the basis for a definition of research relevant to ESD. The NSESD adopted the 1990 Commonwealth Government definition for ESD in Australia as: 'using, conserving and enhancing the community's resources so that ecological processes, on which life depends, are maintained, and the total quality of life, now and in the future, can be increased' (Commonwealth of Australia, 1992, p. 9).

The NSESD described ESD as a new type of development that is sustainable on ecological grounds. There were two major tiers to the NSESD: a goal and three core objectives.[3] The goal provided the major focus of the National Strategy as a whole, while the core objectives suggested the means by which the goal could be achieved on three inter-related fronts. The goals and core objectives are summarized in Box 10.1.

BOX 10.1   GOAL AND CORE OBJECTIVES OF THE
AUSTRALIAN NATIONAL STRATEGY FOR
ECOLOGICALLY SUSTAINABLE DEVELOPMENT

Goal
Development that improves the total quality of life, both now and in the future, in a way that maintains the ecological processes on which life depends.

Core objectives

1. to enhance individual and community well-being and welfare by following a path of economic development that safeguards the welfare of future generations;
2. to provide for equity within and between generations; and
3. to protect biological diversity and maintain essential ecological processes and life-support systems.

Source: NBEET (1995, p. 9).

There are clearly a number of means by which the goal of ESD might be achieved. The National Strategy addressed these as a range of 'challenges' within sectoral issues and inter-sectoral issues, and in terms of the future development of ESD in Australia. As one contribution to the achievement of ESD, the challenge for research and development was described as being 'to provide the knowledge, techniques and technologies needed to achieve the objectives of ESD in all sectors as well as in addressing intersectoral issues' (Commonwealth of Australia, 1992, p. 98). A comparatively simple yet robust definition for categorizing research for ESD was established by focusing on the goal and core object-ives of the National Strategy. Thus research that was relevant to ESD was defined as that which would 'help achieve the Goal of the National Strategy by contributing to the Core Objectives' (NBEET, 1995, p. 9). The principle encapsulated by this definition is that the ESD-relatedness of research can be judged on the basis of whether the research will con-tribute towards the three core objectives. Further, the more of these objectives research contributes towards, the greater is the ESD-relatedness. An important principle in the Australian NSESD is that no single core objective is more important than any other. Although research may be carried out that contributes in varying degrees to these core objectives, it is through the *integration of the objectives* that the goal of sustainability depended. Thus as an informative principle, research that facilitates the integration of these objectives has the greater degree of 'ESD-relatedness'.

Bearing this and the goal and objectives of the Australian NSESD in mind, four categories were used to characterize the degree of relatedness of a specific research project to ESD:

1. *ESD-focused research*. This aims at positively contributing toward the goal described by the NSESD; that is, taking into account the three objectives in its research aims.
2. *Strongly ESD-related research*. This aims to make some positive contribution toward the goal described by the NSESD; that is, taking into account at least two of the objectives in its research aims.
3. *Weakly ESD-related research*. This is likely to make some positive contribution toward the goal described by the NSESD; that is, taking into account only one of the objectives in its research aims.
4. *ESD-neutral research*. This is likely to have neither a positive nor a negative impact on the goal described by the National Strategy for ESD, taking none of the objectives into account in its research aims.

The logic underpinning this approach was that indicators of research for sustainable development should seek to measure the extent to which research is intended to contribute to an *integrated* approach to the life sciences, physical sciences and the social sciences. In essence it was an attempt to reverse the perverse logic inherent in the international systems of indicators that insist research should only contribute to one aspect of the world in which we live (NBEET, 1995). The categories are elaborated further below.

ESD-focused research   The first category in the schema included research that is strongly focused on sustainable development. This category of research included projects that take into account both the physical/biological world and the human/social world in a way that specifically aims to maintain, protect or restore ecological processes. This category, for example, included research into the economic and socio-cultural problems involved in the introduction of a specific ecologically sustainable agricultural practice.

Strongly ESD-related research   Research that is 'strongly related' to sustainable development was defined as projects focused on the physical/biological world with the potential to strongly facilitate action in relation to sustainability; for example, studies of climate change due to global pollution. This category included research focused on the human/ social world and intended to strongly facilitate action in relation to ecological sustainability, such as studies into economic incentives versus regulation for the development of effective industrial waste treatment technologies. Research that takes into account both the physical/biological world and the human/social world in its focus, combined in a way that aims to make some positive contribution to sustainability, could also be included in this category. This would include, for example, a project aimed at developing new engine technologies to reduce greenhouse gas emissions.

Weakly ESD-related research   This category included research focused on the physical/ biological world in a way that would have some positive effect on ecological sustainability. This would include, for example, baseline ecological studies likely at some later stage to contribute to knowledge for environmental management of a particular protected area. This category also included research focused on the human/social world in a way that would have some effect on moving towards ecological sustainability. An example of such research would be a project for the development of new manufacturing techniques with a range of potential applications that *incidentally* have the property of being relatively energy-efficient, or research into attitudes of commuters in cities to determine improvements needed to encourage less use of private cars.

ESD-neutral research   The fourth category in the schema included research that takes into account both the physical/biological world and the human/social world in its focus but with no apparent contribution to ecological sustainability. For example, the development of new materials that may or may not have relative beneficial ecological impact, or the study of the formation of valuable mineral ore deposits, would be included in this group. This 'neutral' category would also include research focused on the physical/biological world with no apparent impact on ecological sustainability. For example, this might include research into the developmental biology of lungfish. It could also include research focused on the human/social world with no apparent impact on ecological sustainability. For example, research into support strategies for sufferers of Alzheimer's disease.

Application of the schema
The application of this method to Australian publicly funded research allowed for an assessment of ESD-related research across core government funding programmes. It also allowed for comparisons between years and across different national socio-economic research objectives and fields of research. Table 10.1 illustrates the outcomes of the analysis for the Australian Research Council's large grants programme for two selected years. Similar output tables were prepared for other programmes and for fields of research and national socio-economic objectives.

Table 10.1   Levels of ESD-relatedness in Australian Research Council (ARC) large grant research investments

ARC large grants for 1993 and 1994	% ESD levels				
	Strongly ESD-focused	Strongly ESD-related	Weakly ESD-related	ESD-neutral	All ESD levels
1993	0.8	12.7	9.2	77.3	100
1994	0.8	19.0	10.7	69.5	100
Total 1993/94	0.8	16.0	10.0	73.2	100

Source: NBEET (1995).

National strategies for economic development in most developing economies are increasingly incorporating policies to develop and manage a strong S&T base. These national strategies are also seeking to achieve development that is sustainable. However, an inherent problem is that the research information systems that monitor ex ante research evaluation remain disconnected from national strategies for sustainable development, not only in the South but also in the industrialized countries that carry out most of the world's research. The Australian case study illustrates one attempt to apply a classificatory schema drawing on, and complementary to, existing international S&T indicators.

Given the importance of research investment for social, economic and technological development it is sensible to create new and innovative ways to link investments and scarce resources to strategies for sustainable development. Developing countries need to make decisions not just about their own investments in science and technology but also about those made elsewhere and imposed through international development programmes. Efforts to capture the impact of global investments in research will depend on a common classification system. Establishing a science and technology classification system that incorporates measures of sustainability can provide one step toward that goal.

The Australian case study described above represented a 'one-off' exercise by the Australian Government to account for research investments in terms of contributions to sustainable development. Although the exercise has not been repeated, it does show that it is possible to take into account potential research investments to sustainable development during ex ante evaluation.

Research policies in developing economies
This section focuses on emerging policy options open to governments to build links between the ecological, social and economic aspects of sustainable development on the basis of outcomes from research. While the previous section emphasized the inter-relationship between research foci from the point of view of those carrying out the research, this section emphasizes this inter-relationship from the perspective of the users of research.

Science and technology policies for economic development
Through the 1970s and 1980s, science and technology policy was adopted as a key tool for achieving economic independence and growth among developing economies. In

Southeast Asia in particular governments relied on science and technology policies as mechanisms for raising levels of innovation to underpin economic growth. Although the rates of growth and public investments in science and technological investments have varied, almost all countries in the region by the late 1980s had clearly articulated science and technology policies directed toward encouraging the development of national technological capability in both private and public sectors. Such policies tended to draw on the experiences of countries that already had strong economic and industrial leverage in the international domain.

Science and technology policy approaches through this period ranged from comprehensive government-planned science to more laissez-faire science approaches, or to the industry-demand approaches characterized by Japan and the Republic of Korea (Hill et al., 1996, p. 169). Such strategies were strongly supported by agencies such as the World Bank and the Asia Development Bank. Indeed these agencies in many cases underwrote the cost of developing the plans.

Two common features, however, can be observed across most national plans. First, in almost all cases there was a strong focus on building an institutional base for steering and carrying out research. Second, the models and ideas that were adopted for developing the institutions were largely borrowed from the successful (and sometimes unsuccessful) models of Europe and North America. In practice, the primary concern of these policies was managing these institutions.

Similar developments occurred in Africa. During the 1970s many African states established national science institutions. Ministries for Science and Technology were set up in Nigeria and Senegal. Ethiopia and Tanzania set up Commissions for Science and Technology, and Ghana, Mali and Sudan established National Research Councils. As Adeboye (1998, p. 168) has pointed out, many of these institutions were created in order to respond to quite specific needs, such as control of crop pests. Once established, however, they provided a central focus for broader scientific development.

In Asia some national policies coincided with quite spectacular economic and scientific growth, while others were more modest. The Republic of Korea maintained an average growth rate in GDP of over 8 per cent throughout the 1980s and, as an indicator of scientific growth, the number of new Korean patents registered in the United States rose from 26 in 1983 to 1486 in 1996. By contrast in the Philippines GDP from 1980–1990 grew at just over 1 per cent per annum and patents registered in the United States remained on average at only four or five per year, with six registered in 1996. While technological development contributed to economic growth in countries such as Japan, Korea, China and Taiwan, in others, such as Indonesia, Thailand and Vietnam, economic performance remained well below the fully industrialized countries. As countries largely acting as subcontractors for large transnationals, this group remained dependent on sustained growth and competitiveness achieved among the former group of nations.

In Mongolia, under the socialist system and with a small population and limited industrial structure, an impressive system of research institutions and universities was established through the 1960s and 1970s. This continued to expand until the collapse of the Soviet system – a system that prevailed from 1926 to 1990 and was dominated by a 'planned economy', in which the state owned all means of production and provided all forms of social service. The backbone of the country's present science system was laid down through these years. The Institute of Sciences and Higher Education, with antecedents in 1921, became the Mongolian Academy of Sciences (MAS) in 1961. The National University of Mongolia established in 1942 provided the platform for development of other state universities such as the Agricultural University (1958), the Medical University (1961), the University of Humanities (1982) and the Technical University (1990). However, like the smaller Asian economies, these institutions remained dependent on the much larger economies of their neighbours. Although the scientific output from Mongolia was impressive through this period, the country's science system was economically unsustainable. This was largely because it was linked to an elite network of Soviet science rather than being embedded in the cultures and practices of local production.

Commenting on early post-independence science policy in India, Krishna (2001) refers to the prevailing approach as one of 'policy for science' because it was the period dominated by the establishment and consolidation of scientific institutions: 'Implicit in the policy-for-science perspective adopted was the view that, once the infrastructure for R&D is created, personnel trained and a set of institutions and universities established, most problems inherent in science to development could be tackled' (Krishna, 2001, p. 183). This policy approach in India, which has remained strong, has acquired the character of what Jain refers to as an 'elite science and technology innovation network' (Jain, 2002, p. 5). The limited extent to which this approach contributed to sustainable development in rural and suburban India created much debate in India through the 1980s (Krishna, 2001). These debates led to alternative approaches in the region, some of which are discussed below.

Policies for innovation and local institutional links

In recognition of 'institutional failures' in delivering widespread benefit for the general population from research investments, new technologies in developing countries led to a new wave of S&T policies through the early 1990s. This flowed from the observation that scientific engineering and technological capability was locked into public sector institutions rather than distributed as skills and expertise in the productive sectors. In the early 1990s in Indonesia, while public investments in science were increasing, only 33 per cent of general expenditure on research and development (GERD) was made by industry and much of that was carried by foreign internationals. In the Philippines the figure was 23 per cent and in Thailand it was only 12 per cent. Thus the outcomes from public sector research had little impact in the productive sectors, while at the same time the public sector found it increasingly difficult to support national S&T capability. A report by the World Bank in 1996 on the Indonesian situation illustrates a general view that increasingly influenced S&T policies in Asian developing economies through the 1990s:

> The public technology institutions . . . lack systems to market their research to firms or to assess the R&D requirements of Industry. Many entrepreneurs are not even aware of the R&D capacity that exists within the institutes. Even when they are sceptical about its relevance to the technology development and competitiveness issues with which they are grappling. (Mans, 1996, pp. 37–8)

Programmes designed to support institutional linkages and the diffusion of technology thus characterized the S&T policy approach during the 1990s. In Indonesia long-term development plans placed an unprecedented level of importance on programmes designed to build closer links between public and private sector enterprises and to promote innovation through research institutes and Technical Service Units (UPTs). The latter were specifically targeted toward raising technological capability among small and medium-sized enterprises (SMEs), particularly in small-scale industrial estates. Similarly S&T policy in Thailand sought to increase comparatively small private sector investments in R&D and increase private sector technical skills development through a new skills development training scheme, through SME development programmes and through enhancing financial incentives for technology upgrading in the private sector.

In China the '863 High-tech Research and Development Programme', started in 1996, targeted applied research for priority areas. These included biotechnology, space technology, information, laser technology, automation, energy and advanced materials, reflecting a major shift from large-scale mission-focused science investment to areas identified as priorities for commercially oriented industrial production. This paralleled a programme started in 1988, the Torch Programme, to commercialize discoveries from institutes and universities and to create new technology enterprises. The Torch Programme was a key initiative in providing technological links for the establishment of 53 New High Technology Zones (NHTZs) across China (see Turpin and Liu, 2002).

Thus science policies were shifting their initial focus from building capability *within* S&T institutions to an alternative approach concerned with building collaboration *between* institutions and productive sectors. The task articulated in many government S&T policy reports through the 1990s was to *manage the links* between institutions and between different sets of activity. Training, research and consulting, previously seen as separate activities, were becoming more intertwined through organizational alliances and practice. Environmental training of engineers, for example, was seen not simply as building education institutions but also as building networks for environmental education. The Network for Environmental Training at the Tertiary Level in Asia and the Pacific (NETTLAP) was designed to enhance the capacity of tertiary institutions in the region to meet the education and training demands associated with 'efforts to achieve sustainable development in the region' (NETTLAP, 1995, p. 1).

In India S&T policies-for-science were giving way to a science-for-policy phase. Thus policies for self-reliance and import substitution, the 'white revolution' in milk cooperatives, the 'green revolution' on food grain production and the development of agroindustrial sectors were underpinned by the 'science effort' (see Krishna, 2001, pp. 183–4). Creating and managing institutional links constituted a key feature of this phase. The policy emphasis was increasingly directed toward the social as well as the technical nature of innovation and toward building networks of 'actors' and 'agencies' to integrate science and markets (Krishna, 1994).

In Mongolia after 'the return to democracy' in 1990, there were close to 100 research institutes. Through the 1990s the government placed a high priority on reforming the public science and technology productive sectors in order to use the science base as a platform for national development. The envisaged role for S&T in Mongolia was clearly articulated as science-for-development: 'Scientific and technological progress is a source for accelerated development of economy and society, and for enriching intellectual wealth, and is a basis for economic security. Therefore, the State shall pay a special attention for development of science and technology' (Government of Mongolia, 1998, Chapter 1, General Provisions, 1.3–1.4). Yet the Mongolian transformation to a market economy has been uneven and the capacity for value-adding to local production and expertise has been limited. The task of stimulating economic development through S&T and 'full process-ing the natural resources and raw material and producing products that have a competi-tive capability in foreign markets' is concentrating the attention of policy-makers (Government of Mongolia, 1998, Chapter 3, Clause 3.2).

Seen through this more recent policy lens, science institutions appeared as only one of many sources of knowledge within a national system of innovation. Bringing S&T policy into a national system perspective meant that S&T could no longer be managed as a sep-arate component of national development. It needed to be formulated, implemented and evaluated in the context of the contributions and the capabilities of a variety of other agencies and institutions.

Science policy for sustainable local development

Examples of competing knowledge systems and the failure of policy to support local development can be observed in many countries (Hewitt and Albu, 1998). The state of the leather industry in Kenya, for example, illustrates a global process through which Kenya was driven by international pressure away from building local technology capacity. Instead, the policy emphasis, driven by economic analyses and structural adjustment poli-cies of international agencies such as the World Bank and the International Monetary Fund, focused on maximizing returns from the production of raw material for export rather than on promoting a domestic value-adding capability. India by contrast intro-duced a policy to ban the export of raw hides and trade in skins and took steps to shift production from raw to semi-processed and finished leather products. The Indian policy specifically sought to increase knowledge acquisition in manufacturing rather than simply increase capacity for trade (Muchie, 2001).

In Kenya, liberalization policies encouraged short-term profit through trade rather than long-term gains through local capacity-building in manufacturing and technology upgrading. The Kenyan government withdrew from supporting local manufacturing with a shift in policy toward import substitution and export promotion. Thus technology development through scientific research in the North carried little long-term sustainable benefit for the Kenyan leather industry. At the same time economic research in the North, if anything, further undermined local attempts to build local technology capacity in the leather manufacturing sector.

In contrast, India, already concerned with building sustainable innovation networks at local community levels, introduced mechanisms to activate S&T-based NGOs (such as the Khadi and Village Industries Commission and the Council for People's Action and Rural Technology). This was a deliberate effort to make a link between the needs of SMEs in the leather industry and R&D institutions and universities. Local organizations and insti-tutions with their inherent 'tacit' and 'codified' knowledge were drawn into innovation networks (Muchie, 2001).

Mozambique experienced a similar dysfunctional policy impact to that of Kenya in the case of cashew production and processing. Policies for maximizing export return from the raw product, again pressed by structural adjustment policies, have predominated over policies for manufacturing. The underlying assumption was that the technological infra-structure was uncompetitive in global markets. The 'cashew controversy', as it has been called, centred on the capacity of the physical infrastructure (essentially technology) with little reference to that of the local context and capability (UNDP, 1998). The latter was simply assumed to be inadequate. The production of cashews in Mozambique involves many forms of knowledge: knowledge about agriculture, industry, processing, trade and transport. All are deeply rooted in local social processes. The capacity to integrate these different forms of knowledge with new technology was largely ignored. This includes integrating crop-pricing policies, industrial and human resource development and choices of technology as well as reconciling the expectations of international agencies and local producers with domestic priorities for development. It is very much a process of integrating social, technological and economic capabilities.

A recognition of the value of building innovation networks has led in some countries to a different way of thinking about S&T policy, one that emphasizes networks and a reconciliation between technology and local capabilities. Evidence suggests that this change is already beginning to take shape in countries such as India, Thailand and Mozambique. For example, in Thailand, the 'One Village, One Product' programme under the responsibility of the Office for SME Development and the Department of Skills Development seeks to consolidate local village production capabilities with grants for technology development and upgrading. In India the National Research and Development Corporation (NRDC) and the Technology Information, Forecasting and Assessment Council (TIFAC) are similarly seeking to match new technologies to local knowledge and experiences to create 'home-grown technologies' (TIFAC, 2003).

In Mongolia an emerging S&T policy action plan is seeking to stimulate innovation and knowledge intensity in 'clusters' of socio-economic development relevant for Mongolia's development plans, rather than simply for the development of science. In India a whole range of initiatives have emerged to create innovation networks that seek to link 'S&T capacities with the aspirations and competences of communities in rural and semi-urban areas – communities that either by design or neglect have been left out of the benefits accruing out of science and its application' (Jain, 2002, p. 14). Thus networks of NGOs, scientists and science institutions are interlinked in local strategies for sustainable development. The Indian National Innovation Foundation, for example, relies now on a large pool of NGOs, high-profile laboratories, the formal business sector, micro-venture management and intellectual property rights legislation to promote collectively synergies between science and local development. These policy-driven networks reflect the growing concern with achieving ecologically and socially sustainable innovation.[4]

Although not yet articulated as a driving force for contemporary S&T policy, this networking policy approach is already beginning to shift national S&T foci in developing economies toward bridging knowledge systems and managing knowledge production, transfer and transmission through education. This is rather different from simply linking S&T institutions or managing links between these institutions and other sectors in the economy. This perspective implies that S&T policy can drive development, not simply through the production and transfer of new knowledge but through its contribution to national knowledge infrastructure.

This suggests a fundamental departure from the policies that characterized earlier waves of S&T policy. It is not the management of institutions, which characterized the first wave of post-war S&T policies, that becomes paramount. Nor is it the management of the links between industry and society, as was the case through the second policy wave, although these of course remain important. Rather, it is a focus on building knowledge networks for linking the social, technological and economic domains that are increasingly characterizing S&T policy in developing countries. For example, the Honey Bee Network set up by the Indian National Innovation Foundation provides incentives for 'scouts' to identify grassroots technological innovations and seeks to develop mechanisms to draw these innovations into the broader science, technology and innovation systems (Jain, 2002).

This presents new challenges for governments. How can they contribute to building creative innovation climates, maintain a balance between development in some of their poorest regions, and maintain development in national knowledge-intensive institutions, such as research institutes, universities, hospitals and industrial enterprises? The recent World Summit on Sustainable Development drew the attention of delegates to the need for a strong level of grassroots enthusiasm, the need to link human resource development to programmes for technology transfer, to communicate more effectively to policy-makers the value of science and technology for achieving sustainable development, and to develop new 'social contracts' between science and society (Dickson, 2002). Putting in place the mechanisms for achieving these broad policy objectives and making them central to national practice is more difficult.

Summary

Although control of the scientific and technological enterprise tends to remain in the hands of developed economies, the dilemma faced by S&T planners in developing countries is not simply to increase their share of the world's S&T resources or scientific output as such, but to derive greater benefits from the global reserve of scientific knowledge and link it, through networks, to local production and services. However, decision-making about national priorities for S&T is inherently difficult. The task is all the more difficult for developing countries because, as the Kenyan and Mozambique cases illustrate, their decisions must be made against the backdrop of powerful international control over the nation's technological choices. There must therefore be wise decisions about mechanisms that will rebalance national autonomy in these choices. Decisions about research and wider resources also involve considerable uncertainty as they deal with knowledge yet to be created or captured.

The country experiences described in this section offer some interesting examples of how national S&T policies in developing economies might be directed toward building sustainable national knowledge infrastructure. Governments within developing economies have been aware that they are dealing with complex environments and variable levels of economic growth. There is also increasing demand to effectively manage the ecological sustainability of the natural environment and the equitable distribution of national resources.

Conclusion

This chapter began by raising the question: to what extent, and how, can public research investments contribute globally to sustainable development? It was noted that outputs and outcomes and the value they carry for countries in the North are often experienced quite differently in the South. Two key issues were addressed. The first concerned the need to develop indicators for assessing and monitoring potential contributions of research to sustainable development. This is essentially a problem for research evaluation.

The second issue concerned the role of national science and technology policies in developing countries and their capacity to capture benefit from global research in order to underpin sustainable local development. This is essentially a problem of policy and practice for dealing with technology acquisition, adaptation and diffusion.

Most governments now require ongoing monitoring and evaluation of national research funding programmes. Evaluation questions, quite reasonably, seek answers about the effectiveness of the programmes in achieving their specific objectives. They therefore emphasize sectoral or programme objectives rather than broad overarching themes such as sustainable development. The difficulty is that there are no indicators in general use for monitoring such thematic objectives. One approach toward developing indicators of research for sustainable development, based on an Australian case study, was presented in this chapter. It was built on the concept of assessing the extent to which research activities seek to achieve an inter-relationship between the ecological, social and economic dimensions of development. Although somewhat simplistic, the schema was sufficiently general to complement prevailing S&T indicators used in many other countries. The applicability of the model depends on the type and availability of information about specific research programmes and projects and the ability of governments to carry out an analysis of the national research effort using this information.

A second theme covered in this chapter concerned the role of S&T policies in achieving local sustainable innovation in the economies of the South. There is growing evidence that innovation networks rather than science institutions are the most effective agencies for steering research investments and capturing benefit for sustainable development. This does not mean that science institutions become less important. Rather, it emphasizes a need to integrate them, through policy objectives and outputs, with policies for local development. Thus S&T policy and evaluation in the South is no longer being supported simply for the benefit of science or scientific institutions or for the industrial and economic potential they carry. Rather S&T policies are being used to promote social networks for sustainable development. As this trend continues the necessity to develop indicators reflecting overall thematic objectives will increase. The earlier section in this chapter suggests a starting point from which more robust and comparable indicators might emerge.

While countries are encouraged by global trends to continue to collect research statistics and develop indicators using UNESCO and OECD methodologies, they should consider the deficiencies of these data in terms of reporting on sustainable development. Developing countries operate in very different economic and industrial environments. It would be wise to selectively use research indicators applicable to different endogenous trajectories for S&T development but integrate these with indicators for monitoring research contributions to sustainable development.

Notes

1. See, for example, Australian Research Council (2003, p. 38).
2. The full story behind this commitment can be followed in a report commissioned by the Australian National Board of Employment, Education and Training (NBEET, 1995).
3. The NSESD states specifically that 'the guiding principles and core objectives need to be considered as a package. No objective or principle should predominate over the others. A balanced approach is required that takes into account all these objectives and principles to pursue the goal of ESD' (Commonwealth of Australia, 1992, p. 9).
4. It is interesting to note that the concern with achieving socially sustainable innovation is also evident in developed economies. Recent research at the Georgia Institute of Technology, for example, is concerned to promote socially sustainable innovation policies (see Cozzens et al., 2003).

References

Adeboye, T. (1998), 'Africa', in *UNESCO World Science Report*, 1998, Paris: Elsevier, 166–81.

Clarke, R. (1985), *Science and Technology in World Development*, Oxford: Elsevier Oxford University Press.

Australian Research Council (2003), *ARC Linkage Program Funding Guidelines*, Canberra: ARC, Commonwealth of Australia.

Commonwealth of Australia (1992), *National Strategy for Ecologically Sustainable Development*, Canberra: Australian Government Publishing Service.

Cozzens, S. and Turpin, T. (guest eds) (2000), Special Edition on 'Evaluation of Publicly Funded Higher Education Research', *Research Evaluation*, **8**(1), 1–3.

Cozzens, S., Kamau Bobb, E., Deas, K., Gatchair, S., George, A. and Ordonez, G. (2003), 'Distributional Effects of Science and Technology-based Economic Development Strategies at State Level in the United States', Working Paper of the Technology Policy and Assessment Center, Atlanta: School of Public Policy, Georgia Institute of Technology.

Deville, A. and Turpin, T. (1996), 'Indicators of Research Relevance to Ecologically Sustainable Development and their Integration with other R&D Indicators in the Asia-Pacific Region', *Chemosphere*, **30**(9), 1777–800.

Dickson, D. (2002), 'Where next after Johannesburg?', in *Scidev.Net: News, Views and Information about Science, Technology and Development*, http://allafrica.com/stories/200409300698.html, accessed 15 November 2005.

Government of Mongolia (1998), *Annex to the Resolution # 55 of the State Great Hural of Mongolia, 1998* (personal translation provided by Ganbat Bulgaa, 2003).

Harger, R., Meyer, F.-M. and Khalil, A. (eds) (1996), Special Edition 'Global Change: Anthropogenic Processes and Indicators for Sustainable Development', *Chemosphere*, **33** (9).

Hewitt, T. and Albu, M. (1998), 'Structural Adjustment, Industrialisation and Technological Capabilities in Africa', *Science, Technology & Society*, **3** (2), 335–64.

Hill, S., Turpin, T. and Spence, H. (1996), 'The State of Science in South-East Asia and the Pacific', in *UNESCO World Science Report*, Paris: Elsevier, pp. 169–89.

Jain, A. (2002), 'Networks of Science and Technology in India: The Elite and the Subaltern Streams', *AI & Society*, **16**, 4–20.

Krishna, V.V. (1994), 'Science Policies to Innovation Strategies', *Knowledge and Policy*, Fall/Winter, **6**(3 and 4), 134–57.

Krishna, V.V. (2001), 'Changing Policy Cultures, Phases and Trends in Science and Technology in India', *Science and Public Policy*, **28**(3), 179–94.

Mans, D. (1996), 'Indonesia – Industrial Technology Development for a Competitive Edge', *Indonesia Discussion Paper Series*, Washington: World Bank.

Muchie, M. (2001), 'Paradoxes of Industrialisation and Unilateral Liberalisation in Africa: A Case of Unrealised Potential of Value-added Leather Manufacture in Kenya', *Science, Technology & Society*, **6**(31), 397–418.

Munro, D.A. (1995), 'Ecologically Sustainable Development – Is it Possible? How will we Recognise it?' in Sivakumar, M. and Messer, J. (eds) *Protecting the Future: ESD in Action*, Wollongong, Australia: Futureworld, National Centre for Appropriate Technology (NCAT) Inc., pp. 49–58.

National Board of Employment, Education and Training (NBEET) (1995), *Australian Research for Ecologically Sustainable Development Commissioned Report No. 38*, Canberra: NBEET.

Network for Environmental Training at the Tertiary Level in Asia and the Pacific (NETTLAP) (1995), *Newsletter of the Network for Environmental Training of Engineers at the Tertiary Level in the Asia-Pacific*, **3**(2), 1–5.

Organisation for Economic Co-operation and Development (OECD) (1991), *The Measurement of Scientific and Technical Activities (Frascati Manual)*, Rome: OECD.

Technology Information, Forecasting and Assessment Council (TIFAC) (2003), *Home-Grown Technology Programme*, www.tifac.org.in, accessed 15 November 2005.

Turpin, T. and Bulgaa, G. (2004), 'S&T Reform in Mongolia: A Challenge during Transition', *Science, Technology & Society*, **9**(1), 129–50.

Turpin, T. and Liu, X. (2002) 'Introduction', in Turpin, T., Garrett-Jones, S., Liu, X. and Burns, P. (eds) *Innovation, Technology Policy and Regional Development: Evidence from China and Australia*, Cheltenham, UK and Northampton, MA, USA: Edward Elgar, pp. 3–13.

Turpin, T. and Martinez, C. (2003), 'Bridging Knowledge Boundaries: A Challenge for S&T Policy in Mozambique', *Science, Technology & Society*, **8**(2), 215–34.

United Nations Development Programme (UNDP) (1998), *Mozambique: National Human Development Report, 1998*, Maputo: SARDC.

United Nations Educational, Scientific and Cultural Organization (UNESCO) (1984), *Manual for Statistics on Scientific and Technological Activities, June 1984*, Paris: UNESCO.

United Nations Educational, Scientific and Cultural Organization (UNESCO) (2000), *World Science Report: 2000*, Paris: UNESCO Publishing, Elsevier.

Victor, P.A. (1991), 'Indicators of Sustainable Development: Some Lessons from Capital Theory', *Ecological Economics*, **4**, 191–213.

11 Technology transfer and uptake of environmentally sound technologies
Steve Halls[*]

Introduction
To be successful, transfer of technology requires more than just the moving of high-tech equipment from the developed to the developing world, or within the developing world. Other requirements include enhanced knowledge, management skills and technical and maintenance capabilities of those receiving the technology. Integrating human skills, organizational development and information networks is also essential for effective technology transfer. Thus, technology transfer is a broad and complex process if it is to avoid creating and maintaining the dependency of the recipient, and if it is to contribute to sustainable and equitable development. The end result for the recipient must be the ability to use, replicate, improve and possibly re-sell the technology. A key element of this wider view of technology transfer is choice. There is no single strategy for successful transfer that is appropriate to all situations. If the transfer of inadequate, unsustainable or unsafe technologies is to be avoided, technology recipients should be able to identify and select technologies that are appropriate to their actual needs, circumstances and capacities. Six main steps in technology transfer have been identified:

1. establishment of cooperative and collaborative partnerships between key stakeholders with the common purpose of enhancing technology transfer;
2. implementation of technology needs assessments;
3. participation in the processes of technology creation, development and adaptation;
4. design and implementation of technology transfer plans and specific actions;
5. evaluation and refinement of the actions and plans; and
6. dissemination of technology information.

Moreover, transferring environmentally sound technologies (ESTs) successfully depends on the potential recipient:

- recognizing and taking advantage of their benefits;
- obtaining information, and having the knowledge and tools to make an assessment and decide on the most appropriate technology option;
- understanding the technologies, especially their operation, responsible use and the systems and infrastructure on which they depend; and
- knowing how to implement and manage technological change successfully.

Encouraging the adoption and use of environmentally sound and, desirably, fully sustainable technologies, thus requires both voluntary approaches and a regulatory framework that nurture innovation and economic, social and environmental accountability.

Enacting policies that lower costs and stimulate a demand for sustainable technologies is necessary in order to achieve the environmental and other benefits that might not otherwise be realized. Furthermore there needs to be greater clarification of existing environmental rules and regulations as well as better coordination and harmonization with international standards.

Context

Technology transfer has the potential to make a significant contribution to sustainable development. This potential and the need for resolute and coordinated actions to realize it, have been recognized at the highest levels internationally. For example, the Rio Declaration on Environment and Development, a key output of the UN Conference on Environment and Development (UNCED, 1992), includes the following: 'States should cooperate . . . by enhancing the development, adaptation, diffusion and transfer of technologies, including new and innovative technologies' (UNCED, 1992, Principle 9).

To operationalize this principle an entire chapter of Agenda 21 was devoted to the transfer of technology and to the related topics of cooperation and capacity building. The chapter formalized a definition of ESTs as technologies that:

- protect the environment;
- are less polluting;
- use all resources in a more sustainable manner;
- recycle more of their wastes and products; and
- handle residual wastes in a more acceptable manner than the technologies for which they are substitutes.

We are now at the threshold of the third generation of environmental technologies – having moved from 'end-of-pipe' technologies to pollution prevention technologies that reduce the environmental 'footprint' of processes, products and services in ways that increase overall efficiency and reduce risks to humans and to the environment. Such technologies emphasize pollution prevention through reduced consumption of raw materials and energy and zero or reduced waste generation. The third generation of environmental technologies will lead to that term becoming redundant to be replaced by 'sustainable technologies', where environmental performance considerations will be fully integrated with economic, social and other operational issues so that the system as a whole is sustainable. Truly sustainable production and consumption technologies will require the development and use of planning, design and management practices that facilitate innovative approaches to the re-use, re-manufacturing and recycling of the limited amounts of 'waste' that cannot be avoided, despite the emphasis on minimizing the consumption of raw materials and energy.

Agenda 21 also emphasized that technology transfer does not just relate to equipment and other so-called 'hard' technologies, but also to total systems and their component parts, including know-how, goods and services, equipment and organizational and managerial procedures (Figure 11.1). Thus, technology transfer is the suite of processes encompassing all dimensions of the origins, flows and uptake of know-how, experience and equipment amongst, across and within countries, stakeholder organizations and institutions (Figures 11.2 and 11.3).

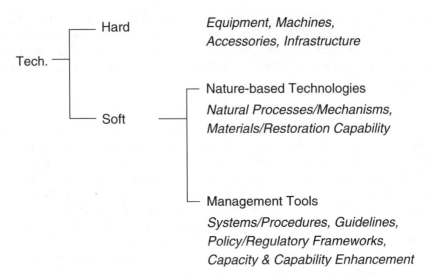

Source: Hay, 2003.

Figure 11.1   Technology typology

Ten years on from Rio, at the World Summit on Sustainable Development (WSSD) held in Johannesburg in 2002, a global stock-take showed that for most communities and countries the anticipated improvements in levels of development and sustainability have proven elusive, with the lack of widespread and effective technology transfer being identified as one of the root causes. Since the 1992 UNCED such revelations and the subsequent soul-searching have resulted in renewed calls for comprehensive but targeted programmes of technology transfer. Two key examples are the Millennium Development Goals and the Plan of Implementation agreed to at the WSSD. Two simple but fundamental questions arise from this brief historic review:

1.   Why has technology transfer failed to fulfil expectations and meet the clearly evident and pressing needs?
2.   What must be done to ensure that the success of renewed international sustainable development initiatives, and those being undertaken at community and national levels, are not frustrated by continued shortcomings in technology transfer?

This chapter will identify and characterize the barriers to successful and effective technology transfer, and provide a roadmap for removing the major impediments to successful technology transfer. Emphasis will be on technology transfer to, between and within developing countries and countries with economies in transition, covering the broad range from community-based technology initiatives to those undertaken as a result of comprehensive national and regional policies. However, it is self-evident that there is no ubiquitous approach to enhancing technology transfer – the suite of measures for addressing barriers and facilitating successful technology transfer, is typically community-specific.

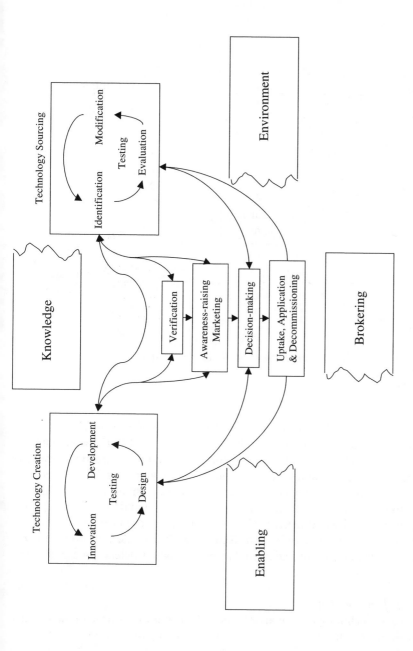

Figure 11.2   The technology transfer process

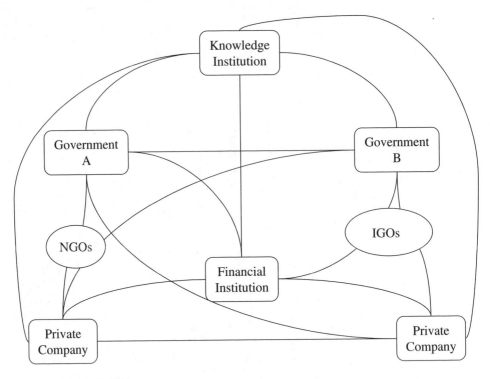

Source: After UNEP, 2001.

Figure 11.3   Some of the key players in the transfer of technologies illustrating the diversity of pathways

Challenges

Challenges are more commonly referred to as 'barriers' – we prefer the former term as it has fewer long-term negative connotations. The discussion of challenges will commence with the supply side of technology transfer (the innovators and developers) and move to the demand side (the recipients and users). Of course, as shown in Figure 11.2, technology transfer processes are neither linear nor simple. Commonly, challenges occur at every node and – due to restrictions on the movement of information and materials – at every linkage. Moreover, the nature and severity of the challenges will be dependent on the prevailing circumstances, varying with the type of technology, its specific application and the characteristics of the technology providers and recipients. Nevertheless some generalizations are possible.

Shortfalls in technology creation

Many of the needs for technologies that arise from implementation of development plans, and from the wish for enhanced sustainability, can be met only through innovation – the design, development and proving of new technologies. New challenges, such as the management of toxic chemicals or the disposal of intractable wastes, often require new solutions rather than a modification of old procedures.

Challenges that need to be addressed if technology creation is not to impede the transfer process include the lack of the following: information relevant to strategic planning and market development; science, engineering and technical knowledge, especially in the private sector; institutions with the mandate and resources to equip people with the requisite knowledge and skills; research, development and testing facilities; technology development and adaptation centres; and joint industry–government planning and collaboration.

Creation of useful and usable technologies is a major factor in ensuring there is the opportunity to make informed and confident choices in technology investment projects.

Underperformance in technology sourcing
Technology sourcing covers those situations where the technology already exists, be it traditional or modern. Transfer of such technologies might start with identifying their existence, evaluating their general suitability and making modifications, as needed. The technologies can then be made available for use in specific applications and locations, where another tier of requirements must be met if the technologies are to prove appropriate for that use.

If technology sourcing is not to impede the transfer process, challenges often need to be addressed, including those related to: ready access to independently verified information on suitable technology options; use of traditional technologies being seen as retrogressive; inability to address the concerns of stakeholders who consider they will be worse off if the technology is introduced; and financial and other disincentives to acquire and use ESTs.

Sub-optimal enabling environments
Figure 11.2 shows that the enabling environment is a key component of the technology transfer process. The enabling environment encompasses many aspects of the context within which the transfer takes place. Shortcomings in the enabling environment reduce the likelihood that the transfer will be conducted in a successful and timely manner.

Stakeholders must have the necessary knowledge and skills to perform the roles and tasks expected of them. High levels of awareness, motivation and empowerment within the public and private sectors and in civil society will help ensure that people, communities and societies are able to adapt continuously to new circumstances and challenges, including developing and modernizing in less wasteful ways than is the current development paradigm, but without losing the sound social and cultural values and practices that underpin their traditional ways of life.

Effective and efficient national and regional systems of innovation, research and development should be in place to facilitate such procedures as adaptation of traditional technologies for use in current settings. Such systems will also contribute to the design, testing and widespread uptake of new technologies suited to local conditions. Networking of organizations and institutions that share a common interest in technology development, diffusion and application will also foster successful transfers of technology.

Networking is just one of many ways in which a functional information management system will facilitate technology transfer. Such a system will also enable needs and practical constraints to be communicated to those involved in designing, developing and providing technologies, including creating a favourable environment for technology transfer. Performance and other data on possible technology options can also be made known to parties contemplating a technology investment, by facilitating the communication and appropriate use of guidelines, codes of practice, standards, verification information and certification procedures. Information management systems will help ensure that key players are aware of both successes and failures in selecting, procuring and applying a technology, allowing systems to be improved. Timely access to quality information is also required if technology investment risks are to be minimized, in part through reductions in both information gaps and uncertainties.

The enabling environment benefits from policy implementation that fosters an appropriate mix of government and private sector investment in ESTs. This can be achieved by implementing policies that address such issues as shortcomings in the financial sector, lack of access to appropriate sources of capital, high or uncertain inflation or interest rates, subsidized or average-cost (rather than marginal-cost) prices for material and energy inputs, high import duties, uncertain stability of tax and tariff policies, investment risk (real and perceived), loss of rights to intellectual property and to productive resources and risk of expropriation.

Appropriate use of regulations, economic incentives and other measures will also facilitate the uptake and operation of ESTs at the community level to the national level and discourage the continued use of environmentally unsound technologies. Moreover, a competitive and open market will encourage ongoing, replicable technology transfers. Financial, insurance and related systems that facilitate investment in ESTs at various levels – community through regional – are also desirable. All institutions must be able to foster and facilitate the transfer and effective operation of ESTs in a transparent, accountable and technically competent manner.

Finally technology transfer will also be assisted by the ability to access, use and benefit from decision support and related methodologies and tools, including those that will help ensure technology investments have equitable, beneficial outcomes that are environmentally sound, socially acceptable and economically viable.

Insufficient and unverified information
In addition to the movement of the technology itself, the transfer process involves a two-way flow of information. For example, technology creation cannot be totally serendipitous – it must reflect, at least in part, the needs and circumstances of the target beneficiaries of the creative effort. Similarly it is essential that those involved in supplying, procuring, financing and insuring technologies are aware of the potential technologies, including their procurement and operational requirements and performance profiles.

Environmentally sound technologies are often considered 'emerging' and hence 'unproven' – there is little confidence in their economic, commercial or technical viability. Such views often relate to a lack of data, information, knowledge and awareness, meaning there is no credible record of performance. Consumer acceptance is thus low and investors, especially those in the private sector, are adverse to financing ventures that appear to carry high risks.

Moreover, available information on the performance and other key attributes of the technology is typically prepared by the technology developer or owner, or their agents, often with little or no accountability for its veracity. The existence of sufficient, reliable information is essential for judging the likely commercial success and other outcomes of a proposed technology investment. A realistic expectation is that the information being used in decision-making is a reasonable representation of reality, with providers of information accountable for reporting in an accurate and meaningful way. Thus the reporting of information must be sufficiently frequent, complete and reliable if stakeholders are to assess on a timely basis the extent to which their expectations of performance are being satisfied.

This requirement has major implications for the manner in which performance information is collected, analysed and communicated. However, uniform reporting measures remain elusive and the variety of approaches to reporting performance information often makes it difficult, if not impossible, to compare technologies and judge their likely commercial viability and environmental and social impacts. This impedes, if not precludes, informed choice. The situation is exacerbated if the information that is available cannot, or should not, be relied upon. An even greater challenge exists for developing countries, given the complexity of factors that influence and determine investment decisions and the environmental, social and commercial successes of a technology intervention.

Choice

A key component of technology transfer is informed choice – the ability of the technology user (country, community, enterprise and so on) to be able to identify and procure the most appropriate (environmentally, economically, socially) technology for a given application in a given locale. Several requirements must be met, including:

- needs well-defined, documented and understood;
- several technology alternatives, all of which are well and reliably characterized in terms of environmental and economic performance and potential social impact;
- rational and functional methods (decision support tools) that facilitate choice of an optimal technology; and
- capability to make the chosen technology fully operational, so that it fulfils its potential and meets the identified needs without detrimental side-effects, including during decommissioning.

Technology needs assessment Technology needs are often poorly characterized and understood, even by the potential users. This failing impedes the effectiveness of others in the technology transfer chain, including innovators, designers, developers, suppliers, financiers, insurers and regulators.

Technology needs assessment is a coordinated set of country-, sector-, enterprise- or community-driven activities leading to the identification of technology needs, evaluation of specific technology options using generally accepted criteria, development of both specific and more integrated technology strategies, and facilitation of the identification, development, selection, transfer, acquisition and uptake of sustainable technologies. The assessment involves all relevant stakeholders in a consultative process to identify the opportunities for, and challenges to, the successful uptake of sustainable technologies and the measures to address these opportunities and barriers through specific and more integrated actions.

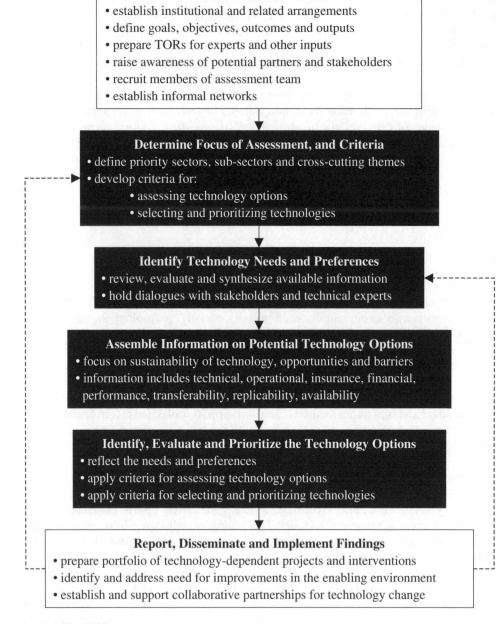

Source: Hay (2003).

Figure 11.4 Key activities in a technology needs assessment

The tangible outcomes of such an assessment are typically a portfolio of technology transfer projects and capacity-building activities to facilitate and accelerate the development, adoption and diffusion of sustainable technologies in particular sectors, enterprises, locations and communities. The key activities are shown in Figure 11.4 and illustrated, using a case study, in Box 11.1.

Certainty

A lack of certainty and the consequential high levels of risk, both real and perceived, are recognized as major impediments to the successful establishment and ongoing operation of a functional market for ESTs. For example, the uncertainty that prevails in markets for ESTs has detrimental repercussions all along the transfer chain, from innovators to users.

The level of risk reduction that occurs in the real world is directly related to the level of performance and effectiveness of the technologies that are purchased or used. Superior performance can manifest itself as reduced operating risk, lower costs and competitive advantage. There are many actions that can be taken to increase the level of certainty and hence reduce risks to technology developers, providers, investors and users. Some are elaborated below.

Development and use of technology performance protocols, criteria and benchmarks
To foster the use of ESTs there is an urgent need for the following credible, comparable and objective information for each candidate technology: technical performance, lifecycle cost, health and safety data, risk, process residuals, regulatory feasibility, future use, natural resource damage and stakeholder concerns. Access to relevant, high-quality performance information is critical to improving the quality of decision-making regarding the selection of ESTs.

Ideally the detailed measurements and assessments required to provide this information will be made in accordance with internationally recognized methodologies or accepted protocols, thereby helping to ensure comparability between information collected by different assessors and/or between commercial variants of the same generic technology. The protocol will involve science-based assessments and should produce statistically valid data sets, with quality control and assurance and application of chain of custody rules. Laboratory testing should be complemented by field trials. If the detailed measurements of performance are made in accordance with standard methods or accepted protocols this will help ensure comparability between information collected by different assessors and/or between commercial variants of the same generic technology.

Such performance assessments are capable of generating large amounts of data. A significant challenge is to identify the key criteria that characterize and distinguish the performance of the technology undergoing assessment. Criteria will be of optimal use to decision-makers and other practitioners if they portray reality and facilitate unequivocal confirmation that a given technology is environmentally sound. However, as yet there is no widely accepted set of technology performance criteria and indicators for evaluating and identifying ESTs. This is recognized as a major impediment to increasing their acceptance and uptake.

Independent quality assurance of performance information

Performance assessments provide standardized information against which individual performance claims can be judged. However, in themselves the procedures do not directly address the concerns and expectations of the end-users with respect to the quality and credibility of the reported information. Where information exists, it may not be credible. This gives rise to the need for additional quality assurance, through verification, whereby independent third parties determine whether the information provided satisfies specific criteria concerning the veracity of the way it is acquired and communicated.

BOX 11.1   TECHNOLOGY NEEDS ASSESSMENT FOR A SMALL ISLAND DEVELOPING STATE

The technology needs assessment process indicated in Figure 11.4 was followed. Criteria used to identify acceptable technologies included:

- improve operational efficiency/effectiveness;
- have social and economic benefits that exceed the social and economic costs;
- reduce vulnerabilities and enhance resilience of social, economic and environmental systems;
- be environmentally sound and compatible with the local culture, society and environment;
- assist people to develop and modernize in less wasteful ways than is the current development paradigm, but without losing the sound social and cultural values and practices that underpin their traditional ways of life;
- facilitate compliance with international agreements such as the UN Framework Convention on Climate Change and the Convention on Biodiversity;
- complement existing technologies and services; and
- be consistent with current capacities for operations and maintenance.

In addition, technology acquisition must involve:

- consultations with all stakeholders;
- recognition of the important roles of traditional knowledge and skills, including endogenous technologies;
- decisions based on comprehensive and credible performance information;
- *either* a tendering process to ensure cost-effectiveness;
- *or* full costing of technology projects funded by donors.

The outcomes of the technology needs assessment are summarized in the following diagram:

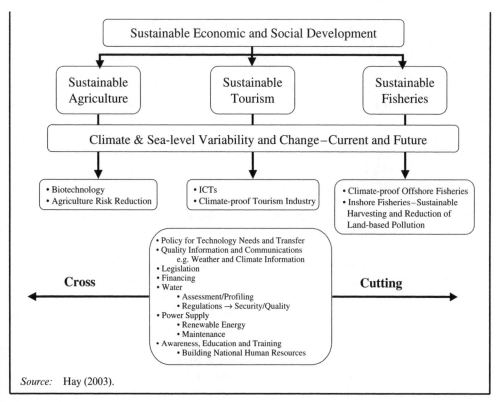

Source: Hay (2003).

Verification is the process of determining that a given technology will produce the results described in a performance claim. This is done through independent, third-party application of guidelines or pre-determined criteria and is substantiated by investigation, statistical analysis and other means. Verification is not an isolated process – it is part of a larger system that includes monitoring, auditing, certification and accreditation. Verification guidelines outline the procedures and information requirements needed to verify a performance claim. A verification system should have:

- *Credibility*. The process should involve credible organizations working in conjunction with internationally recognized bodies that accredit competent organizations to verify and certify.
- *Transparency*. The process should be open and transparent with information shared amongst interested parties.
- *Compatibility*. Verification guidelines should be relevant to national and international applications.
- *Continuous improvement*. The verification system should be designed to accommodate continuous improvement, taking into account new and emerging information and knowledge.

Independent assessments of the veracity of performance claims should follow accepted guidelines and use established criteria. The advantages of third-party assessment and review of technology performance data include enhanced certainty and transparency and increased marketplace acceptance, as well as associated benefits for the proponents of the technology investment, the technology developers, consumers, regulators, insurers and financial investors. Verifying conformity of performance claims against accepted criteria and standards provides stakeholders with an opportunity to assess the real and relative benefits of ESTs, including cleaner production processes and practices. Benchmarking can also be used to improve the usefulness of the evaluation information.

Assessing and verifying conformity to performance standards is usually carried out at the national level, providing an opportunity to reflect at least country-level priorities and bottom lines. In addition, national agencies providing performance assessment services can benefit from strengthened linkages with international organizations that can provide technical assistance in support of their activities. Such linkages are also important in ensuring that country-specific performance assessment protocols, criteria and benchmarks are internationally recognized.

Establishing and strengthening information systems, partnerships and networks
Certainty can also be increased by improving the flow of credible information between technology developers, providers and users (Figure 11.5). Successful technology transfer requires interaction among various stakeholders, many of whom will have different and sometimes conflicting intentions. Free and full flow of targeted and verified information is critical to resolving these differences and allowing greater harmonization of effort. Of all the factors involved in determining the success of technology transfer, the critical element is not public or private sector involvement but the forging of an effective linkage between the generator and the user. The public sector is effective in generating and applying technology when it is also the customer. In contrast, industry is more effective when the technology is intended for use in the wider socio-economic system, including production, distribution, consumption and disposal.

Information to support technology transfer should be demand-driven and results-oriented, supporting key activities in the transfer process such as selecting a specific technology and making an investment decision.

As also indicated in Figure 11.5, monitoring plays an important role in providing information that can improve the success of technology transfer. Operational experience with technologies designated as environmentally sound provides bottom line verification of this description. Monitoring and the effective movement of the resulting information from technology users to developers, assessors and suppliers, provide an important feedback loop that will be the main mechanism for continuing improvement all along the technology transfer chain.

Information flows can be enhanced through the establishment of partnerships involving key players who bring complementary capacities and who can share advantages and gain mutual benefits. One increasingly important example of such relationships is public–private partnerships. These are being seen increasingly as an effective means to leverage public funds, thereby overcoming budget restrictions, while also harnessing the efficiency of the private sector and allowing it to operate more effectively through changes in public policy that create more business opportunities.

The partnerships can take on diverse forms and involve a range of players. Private sector participants can include technology developers, assessors, suppliers, users and

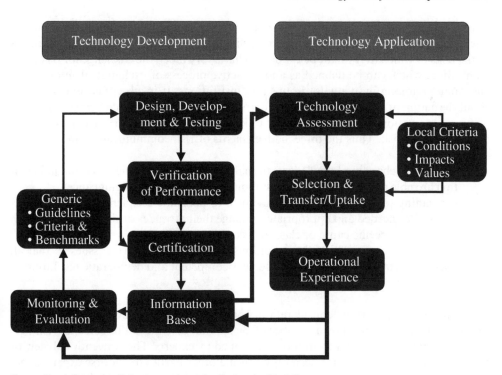

Note: Key information linkages are shown by the heavier black lines.

Figure 11.5 The technology development and application processes

investors, while those from the public sector might be central government departments, agencies, intergovernmental organizations and local government.

Networks can usefully bring together players with similar roles in the technology transfer process who wish to share experiences and information through the exchange of best practices, lessons learned and case studies as well as through protocols, criteria, benchmarks and performance data on specific technologies.

Other networks serve to link stakeholders from different parts of the technology transfer chain as well as those who service the technology transfer process, including advisory centres, specialist libraries, database managers, liaison services and technology intermediaries such as energy and waste service companies and national-level institutions engaged in research, development and assessment.

Communication

Many of the topics that might be covered under communication have been addressed elsewhere as part of discussions related to information access and sharing. However, one important theme that has not been given the attention it merits is risk communication. Previous sections have highlighted the influential role of risk in determining the willingness of potential key players, such as investors, to commit resources to technology development, transfer and uptake. Risk is influential in two ways: (1) there is often a major difference between the perceived and real risk; and (2) perceived risk is typically a major factor in decision-making.

To facilitate technology transfer there needs to be an improved alignment between perceived and real risk, with decision-makers having access to credible and usable estimates of the risks to which they might be exposed. This is very much the challenge for risk communication, which can be defined as an interactive process of exchange of information and opinion among individuals, groups and institutions. It involves multiple messages about the nature of risk and other messages (not strictly about risk) that express concerns, opinions or reactions to risk messages or to legal and institutional arrangements for the management of risk. Thus the three main elements of risk communication are:

- informing (changing knowledge), to make sure that all recipients of the information are able and capable of understanding its meaning and significance;
- persuading (changing attitude/behaviour), to persuade recipients of the information to – as needed and appropriate – change their attitudes or their behaviour with respect to a specific cause or class of risks; and
- consultation – to provide the conditions for a dialogue on risk issues so that all affected parties can engage in an effective, competent and democratic resolution of unacceptable risks.

Perception is an important consideration when communicating about risk. Attitudes towards risk are influenced by a number of biases that often result in personal risk perceptions being very different from those developed by experts. The conventional view of risk communication was that it need only educate other parties to endorse expert judgement concerning which risks are acceptable and which are not. The only challenge was to ensure that other parties were properly informed of the experts' views. This view has now been largely replaced by one that acknowledges that non-experts are also in possession of relevant risk information, thus necessitating a two-way exchange of information between experts and the public. While experts and the public typically view risk in fundamentally different ways, each has something valuable to offer to the understanding and management of risk.

Capacity

This section focuses on processes and measures that help to ensure there is an enabling environment optimized for technology transfer. For technology transfer to be successful, adequate capacity must exist if all stages of the technology development and application process (Figures 11.2 and 11.5) are to operate to an acceptable standard. Circumstances that are supportive of technology transfer include:

- open and competitive market;
- comprehensive and credible specifications on the technology performance;
- financiers who are at least technology-neutral;
- the most cost-competitive technology that also has the most favourable environmental and social performance specifications; and
- policy risks are addressed.

To foster technology transfer developing countries and countries with economies in transition should establish a clear development agenda based on a needs assessment, identify indigenous capacities, coordinate external resources, set up an effective consultation process and develop appropriate monitoring and evaluation systems. While developed countries tend to focus on the private sector and market forces, developing countries tend to stress the role of the public sector and intergovernmental agreements.

Collectively governments have an important role to play in creating enabling environments for the transfer of ESTs. Policy actions can be taken at macro, meso and micro levels. Technology transfer should be integrated into overall national development frameworks as well as broader environmental, economic, social and health policies. Both national and international standards can enhance flows of technology, while risk communication and reduction are important in creating enabling environments (see Certainty).

Commitments

Commitments that will lead to the increased and more effective use of ESTs should be based on cooperation to bring about changes, from the bottom up as well as from the top down. The former might lead to reforms implemented by communities, institutions, organizations, enterprises and sectors in order to facilitate more extensive use of ESTs. Simultaneously the international community can do more to support initiatives at the national and sub-national levels.

Policies and programmes that integrate the elements of capacity building, information and knowledge into comprehensive approaches for EST transfer and cooperation can achieve more than individual actions by themselves, and can contribute to the creation of an innovation culture. This should involve partnerships at all stages of the technology development and transfer process and ensure the participation of private and public stakeholders, including business, legal, financial and other stakeholders within both developed and developing countries.

As an example of the bottom-up approach, and perhaps as a first step at sub-national and national levels, undertaking technology needs assessments (Figure 11.4 and Box 11.1) will lead to identification of technology transfer projects that facilitate and accelerate the development, adoption and diffusion of sustainable technologies in particular sectors, enterprises, locations and communities and to the recognition of capacity-building activities that will enhance the enabling environment (see Capacity).

As noted earlier, a major impediment to the development, transfer and successful uptake of ESTs is the widespread lack of comprehensive and credible information on the performance of technologies. This shortcoming is in part a consequence of the absence of criteria, benchmarks and protocols that must be in place before a technology can be classed as 'environmentally sound'. Addressing this barrier provides a common ground for commitments from both governments and international organizations. At the national level, governments could follow the examples of the Philippine's Department of Science and Technology and Korea's Environmental Management Corporation by establishing national or regional environmental technology verification programmes. At the international level, relevant organizations could do much to facilitate the preparation of the criteria, benchmarks and protocols that could subsequently be adapted for use in the national or regional programmes.

Another nexus is provided by the linkages between industry, commerce and trade organizations operating at international and national levels. These can promote leadership in the application of high standards for environmental performance and create awareness about products, processes and services that use ESTs through means such as eco-labelling, product standards, industry codes and developing environmental guidelines for export credit agencies to promote the transfer of ESTs while ensuring that the transfer of obsolete technologies is discouraged. Governments should also encourage multilateral development banks to account for the environmental consequences of their lending.

Initiatives that can be undertaken at national and sub-national levels with support from regional and international bodies include:

- intensive public education activities to ensure the wider community is committed and able to develop and modernize in less wasteful ways than is the current development paradigm and are willing to continuously adapt to new circumstances and challenges;
- strengthening scientific and technical education institutions in order to help address technology needs;
- discouraging restrictive business practices and promoting open markets and fair competition in EST markets;
- increasing the certainty and responsiveness of legal systems and reducing regulatory risk by reforming administrative law and ensuring that public regulation is accessible to stakeholders and subject to independent review;
- protecting intellectual property rights and using licences to foster innovation;
- encouraging capital flows that support direct investment through the use of specialized credit instruments and capital pools as well as public/private partnerships;
- expanding research and development programmes aimed at improving access to ESTs that are appropriate in developing countries and adaptable to local conditions;
- simplifying and making transparent programme and project approval procedures and public procurement requirements;
- improving systems for the collection, assessment and sharing of specific technical, commercial, financial and legal information.

Addressing the inadequacy of information and decision-support tools used to quantify and qualify the merits of ESTs represents a significant challenge. The effectiveness of ESTs depends on having both broad-based and expert input into their development, adoption and ongoing monitoring. Leverage and synergy through cooperation amongst governments, industry associations, corporations and the financial community is needed for investments in ESTs to occur. At the same time systems for collecting, synthesizing and feeding back information and knowledge on ESTs must be strengthened, deployed widely and maintained. Third-party performance assessment mechanisms such as verification and certification can assist in meeting this need for transparent, credible information on which decisions can be based. Continuous review and improvement will be essential to ensure the establishment of an effective system that is responsive to changing social, economic and political realities.

Conclusions

The foregoing discussion has highlighted the slow uptake of ESTs, provided an analysis of the reasons for this situation and suggested ways in which the situation might be remedied. Key actions that will foster technology transfer include:

- needs assessments, including identification of shortcomings in the enabling environment, with relevant organizations and agencies helping to address these;
- evaluation and strengthening of policies that influence the enabling environment;
- greater communication and interaction between key parts of government;
- intra- and inter-governmental coordination, cooperation and assistance;
- protection of intellectual property rights and legal contracts;
- political support for programmes and institutions that foster technology transfer;
- seed investment programmes to stimulate private sector investment;
- capacity enhancement for major stakeholders;
- delineation of the roles of the private and public sectors in both developed and developing countries;
- economic incentives targeting industries that have the potential to make critical and major contributions to technology transfer;
- ensuring that technology transfer initiatives are compatible with national sustainable development agendas; and
- increasing communication among technology transfer bodies with a view to leveraging limited financial and human resources on issues of common interest, integrating and strengthening regional and country-level activities through information sharing and joint activities, and providing a platform for multilateral approaches and consistency in technology transfer.

Activities that will lead to the increased and more effective use of ESTs should be based in cooperation to bring about changes, from the bottom up as well as from the top down. This calls for partnerships at all stages of the technology development and transfer process, ensuring active and equitable participation of private and public stakeholders, including business, legal, financial and other stakeholders within both developed and developing countries.

Note

* Dr Halls is now Professor of Environmental Science at Murdoch University, Perth, Australia.

References

Hay, J.E. (2003), *Report on Workshop on Technology Needs Assessment and Technology Transfer Projects, Niue Climate Change Project, PHASE II – GF/2010–01–05*.

United Nations Conference on Environment and Development (UNCED) (1992), *Report of the United Nations Conference on Environment and Development, Vol. 1 Agenda 21 (A/CONF.151/26/Rev.1)*.

United Nations Environmental Programme (UNEP) (2001), *Managing Technological Change*, an explanatory summary of the IPCC Working Group III Special Report 'Methodological and Technological Issues in Technology Transfer', Paris: UNEP, Division of Technology, Industry and Economics (DTIE), 20pp.

PART III

ENVIRONMENTAL TECHNOLOGY MANAGEMENT IN BUSINESS PRACTICES

12 Lifecycle assessment
David Evans and Stuart Ross

The evolution of the LCA method

Lifecycle concepts were first developed and applied by industry in the late 1960s and early 1970s. The term originally used to describe this type of study was resource and environmental profile analysis (REPA). REPA enabled assessors to quantify the consumption of resources and energy over the life of a product or service. Throughout the 1970s and 1980s businesses and trade organizations used the information generated by REPA in product design decisions and long-term strategic planning processes.

However, as concern for the state of the environment grew, more questions were being asked about the origin and fate of the materials and energy used across the life of a product or service. People began to have doubts about the ultimate ability of the Earth's natural systems to deal with these wastes (McCormick, 1995) and pressure was placed on manufacturers to reduce the environmental impacts of their products by going about things in different ways. The question arose as to the contribution of less obvious steps in the whole journey of a product from the extraction of raw materials through manufacture to final disposal. This led to the realization that the environmental impacts resulting from a product or service could only be properly understood after a comprehensive assessment in which all process steps from extraction through to disposal had been evaluated. This desire for a broader, systems-based view of environmental problems was the impetus behind work to expand the role of REPA to include impact assessment and careful interpretation of results. This modified form became known as lifecycle assessment (LCA) and first came into use around 1990 (Hunt and Franklin, 1996).

Since developers expanded the role of LCA to include an impact assessment phase, interest in the technique has grown strongly. As a result many increasingly complex systems have been 'assessed', as different stakeholder groups have sponsored lifecycle work aimed at better understanding the impacts of the economy on the natural environment. For example, environmental protection authorities in several European countries have used LCA information for environmental labelling of products. Companies are doing in-house LCAs to support product development, environmental management and marketing. Consumer organizations have used the findings of LCA studies to counsel consumers.

With so much interest in the LCA method, practitioners realized that a single, internationally agreed approach was needed. The task of coordinating and documenting the development of a standard approach was given to the International Standardization Organization (ISO). The ISO 14040 series of standards, which are part of the ISO 14000 series on environmental management, is the result.

The ISO 14040 LCA method

According to the umbrella standard, ISO 14040 – Principles and Framework, LCA is concerned with the material and energy efficiencies of product systems and how their emissions may interact with the environment (AS/NZS ISO 14040, 1998). Stage 1, ISO

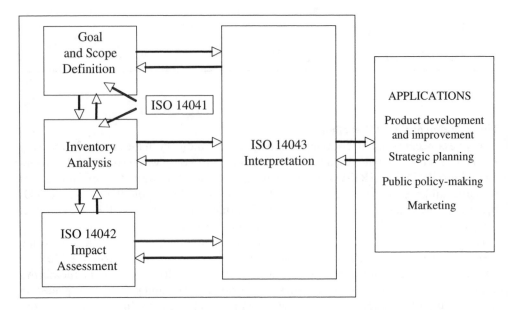

Source: Adapted from ISO 14040 – Principles and Framework (AS/NZS ISO 14040, 1998, Figure 1, p. 4)

Figure 12.1   Relationship between the different stages of the ISO LCA method

14041 – Life Cycle Inventory, is the data collection step. This component of LCA evolved out of REPA. It includes the collection and collation of data describing the material and energy flows into, through and out of the product system (AS/NZS ISO 14041, 1999). Stage 2, ISO 14042, constitutes Life Cycle Impact Assessment. During this stage the aggregated inventory data is assigned to impact categories that represent contributions to significant environmental problems (ISO, 2000a). Stage 3, ISO 14043 – Life Cycle Interpretation, provides advice and guidance on how to interpret the results of the impact assessment phase and how to improve the quality of the study (ISO, 2000b). Figure 12.1 illustrates the relationship between each of these stages.

In this section we review these published standards to show what these three stages are and the relationship between them.

ISO 14041 – Life Cycle Inventory Analysis
The first stage in a LCA, as described in ISO 14041, is to carry out a lifecycle inventory analysis. However, before this can be done, the goal and scope of the analysis must be carefully defined:

> The goal and scope definition phase . . . determines why an LCA is being conducted (including the intended use of the results) and describes the system to be studied and the data categories to be studied. The purpose, scope and intended use of the study will influence the direction and depth of the study, addressing issues such as the geographic extent and time horizon of the study and the quality of the data which will be necessary. (AS/NZS ISO 14041, 1999)

If we are not absolutely clear what the task is and where it starts and ends we will most likely get the wrong answer. To take an example, a German group performed a lifecycle assessment study on resilient floor coverings (Günther and Langowski, 1997). They compared 30 different material combinations on the basis of 'the typical use of 20m² flooring over a period of twenty years'. It would be misleading to base this comparison purely on the *production* of a kilogram of each of the materials used in the different floor coverings because some materials are lighter (less material needed to perform the same function) or more durable (do not need to be replaced so frequently) than others. The introduction to ISO 14041 goes on to say: 'The life cycle inventory phase involves the collection of the data necessary to meet the goals of the defined study. It is essentially an inventory of input/output data with respect to the system to be studied' (AS/NZS ISO 14041, 1999).

This appears to be straightforward, until we begin to reflect on what data might be necessary to meet the goals of the study. The time and cost involved in collecting data is one of the biggest challenges facing users of LCA (Clark and de Leeuw, 1999). Therefore, we do not want to be collecting data that will never be used in the assessment. However, we also do not want to leave out important data or stages in the lifecycle of a product or service system.

The total system will usually consist of extraction of raw materials from the natural environment, transforming these to a useable form, using the transformed materials to make a product or deliver a service, using the product, and finally disposing of it when it is of no further use. ISO 14041 calls these sub-systems *unit processes* (AS/NZS ISO 14041, 1999). The first step in the calculation of the inventory is to draw a system diagram showing all the material and energy flows into and out of all the unit processes. This will show flows into the system from the natural environment, flows of product between unit processes, and flows of wastes out of the system and back into the environment. (An important point is that if any recycling is involved, the recycling system is itself a unit process – or series of unit processes – and it forms part of the total system.) The flows of materials and energy should be expressed in terms of a unit quantity, usually per kilogram of the desired product. All the flows can then be converted later to flows in kilograms required to carry out the desired task. As a check, the laws of conservation of mass and energy must be obeyed for each unit process and for the total system. As an example, Figure 12.2 gives the systems diagram for the production, use and disposal of polystyrene/polyethylene packaging.

Second, we have to be mindful of the partitioning problem. Some of the materials and energy used may be by-products or co-products of the production of other materials not used in our system. For example, when black coal is mined, some methane, a powerful greenhouse gas, is emitted. In the production of polystyrene in Australia one of the main inputs is benzene from the benzene/toluene/xylene (BTX) mixture produced as a by-product of the production of metallurgical coke from bituminous coal in coke ovens. How much of the fugitive methane from coal mining should we debit against the BTX? Several different approaches are discussed in the ISO standards (AS/NZS ISO 14041, 1999, p. 11).

Having decided on how these problems will be dealt with, the final step is to relate this data to the study's functional unit and then aggregate equivalent substances from each unit process that are linked to similar impacts across the lifecycle. Figure 12.3 charts this process.

ISO 14042 – Life Cycle Impact Assessment
ISO 14042 calls the second stage lifecycle impact assessment (LCIA) to distinguish it from lifecycle assessment, which is the whole process consisting of the three stages, and also

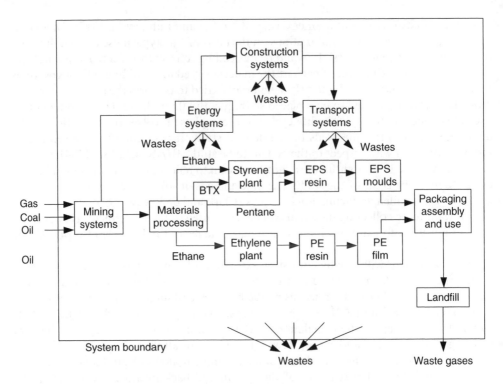

Source: Adapted from Ross and Evans (2003a, Figure 1, p. 563).

Figure 12.2   Systems diagram for the lifecycle of polystyrene/polyethylene packaging used during the transport of white goods

from environmental impact assessment (EIA), the well-known regulatory impact assessment process that looks at the site- and time-specific impacts of particular proposals.

The task of LCIA is to translate the aggregate inventory data (that is, kg of materials or MJ of energy per functional unit) from the previous stage into information on impacts. How is this to be done? According to ISO 14042 (ISO, 2000a) the purpose of LCIA is to assess a product system's lifecycle inventory (LCI) analysis results to better understand their environmental significance. The LCIA stage models selected environmental issues, called impact categories, and uses category indicators to condense and explain the LCI results. Category indicators are intended to reflect the aggregate emissions or resource use for each impact category. These category indicators represent the 'potential environmental impacts' discussed in ISO 14040. The method used to calculate these values from the results of the inventory is shown in Figure 12.4, using the example of the category 'acidification'.

However, before values for category impacts can be calculated, assessors have to decide what categories of impact to include in the assessment. Examples of impact categories that are commonly selected include: abiotic resources; biotic resources; global warming (for example, from greenhouse gases); stratospheric ozone depletion; acidification; eutrophication; photochemical oxidant formation; human toxicity; ecotoxicity; land use.

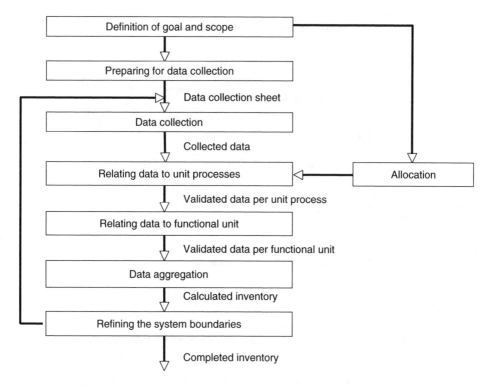

Source: Adapted from ISO 14041 – Goal and Scope Definition and Inventory Analysis (AS/NZS ISO 14041, 1999, Figure 3, p. 8).

Figure 12.3 Simplified procedures for inventory analysis

Next the assessor has to insert the inventory results into these categories in a meaningful way. In the case of acidification, global warming and stratospheric ozone depletion this is straightforward, as the scientific community has worked out the equivalent effects of different chemicals in these two areas. For example, the global warming potential of methane over a 100-year time horizon is 23 times that of carbon dioxide (Watson, 2001). Therefore, the global warming potential of a quantity of methane plus a quantity of carbon dioxide can be obtained in carbon dioxide-equivalents by multiplying the number of kilograms of methane by 23 and adding this quantity to the number of kilograms of carbon dioxide to build up a tally of carbon dioxide-equivalents.[1] However, for impact categories that are site-dependent, such as photochemical smog, there is no easy way of aggregating the contributions of different precursor emissions to these effects (Ross and Evans, 2002).

One can readily see why LCA requires the selection of categories and the relating of the inventory inputs and outputs to be documented in a 'transparent manner' (ISO, 2000b). The lifecycle impact assessors and the readers of the impact assessment might have different views on the importance of different items, and the readers need to be able to see how the assessors obtained their results. Eventually, we end up with burdens expressed as quantities of material, energy or wastes per unit of production/consumption (that is, in terms of the

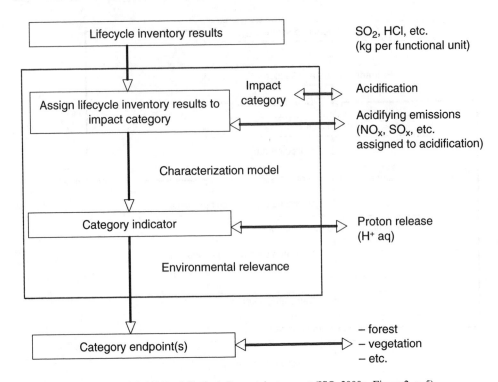

Source: Adapted from ISO 14042 – Life Cycle Impact Assessment (ISO, 2000a, Figure 2, p. 5).

Figure 12.4   The concept of category indicators

chosen functional unit). However, practitioners must use this information carefully when drawing conclusions. This is because, unlike project-level EIA methods, which calculate the actual amount of materials consumed or substances emitted in a given region within a given time, LCA uses material and energy flow data to determine the aggregate quantity of resources used and wastes released, irrespective of the time or place of these events, for an arbitrary unit of production or consumption (Klöpffer, 1996). While this format is satisfactory for comparisons of resource and energy efficiency, it is confusing when attempts are made to predict actual impacts (Ehrenfeld, 1997). In studies where the identification of site-specific impacts is important, assessors need to collect data describing the actual quantities of waste released per unit of time in a particular location so that they can identify single concentrated sources of pollutants and quantify the magnitude of direct impacts. However, LCA cannot do these calculations because the spatial, temporal, dose-response and threshold data needed to complete them is not collected during the inventory phase.

ISO 14043 – Life Cycle Interpretation
According to the introduction to ISO 14043 – Life Cycle Interpretation (ISO, 2000b):

> Life cycle interpretation is a systematic procedure to identify, qualify, check and evaluate information from the results of the LCI and/or LCIA of a product system, and to present them in order to meet the requirements of the application as described in the goal and scope of the study.

Life cycle interpretation includes communication, to give credibility to the results of other LCA phases (namely the LCI and LCIA), in a form that is both comprehensible and useful to the decision maker.

The organization or firm commissioning a lifecycle assessment will generally be doing so for one of two reasons: to inform the general public (usually to correct what they believe to be a misconception) or to determine whether an alternative method for making a product or providing a service is environmentally superior or inferior to the current method. The second of these will provide input to their decision whether to adopt the new method. In this they will be responding to pressures from government, environment protection agencies or the public, including their own shareholders. They must communicate, but the communication must be credible.

ISO 14043 deals with this problem. First, the standard emphasizes that transparency is an essential element of a lifecycle assessment. Where preferences, assumptions or value judgements are involved, these need to be clearly stated by the LCA practitioner in the final report (ISO, 2000b). Then it goes on to describe how to carry out the interpretation (ISO, 2000b).

The lifecycle interpretation phase of an LCA or an LCI study comprises three elements as follows:

- identification of the significant issues based on the results of the LCI and LCIA phases of LCA;
- evaluation that considers completeness, sensitivity, consistency checks;
- conclusions, recommendations and reporting.

The aim of the identification step 'is to structure the results from the LCI or LCIA phases in order to determine the significant issues, in accordance with the goal and scope definition and interactively with the evaluation element' (ISO, 2000b). The aim of the evaluation step 'is to make a synthesis of the study'. We need to get from the identification of the most important inputs, outputs and potential impacts to the conclusions and recommendations. Three checks are applied to do this:

1. *Completeness.* Check that the most important inputs, outputs and potential impacts accord with the goal and scope defined at the outset.
2. *Sensitivity.* Check the reliability of the final conclusions by determining whether they are affected by uncertainties in the data and assumptions made during the study.
3. *Consistency.* Check whether the assumptions, methods and data are consistently applied to all product systems.

The final step is the reporting of results, the drawing of conclusions, and the making of recommendations where appropriate (ISO, 2000b): 'The report shall give a complete and unbiased account of the study, as detailed in ISO 14040. In reporting the interpretation phase, full transparency in terms of value-choices, rationales and expert judgements made shall be strictly observed' (ISO, 2000b, p. 7).

Application of LCA to a typical problem

In this section we illustrate how the LCA process operates in practice by applying it to a typical problem: comparison of the environmental burdens associated with two different options for performing the same task. The problem we have chosen is the comparison of the impacts from two different types of motor car performing the same task. The cars are identical except for their engines, one of which has a traditional cast-iron engine block and the other an aluminium engine block. As aluminium is only one-third the density of cast-iron the engine, and hence the motor car, is much lighter, and less energy is required per kilometre of normal operation. Hence less petrol is required per kilometre and the emissions to the atmosphere from burning the petrol will be correspondingly less. At first sight it appears that the environmental impact of the use of the aluminium engine will be less than that of the cast-iron engine. However, it is not so simple. Aluminium is produced from its ores using an energy-intensive electrolytic process. Iron, however, is produced by a simple chemical reduction process, which is less energy-intensive and produces less environmental impact. Will this lower impact in the engine manufacturing process be sufficient to counterbalance the greater impact in the motor car operating process?

In both cases we have to consider the whole set of processes involved. However, for our comparison we only have to consider the manufacture and disposal of the engines and the petrol required to propel the cars over their lifetime, as all other parts of the cars are identical. Thus we have to consider only the following processes:

- mining the ores for making aluminium and iron;
- smelting these ores to produce the aluminium and iron needed for the two types of engine;
- manufacturing the two types of engine;
- driving the two types of car over the distance expected for their lifetime (this will include making the petrol from the crude oil);
- disposing of the two engines at the end of their useful lives.

The various processes occur at different times and in different places. The engines are manufactured and then installed in motor cars in Japan using raw materials shipped in from Australia. The motor cars are then shipped to the USA and driven there for several years using petrol made in the USA from crude oil imported from the Middle East.

Figure 12.5 is a schematic diagram of the whole system for making the engines and using them to propel the car over its lifetime of ten years. In the aluminium engine option the aluminium is produced by electrolyzing aluminium oxide (alumina) in a solution of molten cryolite (sodium-aluminium fluoride) using expendable carbon electrodes made from pitch and petroleum coke. The alumina is produced from bauxite, a naturally occurring hydrous aluminium-iron oxide ore, by dissolving it in caustic soda produced by electrolyzing common salt. The residue of iron oxide and other impurities, known as red mud, has to be disposed of to landfill. These processes are all carried out in Australia and the aluminium is shipped as metal to Japan. Iron for the iron engine option is produced in blast furnaces in which carbon in the form of coke is used to reduce hematite ore (iron oxide) to metallic iron. The iron oxide and coking coal are both mined in Australia and shipped to Japan. For simplicity, in this figure we have not shown the power station used to generate the electricity required for the electrolytic processes and manufacturing. We have also not shown the flows of energy needed to operate the mining equipment and to transport materials from one process to another, nor the materials and energy used to manufacture all the equipment needed for the various processes, including the mining and transport equipment.

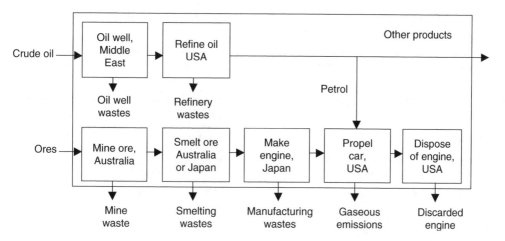

Figure 12.5   Simplified diagram of the system needed to propel a motor car

Having described the processes involved in the two options we can now show how the various steps in LCA would be performed.

Goal and scope
Clearly we have to decide on the type and size of car and the number of kilometres to be driven over its lifetime. We also have to determine a functional unit for the task, for example, driving the car 200 000 km over its life of ten years. The resources and emissions associated with producing the two types of engine and finally disposing of them can then also be expressed on a per 200 000 km basis.

Lifecycle inventory
The system boundary is the boundary between the natural environment and the global economy (see Figure 12.5). Arrows entering the system represent resources being drawn from the natural environment in the form of water, ores, crude oil, natural gas and coal.[2] Arrows crossing the boundary and re-entering the natural environment represent emissions to air, water and land. Between them, these two types of flow depletion of resources and production of emissions, are the environmental burdens of the whole system that we have to identify and quantify in the inventory stage of the LCA process.

Table 12.1 lists the major inputs of materials and energy and emissions from each of the unit processes. Although it is a mandatory requirement of ISO 14041 to collect materials and energy flow information for all of these, it does not require the assessor to list when and where they occur. We will discuss some of the implications of this later.

To avoid overcomplicating this table we have omitted many minor unit processes such as transport within and between countries and construction of the various facilities, including transport equipment (between them these usually require less than 10 per cent of the energy inputs and hence contribute a similar proportion of the air emissions). In practice these would also have to be included in the inventory.

In this table we have merely listed the inputs and emissions: in practice they would have to be quantified by means of appropriate equations. For example, consider the generation

Table 12.1   *Principal resources and discharges for the various unit processes involved in the propelling of motor cars*

Unit process	Resources Drawn from Natural Environment		Emissions to Natural Environment	
	Aluminium engine option	Cast-iron engine option	Aluminium engine option	Cast-iron engine option
Oil well	Crude oil		Hydrocarbons to air	
Oil transport by tanker	Crude oil to provide heavy diesel fuel		CO_2, NO_x, SO_x and smoke from oil tanker	
Oil refinery	Cooling water		CO_2, NO_x and SO_x, hydrocarbons	
Mine	Bauxite ore	Hematite ore	CO_2, NO_x and SO_x from mining machinery	
Processing of ore	Salt for caustic soda; natural gas for process energy		CO_2 and NO_x from burning gas; red mud	
Coal mine	Steaming coal	Coking coal	CH_4 from coal seams; CO_2, NO_x and SO_x from mining machinery; solid and aqueous mine waste, and acid leachate from waste dumps	
Generation of electricity	Coal for boilers; cooling water		CO_2, NO_x, SO_x and fly ash to air from boilers; ash to landfill; cooling water blowdown	
Metal smelter	Crude oil to provide pitch and petroleum coke		CO, CO_2 and fluoride emissions to air	CO, CO_2 and dust emissions to air; slag to landfill
Coke oven				CO_2, dust and hydrocarbons
Transport of ore, coal and metal on high seas	Crude oil to provide heavy diesel fuel		CO_2, NO_x, SO_x and smoke to the high seas from ships	
Manufacture of engine	Natural gas		CO_2 and NO_x	
Operation of engine			CO, CO_2, NO_x and hydrocarbons	
Disposal of engine at end of its life			Recycling or disposal to land	

Table 12.2   Impacts of the two options listed under categories

Category	Aluminium Engine Option	Cast-iron Engine Option
Abiotic resources	Bauxite, natural gas, crude oil, salt, steaming coal and cooling water	Iron ore, crude oil, coking coal, cooling water
Global warming	CO_2, CH_4	CO_2, CH_4
Acidification	NO_x, SO_x, fluorides, acid mine waste water and leachate	NO_x, SO_x, acid mine waste water and leachate
Photochemical oxidant formation	NO_x, hydrocarbons	NO_x, hydrocarbons
Human toxicity	NO_x, SO_x, fly ash, CO, smoke, fluorides	NO_x, SO_x, fly ash, CO, smoke
Ecotoxicity	Fluorides, cooling water blowdown, acid mine waste water and leachate	Acid mine waste water and leachate
Land use	Red mud, coal mine waste, ash	Slag, coal mine waste

of electricity to electrolyze aluminium oxide to produce metallic aluminium. We need to find from the electricity producer the amount of coal in kilograms required to produce one kilowatt-hour of electricity. We can then find from the coal miner the amount of waste water discharged from the mine, the amount of leachate discharged to water bodies from mine waste, the amount of diesel fuel required for the mining operation and the emissions to the air from combustion of this fuel, all brought to the unit basis per kilogram of coal mined and hence per kilowatt-hour of electricity. Downstream we need to know the amount of water drawn from water bodies to provide cooling water for the steam turbines used to generate the electricity, the gaseous discharges from the power station boiler (fly ash, nitrogen oxides and sulphur oxides) and the amount of ash discharged to some disposal site, all per kilowatt-hour of electricity. In addition, we need to calculate the material and energy inputs and emissions associated with the construction of the mining and electricity-generating equipment. By estimating a load factor for this equipment and its useful life we can bring these quantities also to the basis of per kilowatt-hour of electricity produced.

By following through all these connecting equations the quantities per functional unit (here 200 000 km of car travel) can be calculated. This can be achieved most conveniently by entering the connecting equations into a spreadsheet.

Lifecycle impact assessment
The task of the assessor in the lifecycle impact assessment step is to translate each item in the inventories into impacts that can later be interpreted in the form of lifecycle burdens characteristic of the task under consideration, here the propelling of the motor car. As noted earlier, ISO 14042 requires that this task be made manageable by sorting the resource usages and emissions into categories. This is done in Table 12.2 for our two options. Note that three of these categories, biotic resources, stratospheric ozone depletion and eutrophication, are not relevant here. Thus Table 12.2 expresses the results given in Table 12.1 in various categories of impacts rather than for the various unit processes.

Again, note that we have merely entered these resource uses and emissions qualitatively. In practice all the figures from the various unit processes noted in Table 12.1, which are all expressed per 200 000 km of motor car travel, should be added together to give total quantities of each component per 200 000 km of travel.

ISO 14042 recommends that these quantities be collapsed into a single indicator for each category. As noted earlier, this is quite straightforward for global warming, stratospheric ozone depletion (not relevant here) and acidification (see Figure 12.4), but is more problematic for the other categories for which simple characterization models do not exist (Guinée and Heijungs, 1995). This problem is usually overcome by using models that attempt to standardize the impacts of various resources and emissions within categories in terms of their impact on the category endpoint, for example, in the case of the human toxicity category, human health and morbidity researchers have suggested the use of disability-adjusted life years (DALYs) or years of life lost (YLL) (Udo de Haes et al., 1999a, 1999b). As noted in Figure 12.4, the endpoint of this stage of the LCA is a single indicator figure for each category.

Lifecycle interpretation
At this point the assessor has done all the routine assessment work on the two options and now wishes to interpret the results obtained. In our case study the assessor wishes to know which of the two options will have the lower lifecycle impact. Even if we were quite confident of the seven category indicators we obtained in the lifecycle impact assessment step, how do we compare the two options when we still have seven categories of impacts?

One way would be to make pairwise comparisons for each option. Table 12.3 gives an example.

Table 12.3   Comparison of options by categories of impact

Category	Aluminium Engine Option	Cast-iron Engine Option
Abiotic resources	X (less crude oil required)	
Global warming	X (less CO_2 emitted)	
Acidification		X (less coal burnt)
Photochemical oxidant formation	X (less NO_2 and hydrocarbons emitted)	
Human toxicity		X (no fluorides emitted)
Ecotoxicity		X (no fluorides emitted)
Land use		X (no red mud)

Note: The crosses in the cells indicate which option is superior for that particular category of impact.

Thus if all of these categories had equal weights, the cast-iron engine option would be superior to the aluminium engine option. Had the comparison come out say 6–1 in favour of one of the options we might, with some confidence, declare it to be superior. However a close 4–3 count as we have here is not very convincing.

We could attempt to refine this simple comparison by multiplying the indicator figures for each category by a weighting factor indicating, for example, the relative contribution that the category impact makes to the total consumption of a particular resource or the total quantity of a particular pollutant emitted. The weighted values for each category could then be added together to give a grand index. Another approach would be to multiply the category values by weightings purporting to indicate the importance of the particular category to the sustainability of the 'world' environment. These weightings would be obtained by canvassing the views of world sustainability experts (Sangle et al., 1999; Volkwein et al., 1996). A related approach would be for the assessors themselves to eliminate categories that they believed to be of only local or transient importance, and retain only those that are clearly of 'world' sustainability, for example, the depletion of scarce abiotic resources, permanent damage to biotic resources (for example loss of biotic diversity – not an issue in the case study), global warming, acidification and ecotoxicity, especially of the deep ocean.

However, as discussed earlier, whatever approach is used it is a requirement of ISO 14043 that the method used to interpret the results be made explicit and transparent.

Strengths and limitations

The case study permits us to draw some conclusions about the strengths and possible limitations of LCA.

The greatest strength of lifecycle assessment is that, as implied by its name, it examines the whole system contributing to the delivery of a task or product. When making judgements about alternative ways of doing things it is quite easy to make the error of considering only the immediate manufacture of the product or operation of the task, in this case the consumption of petrol when propelling the motor car. If the petrol consumed and the consequent emissions were the only things considered the aluminium engine would clearly be superior, but if the manufacture of the engine was included the situation could be reversed. A striking example is the comparison of plastic and cardboard packaging: plastic packaging is generally more energy-intensive to make than cardboard packaging and is therefore responsible for greater CO_2 emissions and hence more global warming, apparently making it inferior to cardboard packaging. However when we also consider the final disposal of the packaging we find that, because cardboard decomposes anaerobically in landfill, emitting methane, a powerful global warming gas, cardboard packaging is usually inferior for this impact category (Ross and Evans, 2003b, 2003c).

Because of the global sweep of LCA it is suitable for analysis of sustainability problems, whereas environmental assessment methods such as EIA and risk assessment, which focus on local or regional problems, are not (Cushman, 1993; Stahkiv, 1998). It is therefore a powerful tool for policy analysis for governments or for firms responding to government pressures to develop more sustainable ways of doing things. For example, our case study would be a typical response of a car manufacturing company faced with a government requirement to reduce greenhouse gas emissions in order to meet Kyoto Protocol targets.

However, clearly it has limitations. Because there is no requirement in the standards (AS/NZS ISO 14041, 1999) to register times or places where pollutants are emitted, it is quite possible that some of the category indicators will be meaningless. For example, in our case study most of the precursor gases for photochemical oxidants, NO_2 and hydrocarbons, were emitted at different times and in quite different places, and therefore had no opportunity to react with each other to form oxidants.

Moreover, despite the attempts by many workers to combine emissions within a category to form a single aggregated indicator value, in some categories the outcomes are of dubious validity, especially when the emissions occur in places where these pollutants are quickly broken down and do not contribute to a cumulative effect (Ross and Evans, 2002). This problem is compounded when the category indicators are combined to form a single index of environmental burden for the purposes of eco-labelling or for comparisons between different products. The results may have some validity when nearly all of the processes are situated in a single intensely utilized area such as some parts of Europe or the eastern seaboard of the USA but are suspect when they are scattered all over the world, as in our case study.

Thus LCA has pluses and minuses. This is not to say that we have to seek a compromise in which we have to accept some limitations in order to achieve useful outcomes. Rather, we believe that it is advisable to concentrate the use of LCA on those applications where policy outcomes about sustainability are sought. These will concentrate on global categories such as the use of scarce abiotic resources, global warming, depletion of the stratospheric ozone layer and perhaps in some cases regional problems such as acidification. Transient and localized impacts would be better dealt with by more traditional impact analysis such as EIA and risk assessment.

Notes

1. The term carbon dioxide-equivalent is a relative measure of the radiative forcing effect of a gas to that of carbon dioxide with a reference value of 1.
2. One of the arrows crossing the system boundary represents a flow of petroleum products other than petrol from the oil refinery. This of course does not enter the natural environment but flows into other systems within the economy not shown here. This is an example of the partitioning problem mentioned earlier. Some of the emissions from the oil refinery will have to be debited against the petrol and hence the propelling of motor cars, and some will have to be debited against the other useful petroleum products used for other purposes.

References

AS/NZS ISO 14040 (1998), *Environmental Management – Life Cycle Assessment – Principles and Framework*, Australia: Standards Australia.
AS/NZS ISO 14041 (1999), *Environmental Management – Life Cycle Assessment – Goal and Scope Definition and Inventory Analysis*, Australia: Standards Australia.
Clark, G. and de Leeuw, B. (1999), 'How to Improve Adoption of LCA' *International Journal of Life Cycle Assessment*, 4(4), 184–7.
Cushman, R.M. (1993), 'Global Climate Change and NEPA Analyses', in Hildebrand, S.G. and Cannon, J.B. (eds), *Environmental Analysis: The NEPA Experience*, Boca Raton, FL: Lewis Publishers.
Ehrenfeld, J.R. (1997), 'The Importance of LCAs – Warts and All', *Journal of Industrial Ecology*, 1(2), 41–9.
Guinée, J.B. and Heijungs, R. (1995), 'A Proposal for the Definition of Resource Equivalency Factors for Use in Product Life Cycle Assessment', *Environmental Toxicology and Chemistry*, 14(5), 917–25.
Günther, A. and Langowski, H. (1997), 'Life Cycle Assessment on Resilient Floor Coverings', *International Journal of Life Cycle Assessment*, 2(2), 73–80.
Hunt, R.G. and Franklin, W.F. (1996), 'LCA – How it Came About', *International Journal of Life Cycle Assessment*, 1(1), 4–7.

ISO (2000a), *ISO 14042 Environmental Management – Life Cycle Assessment – Life Cycle Impact Assessment*, Geneva: International Standardization Organization.

ISO (2000b), *ISO 14043 Environmental Management – Life Cycle Assessment – Life Cycle Interpretation*, Geneva: International Standardization Organization.

Klöpffer, W. (1996), 'Reductionism Versus Expansionism in LCA', *International Journal of Life Cycle Assessment*, **1**(2), 61.

McCormick, J. (1995), *The Global Environmental Movement*, Chichester: John Wiley & Sons.

Ross, S. and Evans, D. (2002), 'Excluding Site-specific Data from the LCA Inventory: How This Affects LCIA', *International Journal of Life Cycle Assessment*, **7**(3), 141–50.

Ross, S. and Evans, D. (2003a), 'The Environmental Effect of Reusing and Recycling a Plastic-based Packaging System', *Journal of Cleaner Production*, **11**(5), 561–71.

Ross, S. and Evans, D. (2003b), *Life Cycle Assessment of Refrigerator Packaging: Comparison of Expanded Polystyrene-Polyethylene Packaging and Corrugated Cardboard Packaging*, Melbourne: University of Melbourne.

Ross, S. and Evans, D. (2003c), 'Using LCA to Examine Greenhouse Gas Abatement Policy', *International Journal of Life Cycle Assessment*, **8**(1), 19–26.

Sangle, S., Ram Babu, P. and Khanna, P. (1999) 'Evaluation of Life Cycle Impacts: Identification of Societal Weights of Environmental Issues', *International Journal of Life Cycle Assessment*, **4**(4) 221–8.

Stakhiv, E. (1988), 'An Evaluation Paradigm for Cumulative Impact Analysis', *Environmental Management*, **12** (5), 725–48.

Udo de Haes, H., Jolliet, O., Finnveden, G., Hauschild, M., Krewitt, W. and Müller-Wenk, R. (1999a), 'Best Available Practice Regarding Impact Categories and Category Indicators in Life Cycle Impact Assessment', *International Journal of Life Cycle Assessment*, **4**(2), 66–74.

Udo de Haes, H., Jolliet, O., Finnveden, G., Hauschild, M., Krewitt, W. and Müller-Wenk, R. (1999b), 'Best Available Practice Regarding Impact Categories and Category Indicators in Life Cycle Impact Assessment', *International Journal of Life Cycle Assessment*, **4**(3), 167–74.

Volkwein, S., Gihr, R. and Klöpffer, W. (1996), 'The Valuation Step with LCA: Part II – A Formalized Method of Prioritization by Expert Panels', *International Journal of Life Cycle Assessment*, **1**(4) 182–92.

Watson, R.T. (ed.) (2001), *Climate Change 2001: Synthesis Report*, Cambridge: Cambridge University Press.

13 The Natural Step Framework: from sustainability fundamentals to innovation

Joe Herbertson[1] and Christopher Tipler

Introduction

Personal and organizational change are parallel and reinforcing processes. Significant change does not take place in society unless individuals see the need for it and become committed to it. Change, however, is effected through organizations, be they business-, government-, political- or community-based. All of these are important.

From a sustainable development perspective, change in the way we do business is critical. Business is the predominant vehicle by which goods and services are currently delivered to society. It is where most of us work and where enormous human and material resources can be mobilized.

Sustainable development will be built around practices that do not systematically undermine social well-being or the life supporting systems of nature. At the philosophical level, sustainability therefore means a new way of thinking about how individuals relate to each other in society and to the natural systems we all depend on. At the levels of actions and outcomes, business is the arena where most of today's sustainability problems are manifest and most of tomorrow's solutions will need to be found.

If society is to make a successful transition to operating in a sustainable fashion, a common understanding and common language need to emerge about what that implies, at least in principle. Unless innovation and value creation are fundamental to the approach taken towards sustainable development and environmental management, there will be no persistent business case. If sustainability has to rely on some combination of altruism, guilt and compliance, industry momentum will be lost and business initiatives will tend to flounder.

The Natural Step

Working with businesses to create positive role models for sustainable development is a priority for The Natural Step, which was founded in Sweden some 15 years ago by Dr Karl-Henrik Robèrt. The Natural Step now operates in ten countries around the world and its activities combine education, research and consulting functions. An emphasis on principles and systems thinking helps to reinforce its non-adversarial style and consensus-building ethos. The business model assumes that companies, which systematically eliminate unsustainable practices, will be the successful ones. It is an approach of 'enlightened self-interest', which combines the elements of innovation and risk management.

The evolution of The Natural Step Framework and its practical application by pioneering companies is described in detail by Robèrt (2002), with further business case studies given by Nattrass and Altomare (1999 and 2002).

In Australia The Natural Step is working with leading companies in a growing number of important sectors, including resources, manufacturing, chemicals and water services.

Our business clients are typically searching for a coherent sustainability framework, which can (1) place the company's array of activities and reporting systems into a meaningful context or 'story', (2) link sustainable development with value creation and business leadership and (3) engage the workforce.

The theoretical concepts presented in this chapter, which of course draw heavily on the global foundations of The Natural Step Framework, have been reinforced and sharpened in the process of working with our Australian industrial clients.

The sustainability framework

A systems approach

The Natural Step has been instrumental in developing a strategic sustainability framework. The framework represents a coherent and scientifically robust approach to planning in complex systems that can be applied in very diverse situations. It is designed to help organizations move systematically towards sustainability, finding their own distinctive pathways. The flexibility of application is grounded on fundamental principles for a sustainable society that are universally valid. These principles are referred to as the Four System Conditions, which are described in more detail below.

The Natural Step System Conditions, and the scientific analysis that underpins them, make an important contribution to the field of sustainability in their own right. However, what makes them particularly powerful is that they are designed to fit within a 'backcasting' strategic planning process. The backcasting process is driven from a notion of the sustainable future, rather than being projected out from an unsustainable present ('forecasting'). The combination of the System Conditions and the backcasting process within a coherent systems framework is the unique contribution of The Natural Step.

Backcasting from principles

In backcasting, a step-wise methodology, simply referred to as ABCD, links strategic analysis to action. The process can be carried out from a global perspective or from the perspective of a sub-system, such as a company, an industry sector or a region.

The starting point, Step A, defines *in principle* the desired future when the organization is able to fulfil its purpose successfully within sustainability constraints (System Conditions). The sustainable future state then provides the reference point for an evaluation, Step B, of the current state of the organization. The gap between sustainability and current reality (A–B) provides the creative tension for innovation, Step C, where strategies, future options and ideas for new products, processes and business relationships can emerge. In the implementation stage, Step D, the focus shifts to action plans that create an effective pathway towards sustainability that is flexible, robust and economically viable. Effective communication is important throughout the process.

The 'backcasting from principles' methodology provides the compass and map for the sustainability journey, rather than relying too much on the rear view mirror for bearings. This is particularly important when there is a need to drive in a new direction.

As with the systems for managing quality and safety, the ABCD process is cyclical by nature. The ABCD process can be repeated in an organization from time to time as knowledge deepens and early wins build the confidence to stretch further, expanding the time horizon for action, innovating more imaginatively across traditional boundaries and progressively widening the zone of stewardship and business leadership.

In complex situations it can be very difficult to build consensus around advocated solutions to specific problems, when there is no prior shared view of the desired outcome in principle for the system as a whole. Backcasting helps to build consensus at this level and therefore reinforces the non-adversarial style of The Natural Step.

System hierarchy
Dr Karl-Henrik Robèrt (2000) and other global leaders in sustainability (Robèrt et al., 2002) have identified five levels within an effective and comprehensive sustainability planning framework. A modified version[2] of the system hierarchy is presented in Table 13.1. The Natural Step Four System Conditions are Level 2 principles of sustainability designed for backcasting.[3] Backcasting itself is a Level 3 process for shaping development activities.

In this systems hierarchy, what could otherwise be seen as competing sustainability models become complementary and synergistic, since all levels are necessary for strategic analysis, planning and innovation.

At Level 3, for example, models such as Eco-efficiency,[4] Factor 10,[5] Cleaner Production, Industrial Ecology, Zero Emissions and Natural Capitalism provide guiding principles for reducing or eliminating materials flow and practices that are unsustainable because of their intensity or toxicity. Technical strategies for dematerialization and substitution underpin the transition to sustainability by promoting resource productivity, waste minimization, closed-loop cycles, less encroachment on nature, eco-effective design, the use of abundant and degradable materials and a primary focus on services rather than goods.

Also important at Level 3 are the process principles for sustainable development, which include the precautionary principle,[6] transparency, equity, stakeholder engagement, corporate governance and the need for flexible, economically viable pathways to sustainability.

At Level 5 'Triple Bottom Line' reporting systems, the Global Reporting Initiative, Life Cycle Assessments, Ecological Footprint, Factor X (see Note 5) and management systems and standards (such as ISO 14001) are useful tools for monitoring progress, provided of

Table 13.1   Five-level hierarchy in the systems approach to sustainability

Level 1: 'The system'	Society in the eco-sphere	The constitution and boundaries of the overall system: the world we live in and what makes life possible; social and ecological principles
Level 2: 'Success'	Principles of sustainability	Defining the goal: conditions for a state of social sustainability within the eco-sphere; principles for a favourable outcome, the desired endpoint
Level 3: 'Strategies'	Principles of sustainable development	Strategic guidelines for the process to reach a sustainable outcome; technical strategies and process principles that underpin the transition to sustainability
Level 4: 'Actions'	The realm of action	Specific activities, projects and initiatives that contribute towards the goal of sustainability; concrete actions not principles
Level 5: 'Tools'	Metrics and methodologies	Frameworks for monitoring and reporting progress: are the actions being implemented effectively and are the consequent outcomes desirable?

course that they serve the higher level principles and actions and do not become the substitute for strategy.

At present it is quite common for organizations to focus their sustainable development thinking around Levels 3, 4 and 5, without an articulated view of the end goal of sustainability. Given the highly complex systems involved, the most effective approach to sustainability[7] is one based on all five levels, with each level being shaped by the levels above. The Natural Step System Conditions provide practical Level 2 principles that connect the vision of sustainability to the realm of action through the backcasting process. The System Conditions are grounded in Level 1 science pertinent to the sustainability of society within the biosphere.

The System Conditions

For society to be sustainable it cannot afford to undermine systematically the natural systems that make life possible or to undermine systematically people's ability to meet their needs. Four System Conditions have been developed to help identify the separate mechanisms by which systemic ecological and social harm can be done. They are designed to identify unsustainable practices; they have been developed within the constraints of conservation principles, thermodynamics and the laws of nature; they reflect the role of natural cycles and bio-geo-chemical evolutionary processes in making life possible;[8] and they build on an understanding of fundamental human needs and the material foundations of the quality of our personal and social lives.[9] The System Conditions are summarized in Table 13.2.

From this it follows that the strategic objective of organizations is to learn how to fulfil their purpose and prosper without contributing to:

- the systematic build up in nature of substances extracted from the Earth's crust;
- the systematic accumulation in the biosphere of industrial substances foreign to nature;
- the long-term degradation of natural systems through physical impacts on land, water and biodiversity; and
- social, economic and political conditions that undermine the capacity for people to meet their needs.

The System Conditions are a call for creativity since reactions like 'No mining', 'No artificial chemicals', 'No agriculture' are too simplistic. The complex material cycles that underpin society have to be managed in accordance with the System Conditions, and that brings innovation to the core of how the framework is applied.

Sustainability-driven innovation

Given the enormous challenges in creating a stable relationship between human socio-economic activity and natural systems (see Note 5), the path of sustainable development is by necessity one of major innovation and culture change, ultimately on the scale of the Industrial Revolution and the Renaissance. Innovation is necessary in all aspects of the production and consumption cycles that constitute the material foundations of society. Moreover, a deep paradigm shift is needed. Large-scale behavioural change and innovation will flow from a genuine realization that we are an integral part of nature and that we all depend on natural systems for pleasure, vital goods and services and survival.

Table 13.2   The four System Conditions

SC 1	In the sustainable society, nature is not subject to systematically increasing concentrations of substances extracted from the Earth's crust This means substituting certain minerals that are scarce in nature with others that are more abundant, using all mined materials efficiently, and systematically reducing dependence on fossil fuels
SC 2	In the sustainable society, nature is not subject to systematically increasing concentrations of substances produced by society This means systematically substituting certain persistent and unnatural compounds with ones that are normally abundant or break down more easily in nature, and using all substances produced by society efficiently. 'Dilute and disperse' is not a sustainable strategy
SC 3	In the sustainable society, nature is not subject to systematically increasing degradation by physical means This means drawing resources only from well-managed eco-systems, systematically pursuing the most productive and efficient use both of those resources and land, and exercising caution in all kinds of modification of nature, for example, overharvesting and introductions
SC 4	In the sustainable society, people are not subject to conditions that systematically undermine their capacity to meet their needs This means checking whether our behaviour has consequences for people, now or in the future, that restrict their opportunities to lead a fulfilling life; this includes asking ourselves whether we would like to be subjected to the conditions we create (The Golden Rule)

Table 13.3   Five key elements of sustainability-driven innovation

1. Understanding the goal
2. Intelligent constraints
3. Systems approach
4. Engagement
5. Flexible pathways

Not only is innovation required to create a sustainable future, the systems approach described above actually promotes and reinforces innovation by the way the framework is formulated and applied. Five key elements of the framework that facilitate innovation and value creation are listed in Table 13.3.

Supporting data

The Natural Step systems approach to sustainability has provided the context, or 'lens', for an in-depth portfolio analysis of a diverse range of projects (Herbertson, 2004).[10] The projects represented a wide range of commercial drivers, technologies and stewardship focus; the patterns that emerged are believed to be relevant well beyond the resource sector

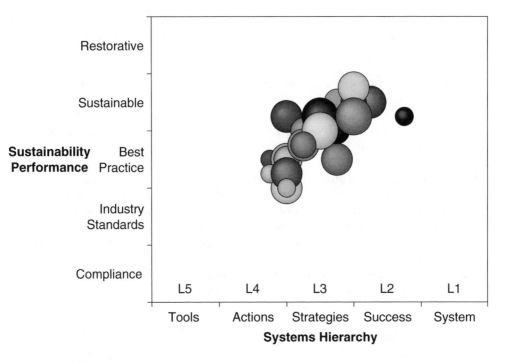

Note: Discs represent the positioning of individual projects.

Source: Herbertson, 2004.

Figure 13.1   Relative sustainability performance of project outcomes in relation to the level of thinking that shaped the project objectives

from which the data was drawn.[11] Some highlights of the portfolio analysis are included in the discussion below to reinforce the observations being made.

Understanding the goal
The innovative process becomes more focused and effective when there is a clear sense of a goal to aim for. Breakthrough innovation implies moving beyond current best practices into progressively unproven territory, which means the goal is to some extent intangible, and best expressed in principles. For sustainable development the System Conditions describe success in principle and provide a Level 2 goal for the innovators (see Table 13.1). The backcasting process provides the creative tension for innovation by highlighting the gap between sustainability and current reality. The gap in many cases will be large and stretch targets will usually stimulate creative thinking.

The portfolio analysis highlights the power of driving projects from a higher-level systems perspective.[12] Figure 13.1 shows that the innovative shift beyond best practice to sustainable outcomes is generally associated with a progressively clearer focus on the desired end game. For instance, striving for a greenhouse-neutral solution is Level 2 thinking whereas improving energy efficiency is Level 3 thinking.[13] In simple terms, you are more likely to get sustainable outcomes if you consciously set out to achieve them. Understanding the goal in terms of fundamental principles enhances the innovative process.

Intelligent constraints
Innovation is borne of constraints. We have already seen this above: operating within the fundamental constraints of sustainability (explicit or implicit alignment with the System Conditions) actually increases the likelihood for innovation beyond current best practice.

The discipline of finding economically attractive solutions is the second powerful constraint that drives and shapes the innovative process (particularly Step D in backcasting).

For a company, there is little point in developing sustainable solutions and yet failing as a business. Sustainability activities need to be aligned with the business as a whole by increasing the capacity to create value relative to ecological impacts. The self-imposed constraints of sustainability and economic viability are intelligent ones for a business to adopt. When these constraints are integrated as a business strategy they can be a powerful driver of innovation.

The portfolio analysis demonstrates how sustainable development can reinforce business success (see Figure 13.2). Innovative projects that move beyond best practice from a

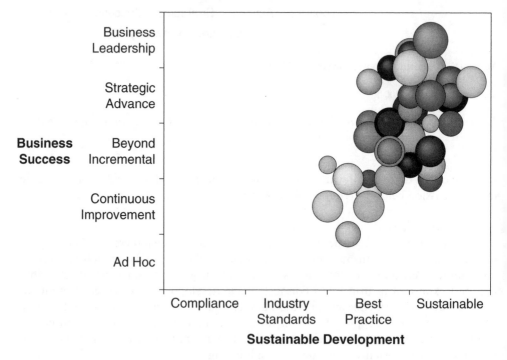

Note: Discs represent the positioning of individual projects.

Source: Herbertson, 2004.

Figure 13.2   The relative potential for projects to contribute to business and sustainability performance

sustainability perspective also tended to be very attractive from a business perspective, well beyond the rather defensive notion of 'licence to operate'.

Systems thinking

The sustainable development objective of greatly increasing the capacity to create value (satisfying human needs in an economically viable fashion) relative to ecological impacts (systemic harm to nature and people) is shown schematically in Figure 13.3, as a generic representation of the material foundations of society. Conceptually sustainable development requires major (by a 'Factor X' [see Note 5]) increases in the ratio of V/I. This can be achieved through a persistent emphasis on service, equity, diversity and viability, accomplished in parallel with significant dematerialization and detoxification of the economy. It is impossible without tremendous innovation throughout our socio-economic systems.

At a more practical level, perhaps the most important benefit of a systems approach to sustainability is the type of innovation it can promote, namely innovation across traditional boundaries. Sustainability is a 'system problem' and therefore needs 'system solutions' that, for instance, exploit regional synergies, streamline value chains, close production and consumption loops, create industrial ecologies and establish new agri-industry linkages.

Implicit in whole systems thinking is taking a multi-disciplinary approach, which brings in new perspectives and stimulates the innovative process further.

The portfolio analysis highlights the power of adopting a whole systems approach, especially when underpinned by root cause, fundamental thinking (see Figure 13.4). With a linear approach, a direct link is perceived between problem and solution, with limited opportunity for intuitive creativity. With systems thinking wider linkages are seen between many coupled factors and issues; system solutions therefore tend to yield multiple benefits. A fundamental approach focuses on root causes and the solutions are therefore lasting ones. The portfolio analysis also confirms that projects shaped by root cause, systems thinking are also good for business (as reflected in Figure 13.2).

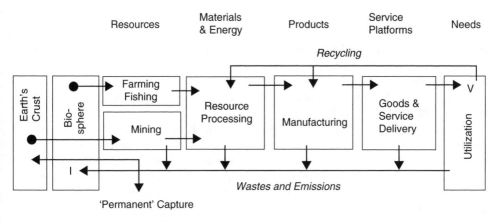

Source: Herbertson and Sutton, 2002.

Figure 13.3 Schematic representation of the material flows that underpin economic activity

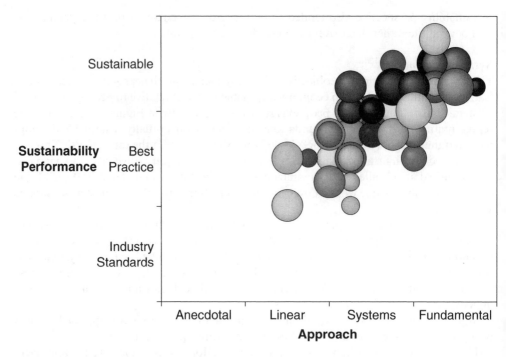

Note: Discs represent the positioning of individual projects.

Source: Herbertson, 2004.

Figure 13.4   Relative sustainability performance as a function of the kind of thinking that underpins projects

Engagement

A follow-on from taking an extended systems view of sustainable development is the need to reach beyond traditional business relationships and engage people in the search for sustainable solutions. This is in part a question of accessing skills and talents and in part about widening the zone of stewardship. Perhaps even more importantly, it is about tapping into the commitment and passion of people who are motivated by the challenge of sustainability ('Doing well by doing good').

Sustainability is about fundamental human needs. It therefore touches profoundly on the human condition, on the prevailing values in society, on the place of individuals in society and on the relationship between man and nature. Innovation is at its most powerful when it can unlock this human factor and connect with human needs for meaning and creativity (see Note 8). Backcasting from fundamental principles has the capacity to engage people in this way.

The portfolio analysis brings out how sustainable development builds on a culture of broader collaboration and extended stewardship. This means engaging with, and learning from, people well beyond the business unit, as shown in Figure 13.5.

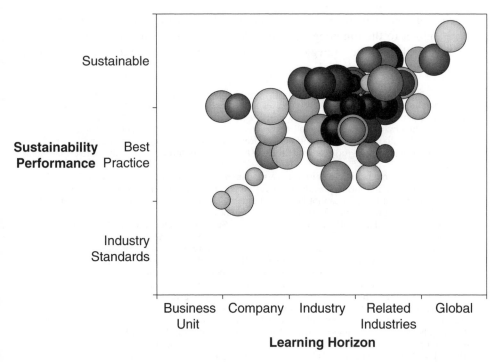

Sustainable

Sustainability   Best
Performance   Practice

Industry
Standards

Business   Company   Industry   Related   Global
Unit   Industries
Learning Horizon

Note: Discs represent the positioning of individual projects.

Source: Herbertson, 2004.

*Figure 13.5   Relative sustainability performance as a function of the extent of
collaboration beyond the immediate business environment*

Flexible pathways
Whilst Step C in the backcasting process creates options and opportunity, Step D builds
the flexible pathways to success. It must be emphasized that successful and sustained
innovation in a business setting will combine persistent incremental improvements with
the occasional breakthroughs. If sustainability is seen as 'blue sky' only, it will be relegated
to the periphery of a business. Early wins build confidence and help justify further innovation and a deepening and widening of the zone of stewardship. The sustainability
imperative does not overcome the need for discipline in capital investment. A robust and
flexible platform will consist of an evolving portfolio of new technology and business
development initiatives that combine a focus on short-term delivery of value with a strategic commitment to sustainable business leadership.

Conclusion
The Natural Step Framework has been developed from the perspective that sustainable
development, business leadership and innovation are inextricably linked. Viable economic
activity that meets human needs and is in harmony with nature will only be possible with
fundamental and widespread behavioural and technological change. The Natural Step

approach is designed to help organizations see through the complexity of the issues involved and get to the fundamental principles that can guide an effective transition to sustainability, which is the strategic context for environmental technology management.

Notes

1. Corresponding author: Dr Joe Herbertson, The Crucible Group Pty Ltd, PO Box 473, Toronto, NSW 2283, Australia, Tel: 02 49598856; mobile 0408 598856, Joe.Herbertson@thecrucible.com.au
2. The system hierarchy presented here follows exactly the same order as Robèrt et al. (2002) but the principles of dematerialization and substitution have been placed in Level 3 rather than Level 2 to distinguish clearly between principles of sustainability (destination) and principles of sustainable development (direction).
3. The Brundtland Commission definition of sustainability (Brundtland, 1987), 'meeting the needs of the present without compromising the ability of future generations to meet their needs', is an aspirational Level 2 principle, although it cannot be applied directly in backcasting exercises without identifying the principal mechanisms by which future generations will be compromised.
4. A good example is The World Business Council for Sustainable Development (WBCSD) framework (DeSimone and Popoff, 1997, p. 47), which defines eco-efficiency as 'the delivery of competitively priced goods and services that satisfy human needs and bring quality of life, while progressively reducing ecological impacts and resource intensity throughout the life cycle, to a level at least in line with the earth's carrying capacity'. The WBCSD then provides seven elements of eco-efficiency: (1) reduce the material intensity of goods and services, (2) reduce the energy intensity of goods and services, (3) reduce toxic dispersion, (4) enhance material recyclability and level of recycling, (5) maximize sustainable use of renewable resources, (6) extend product durability and (7) increase the service intensity of products.
5. Factor X (e.g. Factor 10) and related tools such as the 'ecological rucksack' provide Level 5 metrics for dematerialization. Using these tools, the Wuppertal Institute, the Factor 10 Institute and the Dutch programme Sustainable Technology Development have concluded that the material intensity of the industrialized world needs to be reduced by a factor of 10 or even more to avoid systematic degradation of the biosphere (see, for instance, Schmidt-Bleek, 1994 and Weaver and Schmidt-Bleek, 2000). With these analytical insights the Factor 10 model provides stretch targets for sustainable development and in this respect belongs to Level 3.
6. The precautionary principle is not an excuse for inaction, since doing nothing violates the precautionary principle if Level 2 sustainability principles are being violated, even though the final impacts are not yet obvious or fully understood. ('Lack of full scientific certainty shall not be used as a reason for postponing cost-effective measures to prevent environmental degradation' United Nations Conference on Environment and Development (1992) – this puts the emphasis on innovation to find cost-effective means.)
7. In one study Dunphy et al. (2003) conducted a survey of selected corporations, which showed that among the leaders – who had moved beyond eco-efficiency and adopted a proactive strategic approach to sustainable development ('third wave') – a majority had worked with The Natural Step Framework at some stage in their evolution (private communication, A. Griffiths, Senior Lecturer, UQ Business School, University of Queensland).
8. Bio-geo-chemical evolution and the complex cycles of nature have created the conditions for life on Earth, including sedimentation and detoxification of heavy metals and other toxins in the Earth's crust; decarbonization and oxygen enrichment of the atmosphere; establishing a protective ozone layer; purification of water; the carbon and nitrogen cycles; and progressive biodiversity over time making natural systems robust.
9. There is an infinite array of ways to fulfil or frustrate human needs, reflecting the richness and complexity of human behaviour and the material foundations of society. Fundamental human needs are permanent and should not be confused with the ever-changing platform of material goods and services that address them. Max-Neef (1991) has identified nine fundamental human needs: Subsistence, Protection, Affection, Understanding, Participation, Recreation, Creativity, Identity and Freedom, conditions that systematically undermine people's ability to satisfy these needs are contrary to a sustainable society.
10. A full account of the portfolio analysis, including assumptions, parameters, methodology and results, is being published separately. The study formed the basis of a Masterclass in 'Sustainability and Innovation' at the Inaugural Global Sustainable Development Conference of the Minerals Council of Australia. The Masterclass included practical case studies drawn from the portfolio. Case studies are a powerful way to learn but are by nature anecdotal. For the present purposes the content of individual projects is less relevant than the patterns that emerge from the analysis of the portfolio as a whole.
11. The portfolio selected for analysis contained 60 projects being carried out by two global resource companies, BHP Billion and Rio Tinto. The combined projects represent A$85m in development funds. They

cover projects across many mineral sectors, particularly aluminium, iron and steel, energy and precious minerals. Each project was rated against 13 parameters and then the portfolio was analysed for insightful trends in the data.

12. In the portfolio analysis sustainability performance has been placed in a progression: Compliance (a focus on regulatory pressures and the licence to operate); Industry Standards (operating with accepted good practices); Best Practice (the most advanced solutions commercially proven); Sustainable (within scope of the project, outcomes satisfy the System Conditions); Restorative (fixing past mistakes by others).

13. Reactive responses to problems is Level 4 thinking. It is not unknown for companies to adopt Level 5 measurement systems (for example, Global Reporting Initiative – GRI) without having higher-level clarity. This approach (reversing the system hierarchy) is a very ineffective way to drive sustainable development activity.

References

Brundtland Report (1987), *Our Common Future*, World Commission on Environment and Development: Oxford University Press.

DeSimone, L. and Popoff, F. (1997), *Eco-efficiency: The Business Link to Sustainable Development*, Cambridge, MA: MIT Press.

Dunphy, D., Griffiths, A. and Benn, S. (2003), *Organisational Change for Corporate Sustainability*, London: Routledge.

Herbertson, J. (2004), 'Sustainability and Innovation', Masterclass within the Inaugural Global Sustainable Development Conference, Melbourne, Australia: Minerals Council of Australia.

Herbertson, J. and Sutton, P. (2002), 'Foundations of Sustainable Resource Processing', *Proceedings of Green Processing 2002*, Melbourne, Australia: Australasian Institute of Mining and Metallurgy, pp. 31–8.

Max-Neef, M. (1991), *Human Scale Development: Conception, Application, and Further Reflections*, New York, NY: The Apex Press.

Nattrass, B. and Altomare, M. (1999), *The Natural Step for Business: Wealth, Ecology and the Evolutionary Corporation*, Gabriola Island, BC, Canada: New Society Publishers.

Nattrass, B. and Altomare, M. (2002), *Dancing with the Tiger: Learning Sustainability Step by Natural Step*, Gabriola Island, BC, Canada: New Society Publishers.

Robèrt, K.-H. (2000), 'Tools and Concepts for Sustainable Development – How do they Relate to a General Framework for Sustainable Development, and to Each Other', *Journal of Cleaner Production*, **8**, 243–54.

Robèrt, K.-H. (2002), *The Natural Step Story: Seeding a Quiet Revolution*, Gabriola Island, BC, Canada: New Society Publishers.

Robèrt, K.-H., Schmidt-Bleek, B., Aloisi de Larderel, J., Basile, G., Jansen, J.L., Kuehr, R., Price Thomas, P., Suzuki, M., Hawken, P. and Wackernagel, M. (2002), 'Strategic Sustainable Development – Selection, Design and Synergies of Applied Tools', *Journal of Cleaner Production*, **10**, 197–214.

Schmidt-Bleek, F. (1994), *Wieviel Umwelt braucht der Mensch?* Basel: Birkhauser.

United Nations Conference on Environment and Development (1992), 'Rio Declaration on Environment and Development', http://www.unep.org/Documents.Multilingual/Default.asp?DocumentID=78&ArticleID= 163, accessed 7 April 2006.

Weaver, P. and Schmidt-Bleek, F. (eds) (2000), *Factor 10: Manifesto for a Sustainable Planet*, Sheffield: Greenleaf.

14 Integrating human and ecological factors: a systematic approach to corporate sustainability

Suzanne Benn, Dexter Dunphy and Andrew Griffiths

Introduction

In this chapter we propose an integrated perspective on organizational change, which is designed to progress the corporation to a position of both human and ecological sustainability. Human sustainability is defined here as the development and fulfilment of human needs whilst ecological sustainability is the protection and renewal of the biosphere. The chapter defines key steps along the way to this 'organization of the future' and explains ways of achieving an incremental – or, in some cases, a transformative – transition to the fully sustainable and sustaining corporation. Case studies of both incremental and transformative change are also provided to illustrate how organizations have moved toward ecological sustainability through the development of their human sustainability.

It is this internal driver of ecological sustainability that is our point of departure from theorists of ecological modernization who interpret corporate sustainability as corporate environmentalism. According to this latter perspective, corporations develop more sustainable products and processes in response to external stimuli such as government policy-making or social movements, and according to procedures set by the technical requirements of environmental restructuring (Dryzek, 1997; Hajer, 1995; Mol, 1997). As such, ecological reforms are seen as driven by market forces that have been manipulated by government policy in the form of incentives, subsidies and sanctions or by the increasing importance of social movements in decision-making institutions.

While we agree that international and national governments should continue experimenting with a variety of policy incentives and models of governance to ensure corporate accountability, we argue that a perspective that looks solely to the role of external agents of corporate change misses the crucial relationship between human and ecological sustainability within the individual firm. Recent work has begun to explore this relationship, and has indicated a link between human resource policies and the successful implementation of the Environment Management System (EMS) as well as its maintenance as a strategic business and risk management tool (Daily and Huang, 2001; Dunphy et al., 2000; Wilkinson et al., 2001). This research has concluded that EMS programmes are more successful if factors such as training, empowerment, teamwork and rewards are addressed, and suggests a relationship between human and ecological sustainability in the form of a phase model.

Building upon this work, the argument presented in this chapter is that the organizational move to ecological sustainability is reliant upon human sustainability: the development by the organization of the human capabilities and skills that enable more consistent compliance, and the implementation of eco-efficiency measures and forward planning for sustainability. In other words, we argue that the concepts of human and ecological sustainability need to be merged. Progress towards corporate environmental sustainability is a deliberative process that relates to human decision-making and capabilities. Similarly, corporate social responsibility relates to the capability or capacity of managers to look strategically at the organization's long-term future in the context of both local and global communities.

Drivers of change

An integrated perspective on corporate sustainability is necessary if we are to capture the complex set of corporate responses to the wide array of influences impinging on the contemporary marketplace. These forces have meant a merging of the internal and the external spheres of the firm's interest, resulting in a recognition by management that the firm needs to respond in ethical terms to primary stakeholders (those who are essential to the firm) as well as secondary stakeholders (those who are not essential to the firm but who can influence or are influenced by it). External drivers include the actions of internationally mobilized human rights and environmental activists as well as international and national agreements and regulations concerning environmental protection and social and environmental justice. Internal drivers include management push for business advantage as well as employee awareness and the drive to accumulate new forms of capital.

The need to accumulate new forms of capital is a recent force for corporate change. For instance, companies now trade in symbolic capital. Their interaction with the consumer and with their workforce is symbolic as well as economic (Beck, 1999; Tsoukas, 1999). Many corporations now market themselves as compliant according to standards set by governments, by co-regulatory arrangements between government, industry and non-government associations, or by a range of other non-government associations. The leadership for this change is also coming from managers who recognize the benefits for the firm in participating in the new co-regulatory system. That is, they see reputational rewards for the firm in voluntarily working towards social and environmental standards that are not enforced by government.

Public (and employee) awareness is running ahead of most government and corporate initiatives. For example, Energy Australia's Pure Energy scheme, which offers customers energy derived from renewable sources for a higher price, has not been able to keep up with the demand for clean energy (Peatling, 2003, p. 1).

Knowledge management is drawing attention to the value of an organization's human resources. Motivation, qualifications and commitment, when combined with a significant store of 'corporate memory', are a major asset to the corporation. As prized employees hunt for the firm with a strong sense of values, there are real rewards in becoming an employer of choice. Firms need employees who can give high levels of customer service and who are sufficiently motivated and inspired to achieve the long-range philosophy of the firm. All industry sectors accept the importance of employee loyalty and commitment (Wilkinson et al., 2001).

Shareholders and investors are also looking to more than financial success in the assessment of performance. Their selection of investments increasingly takes into account reputation and performance on the more long-term factors of both social and ecological sustainability. Investors are also placing more value on the human capabilities and commitment that the organization has built. In the new economy the building of knowledge systems, social capital and other strategies designed to increase and sustain human

Table 14.1   A developmental model of human and ecological sustainability

	Human Sustainability	Ecological Sustainability
Level 1 Rejection	Employees and subcontractors exploited. Community concerns are rejected outright	The environment is regarded as a free good to be exploited
Level 2 Non-responsiveness	Industrial relations (IR) a major issue with the emphasis on cost of labour. Financial and technological factors exclude broader social concerns. Training agenda focuses on technical and supervisory training	Environmental risks, costs, opportunities and imperatives are seen as irrelevant
Level 3 Compliance	Human resources (HR) functions such as IR, training, total quality management (TQM) are instituted but with little integration between them	Ecological issues likely to attract strong litigation or strong community action are addressed
Level 4 Efficiency	Technical and supervisory training augmented with interpersonal skills training. Teamwork encouraged for value-adding as well as cost-saving purposes. External stakeholder relations developed for business benefits	ISO 14000 integrated with TQM and occupational health and safety systems or other systematic approaches with the aim of achieving eco-efficiencies. Sales of by-products
Level 5 Strategic proactivity	Intellectual and social capital is used to develop strategic advantage through innovation in products/services	Proactive environmental strategies are seen as a source of competitive advantage
Level 6 The sustaining corporation	Key goals both inside and outside the firm are the pursuit of equity and human welfare and potential	The firm works with society towards ecological renewal

capability are vital to corporate performance. More and more employees have strong expectations for workplace safety and heightened environmental awareness; they are searching for more meaningful work, particularly for work that makes a social and ecological contribution as well as providing an income.

Concerns about developing social capital, too, can explain why companies may wish to support local community participation in corporate decision-making. Social capital is fundamental to the successful working of new organizational forms such as the network organization and communities of practice. Companies are increasingly dependent on employees who can work cooperatively and contribute to the social capital of the organization (Sagawa and Segal, 2000). Yet the dangers of 'over-embedding' (Adler and Kwon, 2002), where the build up of social capital within the organization means mistrust of external stakeholders or external experts, are also increasingly recognized.

In this context, the principles of industrial ecology – of community interconnectedness and cooperation – can be seen as a model for the way forward for a corporation wishing to move towards sustainability (Ehrenfeld, 2000). They provide a framework for new levels of resource productivity and generate new strategic directions. If the point is taken that corporations should serve as moral entities (Starkey and Welford, 2001), then these principles should show us the way. Our developmental model (Table 14.1) indicates how human and ecological sustainability are inter-related developments along this path. At each step of the way, new human capabilities or characteristics of the organization enable further progression of ecological sustainability.

Building capability for sustainability
Moving to compliance

According to the developmental model shown in Table 14.1, the organizations most in need of change are those in the rejection and non-responsive phases. Because expenditure on personal and professional development is kept to a minimum and the community concerns are rejected completely, firms in the rejection phase remain ignorant of the benefits of progressing along the sustainability spectrum. In the non-responsive organization, financial factors so dominate corporate decision-making that often both the human resource and environment factors are taken for granted, usually with the exception of industrial relations (IR) issues. Management policy-making in the non-responsive organization focuses on ensuring that the workforce is easily moulded to the corporate will in order to derive maximum financial return. Barbara Ehrenreich's (2001) work *Nickel and Dimed* provides us with graphic descriptions of the human resources and occupational health and safety (OH&S) policies of many such organizations.

For organizations in these early developmental phases, the move to compliance requires cultural change. Compliance is not just a matter of changing policies and values – it involves enlisting the commitment of all employees and building practical procedures that everyone in the organization can understand. Moving to a risk-free, ongoing position of compliance is not possible without this commitment.

Since moving to compliance is dependent upon both hard and soft variables of change, the human resource function is crucial. A major barrier to the organizational learning required for change may be that employees see the natural environment as a technical issue and difficult to understand. Attaining compliance with environmental laws may require new technical solutions or programmes being implemented to enable cleaner production techniques and improved worker health and safety. These often include operating within specific, technical considerations such as 'consents' or specified standards, eliminating hazardous operations, addressing waste disposal and water quality issues, and complying with planning requirements for new developments. Consequently ongoing HR support may be required if these measures are to be successfully introduced and maintained. Having encountered this challenge, management at KLM, for instance, has developed an awareness-raising course and have instigated a regular staff information day.[1]

As surveillance at all levels of the corporation is not possible, compliance requires the internalization of new standards by people at all levels of the organization. Major litigation costs, for example, often result from employee carelessness. So attitudinal change is essential to ensure respect for all legislative requirements. Values and attitudinal changes are required where employees have not been aware of their responsibilities to environmental legislation as members of the organization. If the collective assumption within the organization has been that exploitation of community groups, employees or the environment is justified by economic benefits to the corporation, the cultural values of the organization are in need of reformulation. Such cultural change cannot be enforced – it only evolves with leadership and over time. Structural change underpins and provides some of the social organization needed for the shift to compliance. However, the enthusiasm and commitment of the workforce at all levels has to be generated. Again management at KLM is now recognizing that new ideas on environmental sustainability need to come from the bottom up.[2] In effect both top-down and bottom-up approaches are needed, neither being sufficient on its own (Stace and Dunphy, 2001).

Community consultative committees can facilitate the internal and external cultural changes required. Placing employees in face-to-face situations with those who will be most affected by their work activity can have a powerful impact on attitudes. Employee empowerment is a key aspect of these initial changes. For instance, if technological developments are to lead to a more sustainable workplace, employees with low standards of qualification must still be able to succeed in the company.

There is a limit to what can be achieved by an ad hoc approach to compliance. A more systematic approach to compliance usually involves making structural changes and developing new coordinating mechanisms to create a systematic, integrated compliance framework. The usual framework for environmental compliance is an Environment Management System (EMS). However the EMS needs to be supported by a range of human resource and community development tools or techniques. These techniques focus on two areas, the management and operation of facilities and the design, manufacture and delivery of products or services. For the latter various laws such as those concerned with the emissions of toxic substances (for example, the Toxic Substances Control Act in the US) must be complied with. At this stage strong adherence to codes of occupational health and safety and well advertised workplace safety precautions are leading areas of change for human sustainability. Progressing this requires an OH&S policy and an OH&S committee with representation from employees and managers.

Compliance requires the implementation of both environmental and human resource reviews. A stakeholder analysis that defines roles and responsibilities, including those of senior management, is a key aspect of the review. For instance, the Dutch investment bank, the ING Group, has identified its core stakeholders as employees, clients, shareholders and society. To ensure compliance, it has established a Code of Conduct and Business Principles as internal regulations applicable to these stakeholders. A compliance organization has been created within ING and an Executive Board member serves as Head of Group Compliance.[3] The relationship between human resource policies and environmental sustainability is reflected in the increasing role played by external organizations in monitoring for compliance. One reason for this shift is the trend toward flatter organizational structures, more participative management styles and the implementation of self-managed teams, task forces and project groups, all of which can make compliance monitoring more difficult.

In sum, establishing a structured, systematic response to compliance requirements depends on implementing learning strategies and appointing change agents or specific change units, such as compliance committees, environmental management committees, human resource and environmental health and safety units, to gatekeep information.

Moving beyond compliance

To go beyond compliance, these responsibilities need to move from the periphery into the core function of the firm (Hoffman, 1997). They need to be integrated into such departments as marketing and product and process design. Importantly these reforms will succeed only if employees recognize that the reforms create value for them. Employee ownership of the changes takes us into the area of cultural change within the context of an overall sustainability plan. The organizational change strategy developed to achieve the sustainability plan must be selected to suit the mission and context of the firm. Too often, environmental managers and human resource managers move in different worlds from strategic business managers. The environmental manager is technically focused and often unaware of business objectives, except in the broadest sense. Similarly the human resource manager's interest may be in training and other intra-organizational needs. As a result, if sustainability rests in their hands alone, sustainability issues can be marginalized or moved off to agendas unrelated to the firm's core business. As a counter to this, human and ecological compliance policies can be integrated into an overall sustainability policy that is an integral part of the firm's business strategy.

Integration is a key concept for both successful business and sustainability. It points to the need for cross-functional relationships and the building of trust between areas of business that have previously been regarded by managers as only loosely connected (Piasecki, 1995). Active relationships can be fostered between line managers and sustainability experts, so that the issues of human and ecological sustainability are not perceived by employees as an add-on, and thus as irrelevant to the core priorities of management. For instance, at KLM, line managers all have sustainability action points and targets.[4]

Limiting staff training to the technical training required for compliance is an add-on approach that is unlikely to bring about either attitudinal changes or allow the development of a culture capable of adopting self-regulation. Restricting skills development to a few key individuals can also lead to disaster, particularly in some industries where toxic chemicals or dangerous technologies are used by others in the corporation. The ING Group, a global banking, insurance and investment company, has attempted to avoid these problems by developing an e-learning compliance course in order to familiarize its 7000 employees in 11 countries with compliance principles.

Sharing knowledge across the institution and developing an integrated policy according to the triple bottom line of social, environmental and economic objectives requires more coordination between internal elements of the firm. A significant step along this road is the merging of human and ecological sustainability policies through the explicit recognition by management of the need for a comprehensive sustainability strategy. Nuon Energy, for instance, the second largest energy producer and largest producer of green energy in the Netherlands, has an Environmental Knowledge Centre that is part of a larger Human Relations Management section.[5] This kind of organizational response can be achieved by encouraging the development of the informal roles that enable inter-unit collaboration; roles such as champion, gatekeeper, idea generator and sponsor, for example (Adler and Tushman, 1997). These informal roles can become ineffective if formalized but thrive when the individuals who adopt them are good networkers and strongly committed to the core value of sustainability.

At the environment-award-winning nickel plant in South Wales, UK, INCO Pty Ltd, the company doctor acts as the 'champion' and represents the company at international conferences and forums. The environmental officer acts as the 'idea generator', regularly attending meetings of business units, negotiating standards with the regulatory agency and ensuring that employees are trained according to ISO 14001 (Certified Environmental

Management) requirements. The role of 'gatekeeper' is played by some senior people who are also trained as auditors for ISO 14001. The 'sponsor' is the plant manager whose idea it was to start the plant down the road to ISO 14000 and who exhibits 'very decisive leadership'.[6]

When a firm makes the move beyond compliance in accordance with standards such as ISO 14000 or industry-specific codes such as Responsible Care, some 'old ways' may still be appropriate if they help the organization achieve compliance with regulatory standards efficiently and effectively. The processes upon which a structured compliance has been built do not necessarily need to be discarded. What must go is the 'machine' metaphor of the closed and highly regulated system. If a structured EMS has been adopted, for instance, it now needs to be re-assessed for its value in delivering the negotiated understandings of 'co-regulation' (where government facilitates a form of self-regulation). With the firm now partly responsible for instigating and monitoring compliance, institutionalizing reflection and implementing feedback become priorities. The whole organization needs to emerge from a legalistic understanding of regulation and begin to support what may seem less tangible ends, such as the capacity of employees to examine their own attitudes towards change.

To this end many managers are recognizing that learning capacity is the true capital of the firm. Dealing effectively with such issues as risk liability is dependent on this capacity. According to de Geus (1997), the successful company encourages the development of a culture that is sensitive to its general environment. Ongoing improvement in ecological sustainability is assured because the organization becomes an open system, capable of learning through interaction among employees and between employees and the wide range of stakeholders. These stakeholders include not only customers, competitors and suppliers but the environment and society (Laszlo and Laugel, 2000). In an organization of this kind, environmental responsibility simply becomes an integral aspect of the way people conduct business.

More organizations are moving into the new structural forms that will characterize the third millennium such as networks, 'spider-webs', the 'infinitely flat' organization, alliances and the virtual organization. In this new order fewer interactions within the organization involve hierarchical relations and more depends on trust. Bureaucratic implementation of procedures will still remain important for some regulatory requirements. Increasingly, however, managers gain credibility from their ability to put in place voluntary codes for healthy human and ecological relations.

Managers are increasingly required to expand the knowledge base of the workforce and institute high performance management practices. Studies demonstrate that these practices (such as reduction of status differences, sharing information and extensive training) can build commitment (Pfeffer and Veiga, 1999). In other words, a culture of voluntarism can only be developed through a shared sense of identity. However, this is increasingly difficult to create given that career paths are ill-defined and unpredictable and more employment is contractual and part-time rather than permanent and full-time. Palmer and Dunford (1997) point out that as a result of these conditions, managers need high-level skills in team-building, empowerment and the development of trust. Because a certain proportion of expert knowledge is tacit, working in self-managed teams requires trust, otherwise the critical knowledge needed to make the new systems work will not be exchanged (Pfeffer and Veiga, 1999).

Developing systems to take the firm to compliance and beyond requires building a baseline level of human and technical competence. Moving voluntarily beyond compliance enables change flexibility as the internal corporate system is modified to address the requirements of external stakeholders. However, voluntary initiatives alone often lack rigorous performance criteria and specific objectives. To ensure rigour, the voluntary phase should be tied firmly to efficiency and strategic objectives.

Moving to the efficiency phase
In the shift to the next stage of efficiency, human resource management becomes resource management (Gollan, 2000). Specific targets include how to identify cost inefficiencies and how to develop 'natural capitalism' by increasing resource productivity and operational efficiency (Lovins et al., 1999). For many corporations, improving performance in terms of efficiency enables and justifies shifting along the spectrum toward ecological sustainability. Improving corporate performance in terms of cost, such as by reducing energy use and waste to landfill, is highly reliant on company-wide awareness and educational programmes. These in turn rely on human resource and quality management systems, and on communication and motivation strategies. In general, developing eco-efficiency requires a number of human resource management measures to be implemented such as highly organized knowledge management and transfer systems, reorganization of line managers to take more responsibility for both reputational and technical risk areas, and change agent development in order to facilitate incremental change.

As pointed out, a well-established learning culture will act as a base for the moves to eco-efficiency. For instance, Carlton & United Breweries (CUB), a major Australian brewer, has been well known for more than ten years for its Lifelong Learning Program, a team-based approach to organizational learning, work practice and problem-solving. This culture change at CUB means that work area team managers are also responsible for health, safety and environment instead of just focusing on production. As a result the role of the environmental coordinator at CUB is to help teams address their particular issues. The company is also moving to a systems approach to gain certification through ISO 14000, as well as to emphasize health and safety, and TQM. The general level of staff awareness has assisted the gains made in efficiency through automation so that the volume of water required for beer production has halved in eight years.[7]

Business advantage can also be obtained by following the principles of industrial ecology. Tracking material and energy flows over the whole producer/consumer cycle reduces the likelihood of 'suboptimal solutions' and 'unintended consequences' (Ehrenfeld, 2000, p. 226). Such a system can be introduced using incremental changes such as those initiated by another environment-award-winning firm, the Panasonic TV Factory situated in Sydney's outer western suburbs. A small group of highly motivated employees, a supportive and enabling corporate culture, and the high social capital of the business networks in this area of Sydney, have resulted in the firm developing ingenious arrangements with a range of local firms for either the sale or free removal of by-products as an alternative to paying for landfill. Major energy-saving initiatives have been included in new warehouse and building sites.

Panasonic employees describe a workplace culture enabling the ready development of such sustainable practices.[8] Television manufacture requires the assembly of many different components and is dependent on an assembly line staffed by skilled employees.

The corporate shift towards ecological sustainability has been built upon a base of human sustainability developed through open workplace communication systems, a policy of ongoing skills development, a customer-driven ideology and an ongoing commitment to the local community. The employee response is to see the company as a long-term employer, demonstrated in the low level of staff turnover.

Many employees have visited Matsushita Training Centres in Malaysia and Japan, and have taken study tours to similar centres in Europe. The company has a policy of continually upgrading skills through accredited courses, setting competency standards and awarding medals. Integration of human and ecological sustainability initiatives has been facilitated through a monthly communication meeting for all employees. The meeting is led by the managing director and addressed by various speakers on a wide range of issues relevant to the business. This commitment to communicate the ideas and strategies driving the firm comes from the example of Matsushita himself, who was noted for his passionate efforts to propagate his ideas. More than 3000 tape recordings of his talks exist as well as numerous in-house newsletters and messages.

Even though the parent company is a multinational (the eighth largest in the world), Panasonic tries to concentrate on the Penrith community. For instance, there is a community welfare committee that makes shop floor decisions about what products are suitable as gifts or donations. Staff assist in the local 'Meals-on-Wheels' programme in paid time. In fact, in 1998 Panasonic was one of only three firms in Australia identified by Volunteering Australia as being genuinely involved in projects that promote community well-being. Overall the community-based activities reflect a relatively seamless 'family' relationship between employees, firm and community.

However, efficiency initiatives alone will not generate the competitive advantage required for long-term high performance. The already lean and trim organization will need to leverage other sources of competitive advantage such as their symbolic capital (Dunphy et al., 2003; McIntosh et al., 1998). The firm's reputation, capability of communicating meaningfully with internal and external stakeholders, and ability to develop intangible assets such as the tacit knowledge of long-term employees all become more important. Panasonic is a case in point here.

The strategic phase
It has been argued that if firms persist with the win–win business logic of natural capitalism – profiting from increasing the productivity of natural resources, closing materials loops and eliminating waste, shifting to biologically inspired production models, providing their customers with efficient solutions and reinvesting in natural capital – they can gain a commanding and strategic competitive advantage (Lovins et al., 1999). It needs to be recognized, however, that building this perspective into an organization requires the development of innovative capacity.

A firm in the strategic phase will have the capability and learning capacity to recognize and develop the skills and organizational culture necessary to innovate in line with the new business standards set by ecological modernization (Hoffman, 1997; Mol and Sonnenfeld, 2000). Hewlett-Packard's Environmental Strategies and Solutions programme for instance showed that sustainability can offer companies a strategic competitive advantage (Preston, 2001). The firm based the programme on the premise that the planet is a closed system that will eventually face limits, placing the firm in a new social and economic situation. In other words, the firm strategically scoped the challenges of a new business environment, developing strategies that would transform potential environmental liabilities such as climate change, resource exhaustion and the energy crisis into competitive advantage (Preston, 2001, p. 29).

More generally, movement to this phase requires management to formulate a strategy that anticipates and scopes future issues and the intended impacts of company operations, products and services (Roome, 1992). In another approach to increasing ecological sustainability through developing human capability, the firm can then focus on redesigning products, processes and service layouts to gain strategic business advantage.

For instance, the Fuji Xerox group of companies has successfully undertaken remanufacturing for many years. However, the sustainability benefits achieved by the Australian plant at Zetland in Sydney make it one of the most successful of the company's plants worldwide. This has contributed to the success of this firm in being awarded the prestigious Museum of Sydney Eureka Prize – Environment Section. The concept of eco-manufacturing involves detailed analysis of 'why things fail', which informs the production of remanufactured products with improvements intended to eliminate future failures. This concept enables business to become more sustainable while making savings and supplying local operators with high-quality locally reprocessed parts.

Eco-manufacturing takes used components and tests them, and then re-engineers and reassembles them into 'new' products while ensuring that the production process and the products do not have adverse environmental effects. However, to produce a re-engineered product as good as (or better than) new while meeting the new sustainability challenges requires addressing some complex technological challenges. For example, the materials of the components may have changed during their first use. These changes may occur due to heat, vibration or some other physical effect of the equipment's operational processes. Thus remanufacturing cannot simply replicate the original production process. In addition in order to qualify as 'eco-manufacturing', the remanufacturing process may have to meet new and higher environmental safety standards.

Fuji Xerox managers describe the work of their Eco Manufacturing Centre as re-engineering and redesigning a product or product component and developing it to equal or better than new condition. This process involves scientifically examining the causes for failure while looking for opportunities to extend the life of the product and in general to improve its performance. These processes also have environmental benefits by reducing the demand for raw materials, energy and waste to landfill. Another major benefit to the business is the acquisition of data about problems that develop in its products over time. That data was previously lost as used products that were defective were sent to landfill. Part of the new remanufacturing/re-engineering process involves analysing the defects in the components that have been returned. This analysis provides information that can be used to improve component design and thereby leads to the production of 'as new' or better remanufactured products. So there are multiple benefits from remanufacturing:

- decreased costs due to recycling (over the year 2000–2001 savings to Fuji Xerox were approximately AUD22.5 million);
- improved design for increased reliability and enhanced performance; and
- savings from import substitution and new export earnings.

Not only are parts renewed or recycled but the technical processes involved in achieving this have been developed to eliminate environmentally damaging emissions, pollution and waste. For example:

- All solvents have been eliminated from the cleaning of parts and components.
- Frozen carbon dioxide (dry ice) is used under high pressure to clean components, a process that creates no liquid wastes or pollutants.
- Environmentally 'neutral' bicarbonate of soda is used under high pressure to remove the old coating from the fuser rollers used in photocopiers, and the spent bicarbonate of soda is then re-used as an industrial water softener.
- The carbon by-product of waste toner (57 000 kg per annum) is extracted and can be used as a combustion agent in steel making.
- All unusable metal parts are sent to Sims Metal to be recycled.
- The company is also undertaking ongoing research and development into ways of reducing all packaging waste through re-use of a range of packaging materials, including plastics. The Eco Manufacturing Centre is working with Collex Waste Management to solve these problems.
- Reduction in energy use is achieved through the implementation of a range of initiatives and monthly tracking to evaluate improvement through these programmes.

The Eco Manufacturing Centre has clearly reached beyond the efficiency phase with these measures. Several cultural and human resource factors at the Zetland plant have enabled this strategic perspective.[9]

A key success factor at the Fuji Xerox Eco Manufacturing Centre in Sydney has been the presence of a small group of skilled, innovative and committed managers willing to listen to staff, customers and other stakeholders. Staff are assigned to teams, with each team being responsible for quality, engineering and production capacity around products or product groups such as print cartridges or lasers. The product-based team structure promotes multi-skilling, enhances communication around problem identification and problem-solving, builds deep expertise and cumulative experience and ensures that improved quality is constantly built into the work process. Managers at Fuji Xerox see this structure as the leading cause of the high level of innovation in the plant. It has also led to a close working relationship between the engineers and the production workers, and joint ownership of the production targets and product quality.

The plant is systematically building the human capabilities of its staff. Staff members are offered a range of developmental opportunities and most have had training in various aspects of 'people management'. Employees are also well remunerated. Staff turnover is low: for example, in 2000, 37 staff were employed on contract work and of these, 28 stayed on to become permanent employees. Clearly Fuji Xerox is seen as an employer of choice.

The shift towards a more enabling and committed culture at the Fuji Xerox Eco Manufacturing Centre has been significant. According to management, the better pay, team-based conditions and considerable team spirit and loyalty, have resulted in the virtual elimination of industrial unrest. The nature of the work at the Eco Manufacturing Plant seems to have generated a sense of worthwhile purpose. Where staff diversity was formerly considered a problem for the organization, cooperative leadership has redefined this issue and the company now benefits from the richness of its cultural mix.

The eco-manufacturing process builds on the quality control systems already in place throughout the company, particularly the ISO 9001 (quality endorsed company), and ISO 14001 (Certified Environmental Management) accreditation and accompanying systematic processes. Gaining and maintaining accreditation with these systems is now company policy for Fuji Xerox plants throughout the world. The Australian plant is both ISO 9001 and ISO 14001 accredited and has in place systems and procedures to ensure that all products leaving the site receive a full quality guarantee. The ISO 14001 Enviro Accreditation is integrated into the company's quality management systems and audits of ISO 9001 also include ISO 14001 progress audits.

Occupational health and safety (OH&S) is viewed as part of the quality process. The director of manufacturing and supply in the Zetland and Mascot offices personally undertook the site inspections for OH&S in order to demonstrate its importance to staff. Similarly, all quality processes including environmental standards have been given great importance in redefining the business. Management's position is that commitment to this standard of environmental accreditation will keep remanufacturing as a core business function. The regular audits associated with the accreditation identify the company's environmental impact and require steps to be taken to solve any identified problems.

The organizational changes at Fuji Xerox illustrate the links between an organizational culture of innovation and one designed to deliver sustainability. Practices designed to enhance human sustainability and social capital within the organization (such as empowerment, teamwork and continuous learning) are linked to the capacity to innovate and escape from rigid models of operation and production.

Arguably, implementing practices that support human sustainability creates an organizational culture that also facilitates product differentiation (Orssatto, 2000). A number of companies have been successful in employing a strategy of environmental product differentiation. While Reinhardt (1999) argues that the success factors of such a strategy include consumers who are prepared to pay more, and an innovation that is unique long enough for a profit to be made, the benefits must be readily communicable. Communication systems are thus a key feature of the human sustainability of the organization.

Corporations face an accelerating rate of change and an increasingly complex society. For these business conditions innovation depends on the cultural and structural characteristics of the organization. Both sets of characteristics are linked to the organization's capacity to engage with sustainability. Cultural factors such as those associated with the learning organization also underpin a culture of precaution. Structural factors such as an internal network culture, employee participation and the capability of developing community partnerships are aspects of human sustainability.

In other words innovation, business concept redesign and human sustainability can be readily linked in a dynamic relationship aimed at delivering long-term business advantage. Importantly such qualities enable the corporation to be more responsive to the external drivers of change. An organization geared to innovation is ready to take up government incentives for ecological modernization. That is, it can readily translate social and moral issues into market issues and can exploit the potentially huge market that ecological sustainability in particular represents. Moreover, such an organization can more critically reflect on the possibilities of new relationships between nature, society and technology that will mark a new, more sustainable age (Hajer, 1995).

The sustaining organization
In order to develop to the stage of the truly responsible and responsive organization the firm will need to make the transition to an organization that functions as an instrument for the fulfilment of human needs and the support and renewal of the biosphere. It will need to develop a codified set of values that are able to be internalized in all employees. We argue that the skills necessary for such an organization will be process-based and include higher-order skills of personal resilience, self-confidence, adaptation and learning, empathy, communication and influence, coaching, negotiation and conflict resolution. The truly sustaining organization will attempt to use its influence so that society as a whole will progress toward a more equitable position both between and within generations.

Hence Nuon Energy, the second biggest energy supplier in the Netherlands, has based its long-term vision on the fact that eventually all energy it supplies will come from renewable sources.[10] It is also working with other energy suppliers and non-government organizations in areas such as Mali, the Sudan and Albania to develop their use of renewable energy resources.[11]

In its relationship with the Landcare movement in Australia, the Fuji Xerox Eco Manufacturing Centre in Sydney displays some of the characteristics of this phase. The 4500 Landcare groups across Australia are part-funded by government but are also dependent upon voluntary labour and corporate sponsorship. Management at Fuji Xerox sees a relationship with Landcare as a means to enable organizational learning around environmental responsibility as well as to deliver business benefits. This corporate–community relationship goes beyond the transactional phase of collaboration to a more integrated relationship (Austin, 2000). The Fuji Xerox managers have taken the unusual measure of attempting to get corporate dollars down to the local Landcare groups, arguing that central administration often takes too great a share of such funds. As well, several senior managers have become Board Members of the Landcare Foundation and have persuaded other companies to sign up as sponsors. The firm also initiated the Olympic Landcare project, which developed a number of solutions to help deliver the first 'green' Olympic Games. These measures are the more ideologically committed measures of the sustaining corporation, an organization that will attempt to persuade other corporations to support the cause of ecological renewal.

Change for sustainability
Incremental and transformational change
For some organizations, the change toward a more sustainable position on the sustainability developmental model will centre on incremental change in human sustainability. For instance for the organization to move from the non-responsive stage to compliance requires incremental change. Small wins can in fact provide the basis for ongoing change (Weick, 1984). The case of Panasonic illustrates this. The informal initiatives of a few employees, encouraged by the enabling corporate culture of Panasonic, gained the efficiency results that encouraged the organization to move towards a more structured EMS.

Key human capability features of incremental change may be identified to include standard project management skills so that progress may be monitored and reported adequately; awareness of external auditing systems and legislative changes; transparent actions of managers and change agents so that problems and challenges are honestly faced and addressed; cross-functional team building; and managers who walk the talk and emphasize innovation.

For other organizations change along the sustainability phase model will be transformational, requiring a redefinition of the company's core business area or its key manufacturing processes, or even its human resource policies and community relationships. For example, a transformative change from the rejection phase to compliance would require a major shift in values and a re-inventing of the corporate image and culture according to a powerful conception of future need.

For some organizations this developmental path will be a slippery one. For example, Orica Ltd (previously the Australian branch of ICI Ltd) attempted what it claimed to be a paradigmatic change in its closed culture, which was fixed in the traditional norms of industrial science, when it established a Community Participation and Review Committee in 1996.[12] The main function of the Committee was to advise on the disposal of 10 000 tonnes of highly toxic organochlorine waste stored at its site on the shores of Botany Bay, Sydney. However, the protracted communications with the local community came to a conclusion as community representatives remained convinced that the firm had colluded with its traditional allies, such as government, professional scientists and corporate power to betray them.[13] The upshot of the Committee's efforts was a mistrustful community and the potential that Orica will suffer a considerable loss in reputation as the results of the negotiations are publicized.

For other organizations, however, transformative change will be enabled because the firm has a more accurate perception of where it is now. Orica failed in its attempt at change because it did not recognize that it faced the classic conditions of the 'risk society' (Beck, 1992, 1999). This is a society in which risk becomes a preoccupation when the uncertain nature of risk causes expert to contradict expert, so that trust and credibility become all important.

Other companies such as Hewlett-Packard and 3M have shown themselves able to continually re-invent themselves. These companies have clearly identified the established cultural characteristics that inhibit profound change and have accepted change as at the heart of their culture (Stace and Dunphy, 2001).

An example of the human capability that can underpin such transformative change is the ability of management to build alternative scenarios for the future. To implement these visions requires other abilities such as the employment of reframing techniques (Palmer and Dunford, 1996; Senge, 1992). In particular, the political frame is useful. Managers need a good understanding of the political issues associated with change (Stace and Dunphy, 2001). Stakeholder analysis will assist in identifying the entrenched interests inside and outside the organization that may organize against the change.

A number of organizational characteristics dependent upon human capability can make useful criteria for the monitoring of transformational change. These include progressive engagement of key individuals and groups; the development of actionable plans; progressive development of skilled change agents at all levels; development of trusting relationships with all stakeholders; and a culture of non-defensive learning (Dunphy et al., 2003, p. 257).

Ad hoc initiatives are important but systematic progress needs competent, qualified professional change agents. Developing a specific sustainability change unit and a cadre of change agents enables attitudinal change, facilitates the building, managing and accounting for social as well as environmental capital and fosters the innovation needed for significant corporate change toward sustainability objectives.

Proactive and collaborative leadership

Proactive leadership is required if the firm is to progress along the spectrum towards becoming a sustaining organization. In our work with organizations we have consistently noted the importance of leadership for collaboration between:

- corporate members of networks and alliances through the identification of negotiated standards of human and ecological sustainability;
- internal environmental and human resource managers as they negotiate performance criteria that integrate their fields of activity;
- external change agents such as government policy-makers, non-government organizations and community groups and the internal change agents of the firm.

Again we quote the example of Orica, where although the standards of health, safety and environment had been re-negotiated internally, the corporate culture remained closed to the recognition of the values of other stakeholder groups. The leaders of the firm throughout the dispute in question did not practise the 'inclusive leadership' required for the firm to take the responsive and responsible position of the long-term high-performance organization.

By contrast there are obvious benefits for a firm that takes a more democratic and collaborative approach to its relationship with community groups. The relationship between Fuji Xerox in Sydney and the Landcare movement in Australia is one example. Managers at the Fuji Xerox Eco Manufacturing Plant in Sydney are also committed to establishing ongoing collaborative relationships with customers, believing that high-quality products and efficient responsive service will result in Fuji Xerox being the preferred supplier of office equipment. Fuji Xerox also offers training to its customers. This includes developing an understanding of the technology and of the company's approach to business. Customers identify talented people in their own organization who are then trained by Fuji Xerox to identify problems, install remanufactured components and ensure that damaged components are returned for remanufacturing.

The close relationship between the nickel manufacturer INCO in South Wales, UK and the regulatory agency also illustrates this inclusive leadership approach. The environmental officer of this environment-award-winning nickel-processing plant keeps a high level of personal contact with the local community. The officer also negotiates standards with the environment agency so that high achievement in one area can be traded off against poor results in another. For instance, INCO's management is currently looking at improving their stack emissions. The firm is pre-empting expected legislative changes and is currently looking at new dust pollution measures. According to INCO's Environmental Officer: 'It's better to be ahead of the change and perhaps even influence the direction of legislation. Our emphasis is on a very transparent, negotiated relationship with the regulator'.[14]

This collaborative approach acts as informal reflexive regulation (Matten, 2001; Orts, 1995), enabling the social learning now recognized by policy-makers as an essential component of compliance. For writers such as Orts (1995) and Fiorino (2001), putting corporate managers in situations that encourage self-criticism is more productive than a prosecutorial approach. In this form of regulation, a regulatory agency that cooperates in the setting of standards with the firm is in effect encouraging the firm to 'self-transform' (Fiorino, 2001). This development of self-awareness or 'reflexivity' is the key factor for INCO in moving incrementally through the sustainability phases, building human capital in the form of skills and capabilities along the way (Stace and Dunphy, 2001).

This company also cooperates with the community within which it is based. Many of the ideas for environmental improvement come from its employees, who are also members of the local community since most employees live locally. Putting these ideas into practice is in INCO's interest since it encourages long-term worker satisfaction and commitment, helping to maintain a well-trained workforce and giving the company a competitive advantage. Cooperation and community are thus vital elements of INCO's success.

Similarly for Panasonic, leadership for collaboration has directed the firm's relationships with the local community, including local business organizations. Indeed with Panasonic and INCO we see the management practices of two companies that, while geographically distant, are otherwise remarkably similar. Their sensitivity to the communities in which they are based shows the similarity of their ideals and is a common factor in maintaining their competitiveness. Panasonic is an admirable example of a firm that is prepared to fund its employees in endeavours that support the community, both through social initiatives and in terms of setting an example of environmental responsibility to industry. It sees itself as having a social function and thus has supported the sustainability initiatives spearheaded by its employees. In Panasonic's case, the champion for the change remains the embedded ethos of Matsushita but this is driven through the firm by committed employees, rather than top management.

Tensions between human and ecological sustainability

This chapter has attempted to draw out the synergies between human and ecological sustainability, arguing for an integrated perspective on corporate sustainability. But are, as many critics have argued, both normative and business cases for integration flawed? Does attempting the simultaneous achievement of human and environmental sustainability inevitably mean a trade-off?

In terms of ideology, for instance, there are tensions between human sustainability, which focuses on the development of employee and community capability, and the ideals of the 'ecocentric' organization that characterize strong sustainability (Kearins et al., 2003; Shrivastava, 1995). Underpinning sustainability with features associated with the development of human capability arguably encourages a 'technological fix', or mechanistic, approach to sustainability. Such an anthropocentric approach puts humans outside the realm of the natural world. It restricts the development of the interconnectedness and reciprocity associated with a more ecocentric perspective (Shrivastava, 1995). Stepwise approaches based on the 'human journey' metaphor can lead to acceptance of imperfection in implementation or weak sustainability (Kearins et al., 2003).

At an operational level there is evidence that while a humanistic approach and investment in employee development produces stakeholder benefits, it may also cause companies to reduce money spent in replacing beyond end-of-pipe with pollution prevention environmental management systems. At a strategic level an emphasis on the principles of natural capitalism (Lovins et al., 1999) is technocentric, shutting out the 'people element' of the organization and thus the capacity to change. Achieving environmental sustainability for instance may also require an overemphasis on Taylorist principles and lead to intensive rather than sustainable work systems (Doherty et al., 2002). Further tensions at the strategic planning level may arise in association with the unbounded nature of environmental risk and the fact that the social perpetrators of the risk may be far distant from the effects (Beck, 1992, 1999).

Conclusion

Our major motivation in applying the phase model to sustainability issues is to examine the relationship between human resource management and corporate citizenship (human sustainability) and ecological sustainability – areas of management that are usually studied in isolation. Despite some tensions between the human and ecological aspects of corporate sustainability, the case studies described here demonstrate that human and ecological sustainability are more readily actioned as integrated processes. In organizations such as Fuji Xerox, INCO and Panasonic, all of which have won awards for the quality of their environmental management, this ecological capability has been built upon and developed through the trusting relationship that exists between staff, management and the communities in which they are based. Each firm has been prepared to invest in the development of social capital. While the motivation for this may be instrumental in that cultivated social capital may be exploited in the future by the firm, the social relations described could also be motivated by the norm of 'generalized reciprocity' (Putnam, cited in Adler and Kwon, 2002, p. 25). As Adler and Kwon (2002) point out, this shared norm of 'I'll do this for you now if you do something for me later' allows the firm to bind with employees and local communities, giving the firm the right to expand in the future (see also Elkington, 1977).

Our research indicates that firms that progress toward ecological sustainability are also investing in the development of human capital. Because their employees are viewed as a long-term investment, the companies have been prepared to put resources into their environmental training and into the time allocated for environmental projects. In particular Panasonic and Fuji Xerox indicate the development of both scientific and normative systems of industrial ecology. Not only are their by-products becoming the raw materials for other organizations, but this is being enabled through the internalized norms of interconnectedness and community (Ehrenfeld, 2000).

In conclusion, if we were to name one key property of the human sustainability agenda that gathers together the various organizational properties that support corporate sustainability, it would be the firm's capacity for social learning. A culture that fosters this property will give the corporation the flexibility, visionary leadership and generative capacity required to move forward in an era where ethics and morality will be increasingly important for societies and the corporations within them. Environmental sustainability can only be effectively created through first building the human capabilities of the organization – no amount of technical capacity can by itself ensure corporate sustainability.

Notes

1. Interview with U. Huiskamp, Environment Manager, KLM, Amsterdam, February 2002.
2. Interview with U. Huiskamp, Environment Manager, KLM, Amsterdam, February 2002.
3. Interview with P. Kroon, Head of Public Affairs, The ING Group, Amsterdam, 5 February 2002.
4. Interview with U. Huiskamp, Environment Manager, KLM, Amsterdam, February 2002.
5. Interview with Hans Nooter, Nuon Energy, Hardewijk, the Netherlands, 7 February 2002.

6. Interview with R. Savage, Environment Officer, INCO, Swansea, South Wales, UK, 14 February 2002.
7. Interview with David Ford, Environmental Coordinator, CUB, Sydney, 18 December 2002.
8. Interviews with Danny Lett, Manager, Personnel and Administration, and A. Holmes, Supervisor, Local Purchasing, Panasonic TV Factory, Penrith, Sydney, 29 August 2000 and November 2002.
9. The authors conducted semi-structured interviews with members of Fuji Xerox staff from the Eco Manufacturing Centre at Zetland, New South Wales. Interviews were conducted with Veera Chippada, Logistics and Warehouse Manager, Gunawan Komardjaja, Project Manager ROS (Raster Optic Scan) Operation, Manwinder Mavi, Project Manager, Bruce Fremann, Production Supervisor, Dan Godamunne, General Manager, Giovanni Calabrese, Operations Manager Cartridge Remanufacturing and Graham Cavanagh-Downs, Australian Director of Manufacturing and Supply and a number of other staff. Promotional and other literature produced by Fuji Xerox and other sources were also examined.
10. Interview with Hans Nooter, Nuon Energy, Hardwijk, the Netherlands, February 2002.
11. Interview with Hans Nooter, Nuon Energy, Hardwijk, the Netherlands, February 2002.
12. Interview with B. Gotting, Site Manager, Orica Pty Ltd, Botany Industrial Park, Sydney, December 2001.
13. Interview with Nancy Hillier, community representative on the Orica Community Participation and Review Committee, Sydney, June 2002.
14. Interview with R. Savage, Environment Officer, INCO, Swansea, South Wales, UK, 14 February 2002.

References

Adler, P. and Kwon, S. (2002), 'Social Capital: Prospects for a New Concept', *Academy of Management Review*, **27**(1), 17–40.

Adler, N. and Tushman, M. (1997), *Competing by Design: The Power of Organizational Architecture*, Oxford: Oxford University Press.

Austin, A. (2000), *The Collaboration Challenge: How Profits and Nonprofits Succeed Through Strategic Alliances*, California: Jossey Bass.

Beck, U. (1992), *The Risk Society*, London: Sage Publications.

Beck, U. (1999), *World Risk Society*, Cambridge: Polity Press.

Daily, B. and Huang, S. (2001), 'Achieving Sustainability through Attention to Human Resource Factors in Environmental Management', *International Journal of Operations and Production Management*, **21**(12), 1539–52.

de Geus, A. (1997), 'The Living Company', *Harvard Business Review*, **75**, 51–9.

Doherty, P., Forslin, J. and Shani, A. (2002), *Creating Sustainable Work Systems*, London: Routledge.

Dryzek, J. (1997), *The Politics of the Earth*, Oxford: Oxford University Press.

Dunphy, D., Benveniste, J., Griffiths, A. and Sutton, P. (2000), *Corporate Sustainability*, Sydney: Allen and Unwin.

Dunphy, D., Griffiths, A. and Benn, S. (2003), *Organizational Change for Corporate Sustainability*, London: Routledge.

Ehrenfeld, J.R. (2000), 'Industrial Ecology: Paradigm Shift or Normal Science?', *American Behavioral Scientist*, **44**(2), 229–44.

Ehrenreich, B. (2001), *Nickel and Dimed: On (Not) Getting By in America*, New York: Metropolitan Books.

Elkington, J. (1997), *Cannibals with Forks*, Oxford: Capstone.

Fiorino, D. (2001), 'Environmental Policy as Learning: A New View of an Old Landscape', *Public Administration Review*, **61**, 322–33.

Gollan, P. (2000), 'Human Resources, Capabilities and Sustainability', in D. Dunphy, Benveniste, J., Griffiths, A. and Sutton, P. (eds), *Sustainability: The Corporate Challenge of the 21st Century*, Sydney: Angus and Robertson.

Hajer, M. (1995), *The Politics of Environmental Discourse*, Oxford: Clarendon Press.

Hoffman, A. (1997), *From Heresy to Dogma*, San Francisco: The New Lexington Press.

Kearins, K., Milne, M. and Walton, S. (2003), 'The Business Journey to Sustainability: Destination Not Defined!' in Problematizing Sustainability Symposium, Academy of Management Meetings, Seattle, USA, August 3–6.

Laszlo, C. and Laugel, J. (2000), *Large-scale Organizational Change*, Boston: Butterworth-Heinemann.

Lovins, A., Lovins, L. and Hawken, P. (1999), 'A Road Map for Natural Capitalism', *Harvard Business Review*, May–June, 145–58.

Matten, D. (2001), 'The Career of the Concept of the Risk Society in Environmental Politics and Environmental Management', Presentation, 9th Greening of Industry Conference, Bangkok, Thailand, January.

McIntosh, M., Leipziger, D., Jones, K. and Coleman, G. (1998), *Corporate Citizenship*, London: Financial Times Management.

Mol, A. (1997), 'Ecological Modernisation: Industrial Transformations and Environmental Reform', in Redclift, M. and Woodgate, G. (eds), *The International Handbook of Environmental Sociology*, Cheltenham, UK and Lyme, USA: Edward Elgar, pp. 138–49.

Mol, A. and Sonnenfeld, D. (2000), 'Ecological Modernisation Around the World: An Introduction', *Environmental Politics*, **9**(1), 3–16.

Orssatto, R. (2000), 'The Ecological Competence of Organizations: Competing for Sustainability', Paper, 16th EGOS Colloquium, Helsinki, Finland, 1–4 July.

Orts, E. (1995), 'Reflexive Environmental Law', *Northwestern University Law Review*, **89**(4), 1227–340.

Palmer, I. and Dunford, R. (1996), 'Reframing and Organizational Action: The Unexplored Link', *Journal of Organizational Change Management*, **9**(6), 12–25.

Palmer, I. and Dunford, R. (1997), 'Organizing for Hyper-competition: New Organizational Forms for a New Age?', *New Zealand Strategic Management*, Summer, 38–45.

Peatling, S. (2003), 'Hold the Coal, Say Customers, and Pass the Wind', *Sydney Morning Herald*, 2 January.

Pfeffer, J. and Veiga, J. (1999), 'Putting People First for Organizational Success', *Academy of Management Executive*, **13**, 37–48.

Piasecki, B. (1995), *Corporate Environmental Strategy: The Avalanche of Change Since Bhopal*, New York: Wiley.

Preston, L. (2001), 'Sustainability at Hewlett-Packard: From Theory to Practice', *California Management Review*, **43**(3), 26–38.

Reinhardt, F. (1999), 'Bringing the Forest down to Earth', *Harvard Business Review*, July–August, 149–57.

Roome, N. (1992), 'Developing Environmental Management Strategies', *Business Strategy and the Environment*, **1**, 11–24.

Sagawa, S. and Segal, E. (2000), 'Common Interest, Common Good: Creating Value Through Business and Social Sector Partnerships', *California Management Review Winter 2000*, **42**(2), 105–23.

Senge, P.M. (1992), *The Fifth Discipline: The Art and Practice of the Learning Organization*, Sydney: Random House.

Shrivastava, P. (1995), 'Ecocentric Management for a Risk Society', *The Academy of Management Review*, **20**(1), 118–37.

Stace, D. and Dunphy, D. (2001), *Beyond the Boundaries*, Roseville: McGraw Hill.

Starkey, R. and Welford, R. (2001), 'Conclusion: Win-win Revisited: A Buddhist Perspective', in Starkey, R. and Welford, R. (eds), *Business and Sustainable Development*, London: Earthscan Publications, pp. 353–7.

Tsoukas, H. (1999), 'David and Goliath in the Risk Society', *Organization*, **6**(1), 499–528.

Weick, K. (1984), 'Small Wins: Redefining the Scale of Social Problems', *American Psychology*, **39**(1), 40–9.

Wilkinson, A., Hill, M. and Gollan, P. (2001), 'The Sustainability Debate', *International Journal of Operations and Production Management*, **21**(12), 1492–502.

15 Using network approaches to engage small and medium-sized enterprises in Environmental Management Systems
Ruth Hillary

Introduction

Small and medium-sized enterprises (SMEs) are of great importance to local, national and European economies, accounting for 99.6 per cent of all businesses and providing 53 per cent of all jobs. There are approximately 20 million SMEs in the European Union (EU) and they represent a huge diversity of organizations working across all economic sectors. At the same time, they exert pressures on the environment such as use of scarce resources, pollution, waste, and so on. More than larger companies, SMEs face a number of obstacles to addressing pressures, such as insufficient resources, too little specific expertise, and a lack of information and awareness.

It is important to bear in mind that SMEs form a very heterogeneous group and that large differences exist between individual companies, both in terms of sector and, very importantly, size. Small and medium-sized enterprises themselves are often not aware, or at least not fully aware, of their environmental impacts, and they are not always well informed about their obligations under environmental legislation.[1] Evidence suggests, however, that there is a correlation between a company's size and its environmental engagement: the bigger the company, the more likely it is to have a proactive environmental policy (e.g. NUTEK, 2003, pp. 9–10).

Environmental Management Systems

Environmental Management Systems (EMSs) can take many different forms. The two formal EMSs are the EU's Eco-Management and Audit Scheme (EMAS) (European Commission, 2001), and the international standard ISO 14001 (British Standard, 1996). The central principle behind both initiatives is one of self-regulation; consequently, participation in ISO 14001 and EMAS is voluntary. Common to both initiatives is the plan-do-check-act (PDCA) model.

The EMAS is based on an EU regulation. It articulates not only what organizations wishing to register to the scheme have to do, but also what administrative systems EU Member States have to establish to allow the scheme to operate. In particular, the EMAS specifies that an organization must undertake an environmental review, establish an EMS and have this audited and reviewed, and produce a public environmental statement (report). All these elements need to be verified by an external environmental verifier accredited for the job. Once approved, EMAS organizations can use the EMAS logo and are listed on a register. The EMAS is clearly focused on environmental performance improvements.

ISO 14001 is an international standard developed by the International Organization for Standardization (ISO). ISO 14001 is a specification with informative annexes, which aims to enable any organization to establish an effective management system for ensuring and demonstrating compliance with that organization's environmental policy and objectives. Certifiers can assess it and, once successful, ISO 14001 certified organizations can publicize their achievement of the standard, but there is no official logo. Organizations can also choose to implement the standard without having it certified by an external certifier. ISO 14001 has many features in common with the EMAS; ISO 14001 also forms Annex 1 of the EU Regulation. However, EMAS contains additional provisions on a number of issues, including legal compliance, continual environmental performance improvement and employee involvement. Moreover, contrary to ISO 14001, the EMAS requires organizations to produce an environmental statement.

A range of additional EMS approaches have been developed and, in general, have arisen for the SME market and in response to SMEs' difficulties with the EMAS and ISO 14001. They include staged EMSs and alternative approaches to environmental management performance improvement. In staged approaches the establishment of an EMS is divided into a number of clearly distinguishable, consecutive steps; an example is the British Standard (BS 8555). Alternative approaches include environment-related logos or labels according to the specific requirements of the scheme in question. Examples include the Norwegian Eco-Lighthouse scheme, the Austrian Ökoprofit model and the Bavarian QuH label for small craft companies. Simplified EMSs are another alternative, which takes some of the elements of an EMS such as EMAS-light, as are customized EMSs designed by a company itself in accordance with its own wishes and needs.

Barriers and drivers

Several studies have attempted to identify the barriers and drivers to the adoption and implementation of EMSs by SMEs in a structured way. One frequently quoted overview study (Hillary, 1999) concludes that *internal* barriers (for example, the lack of human resources rather than financial ones, frequent interruptions in the implementation of an EMS in SMEs, and the lack of information about EMSs and their benefits) are more important than *external* ones (which include high implementation costs, insufficient drivers and uncertainty about market benefits, and lack of good quality consultants and sector-specific guidance). At the same time, the study identifies a large number of benefits that SMEs have experienced in establishing an EMS, such as cost savings, improved environmental performance, market recognition and so on. Many of these benefits are cited by large companies.

Network approaches to EMS adoption

Network approaches, whereby SMEs cooperate with other SMEs, larger companies and/or other actors to implement EMSs, are a means to address the two major barriers SMEs face with EMS adoption: lack of human resources and lack of financial resources. Networks vary in their composition and may be categorized as follows:

- horizontal;
- vertical; and
- multi-stakeholder.

Whilst these categories are useful descriptors, there is potential for some overlap between the network approaches. The following describes a number of different examples in each network category. These have been identified, amongst others, as examples of good practice through an EU-funded 'Best Procedure' (benchmarking) project, which described and analysed different good practice examples of public policies, programmes and initiatives to encourage the voluntary uptake of EMSs in SMEs in Europe.[2]

Horizontal networks

Horizontal networks bring together clusters of SMEs located in the same geographical area. The SMEs are not necessarily from the same sector but their proximity to each other, for example in an industrial district, makes cooperation possible. In relation to the implementation of an EMS, cooperation between companies in horizontal networks can take several forms, examples of which are elaborated below:

- cooperation where cost reductions can be realized, for instance, through joint services and joint consultancy advice as illustrated by the Convoy funding example;
- cooperation aimed at a joint effort to establish an EMS and certification; the Hackefors model, for example;
- cooperation to establish a joint EMS and seek group/territorial verification under one EMAS registration, exemplified in the First Macrolotto in Italy.

Convoy funding

The Convoy funding scheme (Öko-Audit im Konvoi)[3] was introduced by the Ministry for Environment and Transport in the German Bundesland Baden-Württemberg in 2000 and ran until 2004.[4] The main aim of the scheme is to achieve either EMAS registration or, as a first step and for companies with fewer than 100 employees, a less formal ('precursor') version of the EMAS (Vorstufe zum Öko-Audit).[5] For each Convoy group at least 50 per cent of the participating companies must achieve EMAS registration.[6]

One of the innovative design aspects of Convoy funding is that preparation towards EMAS registration takes place in small groups, coordinated by local/regional authorities or Chambers of Commerce. Each group consists of seven to ten SMEs, often from different sectors and of varying size, allowing information sharing and networking. The onus for accessing funds, identifying SMEs to participate and motivating enterprises to succeed is placed squarely on the coordinators or the consultants.[7] The advantage of this approach is that the application for funding is taken care of by the consultant or the coordinator, hence there is little administrative work and lower costs for the single SME.

Hackefors model

The Hackefors model is a network approach to implementing an EMS in clusters of SMEs located in the same geographical area, in the same sector of industry or belonging to the same company group. The model originated in the Hackefors Industrial District in Sweden in 1997 and represented an innovative SME-focused way to tackle EMS implementation.[8]

The Hackefors model is structured in two halves: one half with the parts of the EMS that are common to the group of SMEs (such as meetings, seminars, general procedures and audits), and the other half with parts of the EMS that are specific to each enterprise (such as environmental aspects, objectives and targets, specific procedures and reviews).

In the model each enterprise has an EMS, fulfils the requirements of ISO 14001 and holds its own certificate, and in this way the registrations are like for any other company. However, it is the novelty of the organization that supports the SMEs, as well as the commonly held documentation, that make the Hackefors model particularly interesting for SMEs.

All companies have an environmental manager and together form the EMS group. From this group a steering committee is selected and a central coordinator appointed. The coordinator is responsible for the network and the common parts of the system; approximately 40 per cent of documentation is common to all. The coordinator acts as a hired and shared environmental manager of the group, thereby addressing a key problem SMEs face when seeking to implement EMSs: lack of human resources. The need for a motivated and well-trained environmental coordinator appears essential for the success of the approach.

Participating SMEs work through the stages of ISO 14001 until certification, and are assisted by monthly meetings with homework tasks, training for environmental managers and employees, and dedicated enterprise visits. An external assessor audits the companies each year and all non-conformances – both internal and external – are compiled and distributed to all companies in the group to allow each company to assimilate the results. By coming together they act like a large company purchaser. This enables them to negotiate more cost-efficient certification rates than if they operated individually to the extent that group certification is estimated to be 50 per cent lower than for individual certification.

The first group of SMEs to pass successfully through the Hackefors model were 26 original Hackefors SMEs, in 1999. Unusually, the majority of these enterprises were micro firms.[9] Since that time the Hackefors model – which is commercially run – has been reproduced in 37 different groups and other regions of Sweden, with enquiries for the approach from Greece and Denmark. The model has generated 520 SMEs certified to ISO 14001, including 300 (58 per cent) certificates to micro enterprises, of which 77 were firms with only one employee; 161 (31 per cent) certificates to small enterprises with between ten and 49 employees; and 59 (11 per cent) certificates to enterprises employing more than 50 people.

Territorial EMASs – Italy

This project centres upon an industrial area, the First Macrolotto, at Prato in Italy, which covers approximately 150 hectares and currently hosts 324 enterprises with some 3500 personnel. It is renowned for its innovative efforts to use a network approach to establish a territorial EMAS.[10] A not-for-profit consortium, CONSER, was set up by consensus to provide communal environmental services for the industrial area, such as effective water treatment and recycling and personnel transportation.[11] In 1999, the idea to realize a territorial EMS and achieve registration to the EMAS was launched. This aspect required all enterprises to be viewed as having one EMS covering a geographical location.

Participating businesses are required to make use of communal services and infrastructure, take part in the initial environmental review (that is, to complete questionnaires) and establish an internal structure to cooperate with the consortium, which also coordinates training, education and technical advice. By addressing the environmental impacts of the area as a whole, the overall benefits are much greater than the total of smaller, incremental gains that could be achieved by the individual SMEs. For the participating businesses there are also important social and economic advantages from the organizational structure and service provision, including a centralized EMS 'resource' and access to environmental engineers. A key driver for the SMEs is the fact that they can delegate some of their EMS functions, as well as the provision of support and coordination.

With most of the businesses in the First Macrolotto having less than ten employees involved directly or indirectly with textiles, the project can generally be described as SME-oriented. Progress towards territorial registration to EMAS requirements has been blocked on technical grounds. The Italian Competent Body for the EMAS is insisting that all 324 enterprises develop formal management systems and will not exclude the micro businesses from the registration process.

Aims of horizontal networks

At the heart of horizontal networks is the grouping of SMEs to reap benefits from the grouping, usually in the form of human resource benefits (lack of time, multi-functional nature of staff) and financial benefits, which include not only cost savings, but also marketing opportunities. The horizontal network approach has the added advantages of enabling SMEs to implement and achieve EMSs, tailoring of material to meet local needs with the possibility of developing guides to promulgate the approach, and bringing about some environmental and commercial improvements. Nevertheless, problems can occur with the approach if the starting point of participating enterprises is different and those differences are not accommodated, and if there is not a strong decision-making process.[12]

The Hackefors model has been extensively studied and these studies show the benefits to SMEs of the approach. Interestingly, in both the Hackefors model and the Prato case, the initial network focused on waste and amenity issues before turning to environmental management. Local conditions may shape the exact form of a horizontal network but these do not limit the transferability of each approach to other settings.

The three models illustrate an evolution in the way enterprises tackle EMS implementation. The Hackefors and Prato models are innovative. Whilst the Hackefors approach is being duplicated with some success, the Prato model is yet to secure EMAS registration firmly. There is a continuum from an individual enterprise working solely on its own EMS to individual enterprises implementing an EMS as one large enterprise (no example identified), with the Convoy and Hackefors in between these two poles. In these cases, SMEs come together to gain the benefit of working together. The Prato model is somewhat different. It shifts the centre of the EMS away from the organization altogether and onto a territory. This is a potentially interesting feature for SMEs.

Vertical networks

A vertical network is a relationship in the supply chain between a large company and one or more of its suppliers that, depending on the sector, may be SMEs. Key to the vertical network is the notion of mentoring between the large company and its suppliers.

Greater pressure on larger companies to secure their environmental credentials has stimulated certain large companies to look at their supply chain, which has occurred in the automotive sector. The need for large companies to develop a greater understanding of their supply chains has led to greater cooperation. Blanket demands by customers for their suppliers to adopt an EMS have been found to be counterproductive because many SME suppliers cannot meet such demands.

Supply chain mentoring project

In April 2002 the Easy Access Environmental Management project[13] took a phased approach to EMS implementation of ISO 14001 and the EMAS, which had been developed in the UK's Acorn project, and adapted it to the construction sector.[14] Using the concept of mentoring in the supply chain, CIRIA (see Note 13) approached three of its large company members: AMEC,[15] French Kier Anglia and BAA. Each large company attracted around 15 companies – 43 in total,[16] mostly SMEs – and committed to support their suppliers through the first three phases of the staged EMS approach. CIRIA supported the process with training workshops[17] for each phase and online and telephone support. The three large companies provided remote support and in some cases workshops.

In the sharply tiered supply chain of the UK construction sector, where a small number of larger companies (only around 25) are supplied by a vast number of smaller companies (99 per cent of all construction companies), mentoring appeared to offer a viable solution to promulgating EMSs along the supply chain. The sector's profile means that mentoring offered a practical way of exchanging information, building capacity and developing understanding in the supply chain. The mentoring relationship between the three large companies and their 43 supply chain members was one that developed around trust and shared learning. Supply chain members included designers, specialist and general contractors and materials suppliers.

The phased EMS templates, based on the Acorn method and in line with BS 8555,[18] were adapted to the construction sector's specific legislative and environmental needs and a seventh phase was added to the six-phased model. The seventh phase dealt with integrated management systems and how these could be achieved. The phase seeks to integrate the EMS with health, safety and quality systems. The adaptation was driven by demand within the construction sector and the need for added value for systems implementation.

The large firms did not approach the supply chain as the 'experts' but rather as partners in the process. An understanding developed along the supply chain: the large firms understood what requirements they could request from their suppliers and the suppliers saw the benefit in formalizing their approach to environmental management and concentrating on environmental performance improvements.

Mentoring

Mentoring offers an effective means to engage and support EMS implementation in the supply chain. With increased pressure on the supply chain by customers wishing to secure their environmental credentials, SMEs are likely to respond to this approach. The mentoring approach is largely driven by the private sector with little public sector involvement, except for some cases of funding.[19]

However, the mentoring relationship is not always successful. This was the experience in the Acorn project approach because the characteristics of participants' supply chains meant the relationship between the mentor company and the SMEs was too distant; that is, some of the SMEs were second- or even third-tier suppliers and would not have a direct relationship with the mentor company.

Multi-stakeholder networks

The establishment of EMSs can also be facilitated through networks composed of a variety of both private and public stakeholders: so-called multi-stakeholder networks.

The Performance Bretagne Environnement Plus example below shows how public/private partnership can have mutually beneficial results. It was initiated as part of a broader regional sustainable development strategy, which sought to engage SMEs as one actor able to contribute to the strategy.

Performance Bretagne Environnement Plus

Performance Bretagne Environnement Plus (PBEP),[20] an initiative of the Brittany regional council and the national government, is part of the six-year action programme embedded in the broader sustainable development strategy of Brittany and covers environmental, quality and human resource performance of the region.[21] The PBEP is based on a multi-stakeholder approach involving all players in industry and civil society and public administration including the Europe Commission, the national government (DRIRE,[22] DRAF[23]), the regional council,[24] ADEME,[25] EDF,[26] employers' unions,[27] and industry and trade unions.[28]

The programme is carried out by all these stakeholders with the regional council coordinating the partners and providing funds for the programme. The national government (DRIRE and DRAF) also provides funds and support on regulation information; the EDF provides an energy counsellor for the SMEs; and the industry and trade unions make links between the actors and mobilize SMEs to motivate them to enrol in an EMS.

The PBEP programme clusters SMEs in groups of between ten and 40 enterprises according to specific topics such as ISO 14001. Attention is given to the composition of the groups to make sure each SME is grouped with similar companies so that the learning potential leads to the maximization of environmental and economic benefits. The PBEP offers hands-on assistance to SMEs to achieve an environmental self-diagnosis, with 500 being undertaken to date, and a free four-day training session, as well as counselling for all environmental problems – for example, on environmental regulations, ISO 14001, atmospheric emissions and clean technologies.

A critical factor is the funding of the multi-stakeholder network and its programmes, as shown by the Performance Bretagne Environnement Plus. Viability is dependent on these funds and, in contrast to other forms of networks, the management of very different stakeholders appears to be more time-consuming and potentially more costly. The added value for both stakeholders is the greater understanding of each other's environmental objectives and what is, and what is not, achievable.

Commentary

The examples discussed in this chapter show three types of network approach: horizontal, vertical and multi-stakeholder. There are key success features that are common to all three types of network:

- benefits accrue to all network participants;
- clear goals are articulated and understood by participants in the network; and
- a transparent and effective organizational and leadership structure is established for the network.

First, in all three types of networks, benefits accrue to participants. For the enterprises involved, the motivation for involvement in a network is first and foremost financial, whether directly through cost savings, for example, or indirectly through improved customer relations. For other participants benefits can be varied: for instance, it may fulfil a local or regional authority's need to engage a broader range of actors to help achieve strategic sustainability goals.

Second, there is added value in the network approach because it develops understanding between the partners and engenders a sense of ownership amongst the stakeholders. It is important that the right partners are included in the network, that there are clear goals, and that the expectations of all participants are managed. If goals are not made transparent and mutually agreed – goals may be multiple and different for different stakeholders – then support is not created for the network and participants will lose interest. Poorly formed networks, which take SMEs time but do not bring rewards, appear likely to fail as SMEs quickly lose interest and disengage from the network. The engagement of relevant stakeholders in initiatives is growing, however, a situation that reflects the added value of a more participatory approach by some public authorities, while the work along the supply chain is a reflection of commercial imperatives.

Third, the examples illustrate the importance of effective and focused leadership that has the support and trust of the network members. This is particularly well illustrated by vertical networks where trusted industrial bodies were the hosting and facilitating organizations. Public administrations may not have the trust required to be an effective leader of a network for the business community because of their other functions, including their regulatory function. Therefore, choice of a network leader is a critical factor and requires careful thought and consultation.

Network approaches thus have clear benefits. They do exhibit differences in three key areas, however, as discussed below: the degree of public administration involvement, the relevance of subsidies to their operations, and the broader strategic environmental policy context in which they may be set.

Horizontal networks have the least public administration involvement, followed by vertical networks where involvement may be in the form of funds or through a steering committee. Multi-stakeholder networks have the most public sector involvement. Indeed, multi-stakeholder networks are more likely than not to be initiated by public bodies, which seek to engage with other stakeholders, particularly the industrial sector.

Funding appears critical to the operation of multi-stakeholder networks. In the case of the vertical network, funding was necessary for the project to proceed because it was testing a model of mentored supply chain EMS implementation, which could then be promulgated in the construction sector. Horizontal networks could either be self-funding or accessing subsidies for specific aspects of the tasks within the network; training and purchase of consultancy services, for example.

Multi-stakeholder networks' goals are set within broader public strategic sustainable goals, whereas for vertical networks this may be only indirectly the case. For example, financial support for CIRIA's construction sector project was consistent with the UK Government's policy paper, which highlighted the construction sector as a key sector for sustainable development initiatives. In contrast, horizontal networks' aims are not set within such public goals.

Conclusions
Network approaches address many of the internal barriers SMEs face when implementing EMSs. They offer SMEs the opportunity to share experiences, costs and support to

enable a more collaborative approach to EMS implementation. When other stakeholders are involved, such as public administrations and regulators, networks allow all actors (including SMEs) to understand the environmental challenges they face and the expectations other stakeholders have of them.

However, networks are not a panacea for the problem of SMEs' low participation in EMSs, such as EMASs and ISO 14001. Questions about the applicability of the formal standard to SMEs still apply. The degree to which these require a formalization of management structures for environmental issues is contrary to many SMEs' informal business management structures. Hence the proliferation of 'light' versions of EMSs and approaches such as networks to enable SMEs to access the benefits of tackling environmental performance in a systematic way.

Notes

1. According to a survey published in January 2003 by NetRegs, a joint initiative between UK environment agencies and small business, 86 per cent of all SMEs do not consider their activities harmful to the environment and only 18 per cent are able to mention any environmental legislation that applies to them. On a more positive note, the study also shows that most larger SMEs have undertaken practical measures to reduce harm to the environment.
2. The author was the consultant on the study and this chapter draws from material in the report: *Public Policy Initiatives to Promote the Uptake of Environmental Management Systems in Small and Medium-sized Enterprises, January 2004*, http://europa.eu.int/comm/enterprise/enterprise_policy/best/best_projects_2002/index.htm.
3. http://72.14.203.104/search?q=cache:HC77wi_UdCkJ:www.uvm.baden-wuerttemberg.de/servlet/is/233/+%C3%96ko-Audit + im + Konvoi&hl = en&gl = au&ct = clnk&cd = 4
4. From 2000 to 2001, €500 000 was spent on Convoy projects. Funding has been extended until 2004 but it is unclear whether there will be further funding beyond 2004.
5. This 'precursor' version consists of the following elements: an environmental check; the development of a basic environmental programme setting out the most relevant environmental measures, responsibilities and deadlines; documentation of the SME's environmental performance, its environmental management and its future tasks; and a check of the documentation by an approved EMAS auditor. There is no official label for this light EMAS version because the intention is that it is upgraded to an EMAS within three years.
6. Since October 2000, 30 Convoy projects with over 200 participants have been completed or are still under way; and 35 to 40 organizations have already achieved EMAS registration.
7. Costs for consultants are funded up to 60 per cent (up to a maximum of €3000 per SME); for SMEs with fewer than 100 employees, up to 80 per cent (up to €4000) can be funded.
8. Interestingly, the members were initially told that they would not be able to achieve a collective EMS and this negative response spurred them on to develop their EMS.
9. Sixteen employed less than ten people, seven employed between 11 and 50 people and three employed between 51 and 80 people.
10. http://www.macrolotto-prato.com/.
11. Other further projects under way include waste management and a co-generation power plant.
12. These problems were encountered in a Finnish project, financed by the National Technology Agency, where a small industrial estate was deemed to be the entity to pilot a group EMS for 12 companies.
13. The project is managed by CIRIA (Construction Industry Research and Information Association) and funded by the UK's Department of Trade and Industry. CIRIA is a not-for-profit member-based research organization undertaking research and developing tools to meet the needs of its members. The project has a multi-stakeholder project management committee including industry, government and NGOs. Representatives are Government Office for the East of England, White Young Green Environmental, Davis Langdon Consultancy, Haswell Environment Agency, AMEC, CIRIA, French Kier Anglia, BAA, Civil Engineering Contractors Association, The Acorn Trust, Groundwork UK, University of Bradford, House Builders Federation, Envirowise, Federation of Master Builders, University of Bradford, AMEC, Constructing Excellence, Small Business Service, Institution of Structural Engineers, BM TRADA Certification Ltd, Environment Agency, Sheffield Hallam University.
14. www.ciria.org.uk.
15. AMEC's suppliers came from Wales, South East England, London and Anglia.

16. The suppliers included building contractors, civil engineering contractors, specialist contractors, architects, designers and construction production manufacturers and suppliers.
17. Three workshops included securing commitment and establishing baseline; identifying compliance with legal, customer and market requirements; and developing environmental management programmes.
18. A new British Standard on the phased implementation of EMSs with performance evaluation.
19. EU Epicentre project is currently funding the supply chain vertical network approach based on the Acorn approach. However, in the case of Ireland it has been difficult to identify mentor companies, or large firms willing to participate.
20. http://www.region-bretagne.fr/.
21. Other performance initiatives include: Performance Bretagne Qualité Plus, Performance Bretagne PSA Peugeot Citroën Plus, Performance Bretagne Ressources Humaines Plus, http://www.cordis.lu/bretagne/fr/services_enterprises.htm.
22. Directions Régionales de l'Industrie de la Recherche et de l'Environnement, www.drire.gouv.fr.
23. Direction Régionale de l'Agriculture et de la Forêt.
24. http://www.region-bretagne.fr/.
25. ADEME (Agence de l'Environnement et de la Maîtrise de l'Energie) is a state-funded public industrial and commercial whose activity is supervised by the French government ministries in charge of research, the environment and energy. ADEME came into existence on 1 January 1992. The agency's mandates cover conservation of energy and raw materials, promotion of renewable energy resources, promotion of clean and energy-efficient technologies, waste minimization, disposal, recovery and processing for economic value, prevention and reduction of air pollution, mitigation of noise pollution, and prevention and remediation of soil pollution, /www.ademe.fr.
26. Electricité de France, http://www.edf.fr/.
27. Mouvement des Entreprises de France, http://www.medef.fr/.
28. Chambres de Commerces et d'Industries de Bretagne, http://www.bretagne.cci.fr/.

References

British Standard (1996), EN ISO 14001:1996, *Environmental Management Systems*, London: British Standard Institute.
European Commission (2001), 'Regulation (EC) No 761/2001 – Allowing Voluntary Participation by Organizations in a Community Eco-Management and Audit Scheme (EMAS)', *Official Journal*, L114, Brussels.
Hillary, R. (1999), *Evaluation of Study Reports on the Barriers, Opportunities and Drivers for Small and Medium Sized Enterprises in the Adoption of Environmental Management Systems*, London: Department of Trade and Industry.
NUTEK (2003), *Environmental Work in Small Enterprises – A Pure Gain?*, **R 2003**, 7, Stockholm.

16 Green marketing and green consumers: exploring the myths

John Connolly, Pierre McDonagh, Michael Polonsky and Andrea Prothero

Introduction

Green marketing is often incorrectly associated with superficial 'green hype' slogans such as 'new, improved and friendly to the natural environment'. Organizations (for-profit, non-profit and governmental) using these types of 'claims' frequently try, with varying degrees of success, to associate themselves and/or their products with an environmental image without substantially improving the environmental characteristics of the organizations' practices and/or products. While such inappropriate behaviour is on the decline, its early use has caused irreparable harm to genuine environmentally responsible marketing actions, as many consumers are generally sceptical of all green marketing activities (Crane, 2000; Connolly and Prothero, 2003; Crude, 1993; Davis, 1994; Dolan, 2002; Mendelson and Polonsky, 1995; Prothero and Fitchett, 2000).

At the same time, there is still increased pressure from NGOs and special interest groups (Crane, 2000) to influence firm behaviour. For example, in 2003 the WWF (Worldwide Fund for Nature) attacked petroleum giant BP by taking out a full-page advertisement in the *Financial Times* criticizing BP's planned pipeline. Companies that introduce misleading advertising are also increasingly being caught out by consumers, government and special interest groups. For example, Earth Day Resources* (EDR) publishes a list of organizations that make misleading environmental claims dubbed 'Greenwash', and defined as: 'Disinformation disseminated by an organization so as to present an environmentally responsible public image' in the *Concise Oxford Dictionary*. 'Greenwashers' that have been listed on the EDR website include Ford, General Motors, Kraft Foods and Herbal Essences.

Companies that introduce 'ethical' activities, including green actions, can in fact be profitable. In 2003 the Co-operative Bank, in conjunction with the New Economics Foundation and the Futures Foundation in the UK, published an Ethical Purchasing Index suggesting that whilst the UK economy grew by only 1 per cent in 2002, the sale of goods marketed as 'ethical' in some way increased by 13 per cent to £6.9 billion pounds sterling. At the same time the report suggests that consumer boycotts of companies that behave inappropriately cost firms approximately £2.6 billion pounds a year.

Green marketing encompasses much more than simple 'marketing hype' (Menon and Menon, 1997; Polonsky and Rosenberger, 2001). Although there is no universally accepted definition of green marketing, accepted elements include voluntary exchange between organizations and consumers that achieves both parties' objectives, while attempting to minimize these exchanges' negative environmental impacts, and ensuring that company activities are sustainable. McDonagh (1998) argues that green marketing is just one indicator of a process of sustainable communication, where sustainable communication is a complex social process of unravelling and eradicating alienation between an organization and its public.

Green marketing involves more than just promoting a product's or organization's green attributes or activities. It requires that organizations carefully evaluate the very nature of organization–consumer exchange. This may require a complete paradigm shift in organizational thinking to address sustainable behaviour (Kilbourne et al., 1997). As such it may involve radical changes in marketing processes, which may go as far as asking whether consumers need to actually own products, or there are other ways of delivering want-satisfying capabilities (Peattie, 1999). Take the example of Toyota, which has been working with the Japanese government to develop a trial programme where people purchase transportation without owning automobiles. Instead, people buy access to a fleet of electronic automobiles that can be used to travel short distances to shops and/or connect with traditional public transportation, which is also accessed as part of the transportation package. In this way green marketing has removed the need for individual ownership of cars, while still allowing people to have their core need for transportation met. Toyota therefore is 'seizing the day' and adapting corporate activities to take advantage of environmental opportunities, rather than waiting to be pushed by regulation or other external pressures. Companies that produce greener products are also aided by various government initiatives, for example, the Toyota Prius, which has an environmental engine, is exempt from the recently introduced congestion charge in London. Indeed, there are even green consumer groups such as the NESEA (Northeast Sustainable Energy Association) in the USA organized specifically for green car enthusiasts. In some cases companies that are usually competitors even work together, for instance, a number of chocolate manufacturers globally have joined forces with human rights groups to eliminate the use of child labour in developing nations.

Why green marketing?
Unfortunately, many organizations do not incorporate environmental activities into their core evaluative criteria (Kilbourne, 1998). When these organizations undertake green marketing activities, they do so to increase profits, return on investment, market share and so on, rather than necessarily to achieve a competitive advantage by enhancing the sustainability of firm activities. Some organizations do 'go green' for more altruistic reasons, however, this growing number of organizations are in the minority (Drumwright, 1994).

Understanding organizational motivations for 'greening' is important, for it shapes how an organization undertakes its specific green marketing activities. For example, organizations that undertake green initiatives simply to comply with regulations may choose to undertake minor changes to their production processes for the sole purpose of avoiding 'prosecution'. Organizations that choose to target niche green market segments may be required to modify significantly their environmental activities across a range of functional areas. While many organizations see environmental regulations introduced by governments as a constraint, more forward-thinking organizations regard them as an opportunity to maintain a sustainable competitive advantage (McDonagh, 1998; Porter and van der Linde, 1995).

One such organization is Toyota, which has established an 'eco-technologies' division that is developing technologies designed to exceed existing emission and fuel efficiency regulations. Thus Toyota has chosen to be an environmental leader rather than an environmental follower based on the realization that environmental issues must be considered and addressed today if organizations are to continue operating into the future.

In this way Toyota is reshaping its direction and designing more fuel-efficient traditional engines as well as developing a range of new automobiles, including its electric and hybrid electric-combustion automobiles, which may ultimately replace petrol engines (Toyota, 1997). Similarly, Honda is planning to bring out a green version of its popular Accord model, which will be a hybrid gas-electric vehicle. Indeed, at the Tokyo Motor Show in 2003, clean cars were at the forefront of company displays as the car of the future, with the focus on fuel-cell and hybrid cars that do not use fossil fuels. Even governmental bodies have seen the need to gain competitive advantages through greening their activities. In Australia the Newcastle City Council has adopted the mission of becoming one of the greenest councils in Australia as they believe this will not only improve the general quality of life, but will draw new industries and technologies to the region (Newcastle City Council, 1999).

In some cases organizations undertake greening initiatives as a reaction to consumers seeking out new greener products/services. Adopting a demand-driven approach unfortunately assumes that the customer is always right (in other words that customers know which environmental improvements are most appropriate), which may not be the case. Customer- or market-oriented greening can take a number of different forms, such as organizations producing new products that are not as environmentally harmful alongside their 'traditional' goods. In other cases organizations modify products to address particular consumer concerns. McDonalds employed this strategy, for example, when it replaced polystyrene clamshell packaging with waxed paper as a direct response to consumer concerns over chlorofluorocarbons (CFCs) produced in the production of polystyrene (Gifford, 1991). Unfortunately it is unclear whether waxed paper is in fact less environmentally harmful than new generation CFC-free polystyrene. This example highlights the fact that responding to customers' wants may not result in the best environmental outcome (Oleck, 1992). Any analysis of the environmental impacts of products or processes is complex, and may require detailed lifecycle analysis (LCA) to identify the most environmentally responsible course of action. Importantly this may mean that the alternative selected does not appear to be the least environmentally harmful, at least on the surface (that is, to consumers).

Competitive pressures may also force organizations to introduce green marketing activities, sometimes triggering a chain reaction within an industry. This occurred when consumers became concerned with the numbers of dolphins being killed by tuna boats in the late 1980s, for example. In response, Starkist, one of the major tuna marketers in the US, announced it was going 'dolphin-free', a move that forced its competitors to follow suit or lose market share (Anonymous, 1991). In another example when a laundry detergent manufacturer in Australia introduced a concentrated formula, others quickly followed suit in the fear they would be left behind (Anonymous, 1990). The introduction of CFC-free refrigerators by a German manufacturer was quickly followed by other manufacturers, despite initial opposition to this innovation. This brought about strategic change across the industry and required a complete shift in market practices. In some cases, however, changes to organizational 'environmental' behaviour may only be tactical in nature. The introduction of detergent concentrates in the example given above did not cause manufacturers to shift corporate environmental thinking or behaviour substantially but rather simply meant that they added a 'green' product alongside their less environmentally friendly alternatives.

Suppliers and downstream customers may also exert pressure on organizations to change their environmental behaviour. In the Starkist tuna example suppliers were required to modify their fishing behaviour or face losing Starkist as a major customer (Rice, 1990). The international standards of ISO 14000 are a significant influence too, as firms complying with these requirements are required to evaluate the environmental performance of their suppliers. In this way complying organizations pressure their suppliers to meet appropriate environmental standards, which in turn pressure their suppliers to comply. The implications of these conditions for national and international trade are exemplified by the actions of IBM and Volvo, which require that all their suppliers worldwide have appropriately approved environmental management systems in place (Zuckerman, 1999).

Together these motivators suggest that changes can occur as a result of external pressures on the organization. However, this is not always the case and there are a number of situations where changes are internally motivated on financial or philosophical grounds. The cost perspective is driven by the realization that there are savings associated with greening activities, which may give organizations a competitive advantage in the marketplace (Porter and van der Linde, 1995). Organizations operating in a less environmentally harmful way typically increase efficiency and reduce costs by using fewer materials and producing less waste and pollution. For example, more efficient fan/blower motors, such as the new motor vector drive installed by the Greenville Tube Company into its tube-making process, use less electricity and can improve the production process. The company has since realized a US$75 000 saving per year from the reduction in wasted material as well as a 30 per cent saving on their energy bill. In addition the process now runs for less time, introducing fuel and maintenance savings that also reduce polluting emissions (Anonymous, 1999).

Greening may also be internally motivated through a 'responsible' organizational philosophy where environmental issues are given equal weighting to financial objectives. Such philosophies are often transferred from an organizational founder down to management, which are then integrated into all organizational activities (Drumwright, 1994). For example, the founder of Blackmores, an Australian firm that produces and markets nutritional supplements and health care products, had incorporated environmental issues as a core focus of the company's activities before it was fashionable to address these issues:

> In 1967, long before environmental issues were a concern, Blackmores' founder, Maurice Blackmores, spoke of the importance of conservation and the effect of pollution on the environment when he stated: 'If man persists in ignoring or defying the recycling laws of Nature he will not avoid pollution, malnutrition or starvation . . . Nature does not know how to handle pollution or preserve the balance of nature in the face of it'. (Blackmores, 1999)

Whereas in the Blackmores example environmental values were adopted because they were part of top management's philosophy, it is also possible that a concern for the natural environment develops slowly over time, to become eventually a centrepiece of organizational philosophy. Such concern is evident in a statement by Toyota's president Hiroshi Okuda who wants Toyota to have a strategy that promotes 'harmonious growth in the next century, balancing the needs of people, society and the global environment' (Toyota, 1997, p. 3). Thus global organizations may integrate environmental issues into overall organizational philosophy because they believe it is the 'right thing to do' *and* because it is a sustainable practice. Importantly, such acts can also be profitable. The Co-operative

Bank in the UK, for instance, set record profits in 2002 whilst being voted in the *Sunday Times* one of the best companies to work for.

Levels of green marketing

A number of catchphrases – including green marketing, sustainable marketing and responsible marketing – have been used in the green marketing area. As noted earlier, however, use of these terms requires that firms satisfy organizational objectives and consumer needs while minimizing environmental harm and ideally making environmental improvements. Thus organizations must ask themselves: how can organizational and the consumer's objectives be achieved in more environmentally responsible and sustainable ways? The ramifications of such an approach are substantial and require that organizations think globally about their activities. The motivations for undertaking green marketing activities discussed in the previous section will most likely impact on the strategies and tactics that are used to address environmental issues.

True green marketing entails that environmental issues are a strategic organizational focus, as are issues such as quality or customer service. Thus changes in the organizational mindset as well as organizational behaviour are likely to be required. Menon and Menon (1997) have suggested that greening marketing activities can occur at three levels within organizations:

- *Strategic greening* requires a substantial and fundamental change in organizational philosophy. The Australian firm CarLovers achieved such change when it designed its entire car washing process on a closed-loop system where water is recycled, ensuring that environmental issues are a core part of business philosophy and practice.
- *Quasi-strategic greening* requires a substantial change in practice. For example, some hotels promote the fact that they are trying to minimize their environmental impact. In an attempt to reduce water consumption associated with washing linen, these hotels ask guests to indicate when they want their towels washed by leaving them on the floor or in the bathtub.
- *Tactical greening* involves shifts in functional activities such as promotion. This occurs, for example, when water authorities encourage consumers to behave in a more responsible fashion in times of drought. In this case there is little, if any, change in organizational activities, other than a short-term reallocation of promotional resources.

These three levels of green marketing can be used to identify the amount of change required by an organization and may reflect the degree of 'commitment' to various environmental objectives. Take the example of a jeans manufacturer who in the early 1990s promoted their commitment to donate a proportion of each sale for planting trees. Such a tactical activity might be viewed with intense scepticism as there is no logical link between these activities (that is, between manufacturing jeans and planting trees). However, a similar programme by a paper company that explains the links between its activities and the natural environment – the specific environmental issue being addressed as well as how this programme will assist in improving the environment – would most likely be seen as an appropriate tactical activity.

On the other hand, any strategic greening activity would require an extensive long-term financial investment as well as a shift in organizational behaviour and mindset. As such, effectively implemented strategic greening would rarely be seen as superficial. Take, for example, the German furniture manufacture Wilkhahn. This company has undertaken substantial environmental activities that shape its fundamental philosophy *and* activities, including the design of long-lasting products that either use recycled materials or minimize its use of virgin products, are produced in an environmentally designed factory, and are ergonomically designed as well (Wilkhahn, 1996). Indeed, according to the *Wall Street Journal* (Mechling, 2003) 'green is the new black' for furniture design in the USA. Quoting examples of furniture made of recycled materials such as newspapers, wax and wood, the article suggests that in 2002 Americans spent US$2 billion on natural furnishings for the home, with a single coffee table selling for up to $14 000 US.

Green marketing strategies and tactics

Popular ideas about green marketing tend to incorrectly focus on promotion or product design activities. While these aspects are important, green marketing is an integrated approach that continually re-evaluates how all types of organizations (for-profit, non-profit and governmental) can better achieve their objectives and consumer needs while reducing the long-term environmental harm of these activities. As such, green marketing requires ongoing improvements across a diverse range of activities (McDaniel and Rylander, 1993) and has evolved since its initial conception within the mainstream business community in the early 1970s (Kilbourne and Beckmann, 1998).

In addition to the preceding discussion of the various forms green marketing can take and what it may involve in practice, a range of related activities may be identified (Polonsky and Rosenberger, 2001). These are summarized in Table 16.1. The next section moves on to explore the green consumer in more detail.

Green consumers

It seems that in the twenty-first century consumption plays a significant part in our lives and much is written about it both in academia and in the popular press; we now even have television programmes such as *Rich Girls* on MTV, viewers of which spend significant amounts of air time watching rich girls buy things! Thus it seems fair to say 'we are what we consume'. What then of green consumers? Are they defined by what they don't consume? It seems that is not necessarily so. As with green marketing, there are no agreed definitions as to who are green consumers; instead it would be more appropriate to suggest various shades of green consumption, by people from all walks of life who cannot easily be segmented into particular target groups. It seems there are very many reasons why people in their own opinion are green. The 1990s saw incredible publicity for green products and environmentally friendly consumers. The commodification of green consumer information was also accompanied to a degree by its segmentation as evidenced by the variety of books on the subject: *The Green Consumer Guide: From Shampoo to Champagne, How to Buy Goods That Don't Cost the Earth* (Elkington and Hailes, 1988); *The Green Consumer's Supermarket Shopping Guide* (Elkington and Hailes, 1989); *The Young Green Consumer Guide* (Elkington et al., 1990, 1997); *Green Earth Resource Guide: A Comprehensive Guide About Environmentally-Friendly Services and Products: Books, Clean Air, Clean Water, Eco-Tourism* (Gorder, 1991); *The Green Consumer Supermarket*

Table 16.1   Green marketing activities

Activity	Explanation
Targeting	The firm undertakes activities that target a specific segment of consumers. This would include developing a new business unit or brand as well as possibly designing promotions targeting specific types of consumers
Green design/New product design (NPD)	The firm has either redesigned its product and production process or has integrated environmental principles in the development of all its activities
Green positioning	The firm explicitly promotes the product or the firm's environmental performance as being a core component of company activities
Green pricing	In developing prices the firm moves beyond simply discussing out of cost price to identify the price of utilizing the good from cradle to grave
Greening logistics	Firms design logistics systems that minimize the utilization of resources that will frequently include reverse logistic systems that move packaging and/or used products back to the firm
Marketing waste	The firm either seeks to improve efficiency and thus reduce 'waste' or seeks to identify viable consumers of waste products
Green promotion	The firm includes information in its promotional activities that mention environmental benefits of products in addition to other activities. In some cases these may be undertaken simply to deal with negative publicity
Green alliances	A company develops joint activities of varying types that have some environmental linkage. These may include co-branding activities or more substantive programmes where parties work together to deal with salient environmental issues

Shopping Guide (Makower et al., 1991); *The Green Supermarket Shopping Guide* (Wasik, 1993); *Mother Nature's Shopping List: A Buying Guide for Environmentally Concerned Consumers* (Shook, 1995).

Are such books or guides, where they are promoted by environmentalists or commercial enterprises, little more than recipes for selecting green options? Whilst this may be the case, such guides may also be seen as a response to the increasing reflexivity of the self in late-modernity. If, as Giddens (1991) contends, the self is a reflexive project for which the individual is responsible, information about ways to live and lifestyles to follow take on increasing significance. Thus, the proliferation of self-therapy and self-help books. Green guides may well be part of this condition.

As a result of concerns about misleading advertising claims, the gloss has certainly dulled over green consumption in recent times. Coupled with high mark-ups on goods such as organic foods and fairly traded products, the green consumer literature is not always positive. Similarly, those consumers who say they would like to buy greener products if they were more readily available do not necessarily do so. Consequently our understanding of green consumption is definitely blurred.

In addition, green consumption needs to be understood against the backdrop of changing social conditions. The shifts to globalization, individualization, reflexive modernization and the 'risk society' associated with the work of Beck (1992, 1999), Beck et al. (2003) and Giddens (1991, 1994) has resulted in a new understanding and conceptualization of society, ecological issues and consumption. Consequently the most recent literature addressing green/ethical consumption (see Connolly and Prothero, 2003; Halkier, 1999, 2004; Micheletti, 2003; Micheletti et al., 2004) is placed within the context of these changing social boundaries.

The market system is now installed not only further across the globe than ever before, but also within ever more aspects of society. Thus shopping, consumption and consumers, to paraphrase Miller (1995), are now at the vanguard of contemporary society. A consequence is the increasing assemblage of shopping and morality. The perceived personal responsibility for global environmental problems (Micheletti, 2003), allied with the intrusion of global ecological dilemmas into people's lives (Giddens, 1991), has led to a politics of the personal (Giddens, 1991) in which green and ethical consumption often plays a significant role.

Consequently there are many reasons why consumers engage in green activities. For some it may be for health care reasons for themselves and their family, to protect the environment or to protect wildlife; for others it may simply be because they feel such activities are the right thing to do. As such there are many types of activities that could be classed as green in one way or another. They range from buying organic foods, not buying genetically modified foods, being vegetarian, purchasing ethical investment products and buying recycled paper products, to trying simply to consume less of everything.

However, despite a lack of being able to define green consumers one issue is clear: the dramatic and ever-increasing levels of consumption that have accompanied consumer culture have led to a variety of individual consumers who question their own lifestyles and consumption behaviour. Marketing and consumer research has attempted to examine these consumers in minute detail. Different terms have been used to described or categorize consumers of different persuasions by marketers and consumer researchers over the years, as detailed in Table 16.2.

Various studies have examined issues surrounding the interaction between consumers and the natural environment and its constituents (animals), quite often for the commercial purposes of new product development, new targets and segments, developing green communications campaigns and so on. Despite the commercial interest of such studies, they also draw attention to consumers who attempt to adjust their consumption behaviour and lifestyle, many for environmental and ethical reasons. Shaw and Newholm (2002) point out that awareness of environmental and ethical issues such as environmental degeneration and animal welfare inevitably raise individuals' awareness of their own levels of consumption. One of the current difficulties with green consumer research is the conceptualization of such consumers where green consumption tends to be equated with green purchasing or disposal decisions (Minton and Rose, 1997; Ottman, 1998; Peattie, 1999). As such it does not address the fundamental lifestyle issues that underpin decisions to completely change one's consumption activities and thus to become what Ottman (1998) describes as a true-blue green consumer.

Since the late 1980s and early 1990s, when the environmental crisis hit home to large numbers of consumers, the notion of a green consumer has become an increasingly

Table 16.2   Categorizations of green consumers in the literature

Consumer categorization	Authors
Responsible consumers	Fisk 1973
The socially conscious consumer	Anderson and Cunningham, 1972; Antil, 1984; Brooker, 1976; Mayer, 1976; Osterhus, 1997; Webster, 1975
Ecologically concerned consumers	Balderjahn, 1988; Bohlen et al., 1993; Kinnear and Taylor, 1973; Kinnear et al., 1974; Roberts and Bacon, 1997; Schwepker and Cornwell, 1991
Environmentally concerned consumers	Berger and Corbin, 1992; Follows and Jobber, 2000; Murphy et al., 1978
Environmentally conscious consumers	Dembkowski and Hanmer-Lloyd, 1994
Downshifters	Schor, 1991, 1998
Conserving consumer	Pickett et al., 1993
Humane consumers	Ger, 1997
Ethical simplifiers	Shaw and Newholm, 2002
Ethical consumers	McDonagh, 2002; Shaw and Clarke, 1999; Shaw and Shiu, 2002; Shaw et al., 2000; Strong, 1997
Green, ethical and charitable consumer	Schlegelmilch, 1994
Green consumers	Diamantopoulos et al., 2003; Laroche et al., 2001; Prothero, 1990; Roberts, 1996; Schlegelmich et al., 1996; Shrum et al., 1995
The voluntary simplifier	Craig-Lees and Hill, 2002; Etzioni, 1998; Rudmin and Kilbourne, 1996; Shama, 1985

accepted phenomenon. From a marketing perspective efforts to identify environmentally conscious (green) consumers can be traced back to the 1970s (Anderson and Cunningham, 1972), with numerous consumers environmentally, ecological or socially conscious. Despite the various labels, the central theme running through these definitions is that of consumers who consider the environment to be important and who respond by evaluating their purchasing decisions to take this into account. At the same time green products can also be regarded as possessing a set of attributes that are in some sense ethical (Crane, 1997). In their book, *The Green Consumer Guide*, Elkington and Hailes (1988) describe the green consumer as one who avoids products that are likely to endanger the health of the consumer or others; cause significant damage to the environment during manufacture, use or disposal; consume a disproportionate amount of energy; cause unnecessary waste; use materials derived from threatened species or environments; involve unnecessary use or cruelty to animals; adversely affect other countries. Yet while many of these attributes might be considered ethical, Shaw and Clarke (1999) distinguish ethical consumers by their concern for deep-seated problems, such as those of the Third World, in addition to environmental problems. Furthermore, Shaw and Newholm (2002, p. 168) state that 'the inextricable link between consumption and ethical problems, such as environmental degeneration and fairness in world trade, has resulted in the emergence of a group of consumers commonly referred to as ethical consumers'. It is hardly surprising then that the notion of a 'typical green consumer' continues to be elusive (Ottman, 1998). Consequently it may well be impossible to categorize green consumers into one segment to be treated differently from other mainstream consumers. As discussed earlier, green consumers themselves differ in what they believe green consumption to mean and thus what they identify as a green product.

Two areas of green consumption based in ethical concerns can be expanded upon in more detail: ethical consumption reflected in product purchases on the one hand and the choice by individuals to adopt a lifestyle of voluntary simplicity on the other.

Ethical consumers

According to Strong (1996, p. 5), ethical consumerism incorporates the principles of environmental consumerism as well as 'buyer behaviour that reflects a concern with the problems of the Third World, where producers are paid low wages and live in poor conditions simply to produce cheap products for western consumers and profits of multinational companies'. As with the range of definitions on green consumers, Harper and Makatouni (2002, p. 289) posit that:

> being an ethical consumer means buying products which are not harmful to the environment and society. This can be as simple as buying free-range eggs or as complex as boycotting goods produced by child labour. Products which fall into the ethical category include organic produce, fair trade goods, energy efficient light bulbs, electricity from renewable energy, recycled paper and wood products with Forest Stewardship Council approval.

Of central concern to ethical consumers are such issues as fair trade, worker exploitation and environmental impacts in Third World producer countries. Raynolds (2002) points out that the growth of fair trade labelled food products is also intertwined with green consumption issues.

Fair trade is a sub-set of ethical consumerism (Bird and Hughes, 1997). The growth of ethical consumerism provides the main driver behind the development of a fair trade market. Fair trade networks support and are supported by elevated concerns amongst western consumers over global ethics and the rise of 'ethical consumption' practices where the social relations embodied in particular commodities increasingly shape product choices (Raynolds, 2002). Yet, according to McDonagh (2002), no universally accepted or authoritative definition of fair trade exists. He puts forward the following text by the Fair Trade Federation and a summation of its principles as a means to understand the concept:

> 'fair trade' means that trading partnerships are based on reciprocal benefits and mutual respect; that prices paid to producers reflect the work they do; that workers have the right to organize; that national health and safety, and wage laws are enforced; and that products are environmentally sustainable and conserve natural resources. (McDonagh, 2002, p. 643)

Fair trade principles include (McDonagh, 2002):

- to benefit the artisans with whom they work, not to maximize profits;
- to work with producer cooperatives that employ democratic principles;

- to encourage producers to reinvest their profits in their communities;
- to shift processing and packaging activities to the Third World where possible, so that as much work as possible will remain with producer countries;
- to encourage cooperation amongst fair trade organizations;
- to evaluate individual products in a manner analogous to a social audit for a company.

According to Raynolds (2002) fair trade labels strive to re-establish consumer trust in the origins and content of their food, attesting that items have been produced outside the agro-industrial system responsible for recent food scares and widespread environmental degradation. From a marketing perspective, numerous initiatives are in place to promote and extend fair trade products in western markets (for examples see Bird and Hughes, 1997; Goodman and Goodman, 2001; McDonagh, 2002). McDonagh (2002) contends that the use of anti-slavery and fair trade labels (formal labelling initiatives) permit consumers to engage in a variety of fair trade symbolic actions. One of the more promising marketing activities employed in this regard is the enactment of the life histories of the producers of fair trade products in labelling and promotion. These narratives thus allow often-substantial geographical distances to be transversed. According to Raynolds (2002), by embedding commodities with social and environmental information, fair trade labels function as a mirror for the consumer in securing the benefits of self-expression and positive social identity. Both McDonagh (2002) and Raynolds (2002) contend that the consumption of fair trade labelled commodities with high symbolic value offers an important opportunity for consumers to identify themselves as socially and environmentally conscious individuals.

Another part of this rather diverse yet interrelated family is organic production and consumption. According to Browne et al. (2000), sustainability and organics are closely related, with many of the goals of sustainable agriculture corresponding closely to those of organic agriculture. Although different in origin from the ethical debate, the principles of organic agriculture are wide ranging and include concerns for safe food production, the environment, animal welfare and issues of social justice. It encompasses the conservation of non-renewable resources and issues of environmental and social responsibility, and states that sustainability lies at the heart of organic farming. Increasingly, according to Browne et al. (2000, p. 72), 'ethical and organic trading are beginning to overlap. An increasing number of fairly traded goods are also organic and the organic movement is moving towards including social rights and fair trade in standards'. Indeed, for one of the authors, a regular dilemma at the local supermarket is whether to buy the fairly traded or organic bananas.

A more recent development in the area of organics is the convergence of organic production under the auspices of the fair trade movement (Browne et al., 2000; Goodman and Goodman, 2001). The products involved are not only produced through environmentally friendly production methods but are also distributed via socially responsible trade, which creates more equitable and favourable conditions for increasingly marginalized small-scale producers in the global South (Goodman and Goodman, 2001).

Fair trade purchases, together with the variety of other ethical purchasing options such as green goods and organic produce, favouring small stores or local production is seen as the positive action of consumers to 'buycott' particular goods and services. In contrast to boycotts that often aim to punish firms for past misdeeds, buycotts are seen to reward firms for their activities (Friedman, 1996). These buycott activities could be regarded as contributing to improved sustainability through the support of more environmentally benign practices by Third World producers or conversely the lower environmental impact of local production (Shaw and Newholm, 2002).

While green and ethical consumerism has been the focus of much consumer research, the rejection or limiting of consumerism has become of increasing interest to marketers and consumer researchers (Craig-Lees and Hill, 2002; Etzioni, 1998; Higgins and Tadajewski, 2002; Rudmin and Kilbourne, 1996; Rumbo, 2002; Schor, 1991, 1998; Shaw and Newholm, 2002; Zavestoski, 2002). Most notable in this search for consumptive resistance is the voluntary simplicity movement, which has been termed 'countercultural, potentially subversive' and 'mainstream' (Maniates, 2002, p. 199).

Voluntary simplicity

Voluntary simplicity can be described as a system of beliefs, and as a practice. It is centered on the idea that personal satisfaction, fulfilment and happiness result from a commitment to the non-material aspects of life (Zavestoski, 2002). For Etzioni, 'voluntary simplicity refers to the choice of free will – rather than by being coerced by poverty, government austerity programs, or being imprisoned – to limit expenditures on consumer goods and services, and to cultivate non-materialistic sources of satisfaction and meaning' (Etzioni, 1998, p. 620). Although voluntary simplicity is often associated with the '60s generation (Zavestoski, 2002), its origins can be traced back through various ancient customs, such as Hebraic, Chinese Taoist and Greek traditions (Rudmin and Kilbourne, 1996). In the context of the contemporary voluntary simplicity movement, Rudmin and Kilbourne (1996) contend that present-day advocates of voluntary simplicity put stewardship of the global ecology at the forefront of their argument and agenda. They maintain that the goal of voluntary simplicity in the modern age is not much different from that professed by the Ancient Greeks, Christians or Romantics; that is, to be a better, more complete individual.

Zavestoski (2002) claims that one explanation for the recent interest in the ideas of voluntary simplicity is that people experiencing unhappiness and discontent are linking these feelings to the media and to culture-driven messages to consume increasing amounts of goods at greater rates. According to Craig-Lees and Hill (2002, p. 191), certain themes are associated with voluntary simplicity:

- the element of free choice to lead a simple life;
- material consumption is reduced but a life of poverty is not required;
- access to resources such as wealth, education and unique skills that could be traded for high income;
- control and personal fulfilment;
- values such as humanism, self-determination, environmentalism, spirituality and self-development.

Maniates (2002) notes that voluntary simplicity is also referred to as simple living, downsizing, downshifting or simplifying. For Shama (1985) voluntary simplicity refers to a lifestyle of low consumption, ecological responsibility and self-sufficiency. Etzioni (1998) takes a more precise view, describing three categories of voluntary simplifiers: down-shifters, strong simplifiers and holistic simplifiers. If constituted on a large scale, Etzioni believes that society's ability to protect the environment would be significantly enhanced. Voluntary simplifiers are more likely to recycle, build compost heaps, and engage in other civic activities that indicate stewardship toward the environment because simplifiers draw more satisfaction from nature than from conspicuous consumption. There is also, he believes, a converse correlation: as people become more environmentally conscious and committed, they are likely to find in voluntary simplicity a lifestyle and ideology compatible with their environmental concerns. A similar position is taken by Shaw and Newholm (2002), who contend that voluntary simplicity may be demonstrated among consumers whose behaviour includes some ethical consideration of the environmental and social impact of their consumption choices. These consumers are termed ethical simplifiers. In addition, Shaw and Newholm (2002) maintain that although ethical consumers may be concerned about consumption levels per se, radical anti-consumerism may not be an option for them in a society that requires or demands some level of consuming. Important decisions for these consumers therefore surround the issue of whether to consume with sensitivity through the selection of more ethical alternatives or whether to reduce levels of consumption to a more sustainable level through voluntary simplicity.

We cannot assume, however, that voluntary simplicity and green or ethical concerns go hand-in-hand however. Although interest in voluntary simplicity for many has a clear environmental basis, these individuals may constitute a very small portion of the increasing number of people interested in voluntary simplicity (Zavestoski, 2001). This would indicate that large numbers of people are involved in the movement for other reasons. As such Zavestoski (2001, p. 186) contends that: 'The potential for massive voluntary changes in consumption practices, not for altruistic or environmentally motivated reasons, but for more inwardly orientated reasons must be explored'. Brand (1997) makes a similar claim, namely that the eco-balance of the lifestyle of a number of older, immobile, traditionally thrifty people who show no specifically pronounced environmental consciousness is in many cases better than that of the environmentally conscious.

There is a converse opinion on the ability of voluntary simplicity to address environmental problems. Maniates (2002), echoing the criticism levelled at green and ethical consumers for their individual focus and lack of collective political action, maintains that voluntary simplicity is taking a similar path:

> Locked into a rhetoric of the individualisation of responsibility, it propagates the all-too-familiar 'plant a tree, save the world' environmental mentality. This mentality imagines consumers to be immune to the marketer's ability to tap into environmental concern to sell a host of environmentally *un*friendly products, and draws attention away from inequalities in power and responsibility that occupy the centre of the environmental crisis. Ironically, the VSM [Voluntary Simplicity Movement] risks aiding and abetting the very cultural and political forces it philosophically opposes. (Maniates, 2002, p. 233)

Significantly, and not unlike what has occurred within the areas of green marketing and green consumer research, research on voluntary simplicity by marketers has tended to focus on two distinct areas (Rudmin and Kilbourne, 1996): identifying new market segments for exploitation and examining how values of voluntary simplifiers might be manipulated for the support of national policies of energy conservation (for examples see

Rudmin and Kilbourne, 1996, p. 166). In fact this research focus continues with Craig-Lees and Hill's (2002) examination of the segmentation viability of the voluntary simplicity cluster, for example. As Maniates (2002) illustrates, commodification has already made moves in the area of voluntary simplicity, with various books, cassettes and magazines on the subject. He describes how the magazine *Real-Simple*, which is published by AOL Time Warner, has 20 pages of advertising before the table of contents. Moreover, 116 out of a total of 216 pages are devoted solely to advertising. In a similar vein, Satterthwaite (2001) describes how for ten dollars you can buy the 'Simple Living and Earth Saving Action Kit'. The danger in this for Miles (1998) is that any movement against consumerism is subsumed within the system as yet another market niche. Similarly Rumbo (2002) contends that although resistance by environmentally and politically motivated consumers can effect change, marketers also strengthen the consumerist hegemony by absorbing criticisms and converting such resistance into reasons for consumption. Despite this, green consumption, ethical behaviour and voluntary simplicity have the potential to function as a progressive step towards environmental sustainability. Yet it is far from inevitable, given the demonstrated ability of marketing to co-opt consumer movements, as critics such as Miles (1998), Maniates (2002), Rumbo (2002) and Satterthwaite (2001) have pointed out. Therein lies the fulcrum of the issue.

Conclusions

This chapter has shown that within the literature there is a lack of clear definitions for both green marketing and green consumers as both mean very different things to the individuals buying products and services and the organizations providing them. Instead of focusing on definitions, however, this chapter has introduced the reader to the various issues of importance in our understanding of both green marketing and the green consumer, with particular focus in the green consumer literature on ethical consumers and voluntary simplifiers. It seems that at the beginning of the twenty-first century the green marketing fad of the early 1990s is very much alive and well and is certainly not likely to slip-slide away in the foreseeable future (Prothero, 1998). Green markets are growing all the time, as it is now more difficult to get away with green washing claims. Similarly, the green marketing strategies of organizations can easily be tied in with the company's wider socially responsible activities, so one can now argue that we are moving towards a time where social responsibility is regarded as part of the mainstream activities of all organizations and not just a small minority. Coupled with an increase in the number of consumers who are questioning their consumption on many different levels, including from a green perspective, it seems fair to say that as the numbers of these consumers continue to grow, so too will the green marketing strategies of organizations. Thus these strategies will need to recognize that a complete paradigm shift in organizational thinking is required if the company wishes to be successful, both now and in the future.

Note

* Now The Green Life, see http://www.thegreenlife.org/.

References

Anderson, T.W. and Cunningham, W.H. (1972), 'The Socially Conscious Consumer', *Journal of Marketing*, **36**, 23–31.

Anonymous (1990), 'Green Cleaners', *CHOICE* (September), 10–14.
Anonymous (1991), 'Spurts and Starts: Corporate Role in '90s Environmentalism Hardly Consistent', *Advertising Age*, **62**(46), GR14-GR16.
Anonymous (1999), 'A Challenge from the US', *Australian Energy News*, Issue 12 (June).
Antil, J.H. (1984), 'Socially Responsible Consumers: Profile and Implications for Public Policy', *Journal of Macromarketing*, **4**(2), 18–39.
Balderjahn, I. (1988), 'Personality Variables and Environmental Attitudes as Predictors of Ecologically Responsible Consumption Patterns', *Journal of Business Research*, **17**(1), 51–6.
Beck, U. (1992), *Risk Society: Towards a New Modernity*, Newbury Park, CA: Sage.
Beck, U. (1999), *World Risk Society*, Cambridge: Polity.
Beck, U., Bonass, W. and Lau, C. (2003), 'The Theory of Reflexive Modernization: Problematic, Hypotheses and Research Programme', *Theory Culture and Society*, **20**(2), 1–33.
Berger, I.E. and Corbin, R.M. (1992), 'Perceived Consumer Effectiveness and Faith in Others as Moderators of Environmentally Responsible Behaviours', *Journal of Public Policy and Marketing*, **11**(2), 79–89.
Bird, K. and Hughes, D.R. (1997), 'Ethical Consumerism: The Case of "Fairly-traded" Coffee', *Business Ethics*, **6**(3), 159–67.
Blackmores (1999), http://www.blackmores.com.au/content/co_enviro.htm, accessed 31 May 2005.
Bohlen, G.M., Schlegelmilch, B. and Diamantopoulos, A. (1993), 'Measuring Ecological Concern: A Multi-construct Measure', *Journal of Marketing Management*, **9**(4): 415–30.
Brand, K. (1997), 'Environmental Consciousness and Behaviour: The Green of Lifestyles', in Redclift, M. and Woodgate, G. (eds), *The International Handbook of Environmental Sociology*, Cheltenham, UK and Lyme, USA: Edward Elgar, pp. 204–17.
Brooker, G. (1976). 'The Self-actualising Socially Conscious Consumer', *Journal of Consumer Research*, **3**, 107–12.
Browne, A.W., Harris, P.J.C., Hofny-Collins, A.H., Pasiecznik, N. and Wallace, R.R. (2000), 'Organic Production and Ethical Trade: Definition, Practice and Links', *Food Policy*, **25**, 69–89.
Connolly, J. and Prothero, A. (2003), 'Sustainable Consumption: Consumption, Consumers and the Commodity Discourse', *Consumption, Markets and Culture*, **6**(4), 275–91.
Craig-Lees, M. and Hill, C. (2002), 'Understanding Voluntary Simplifiers', *Psychology and Marketing*, **19**(2), 187–210.
Crane, A. (1997), 'The Dynamics of Marketing Ethical Products: A Cultural Perspective', *Journal of Marketing Management*, 13, 561–77.
Crane A. (2000), 'Facing the Backlash: Green Marketing and Strategic Reorientation in the 1990s', *Journal of Strategic Marketing*, **8**, 277–96.
Crude, B.J. (1993), 'Consumer Perceptions of Environmental Marketing Claims: An Exploratory Study', *Journal of Consumer Studies and Home Economics*, **17**(3), 207–25.
Davis, J.J. (1994), 'Consumer Responses to Corporate Environmental Advertising', *The Journal of Consumer Marketing*, **11**(2), 25–37.
Dembkowski, S. and Hanmer-Lloyd, S. (1994), 'The Environmental Value-Attitude System Model: A Framework to Guide Understanding of Environmentally Conscious Consumer Behaviour', *Journal of Marketing Management*, **10**(7), 593–603.
Diamantopoulos, A., Schlegelmilch, B., Sinkovics, R. and Bohlen, G. (2003), 'Can Socio-demographics Still Play a Role in Profiling Green Consumers? A Review of the Evidence and an Empirical Investigation', *Journal of Business Research*, **56**(6), 465–80.
Dolan, P. (2002), 'The Sustainability of "Sustainable Consumption"', *Journal of Macromarketing*, **22**(2), 170–81.
Drumwright, M.E. (1994), 'Socially Responsible Organizational Buying: Environmental Concern as a Noneconomic Buying Criterion', *Journal of Marketing*, **58**(3), 1–19.
Elkington, J. and Hailes, J. (1988), *The Green Consumer Guide: From Shampoo to Champagne, How to Buy Goods That Don't Cost the Earth*, London: Gollancz.
Elkington, J. and Hailes, J. (1989), *The Green Consumer's Supermarket Shopping Guide*, London: Gollancz.
Elkington, J., Hailes, J. and Hill, D. (1990), *The Young Green Consumer Guide*, London: Gollancz.
Elkington, J., Hailes, J. and Hill, D. (1997), *The Young Green Consumer Guide*, London: Penguin.
Etzioni, A. (1998), 'Voluntary Simplicity: Characterization, Select Psychological Implications, and Societal Consequences', *Journal of Economic Psychology*, **19**, 619–43.
Fisk, G. (1973), 'Criteria for a Theory of Responsible Consumption', *Journal of Marketing*, **37**(2), 24–31.
Follows, S. and Jobber, D. (2000). 'Environmentally Responsible Purchase Behaviour: A Test of a Consumer Model', *European Journal of Marketing*, **34**(5/6), 723–46.
Friedman, M. (1996), 'A Positive Approach to Organized Consumer Action: The "Buycott" as an Alternative to the Boycott', *Journal of Consumer Policy*, **19**, 439–51.
Ger, G. (1997), 'Human Development and Humane Consumption: Well-being Beyond the Good Life', *Journal of Public Policy and Marketing*, **16**(1), 110–25.

Giddens, A. (1991), *Modernity and Self-identity*, Cambridge, Polity Press.

Giddens, A. (1994), *Beyond Left and Right: The Future of Radical Politics*, Cambridge: Polity Press.

Gifford, B. (1991), 'The Greening of the Golden Arches – McDonalds Teams with Environmental Group to Cut Waste', *The San Diego Union*, 19 August, C1, C4.

Goodman, D. and Goodman, M. (2001), 'Sustaining Foods: Organic Consumption and the Socio-ecological Imaginary', in Cohen, M. and Murphy, J. (eds), *Exploring Sustainable Consumption. Environmental Policy and the Social Sciences*, London: Pergamon.

Gorder, C. (1991) *Green Earth Resource Guide: A Comprehensive Guide About Environmentally-Friendly Services and Products: Books, Clean Air, Clean Water, Eco-tourism*, Mesa, AZ: Blue Bird Publishing.

Halkier, B. (1999), 'Consequences of the Politicization of Consumption: The Example of Environmentally Friendly Consumption Practices', *Journal of Environmental Policy and Planning*, 1(1), 25–41.

Halkier, B. (2004), 'Consumption, Risk, and Civic Engagement', in Micheletti, M., Follesdal, A. and Stolle, D. (eds), *Politics, Products, and Markets*, New Brunswick: Transaction Publishers, pp. 223–44.

Harper, G.C. and Makatouni, A. (2002), 'Consumer Perception of Organic Food Production and Animal Welfare', *British Food Journal*, 104(3/4/5), 287–99.

Higgins, M. and Tadajewski, M. (2002) 'Anti-corporate Protest as Consumer Spectacle', *Management Decision*, 40(4), 363–71.

Kilbourne, W. (1998), 'Green Marketing: A Theoretical Perspective', *Journal of Marketing Management*, 14(6), 657–77.

Kilbourne, W. and Beckmann, S. (1998), 'Review and Critical Assessment of Research on Marketing and the Environment', *Journal of Marketing Management*, 14(6), 513–32.

Kilbourne, W., McDonagh, P. and Prothero, A. (1997), 'Sustainable Consumption and the Quality of Life: A Macromarketing Challenge to the Dominant Social Paradigm', *Journal of Macromarketing*, 17(1), 4–24.

Kinnear, T.C. and Taylor, J.R. (1973), 'The Effect of Ecological Concern on Brand Perceptions', *Journal of Marketing Research*, 10 (2), 191–7.

Kinnear, T.C., Taylor, J.R. and Ahmed, S. (1974), 'Ecologically Concerned Consumers: Who are They?', *Journal of Marketing*, 38, 20–24.

Laroche, M., Bergeron, J. and Barbaro, G. (2001), 'Targeting Consumers Who Are Willing to Pay More for Environmentally Friendly Products', *Journal of Consumer Marketing*, 18(6), 503–20.

Makower, J., Elkington, J. and Hailes, J. (1991), *The Green Consumer Supermarket Shopping Guide*, London: Gollancz.

Maniates, M. (2002), 'In Search of Consumptive Resistance: The Voluntary Simplicity Movement', in Princen, T., Maniates, M. and Conca K. (eds), *Confronting Consumption*, London, MIT Press, pp. 199–236.

Mayer, R.N. (1976), 'The Socially Conscious Consumer – Another Look at the Data', *Journal of Consumer Research*, 3(2), 113–15.

McDaniel, S.W. and Rylander, D.H. (1993), 'Strategic Green Marketing', *Journal of Consumer Marketing*, 10 (3), 4–10.

McDonagh, P. (1998), 'Towards a Theory of Sustainable Communications in Risk Society: Relating Issues of Sustainability to Marketing Communications', *Journal of Marketing Management*, 14(6), 591–622.

McDonagh, P. (2002), 'Communicative Campaigns to Effect Anti-slavery and Fair Trade', *European Journal of Marketing*, 36(5/6), 642–66.

Mechling, L. (2003), 'The Home Front: "Green" Furniture's New Terrain; Eco-friendly Homes Go Chic with High End Designs; A $1200 Bamboo Bench', *Wall Street Journal*, 12 December, W12.

Mendelson, N. and Polonsky, M. (1995), 'Using Strategic Alliances to Develop Credible Green Marketing', *Journal of Consumer Marketing*, 12 (2), 4–18.

Menon, M. and Menon, A. (1997), 'Enviropreneurial Marketing Strategy: The Emergence of Corporate Environmentalism as Marketing Strategy', *Journal of Marketing*, 61(1), 51–67.

Micheletti, M. (2003), *Political Virtue and Shopping. Individuals, Consumerism and Collective Action*, New York: Palgrave Macmillan.

Micheletti, M., Follesdal, A. and Stolle, D. (eds) (2004), *Politics, Products, and Markets*, New Brunswick: Transaction Publishers.

Miles, S. (1998), *Consumerism as a Way of Life*, London: Sage.

Miller, D. (1995), 'Consumption as the Vanguard of History', in Miller, D. (ed.), *Acknowledging Consumption*, London: Routledge, pp. 1–57.

Minton, A.P. and R.L. Rose (1997). 'The Effects of Environmental Concern on Environmentally Friendly Consumer Behaviour: An Exploratory Study', *Journal of Business Research*, 40(1), 37–48.

Murphy, P.E., Kangun, N. and Locander, W. (1978), 'Environmentally Concerned Consumers: Racial Variations', *Journal of Marketing*, 42(4), 61–6.

Newcastle City Council (1999), *State of the Environment Report 1998/99*, Newcastle, NSW Australia: Newcastle City Council.

Oleck, J. (1992), 'The Great Clamshell Debate', *Restaurant Business*, **91**(16), 68–72.
Osterhus, T.L.(1997), 'Pro-social Consumer Influence Strategies: When and How do They Work?', *Journal of Marketing*, **61**(4), 16–29.
Ottman, J. (1998), *Green Marketing: Opportunity for Innovation*, Lincolnwood: NTC Business Books.
Peattie, K. (1999), 'Rethinking Marketing: Shifting to a Greener Paradigm', in Charter, M. and Polonsky, M.J. (eds), *Greener Marketing: A Global Perspective to Greening Marketing Practice*, 2nd Edition, Sheffield, UK: Greenleaf Publishing, pp. 57–71.
Pickett, G.M., Kangun, N. and Grove, S. (1993), 'Is There a General Conserving Consumer? A Public Policy Concern', *Journal of Public Policy and Marketing*, **12**(2), 234–43.
Polonsky, M.J. and Rosenberger, P.J. III (2001), 'Re-evaluating to Green Marketing – An Integrated Approach', *Business Horizons*, **44**(5), 21–30.
Porter, M.E. and van der Linde, C. (1995), 'Green and Competitive: Ending the Stalemate', *Harvard Business Review*, **73**(5), 120–34.
Prothero, A. (1990), 'Green Consumerism and the Societal Marketing Concept: Marketing Strategies for the 1990s', *Journal of Marketing Management*, **6**(2), 87–103.
Prothero, A. (1998), 'Editorial – Green Marketing: The "Fad" That Won't Slip Slide Away', *Journal of Marketing Management*, **14**(6), 507–12.
Prothero, A. and Fitchett, J.A. (2000) 'Greening Capitalism: Opportunities for a Green Commodity', *Journal of Macromarketing*, **20**(1), 46–55.
Raynolds, L.T. (2002), 'Consumer/Producer Links in Fair Trade Coffee Networks', *Sociologia Ruralis*, **42**(4), 404–24.
Rice, F. (1990), 'How to Deal with Tougher Customers', *Fortune*, **122**(14), 38–48.
Roberts, J.A. (1996), 'Green Consumers in the 1990s: Profile and Implications for Advertising', *Journal of Business Research*, **36**(3), 217–31.
Roberts, J.A. and Bacon, D.R. (1997), 'Exploring the Subtle Relationships between Environmental Concern and Ecologically Conscious Consumer Behaviour', *Journal of Business Research*, **40**, 79–89.
Rudmin, F.W. and Kilbourne, W.E. (1996), 'The Meaning and Morality of Voluntary Simplicity: History and Hypotheses on Deliberately Denied Materialism', in Belk, R.W., Dholakia, N. and Venkatesh, A. (eds), *Consumption and Marketing: Macro Dimensions*, Cincinnati, Ohio: South-Western College Publishing, pp. 166–215.
Rumbo, J.D. (2002), 'Consumer Resistance in a World of Advertising Clutter: The Case of Adbusters', *Psychology and Marketing*, **19**(2), 127–48.
Satterthwaite, A. (2001), *Going Shopping: Consumer Choices and Community Consequences*, New Haven and London: Yale University Press.
Schegelmilch, B. (1994), 'Green, Ethical and Charitable: Another Marketing Ploy or a New Marketing Era', in Baker, M.J. (ed.), *Perspectives on Marketing Management*, **4**, London: Wiley, pp. 55–71.
Schlegelmilch, B.B., Bohlen, G.M. and Diamantopoulos, A. (1996), 'The Link Between Green Purchasing Decisions and Measures of Environmental Consciousness', *European Journal of Marketing*, **30**(5), 35–55.
Schor, J. (1991), *The Overworked American*, New York: Basic Books.
Schor, J. (1998), *The Overspent American*, New York: Basic Books.
Schwepker, C.H. and Cornwell, T.B. (1991), 'An Examination of Ecologically Concerned Consumers and Their Intention to Purchase Ecologically Packaged Products', *Journal of Public Policy and Marketing*, **10**(2), 77–101.
Shama, A. (1985), 'The Voluntary Simplicity Consumer', *Journal of Consumer Marketing*, **2**, 57–64.
Shaw, D. and Clarke, I. (1999), 'Belief Formation in Ethical Consumer Groups: An Exploratory Study', *Marketing Intelligence and Planning*, **17**(2), 109–19.
Shaw, D. and Newholm, T. (2002), 'Voluntary Simplicity and the Ethics of Consumption', *Psychology and Marketing*, **19**(2), 167–85.
Shaw, D. and Shiu, E. (2002), 'An Assessment of Ethical Obligation and Self-identity in Ethical Consumer Decision-making: A Structural Equation Modelling Approach', *International Journal of Consumer Studies*, **26**(4), 286–93.
Shaw, D., Shiu, E. and Clarke, I. (2000), 'The Contribution of Ethical Obligation and Self-identity to the Theory of Planned Behaviour: An Exploration of Ethical Consumers', *Journal of Marketing Management*, **16**, 879–94.
Shook, M.D. (1995) *Mother Nature's Shopping List: A Buying Guide for Environmentally Concerned Consumers*, New York: Citadel Press.
Shrum, L.J., McCarty, J.A., Lowrey, J.A. and Tina, M. (1995), 'Buyer Characteristics of the Green Consumer and Their Implications for Advertising', *Journal of Advertising*, **24**(2), 71–83.
Strong, C. (1996), 'Features Contributing to the Growth of Ethical Consumerism', *Marketing Intelligence and Planning*, **14**(5), 5–13.
Strong, C. (1997), 'The Problems of Translating Fair Trade Principles Into Consumer Purchase Behaviour', *Marketing Intelligence and Planning*, **15**(1), 32–7.

Toyota (1997), *Care for the Earth*, Japan: Toyota Motor Corporation.

Wasik, J.F. (1993) *The Green Supermarket Shopping Guide*, New York: Warner Books.

Webster, F.E. (1975), 'Determining the Characteristics of the Socially Conscious Consumer', *Journal of Consumer Research*, **2** (December), 188–96.

Wilkhahn (1996), 'Wiklhahn Awarded the German Ecology Prize 1996', Germany: Wilkhahn.

Zavestoski, S. (2001), 'Environmental Concern and Anti-consumerism in the Self-concept: Do They Share the Same Basis?', in Cohen, M. and Murphy, J. (eds), *Exploring Sustainable Consumption: Environmental Policy and the Social Sciences*, London: Pergamon, pp. 173–89.

Zavestoski, S. (2002), 'The Social-Psychological Bases of Anticonsumption Attitudes', *Psychology and Marketing*, **19**(2), 149–65.

Zuckerman, A. (1999), 'Using ISO 14000 as a Trade Barrier', *Iron Age New Steel*, **15**(3), 77.

17 Business, environmental management and the triple bottom line
Steven Schilizzi

Introduction

It would be a provocative statement these days to assert that 'triple bottom line' (TBL or 3BL) accounting is just empty rhetoric. As Norman and MacDonald (2004, p. 2) point out: 'Organisations such as the Global Reporting Initiative and AccountAbility have embraced and promoted the 3BL concept for use in the corporate world. Companies as significant as AT&T, Dow Chemicals, Shell, and British Telecom have used 3BL terminology in their press releases, annual reports and other documents.' A KPMG survey (Kolk et al., 2002) shows 45 per cent of the top 250 companies worldwide now publish a separate corporate report containing details of environmental and/or social performance. If one checks the number of 'Google' hits for the expression 'triple bottom line', one can see, as shown in Figure 17.1, how it has spread throughout the business world like wildfire.

Norman and MacDonald (2004), who strongly criticize this new linguistic fad, claim that this popularity is due to the confusion the expression maintains. They claim that behind the language of rigorous numerical accounting there lies only a vague commitment to environmental and social concerns. Although their point is valid, it also appears somewhat superficial and seems to ignore the prospects permitted by new developments in the field of financial accounting and modelling.

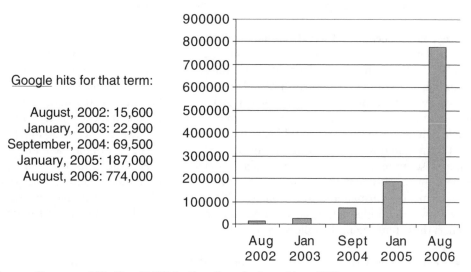

Google hits for that term:

August, 2002: 15,600
January, 2003: 22,900
September, 2004: 69,500
January, 2005: 187,000
August, 2006: 774,000

Source: Norman and MacDonald (2004) at http://www.businessethics.ca/3bl/).

Figure 17.1 'Triple bottom line' rhetoric is spreading like wildfire

The purpose of this chapter is to provide a summary overview of what these prospects are and, if the rhetoric of TBL does indeed signal some degree of commitment towards environmental and social concerns, to show how these concerns can translate into specific decisions, actions and reporting. But first of all, what do we actually mean by TBL? Can it be operationalized and if so, how?

What is TBL?

Triple bottom line is an expression referring to three key dimensions of business performance: financial, environmental and social. In the private sector, financial performance reflects success in the marketplace and stewardship to shareholders. Environmental performance reflects compliance with government regulations and stewardship to a growing class of customers. Social performance reflects stakeholder management or partnership – in particular the workforce and local and neighbouring populations. One can thus interpret TBL as an enlargement of the stakeholder base to which the firm is held accountable, in line with the general 'stakeholder approach' first highlighted by Freeman in 1984.

The term 'triple bottom line' dates back to the mid-1990s, when management think tank AccountAbility coined and began using the term in its work. The term found public currency with the 1997 publication of the British edition of John Elkington's *Cannibals with Forks: The Triple Bottom Line of 21st Century Business*. It then took only a few years, as mentioned above and shown in Figure 17.1, for the term to spread widely in the business world. Those who use the language of TBL are part of a much larger movement sometimes identified by the acronym SEAAR: social and ethical accounting, auditing and reporting. They have produced a variety of competing standards and standard-setting bodies, including the Global Reporting Initiative (GRI), the SA 8000 from Social Accountability International, the AA 1000 from AccountAbility as well as parts of various ISO standards, in particular the ISO 14000 family of norms.

Current environmental and social accounting efforts

Within the current practice of business accounting there is strictly speaking no such thing as TBL. As Norman and MacDonald (2004) amply demonstrate, it is not possible, strictly speaking, to account for the environmental and social impacts of business activities in the same way as financial performance. The reason is simply that there is no unique standard of measurement that can aggregate all environmental and social dimensions each into a single number or 'bottom line'. While financial revenues and expenses can add and subtract from each other using a single monetary accounting unit, environmental and social impacts seem to be inherently multi-dimensional concepts. Trade-offs between each dimension are far from obvious and will be subject to value judgments and interpersonal disagreements. Norman and MacDonald (2004) provide illustrative examples by asking how we would decide on the *net* social welfare impact of good versus bad contributions such as those figuring in the following pairs:[1]

- *Pair 1*. A generous family-friendly policy that includes extended maternity leave as well as part-time and job-sharing provisions for women returning to the firm after maternity leave but also three sexual harassment suits against it in the past year.

- *Pair 2*. An environmental control policy for its overseas contractors that is audited by an international environmental NGO but also a spotty record of pollution control and reporting of its own operations in a less developed country.
- *Pair 3*. A charitable donation equal to 2 per cent of gross profits but also a conviction for toxic chemical releases into the local river system.

The question also extends to comparing good to good and bad to bad as in these two examples:

- Would a firm do more social good by donating one million dollars to send under-privileged local youths to college or by donating the same amount to the local opera company?
- How should we compare a company-wide programme to train personnel in appropriate environment, health and safety behaviour and an investment aimed at gradually increasing the degree of waste recycling from the company's operations?

If TBL accounting is to mean more than just a more or less vague commitment to environmental and social concerns, then it is important to understand the potential and limitations of current environmental and social accounting efforts. We shall then see that to be able to integrate these dimensions into serious accounting procedures nothing less than an overhaul of standard accounting practices is necessary.

If we focus on the environmental leg of TBL,[2] what we have first is what Schaltegger and Burritt (2000) call 'environmentally related financial accounting' (ERFA). This includes the cost accounting of ongoing environmental management activities and the financial accounting of environmentally related investments. Neither is simple for both accountings face the long-standing problem of cost allocation. How are we to know how much of an investment was 'environmentally related'? Replacing old polluting equipment with a new cleaner one could have happened irrespective of pollution considerations if the equipment needed replacing anyway – because it had become obsolete or dysfunctional. Different rules can be, and have been, imagined but without standardization they are bound to appear to some degree arbitrary.

Secondly we have 'ecological accounting' or 'environmental impact accounting'. While ERFA is done in monetary terms, ecological accounting is done in physical units, such as tonnes of CO_2 or SO_2 emitted per year, kilolitres of waste water into water bodies, hectares of land disturbed through logging, mining or clearing and so on. Thus ERFA examines the impacts of environmental management on the firm's financial status; ecological accounting examines the impacts of the firm's activities and products on the environment. Schaltegger et al. (1996) develop the concept of environmental impact added. An obvious problem is consolidation and aggregation. How do you add up different physical quantities? One (partial) solution is to use what we may call 'functional aggregation', a procedure explained at length in Lesourd and Schilizzi (2001, p. 118). An example comes in terms of global warming potential (GWP) where for example 1 kg of nitrous oxide is equated to 310 kg of CO_2. However, different environmental functions – such as GWP and soil and water acidification potential – cannot readily be aggregated.

Another aspect of ecological accounting appears when the focus shifts from activity-based to product-based accounting. The former leads to so-called 'eco-balance accounting' where a given economic unit uses an accounting framework in the form of an input–output materials balance to track all potential pollutants and environmental impacts. On the input side you have natural resource use and on the output side you have emissions to land, water and air as well as biological disturbances. A synthetic table to bring together all such data is proposed in Lesourd and Schilizzi (2001, p. 180). Product-based environmental accounting has led to so-called product 'lifecycle assessment' (LCA). Here the firm tries to account for the environmental impact of its products in terms of manufacturing, packaging, use and final disposal. Interestingly this apparently technical issue has correlated with an intense debate on the firm's responsibilities: just how far do its responsibilities extend beyond its field of direct control? Instructions on packages about how to use and dispose of the product have become a focus of attention. Unfortunately the legal aspects have far outperformed the technical ones: LCA is still a controversial technique that lacks the necessary level of standardization, partly because it is so hard to standardize!

Ideally we would like to integrate both types of environmental accounting, as well as any system of social accounting, with financial accounts. We are a long way off. There are efforts, in particular in the German-speaking and Scandinavian countries, to implement eco-financially integrated accounts, stemming from Müller-Wenk's seminal work on 'environmental impact points' in 1978. These countries have mainly focused on integrating cost accounting and ongoing environmental impacts. Recently Schaltegger and others, mainly in Switzerland, have tried to link financial accounting to 'environmental investments' (and disinvestments) but the reality of potential liabilities – both environmental and social – still eludes current solutions.

New prospects from financial accounting

Triple bottom line accounting – translating environmental and social liabilities (and assets) into financial terms – can become a reality however, provided current accounting techniques and frameworks evolve in a specific manner.

Technically only financial performance is measured in a clearly regulated and quantitative way. Environmental accounts do exist but they are not (yet?) regulated and are not easily comparable across sectors or companies, or between countries. And social accounting is still in its infancy. Most importantly there is no accepted framework to bring all three dimensions consistently together.

In this context what is TBL accounting supposed to mean? Reminiscent of the use of the term 'sustainability', it signals a willingness to show concern for the three dimensions of business performance as opposed to a single-minded focus on 'just profits'.

At worst TBL may be seen as a spin doctor's exercise in PR (public relations). At best however, it signals a genuine desire to perform well in all three areas – motivated by what customers, consumers, government and society at large think and feel because, very simply, upsetting customers, consumers and government is likely to bring on extra costs. This may manifest itself as reduced market share, new and more stringent regulations or loss of access to key resources, leading to a likely fall in the value of company shares. In the process a company will lose some of its clients' trust or loyalty and suffer image and reputation damage: in other words, a loss of what is known as 'social capital'. There is growing evidence that good reputation in terms of environmental and social management pays when it comes to investors' choices on the stock market, even if the link is more of a correlation than a clear causal relationship. For example, stock indexes such as the

Domini 400 Social Index have slightly outperformed the Standard and Poor's 500 Index over the last decade or so, and this generalizes to most 'sustainable business' indexes (Lesourd and Schilizzi, 2001, pp. 232–9).

Negative environmental and social impacts then are materially important to a company insofar as they are likely to lead, sooner or later, to higher costs or liabilities. Indeed disgruntled stakeholders constitute just such a liability. This liability is a contingent liability however: it eventuates only if government or civil society takes action. The same can be said of positive environmental and social impacts: they only lead to contingent benefits or assets for the private firm. As a result a proactive, strategic attitude is needed that clashes in many ways with the standard, backward-looking and book-keeping approach to business accounting. But then standard accounting, and financial accounting in particular, is, as Kierkegaard (1997) and others have increasingly pointed out, in crisis. There is as yet no accounting system that reliably predicts bankruptcy in a timely manner.

Quantification: just technicalities or ethics?

The crux of the matter is that it is very difficult at present to measure and quantify a contingent liability. It is in the future and it is uncertain: financial losses may never eventuate or they may be huge. This redefines what TBL accounting should be all about if it is to mean anything real.

What initially started out as a philosophy of social duty must now be seen as a technical challenge. Currently there is no easy way to translate environmental and social liabilities into financial terms. And yet, such liabilities are as real as any other. All that is needed is to pin a number on them.

The good news is that there is hope. It should be possible to develop techniques to estimate and quantify such liabilities. The bad news is that it is not easy, at least not yet. The solution to the problem should come from bringing together insights from financial economics and from environmental economics: by combining option valuation techniques from the former with non-market valuation techniques from the latter. The approach holding most promise, it seems, is that of real options valuation. The standard reference to this approach is Dixit and Pindyck's (1994) seminal book, *Investment under Uncertainty*; but the sophistication of the techniques proposed have led other authors to propose simplified but practical implementation methods (see in particular Copeland and Antikarov, 2001). The next step should be to include consideration of unpriced capital assets, whether natural or social.[3]

In the public sector, because of government's responsibilities and mandate, TBL accounting can, as exemplified by the Australian Government (Environment Australia, 2003), involve three areas of separate accounting[4] but in the private sector it can only mean an end result in terms of financial outcomes. From the company's point of view implementing new accounting techniques that reflect TBL concerns will also reflect stakeholders' level of satisfaction. And doing so, the firm can yet remain faithful to Milton Freedman's much cited quip: 'the business of business is business'.

Will ethics disappear behind technicalities? No. In considering an uncertain future the rate at which decision-makers discount future financial impacts and their willingness to take risks when others' interests are at stake, both involve ethical aspects. As soon as others' interests are at stake, ethics are involved. The ethical content of decisions may be seen to vary as a proportion of total interests that are others' interests – those that affect stakeholders other than the firm itself or its shareholders. The temptation always exists to discount those future impacts that affect others and to risk negative backlash in the hope that victims will not react. Knowledge of the high transaction costs of concerted action, legal action and political action can be used by firms to push through environmentally and socially disruptive decisions. It may then be up to government, the legal system or consumer organizations to influence managers' risk attitudes and approaches to discounting. There seems to be as yet an unfathomed link between risk attitudes, discounting of the future and equity considerations.[5]

Conclusion
That the concept of a 'triple bottom line' is controversial, especially for the private sector, should itself be uncontroversial. Echoing Norman and MacDonald's critiques, taken at face value, TBL is at worst a spin doctor's exercise in public relations and at best little more than a metaphor to signal a company's willingness to show concern for the three dimensions of business performance, as opposed to a single-minded focus on 'just profits'. It certainly cannot designate three columns in an accounting exercise but then nobody really claims TBL to be a real accounting exercise. Rather the question is how can such concerns be operationalized in a way that is meaningful to a private company?

A common misperception is to oppose a TBL-minded company to a hard-nosed 'just profits' attitude. A better opposition would be between short-term and long-term profits, and between surer but lower profits and potentially higher but more uncertain gambles. Negative environmental and social impacts create liabilities and these are as real as anything that figures in an accountant's ledger. The difference is that, much like goodwill and social capital, they are much harder to pin down and measure. Such a liability is, from the firm's point of view, ultimately a financial liability, even if it first materializes as a legal or social liability. This liability will appear in broad daylight if the firm is taken over by another firm, given that it will discount the bought-out firm's value accordingly. In other cases, however, new, forward-looking accounting techniques are needed. These will very likely stem from the developments of real option valuation methods augmented with non-market valuation techniques; in other words, from the cross-fertilization of two fields initially as far apart as financial and environmental accounting but which are quickly converging in academic spheres.

Instead the ethics in this new perspective focus on managers' discounting of future costs and benefits and on their willingness to take risks, especially when other stakeholders' interests are at stake. Although new accounting techniques might allow environmental and social liabilities to enter a firm's financial decision horizon, they will by no means ensure that such liabilities are acted upon. Businesses will continue to vary widely in their behaviour.

Although TBL is not an accounting exercise in itself, the rhetoric may well have contributed in bringing about new accounting developments that will address, to some degree at least, society's three concerns of economic, environmental and social performance without asking of business anything other than 'doing business'. Of course, the way one best does business evolves over time.

Notes
1. We have replaced some of the social examples by environmental equivalents.
2. Much of what can be said about environmental accounting carries over to social accounting as well.

3. See Figge (2005).
4. However, advocates of 'sustainability accounting' – as opposed to TBL accounting – emphasize the need to integrate the three aspects, as Sweden has demonstrated using the following headings: efficiency; contribution and equality; adaptability and values; and resources for coming generations.
5. See Pannell and Schilizzi, ch. 11, for insights into this triple link.

References

Copeland, T. and Antikarov, V. (2001), *Real Options: A Practitioner's Guide*, New York: Texere, Monitor Group.

Dixit, A.K. and Pindyck, R.S. (1994), *Investment Under Uncertainty*, New Jersey: Princeton University Press.

Elkington, J. [1997] (1999), *Cannibals with Forks: The Triple Bottom Line of 21st Century Business*, Oxford: Capstone.

Environment Australia (2003), *Triple Bottom Line in Australia. A Guide to Reporting Against Environmental Indicators*, Canberra: Department of Environment and Heritage (June).

Figge, F. (2005), 'Value-Based Environmental Management from Environmental Shareholder Value to Environmental Option Value', *Corporate Social Responsibility and Environmental Management* **12**(1): 19–30.

Freeman, R.E. (1984), *Strategic Management: A Stakeholder Approach*, Boston, MA: Pitman.

Kierkegaard, H. (1997), *Improving Accounting Reliability: Solvency, Insolvency, and Future Cash Flows* (original title: *Dynamical Accounting*), Westport, London: Quorum Books.

Kolk, A., van der Veen, M.L., Hay, K. and Wennink, D. (2002), *KPMG International Survey of Corporate Sustainability Reporting 2002*, Amsterdam: KPMG Global Sustainability Services.

Lesourd, J.-B. and Schilizzi, S. (2001, 2003 2nd Edition), *The Environment in Corporate Management. New Directions and Economic Insights*, Cheltenham, UK and Northampton, USA: Edward Elgar.

Müller-Wenk, R. (1978), *Ökologische Buchhaltung* (*Ecological Bookkeeping*). Frankfurt: Campus (the original 1972 text is a mimeo from St Gallen, Switzerland).

Norman, W. and MacDonald, C. (2004), 'Getting to the Bottom of "Triple Bottom Line"', *Business Ethics Quarterly*, **14**(2) (April).

Pannell, D.J. and Schilizzi, S. (2006), *Economics and the Future: Time and Discounting in Private and Public Decision Making*, Cheltenham, UK and Northampton, USA: Edward Elgar.

Schaltegger, S. and Burritt, R. (2000), *Contemporary Environmental Accounting. Issues, Concepts and Practice*, Sheffield: Greenleaf.

Schaltegger, S., Müller, K. and Hindrichsen, H. (1996), *Corporate Environmental Accounting*, Chichester: John Wiley & Sons.

18 Eco-management and Audit Scheme in sustainable tourism
Giulio Querini and Carmen Bizzarri[1]

Environmental quality in the tourism sector

Assessing environmental quality and taking the consequent decisions is probably one of the most ancient jobs undertaken by humans; indeed human survival and prospects for progress have always depended on the supply of natural resources. Only recently, however, has this ancient vocation been enriched by sophisticated technical instruments, so that it can be used to resolve complex economic and technological problems.

Through an assessment of environmental quality we may analyse the impact of human activities from the viewpoint of both public and private interest. In both cases the basic goal in pursuing environmental quality is to reduce the 'rate of consumption' of environmental resources in the various processes of production, with the adoption of 'appropriate technologies'. In other words, it is a question of reducing the waste of natural and environmental resources that are insufficiently reported, due to 'market failure'. Defined in these terms, environmental quality is a substantial, objective characteristic that is intrinsic to the process of production of goods and services, and has very precise connotations from an economic and technological point of view. It is not therefore, as we are often led to believe (especially in the services sector), simply a 'marketing expedient' aimed at making a company's image more attractive to the client.

'Objectivity of environmental quality' in the tourism sector should be regarded as being intrinsic to the various phases of the supply of tourism services. Indeed the sustainability of the 'rate of consumption' of the various natural and environmental resources varies according to the fragility of the each context in which the tourism enterprise takes place.

What is more, reducing the waste of natural and environmental resources will have a different 'benchmark', depending on the degree of scarcity of these resources in the various 'lifecycle' phases of the production of the tourism-related good or service. In the same fashion, the goal of waste reduction could differ according to the type of tourism. For example, in the case of tourism in protected areas, the chief scarce resource to be conserved will be biodiversity; in the case of seaside resorts on the other hand the main concern will be the reduction of pollution deriving from land-based sources of pollution.

As far as sustainable tourism is concerned, it would seem appropriate to define the special meaning assumed by the concept of sustainability in this context. Indeed it has a double connotation, which is on the one hand *ecological* – that is, it refers to the conservation of equilibrium among all the aspects of the natural environment (flora, fauna, hydro-geological resources, climate, landscape and so on) – and on the other hand *anthropological*, referring to the continuing possibility for successive waves of tourists to enjoy this environment, chiefly in its natural aspects but also in its cultural ones.

It is evident, at least for the economist, that the two aspects of sustainable tourism – ecological and anthropological – are strongly inter-related. Indeed the main effect of the degradation of the more fragile elements of the natural environment, especially if this is irreversible, is a halt in the development of the tourism industry, with marked economic and social consequences. In addition to the loss of financial and human resources, this situation of underdevelopment and impoverishment may go on to provoke a loss of interest in conservation and in the full appreciation of those natural and environmental resources that are the object of the tourist's attention.

In this respect it should be stressed that it is precisely this interaction between environmental degradation and loss of economic profitability that can be found at the root of the well-known phenomenon of the 'lifecycle' of tourism activities. Indeed, this cycle takes place in localities that are highly rated from the point of view of landscape and culture, with a transition from luxury tourism to mass tourism within a varying space of time. As the short-sightedness of public officials and private operators prevails, the loss of quality – in terms of both tourist-customer and natural environment – might not be compensated by the quantity of hotels, tourist groups, campsites, nightclubs and the like.

The lack of sustainability of this mass-tourism-oriented strategy soon emerges in its economic and environmental consequences, which are both negative: we face what is a real 'vicious circle', loss of economic profitability and growing environmental degradation.

In the situation of a fall in prices, there is a certain degree of deterioration of tourism services and of environmental quality, and the elasticity of the demand for these services is low, with a consequent drop in the profitability of the whole tourist sector.[2] Alongside the congestion caused by the influx of a huge number of tourists is an increasing degradation of the environment, and what is more, transport and catering activities also reach levels that are incompatible with efficient economic management. In extreme cases, the entire region, once an 'island paradise' justifying elite tourism, is affected by the phenomena of 'tourist desertification' with regard to the tourism sector, resulting in serious, irreversible environmental degradation and the failure and retreat of the better-quality operators.

In assessing the environmental sustainability of activities in the tourism sector, a cross-sector analysis is needed and must refer to all stages of the production process:

- the production of primary products for food (farming);
- the supply of the industrial equipment needed in order to provide tourism services (manufacturing);
- the building of infrastructure for the accommodation and mobility of tourists (construction); and
- the organization of activities aimed at rendering the supply of tourism services economically profitable for the tourism company (management).

The attainment of high standards of environmental quality in all the various phases of the production of tourism services, the so-called lifecycle assessment (LCA), sets down the prerequisite for achieving a substantial and lasting environmental sustainability in all tourism activities. It is important to note that this goal – which cannot be divorced from considerations of economic or financial profitability for the company – requires substantial changes in behaviour on the part of all the operators involved (Huybers and Bennett, 2002, pp. 43–4).

The transformation of the modus operandi in the tourism sector seems to be rendered easier when it is aimed at the implementation of an EMS (Environmental Management

System), certified by a body that is both authoritative and neutral. This body usually performs a double role: on the one hand, it suggests that the business should introduce those innovations internally, which might bring it closer to the elimination of wastes and inefficiencies, thus helping to improve the level of environmental quality; on the other hand it provides external documentation and certification (for the benefit of the various stakeholders) of the effective implementation of procedures aimed at pursuing higher levels of environmental quality. The procedures for the implementation of an EMS generally involve four stages:

1. Application includes the call for application, the consulting services for applicants and the types of required documents.
2. Verification includes the question of third-party assessment, the examination of documents, on-site visits, fees and post-certification control.
3. Certification includes the question of one-level or multi-level awarding, validity of the certificate and the awarding ceremony.
4. Marketing includes the type of information and marketing activities, and the channels for distribution, such as the Internet (WTO, 2002, p. 64).

The success of this procedure in the tourism sector depends on several different variables. In most cases, however, the essential question is that the company should perceive that the accrued benefits are real and tend to grow over time. The range of benefits that might be attained concern the reduction of costs due to the elimination of wastes, entitlement to benefits from public authorities and access to public funding, and the acquisition of advanced know-how already present in other tourist enterprises adhering to the same EMS.

As far as contents are concerned, an eco-audit would cover aspects of environmental management, including:

- the impact of company activity upon the surrounding physical environment, including noise and aesthetics;
- the use of energy, water and commodities;
- waste management and waste water schemes;
- impacts of the company's purchasing policies;
- communication of environmental practices to its staff; and
- their level of environmental training.

Of course, the eco-audit must be regarded as an integral part of a wider-ranging EMS. Crucial in any case is full confidence in the technical proficiency and professional seriousness of the certifying body conducting the EMS (Holden, 2000, p. 147; Taschwer, 1998, pp. 22–3).

Reporting and auditing of environmental quality
The implementation of an EMS is both beneficial and relevant to the tourism company because it allows two converging objectives to be pursued:

1. It encourages the pursuit of technological and managerial improvement in a systematic and continuous way, aiming at a high benchmark value of environmental quality.

2. It allows the stakeholders to be informed about the ongoing effort to raise the level of environmental quality, thus improving the company's commercial image as well as that of its brands and products.

The first objective results essentially in a reduction of costs, the second in an increase in company profits and market share.

The improvement of the ecological image of the company doesn't follow directly from the pure and simple communication to the stakeholders that the company has implemented an EMS. This type of communicative message should also materialize in a logo or icon, well recognizable by the potential consumers of the goods or services that the company offers to the market.

Such labelling procedures, which fundamentally constitute global standards, have critical economic and political importance in the ongoing process of the globalization of trade. It should be noted that while in the past such standards focused on the products, today they focus mainly on various aspects of the productive process: technological choices are in fact influenced by elements such as health, job safety, the recycling of materials, as well as the protection of the environment. Productive standards constitute an essential component of the new rules that govern international exchanges and their compliance to them plays a fundamental role, evermore destined to spread to new operators.

Standards are agreed rules that are codified and used to assess performance. They can apply to individual companies and their supply chain (company codes of conduct), are generic to particular areas (for example, environmental concerns), or relate to particular industries. They can be self-monitored, independently audited and certified, or publicly regulated. The policy challenges on standards are four-fold: in setting standards, in monitoring compliance, in providing assistance on achieving compliance and in sanctions on non-compliance (Institute of Development Studies, 2003, p. 4).

A positive contribution may come from governments that provide the infrastructures for effective auditing and monitoring as well as the technical and financial assistance on achieving compliance. Moreover, governments may play an essential role in preventing illicit and fraudulent behaviours by implementing exemplary sanctions.

The effectiveness of this communication tool in the tourism sector will depend upon the user's degree of environmental sensibility, which presumably varies with the type of tourism: for example, it will be higher in the case of individual tourism in national parks; it will be lower in the case of the mass tourism of seaside resorts.

It is interesting to note some of the characteristics of the icons or logos that can be used at an international level to certify tourist services. Outside Europe, 85 per cent of the eco-labels focus on all types of accommodation, only 15 per cent exclusively concentrate on hotels. The title of the eco-labels reflects their mission: approximately 50 per cent include the word 'green' or 'environment'; 25 per cent incorporate 'eco' or 'eco-tourism'; and the remaining 25 per cent use other terms. Based on these titles, 73 per cent require good environmental performance and 27 per cent include environmental and socio-cultural and economic criteria (WTO, 2002, p. 61).

Both the number of potential participants and the operability of voluntary systems show great potential for expansion. Nevertheless the options of tour operators as well as individual tourists are still decidedly limited: in fact it is worth noting that the rate of certified tourist services on the global scale is only 11 per cent of the total. Consequently, many voluntary initiatives for sustainable tourism (VISTs) have great difficulty in surviving and have a strong need for cooperation between themselves. Indeed the biggest difficulty, as noted by the World Tourism Organization (WTO), is 'the uneven spread of labels across different areas of the supply chain' (WTO, 2002, p. 13).

As far as the certifying body is concerned, one-third of the labels come from the private sector, another third from NGOs, and the last third come from public certifying bodies. Among the objectives pursued by certifying bodies, a recognizable trademark is considered the most important: a label is generally believed to have achieved reliability only when it covers at least 10 per cent of the total tourist services market.

An important contribution to trademark labels comes from voluntary initiatives for sustainability (VISs) as well as from other sectors, in particular those strictly associated with tourism, like catering, transportation, cultural initiatives, resource management and so on. At an international level, the tendency prevails to create certifying systems that cover the entire lifecycle of the tourist product, in order to create economies of scale through the mutual recognition of labelling systems.

The level of lifecycle considerations is determined by the number of performance criteria as an indicator: the more detailed the criteria for green purchase, for production, consumption and waste impacts at all relevant services at the product group (hotel, beach, marina, whole destination or tour package), the better. According to this indicator about 50 per cent have a high level, 25 per cent a medium level and 10 per cent a low level. The remaining 15 per cent did not provide reliable or enough information on this issue. High lifecycle considerations are given especially by those eco-labels that are led or supported by governmental organizations. About 50 per cent of the eco-labels regularly update their criteria, at least every three years; 15 per cent every two years; and 5 per cent annually (WTO, 2002, p. 63).

The potential role of the VIST system can be increased through the following measures:

- intensified cooperation with analogous initiatives already successful in other sectors, with the aim of ensuring the environment quality of products and processes: biological food, sustainable transportation, alternative energy sources, energy saving, eco-compatible management of water resources, the prevention and recycling of waste;
- promoting a wide and transparent connection network, with labelling systems that ensure not only environmental quality, but also the safety and social sustainability of the production of goods and services;
- the introduction of an extensive valuation method for new types of tourist activities and categories of tourist facilities but only if compatible with management limits and the monitoring of auditing bodies; and
- the standardization of eco-labels and standard criteria for issuing certificates, even in those areas that substantially differ from each other.

The advantage of economies of scale resulting from a reduction in the number of eco-labels and subsequent standardization could, however, enter into conflict with the need to pay particular attention to specific variables indicated on the reporting certificates, which will depend on the variety of users and the kind of tourism. In fact, it is probably appropriate that each tourism type (seaside, mountain, ecological, religious) is given a corresponding specific label showing the characteristics of the services provided, and bearing in mind the particular needs and demands of the tourist.

Such variables could also be ultimately divided into levels of quality similar to the 'stars' system that distinguishes the different hotel and restaurant categories. Likewise, in order to categorize the information for the tourists in a particular way, the WTO has put forward a proposal to create different levels of certification with the same label depending on the size of the company (WTO, 2002, p. 63).

Hence, with respect to standardization, the huge variety of eco-labels for the type of tourism and the quality of services provided has produced both advantages and drawbacks. In fact, with regard to the latter, we can point out the difficulty of compliance on the part of the weaker tour operators, especially small and medium-sized enterprises (SMEs), operating mainly in developing countries.

What is more, the proliferation of standards makes compliance difficult and costly. Failure can mean losing access to key markets. Yet compliance costs can be high in terms of auditing charges, improving production and management. Many standards require local stakeholder participation to ensure that specific groups of workers do not become excluded. The plethora of standards raises costs for firms and global buyers in developing countries that can create more confusion on the part of the consumers (Institute of Development Studies, 2003, p. 4).

The setting of standards, which establishes social conduct and ethical principles, can prove useful not only for workers, but also for tourist operators: in fact, these measures tend to increase competition between companies and facilitate access to global markets. Nevertheless, problems do arise in the developing countries, especially concerning SMEs, in that the costs involved in conforming to world tourist standards are particularly high, and it is more difficult to take into account the demands of a specific local and social context that often faces different categories of tourist operators.

The particular needs of developing countries, including the capacity building of various stakeholders, should be given special attention. Southern initiatives and joint North–South partnership projects are an excellent tool for exchanging and disseminating best practice, as long as care is taken to allow for regional and cultural differences during their implementation (WTO, 2002, p. 13).

When confronted with a variety of eco-labels, the weak tourist operators often seem to be disoriented and this disorientation tends to impact on the tourists themselves. In fact, the tourists are more and more inclined to trust recognized trademark labels: moreover, the modern tourist, unlike the traveller of the past, is generally reserved and poorly motivated to find pleasure in the self-discovery of a place with an unexpected high level of environmental quality. Furthermore, with regard to small and weak operators, the variety of eco-labels requires full and precise monitoring work, which is difficult and costly for certifying bodies.

About 60 eco-labels, the majority of which are for accommodation, have been operating worldwide for years and have certified thousands of hotels, campsites and beaches. Most of them operate in Europe and many require a relatively high environmental performance level but not all can guarantee a high degree of reliability owing to factors such as the lack of on-site control. Only a few can really offer good consumer operations with a long list of certified hotels and campsites in one country (ARPA, 2002, p. 6).

The poor operating effectiveness of the majority of eco-labels arises from the fact that they are not well known to the tourists and what is more they lack coordination amongst themselves. The eco-label situation in Europe is particularly complex: in actual fact in the year 2000 there were 40 regional, national and international labels. Progress towards an increasingly homogeneous and unified European market makes this large variety of labels poorly recognizable and creates disorientation among the consumers and the tourist service suppliers.

Eco-labels can be considered successful when they certify really good – 'better than non-certified' – and enough products in their countries to create consumer options; they generally have a reliable verification procedure and they are usually well-recognized and accepted by the consumer. The VISIT (Voluntary Initiatives for Sustainability in Tourism) project, with the support of the European LIFE (the Financial Instrument for the Environment) programme, intends to demonstrate how tourism eco-labels in Europe can be an effective instrument in moving the tourism industry and consumers towards sustainability. In the year 2000, VISIT agreed to found a common international umbrella association for tourism eco-labels and other voluntary initiatives for sustainable tourism. The main tasks were to control and develop VISIT standards for eco-labels, to continue joint promotion and marketing activities on a European level, to monitor their environmental effects, to strengthen the quality of their criteria and procedures, and to seek other complementary tools and initiatives (for example, the Audit Scheme, and the European eco-labelling system EU Flowers).

Recently there has been a lively ongoing debate on the role the Internet can play in the implementation of Eco-management and Audit Schemes (EMASs). On the one hand, optimists maintain that this tool can play a positive role in the dissemination of information on the high standards of environmental quality achieved by tourist operators, and can thus contribute to the diffusion of best practices. On the other hand, pessimists believe that the Internet, by rapidly pouring out enormous amounts of information on the environmental quality of places and operators, weakens the net-user's ability to make critical selections, creating confusion and alienation on the part of the tourist operators and among the tourists themselves.

To throw light on the problem of tourist sensitivity towards environmental quality as a criterion for purchasing tourist goods and services, many empirical researches have been carried out on the importance of eco-labels in destination options, as well as on the different uses of tourist services in Europe. Another aspect under analysis is the willingness of the tourist operators to collaborate with the labelling procedure.

The results so far are that for consumers environmental quality is not the main decision-making factor (such as prize, sun or security) but environmental qualities are desired as a 'quality plus' factor. For example, 50 per cent of Germans only take their holidays where the environment is still intact/unspoiled: clean beaches, clean water, no rubbish and pollution, protected nature, no noise (ARPA, 2002, pp. 5–6).

Research carried out in Italy in 2000 (ANPA-ISTAT, 2001, p. 12) found that the most important criterion for tourist destinations dependent on environmental variables was the 'proximity of the structure to areas of particular environmental interest', which was indicated as 'very' or 'quite' important in 45.5 per cent of the holidays, and especially taken into high consideration on seaside holidays (70.4 per cent) or mountain holidays (69.3 per cent); whereas it had markedly less importance on business trips (17 per cent). The management of the service provided by the structure that respects the environment appeared to be quite an important element, considering it had a determining influence on the choice of the receiving structure in 33 per cent of the cases. Regarding this decision-making principle of choice, women seemed to be a little more sensitive (35.9 per cent), as did people between the ages of 25 and 44 (34.2 per cent).

With reference to the decision-making principles of the people interviewed in the survey, including the price, the proximity to places of specific interest and the quality of services, environmental quality was not in a priority position. Nevertheless, it was just as important as other criteria, especially for younger tourists and for those coming from densely urbanized areas. In fact, 81.7 per cent of the people interviewed believed that a hotel or a receiving structure that put into practice measures for the defence of the environment also improved the overall quality of the services offered. Furthermore, 87.4 per cent thought it 'very' or 'quite' important for hotels and receiving structures adopting such measures to be recognized with an environmental quality trademark certified by a national public authority or by the European Union (ANPA-ISTAT, 2001, pp. 12–15).

Similar research investigating the environmental quality sensitivity of Italian tourists travelling in India is currently being carried out by Tisdell (2001, pp. 168–79), the results of which could also prove to be of considerable importance in a wider international context. The main aim of the above-mentioned research is to analyse the trade-off between the pleasure that Italian tourists gain from the high 'quality' environmental and cultural resources of the places they visit, in contrast to the socio-cultural ecological degradation they may cause when they come into contact with those environments in India.

The role of the Environmental Management Audit Scheme
Tourist companies operating within the European Union are conforming more and more to the norms regulating the environmental quality of their products and services. On an international level, quality standards such as ISO 14000 have long been established. Likewise the European Union has prepared its own set of regulations that vary according to the different environmental objectives to be achieved. From the point of view of public interest, environmental impact assessment (EIA) in its two-fold 'project' and 'strategic' versions is particularly suitable as an instrument for evaluation of environmental quality. If instead the goal is to measure the impact that environmental quality can have on a company's profits, more suitable instruments would include the corporate environmental report, eco-labels and the Eco-management and Audit Scheme.

The main new additions made to the EMAS in 2001, which update the inadequate 1993 version, concern the possibility of certifying not only individual industrial sites but also more complex networks and organizations. Furthermore, EMAS-2 is no longer limited to the certification of manufacturing companies but can also be implemented by businesses operating in diverse sectors such as tourism, transport, construction, local public utilities, forestry and so on. The application of the EMAS to the tourist sector has had a positive start, especially since environmental quality has become an important option for a variety of tourists. However, a problem occurs with the fact that the EMAS is an operative scheme for environmental management that is quite rigid and homogeneous at the EU level, consequently the resulting standard guidelines are not always simple to implement in situations where there are different types of tourism, and especially in cases where natural and environmental resources are particularly vulnerable and scarce. In any case, the objectives of the EMAS come from a range of priorities that vary from case to case; examples include reduction in energy and water wastage, keeping waste pollution to minimum rates and safeguarding biodiversity.

As a number of environmental management system tools exist at an international level, it is clear that the public authorities that decide to implement one of these tools compare both costs and benefits in terms of their particular aims (Eiderstrom, 1998, pp. 24–5). In the tourist sector, big organizations like hotel chains, integrated structural areas (national parks and transport systems for example) especially benefit from the EMAS, whilst other more flexible and less costly schemes are suitable for SMEs, which need to pay particular attention to administrative and financial costs.

Most enterprises agree that one great advantage of the EMAS – being a voluntary compliance mechanism – is that it specifically certifies 'good environmental conduct', which in turn avoids scrupulous state controls. In other words, it becomes commonly assumed that certified enterprises respect the standards and norms imposed by current regulations. Moreover, there is the added advantage of benefiting from good relations with public authorities and institutions. The WTO has effectively noted that cooperation with private companies and NGOs is particularly fruitful.

Those voluntary initiatives for sustainable tourism where private and public operators cooperate seem to be the most successful. In fact these VISITs are the most transparent and have achieved a great deal of recognition both socially and environmentally. In cases where they are combined with professional marketing, they demonstrate a high degree of effectiveness. Furthermore, VISITs appear to be the best tool for overcoming conflict between compulsory state regulations and voluntary schemes. VISITs can both pre-empt and complement the regulation of industry, and are doing their best to avoid and eradicate 'greenwashing'. Their experience of defining best practice in the marketplace offers an excellent opportunity to develop regulations to suit and stimulate best practice (Altham and Guerin, 1999, pp. 65–6; WTO, 2002, p. 13).

In addition, the more the regulations at the basis of the EMAS spread among companies, the more it becomes a reference point for public norms and authorities. The latter will eventually adopt the EMAS, making monitoring procedures and environmental quality standards obligatory. It must be noted that proactive tourist operators implement the regulations voluntarily. In accordance with this inclination, the European Union is encouraging a wider range of EMAS applications, which will result not only in a higher level of environmental quality, but also in more rigorous standards that safeguard the economic and social quality of products and processes.

Another direction in which the EMAS is expanding, along with most environmental management schemes, involves combining the certification of the services offered by the tourist facilities with the certification of the territory to which tourists flock. In reality, a tourist facility cannot consider itself isolated from the social and territorial context in which it operates: external economies and diseconomies interfere with the efficiency of the tour operator, which can in turn either improve or degrade the local environment.

The typical European tourism product depends to a large extent on the sustainable development of destinations. The vast majority of tourists are looking for intact nature, beautiful landscapes and a rich cultural heritage. They want a clean, healthy environment and a friendly climate. In fact there is a strong link between sustainability and quality. Most issues, such as low noise, less traffic, clean air and water, rich biodiversity, intact landscapes are the focal point of sustainability strategies and crucial to the quality of destinations. If we want to analyse the effects of tourism on sustainability, we have to consider the entire lifecycle of tourism, which consists of three phases: 'access and return travel', 'stay at the destination' and 'activities at the destination' (ARPA, 2002, p. 10).

Taking into account the variety of impacts, the application of the EMAS to the tourist industry presents considerable difficulties inasmuch as it requires sound coordination in the standards applied to the companies operating in the above three phases. It is therefore necessary to standardize the criteria for EMAS compliance especially when considering the following problems:

- The cost of technological innovations resulting from EMAS implementation. The long-term benefits at the rate of interest on the current financial market must be higher than the total cost of current EMAS implementation.
- The evaluation of the environment's carrying capacity for tourists, bearing in mind some important characteristics: its seasonal nature, the precarious working conditions, and the over-dependence of the local economy on tourism.
- The negative environmental impacts induced by greater mass tourist inflows. It is necessary to evaluate not only the impact of tourism on the means of transport systems, but also the environmental degradation following the building of new infrastructures for the incoming tourists.
- The environmental monitoring of the effects of tourist activities on critical natural resource shortages, which characterize the local environment. Depending on the type of tourism, monitoring of energy, water resources, solid waste management, agricultural land, biodiversity and the landscape may be required.

In order to take all these elements into account, the EMAS today is moving towards involving whole networks of productive structures, the so-called 'districts', both at sectoral and territorial level. This wider and more complex system of evaluating and monitoring environmental quality forces the auditors who intend to collaborate with EMAS implementation to face a series of indicators that are important for the analysis of the overall environmental quality of the 'firm–territory' system. The characteristics of these indicators, which are still the subject of debate in economic theory and among tourist operators themselves, can be summarized as follows:

- *Quantifiability*. There is a need for reliable and readily accessible data that allows the identification and quantitative processing of the indicators.
- *Relevance*. The selected indicators must have a high degree of significance in terms of the environmental sustainability of the development under way.
- *Transparency*. On the basis of the 'principle of subsidiary', all the decisions relative to sustainable development – and therefore also to EMAS implementation – must be taken with the participation of citizens. For this reason the selected indicators must be easy to understand for all who participate in local and regional planning.
- *Operating effectiveness*. Since the debate between the social forces concerned must lead to innovative choices about public and private investments, the indicators must be practically usable in order to modify behaviour towards a high level of environmental sustainability.

Each of these indicators should not be used separately: the effectiveness of such a method of environmental policy is conditioned by the fact that they are used like a network and a composite system (ARPA, 2002, p. 10).

The sustainability of tourist services – both of those already existing and those planned – must therefore be evaluated in the context of existing planning systems for the natural and environmental resources of the local area, with particular attention given to the protection of scarce resources, or those that are more vulnerable to a potential increase in the inflow of tourists.

Furthermore in this context the EMAS must assume not only a defensive protective role but also a wider proponent role: the new objective is the identification of relatively more abundant and under-used natural environmental resources that can be enhanced by means of an increase in tourist activities. The particular local characteristics can be exploited by tourist companies – or better still by a network of enterprises or local public bodies – for the purpose of rationalizing the production system, leading to a reduction in waste and achieving economies of scale. A further important objective can be that of creating a true position of 'rent' – even if only imaginary but equally effective – deriving from the non-reproducibility of tourist services offered by each local tourist complex.

Conclusions

The implementation of EMAS on the part of European companies operating in the tourism sector involves relevant advantages both in terms of economic profitability and protection of the environment. These advantages are comparable to those that might be obtained through other systems of quality certification, which are gradually spreading throughout the world. The main advantage of the European Union systems (not only the EMAS but also eco-labelling and other VISITs) lies in their standardization: homogeneity of the issuing procedures and uniformity of certification allows for greater transparency of content and a broader degree of recognition among tourists.

The recent evolution of the EMAS in the direction of the certification of company networks and local authorities is a very positive contribution towards the environmental sustainability of the tourism industry. The positive aspects effectively fall into two groups. On the one hand, the EMAS analyses the entire lifecycle of tourism-related products, favouring the coordination of VISITs between private business and local authorities. On the other, it makes it possible to take account of the specific nature of the various categories of tourism (seaside resorts, cultural-urban, winter-skiing, nature and so on) as well as the 'critical fragilities' (water resources, landscape and biodiversity) of the various local situations.

One question remains open, a question of great importance for both companies and those responsible for environmental policy at national and local levels: what is the role of the EMAS, which is now only voluntarily implemented by European companies, with regard to the allocation of financial incentives and administrative benefits? We predict that the EMAS will probably be transformed in the near future from a voluntary tool of environmental management to a compulsory procedure imposed by law on most European businesses: a prospect feared by some and welcomed by others.

Notes

1. This chapter reports the results of researches financed by the Italian Ministry of University and Research (MIUR) and the University of Rome 'La Sapienza'. The first section is attributed to Prof. G. Querini, the

second and third are attributed to Dr C. Bizzarri. This chapter was translated into English by Dr Ersilia Incelli, English Language Lecturer.

2 For example, if a hotel cuts its prices by 20 per cent, the number of clients will only increase by 10 per cent, so profits will still fall.

References

Altham, W. and Guerin, T. (1999), 'Where does ISO 14001 Fit into the Environmental Regulatory Framework?', *Australian Journal of Environmental Management*, **6**, 86–98.

ANPA-ISTAT (Agenzia Nazionale per la Protezione dell'Ambiente – National Agency for the Protection of the Environment/Instituto Nazionale di Statistica – National Statistics Institute) (2001), *Domanda turistica e qualità ambientale: l'applicazione del marchio europeo di qualità ambientale nel settore del turismo*, Roma.

ARPA (Agenzia Regionale Prevenzione e Ambiente – Regional Agency for Environmental Prevention) (2002), *Voluntary Initiatives for Sustainability in Tourism*, Ravenna, 27/28 Sept.

Eiderstrom, E. (1998), 'Ecolabels in E.U. Environmental Policy', Working Paper No. 98/20, European University Institute.

Holden, A. (2000), *Environment and Tourism*, London and New York: Routledge.

Huybers, T. and Bennett, J. (2002), *Environmental Management and the Competitiveness of Nature-based Tourism Destinations*, Cheltenham, UK and Northampton, MA, USA: Edward Elgar.

Institute of Development Studies (2003), 'The Cost of Compliance', *Policies Briefing*, 18 May, pp. 1–4.

Taschwer, K. (1998), 'Environmental Management Systems: The European Regulation', Working Paper No. 98/21, European University Institute.

Tisdell, C. (2001), *Tourism Economics, the Environment and Development: Analysis and Policy*, Cheltenham, UK and Northampton, MA, USA: Edward Elgar.

World Tourism Organization (WTO) (2002), *Voluntary Initiatives for Sustainable Tourism*, Madrid: WTO.

PART IV

MEASURING ENVIRONMENTAL TECHNOLOGY MANAGEMENT

19 Measuring the true productivity gains from environmental technology improvements
Robert Repetto

Why is it important to measure productivity growth accurately?

Productivity largely determines the level of real incomes because it represents the efficiency with which firms, industries and the entire economy transform inputs into outputs. In turn the productivity growth rate influences how fast real incomes can rise. In the United States for example, rapid productivity growth fuelled the economic boom in the 1950s and 1960s; slow productivity growth in the 1970s and 1980s inhibited real wage increases; and rapid productivity increases in the late 1990s permitted rising real wages. If the availability of goods and services had been limited entirely by the gradual increase in the labour force and capital stock, then US living standards today would be much impoverished.

The effects of productivity are also felt at the level of the individual firm, where productivity growth rates exceeding those of rival firms should result in increased market share and profitability. Accordingly, for business managers, productivity growth within the firm reflects organizational efficiencies achieved in production; for engineers and technicians, it measures the success of productive innovation.

Productivity can be measured in different ways: labour productivity measures output per worker; multifactor productivity, a broader indicator, measures the productive efficiency of labour, capital and other inputs in combination. Either way productivity is a key indicator of technological and organizational efficiency.

Naturally the marked decline in the US productivity growth rate first observed in the 1970s caused widespread concern. After decades of rapidly increasing prosperity this sudden decline cast doubt on the economy's ability to provide rising living standards. Multifactor productivity in the private sector, which had been growing at an average annual rate of 2.2 per cent throughout the period 1948–73, remained static between 1973–79; worse still, private non-farm business productivity showed an absolute decline of −0.1 per cent per year for the 1973–83 period, implying that the same inputs produced less output at the end of the period than had been the case ten years earlier. Reinforcing these concerns, productivity growth rates remained persistently low throughout the 1980s and early 1990s, as shown in Table 19.1 (Bureau of Labor Statistics, 2005).

This abrupt slowdown prompted an outpouring of studies that sought to identify the cause and provide a basis for corrective policies. No such 'silver bullet' was ever found (Gordon, 1981; Munnell, 1990, p. 3). Instead it was generally accepted that various factors were responsible. Economists have pointed to several factors, including lower pay-offs from research and development (R&D) expenditures; reductions in public spending on core infrastructure – a necessary (though often overlooked) requisite for private productivity gains; and the growing importance of the service industries, whose productivity improvements are difficult to measure (Baily, 1986; Fischer, 1988; Munnell, 1990).

Table 19.1   Trends in private business productivity growth by period (annual average percentage changes)

1948–73	1973–90	1990–95	1995–2000
2.1	0.5	0.7	1.3

Source: Bureau of Labor Statistics, 2005.

In addition three 'shocks' to the economic system were identified as important. First it was posited that productivity probably declined in reaction to the sharp increases in fuel prices during the 1970s (Baily, 1981; Gordon, 1981). Second, as the baby boomers came of age and women's labour force participation rates increased, there was a large influx of inexperienced labour into the workforce. Both groups' lack of work experience and the strain of absorbing these new entrants meant that overall labour productivity fell (Baily, 1986).

Finally, it has been claimed that the cost of complying with environmental regulations introduced in the early 1970s, including the Clean Air Act and Clean Water Act, required industries to divert investments toward the installation of costly abatement technologies and raised production costs. According to some studies, this shock may have been responsible for up to half of the productivity decline observed in pollution-intensive industrial sectors.

This latter finding however, is an artefact of a basic flaw in the way productivity is measured – a methodology that counts the cost of environmental protection but ignores the cost of environmental degradation. This problem in productivity measurement has led to serious misunderstandings about the effects of environmental protection on the economy. This chapter explains the flaw in productivity measurement, shows how it distorts the measured impact of environmental improvement on the economy and proposes a better way of measuring productivity growth. Case studies of the electric power and pulp and paper industries illustrate what difference this methodological change would have on the record of productivity growth.

It is argued that a better approach to productivity measurement would also provide business executives with a useful way of integrating their environmental impacts into management accounting frameworks. Just as EVA (economic value added) measures have enabled managers to track the efficiency with which they use their capital assets, the approach explained here – which could be called EEVA (environmental economic value added) – enables managers to track the efficiency with which they use environmental assets.

How is productivity currently measured?
Until 1983 productivity figures produced by the Bureau of Labor Statistics were expressed in terms of output per hour worked – a simple measure of labour productivity. Now the Bureau of Labor Statistics uses a more sophisticated measure of labour productivity that distinguishes many different categories of labour. It is computed as the ratio of an index of outputs weighted by their respective market prices to an index of various categories of labour services weighted by their respective costs of employment. After adjustments for inflation the change in this index over time is taken as the measure of labour productivity growth.

One problem with this labour productivity indicator is that it reflects not only changes in technology and the reallocation of labour to higher-valued occupations, but also changes in the availability of capital per worker – a result of capital accumulation rather than of improved efficiency. For this reason the Bureau of Labor Statistics followed the lead of academic economists and has introduced a broader measure of multifactor productivity (or total factor productivity) to measure the efficiency with which other inputs are used, which includes capital and materials as well as labour. This indicator includes capital and materials used in production, along with labour, in the index of inputs. Each of these factors of production is made up of different constituent inputs weighted by their respective costs to the firm. If industries exhibit constant returns to scale and input markets are competitive, the contribution made by the increasing use of each factor to the growth rate of output can be determined. The remaining change – defined as multifactor productivity growth – represents the increase in output that cannot be explained by mere increases in inputs. It can instead be attributed to technological and organizational advance.

Much time and energy has been devoted to improving the methodology and the data used to calculate these productivity indicators. The measurement of labour inputs now distinguishes between categories of labour whose effect on productivity differs because of educational attainment or accumulated experience. The measurement of capital services takes account of the age and relative efficiency of plant and machinery. Finally the measurement of output has been improved by distinguishing quality improvements along with quantitative increases in the output of goods and services.

To be sure, numerous difficulties remain. Measuring output is still problematic in service industries, such as the legal profession and banking, where the nature of the end product is hard to define or may change from year to year. On the methodological side the index used may imply unrealistic assumptions about the production process. Despite these remaining problems in most respects productivity measurement over the last 15 years has become more sophisticated and informative. Unfortunately in dealing with environmental issues, little progress has been made and serious problems remain.

The conventional approach to assessing environmental protection's impact on productivity

The current methodology leads almost inevitably to the conclusion that environmental protection reduces productivity growth. Though this perception is reinforced by extensive empirical work, it is basically an artefact of the methodology now being used and is not necessarily correct. Environmental regulations have induced firms to reduce emissions by altering production processes, mainly through installing pollution abatement equipment (for example, exhaust gas scrubbers and waste water treatment plants). Purchasing inputs whose main function is to curb pollution has raised input costs with no corresponding increase in marketed outputs. Thus since the productivity measure gives industries no 'credit' for reducing emissions, however damaging, measured productivity has been depressed.

Even under more flexible regulatory systems that allow firms to decide for themselves how best to meet standards, the result is much the same: the means of pollution abatement may be different and perhaps more cost-effective, but qualitatively the result will still be higher input costs with no offsetting rise in production. For example, when faced with lower sulphur dioxide emissions standards in 1992, most electric power utilities reacted by switching to low sulphur coal as the cheapest way to meet the requirements. Since low sulphur coal costs more but produces no more energy, measured productivity suffered.

Only if steps taken to reduce emissions actually reduce production costs or raise the value of saleable outputs sufficiently would environmental protection measures have a positive effect on productivity as currently measured. Sometimes this is of course possible, especially when firms solve emissions problems by fundamentally redesigning their products or production methods (Schmidheiny, 1992). Many firms have reported such successful experiences and environmental regulation can therefore sometimes be a useful prod to induce firms to rethink their long-embodied operating systems (Porter, 1990 and 1991). However, such cost-reducing examples are generally regarded as the exception rather than the rule (Oates et al., 1993). Were they typical, profit-maximizing firms would seek out such cost-saving opportunities even in the absence of environmental regulation. Typically the current methodology will find that environmental protection measures lead to lower productivity.

Environmental protection has been found to retard productivity growth in econometric studies dating from the 1970s. An early study of the private sector between 1972 and 1975 (Denison, 1979) concluded that 16 per cent of the decline in productivity growth could be attributed to the imposition of environmental regulations. Christainsen and Haveman (1981) reported that 8–12 per cent of the decline in labour productivity in the manufacturing sector between 1973 and 1975 was due to regulation. For the period 1973–78 two studies found that regulation was responsible for 12 per cent of the downturn in multifactor productivity for manufacturing industries (Gray, 1987; Norsworthy et al., 1979).

A more recent study of five US manufacturing industries (paper; chemicals; stone, clay and glass; iron and steel; and non-ferrous metals) attempted to measure separately the direct effect of having to buy abatement capital and the indirect effect of having to alter input combinations to accommodate the new equipment (Barbera and McConnell, 1990). After comparing annual average productivity growth figures for the period 1960–70 and 1970–80, the authors concluded that between 10 per cent and 30 per cent of the observed decline could be attributed to the overall effect of the introduction of abatement capital, of which one-half represented a direct effect. Other studies have considered the productivity effects felt in countries in addition to the US, such as Canada and Germany (Conrad and Morrison, 1989) and Japan and Germany (Nestor and Pasurka, 1993). With some exceptions, these findings have mirrored those in the US.

Studies of the electric utility sector that faced some of the most costly abatement requirements have estimated greater productivity losses than in manufacturing industries. For the period 1973–79, 44 per cent of the recorded productivity decline in this sector was attributed to environmental regulation, equivalent to a reduction in annual productivity growth of 0.59 percentage points (Gollop and Roberts, 1983). For the shorter period 1975–79, when the industry was attempting to comply with EPA air quality standards, the estimated impact was an even larger 0.88 percentage points per year. Crandall (1981) also found evidence for the relatively heavy burden regulation imposed on the electricity generating sector.

The effect of regulatory restrictions on the technology used in 100 steam electric plants in 1975 was estimated by Fare et al. (1986) who found that these restrictions ' "cost" an average of roughly 16 million kilowatt hours in lost potential output for each plant'

(Fare et al., 1986, p. 184). This was in addition to the monetary costs of outlay on pollution abatement equipment. Furthermore the authors claimed that these costs almost certainly increased after 1975 as federal regulations were implemented.

A study of individual steel-making plants in the United States for the period 1979–88 found that each additional dollar in environmental operating expenditures per ton of output ultimately raised marginal costs by US$7–12 per ton (Joshi, 1997). Gray and Shadbegian (1993) measured regulatory impact on plant-level productivity for pulp and paper mills, oil refineries and steel mills and found a similar effect. Their initial study estimated that for every US$1 of compliance expenditure there was a further US$3–4 cost increase from indirect effects. However, in follow-up work they found a much smaller indirect effect (Gray and Shadbegian, 1995).

In another study of this kind Robinson (1995) estimates statistically the effects of environmental protection expenditure on productivity growth across 445 US manufacturing industries, finding a significant negative effect. He concludes that 'the productivity-reducing burden of past regulation haunts the future of environmental . . . policy' (p. 414).

These studies differ in their analytical approach, the time period and the industries under observation but all conclude that the response to environmental regulation has impeded productivity growth. Whether intended or not, the inevitable consequence of this consensus has been to strengthen the impression that environmental protection hinders economic growth and reduces living standards. Unfortunately, however, this conclusion is an artefact of the assumptions underlying the definition of productivity. A more reasonable definition would lead to different conclusions.

Why are studies using conventional methodologies biased against environmental protection?

The productivity measure used in all of the productivity studies cited above rests on an incomplete depiction of industrial processes. Basically, industries transform material and energy inputs into marketed outputs. These transformations conform to physical laws, including the conservation of matter and energy, which dictates that all the raw materials drawn into an industrial process re-emerge in some form. An industrial engineer can lay out a materials and energy balance for any industrial process and show where all the inputs go, some to product and some to waste streams. In the words of Sesame Street's Big Bird, 'You can't make nothing out of something'.[1]

For example, a typical 500 megawatt coal-fired power station produces not only 3.5 billion kilowatt-hours of electricity per year, the measured 'output', but also 5000 tons of sulphur oxides, 10 000 tons of nitrogen oxides, 500 tons of particulate matter, 225 pounds of arsenic, 4.1 pounds of cadmium and 114 pounds of lead as well as trace amounts of other minerals embedded in the coal. All of the 1.5 million tons of coal burned each year in the plant for energy ends up as ashes, emissions and other waste products, including more than a million tons of carbon, virtually all of which is emitted as carbon dioxide.[2] The plant also generates a good deal of waste heat, which is usually dispersed in cooling waters. The conservation of matter and energy dictates that along with useful outputs industrial processes also inevitably generate residual outputs that are potentially damaging when released to the environment.

When industrial production is considered in its entirety like this it is obvious that in physical terms inputs and outputs must grow at the same rate. The right question then is

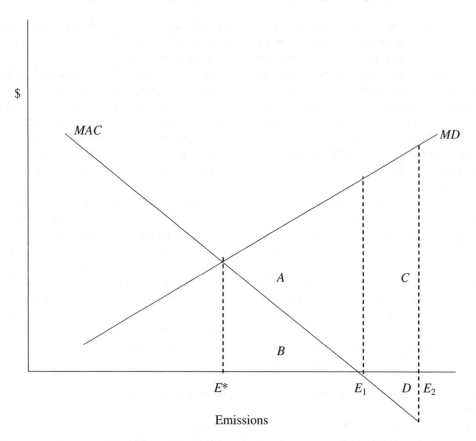

Figure 19.1   Efficient pollution abatement and conventional productivity measurement

whether industrial processes transform these inputs into outputs of greater value, recognizing that some outputs are valuable when sold and that others are damaging when released. Conventional productivity measures generate differential growth rates for inputs and outputs only by ignoring an entire class of outputs, those that are a nuisance to society and therefore unsaleable. The productivity index counts the electricity that is produced but ignores all the other, less desirable, outputs of the process that are nonetheless significant in economic terms. The result is an incomplete and misleading indicator – one that misrepresents the underlying process and provides a misleading indicator of efficiency.

The resulting measure of efficiency is not only inconsistent with the laws of physics, it also disregards the concept of efficiency used when economists evaluate environmental protection measures explicitly. This basic concept of efficiency in environmental economics is portrayed in Figure 19.1.

In the diagram two curves represent the costs of pollution damage and the costs of pollution abatement at different levels of emissions. Specifically the marginal damage (*MD*) curve measures the damage in monetary terms caused by the last (or 'marginal') unit of pollution. Though expressed in monetary terms, these damages might include illness among the exposed populations, degradation of natural resources that makes them less valuable to users, damages to buildings and materials from exposure to pollutants and other environmental impacts. If all units of pollution were equally harmful, this line would be horizontal. Instead the *MD* curve slopes upward, reflecting the tendency for an additional unit of emissions at low levels of pollution to cause proportionately less damage than an additional unit at high levels. At low levels the effects of emissions may be negligible because of the natural capacity of the receiving environment so that the addition of an extra unit results only in limited damage. However, as emissions increase, it becomes more likely that critical thresholds are passed and the addition of an extra unit has greater impact.

The marginal abatement cost (*MAC*) curve shows the costs to the firm of removing the final unit of pollution. Again this varies according to the initial level of emissions. At high levels of emissions the cost of removing a unit of pollution should be low, or even negative, if the firm can save materials or reduce costs through housekeeping improvements. If little or no pollution reduction is presently taking place, relatively cheap and easy abatement options are likely to be available. However, as overall emissions levels are reduced it becomes harder and more expensive to find ways to reduce emissions further. The *MAC* curve reflects the extra input costs to the firm of different abatement options, assuming it will implement the least expensive ones first.

From the standpoint of the economy as a whole the efficient amount of emission reductions occurs at the intersection of the two curves where the marginal damages from pollution are equal to the marginal costs of reducing pollution. At higher levels of emissions the costs of reducing pollution by a unit (equal to *MAC*) are less than the damage costs associated with this unit (equal to *MD*). Hence efficiency increases if this unit is removed. If emission levels are reduced further than this the high costs of removing these units are not justified by the meagre reduction in damages. Efficient emissions reduction efforts are those that balance the incremental costs of pollution control against the incremental costs of pollution damage.

From the standpoint of the individual private sector firm this level of emission control would not be efficient unless the environmental damage costs were 'internalized' and reflected as private costs. Partial internalization might result from pollution charges and fees, from liability to private lawsuits for damages or reputational losses but internalization of damage costs is likely to be incomplete. Therefore private sector firms rarely achieve economically efficient levels of environmental control in the absence of regulatory requirements.

The current methodology of productivity measurement completely ignores this efficiency criterion. Referring again to Figure 19.1, consider a firm that generates E_2 emissions in producing its marketed output. The diagram depicts a situation in which emissions are so high that both emissions and input costs can be reduced simultaneously. Using conventional productivity accounting, a reduction from E_2 to E_1 would be recorded as a productivity increase corresponding to area *D*, the total input cost saving. However, this understates the true efficiency gain because it ignores the reduction in environmental damage from lower emissions, equal to area *C*.

Worse still, cutting emissions further from E_1 to E^*, which would actually maximize efficiency, would cause measured productivity to *fall* (by an amount equal to area *B*) because the firm incurs abatement costs with no offsetting increase in revenues. Despite this reduction in measured productivity, economic efficiency actually would rise by an amount equal to area *A*, the amount by which the avoided costs of environmental damages exceed the costs of abatement. As an evaluative measure the conventional productivity indicator is seriously misleading because environmental protection measures that actually improve economic efficiency can be recorded as lowering productivity.

A recent EPA study prepared at the request of the US Congress provides strong evidence that this distortion has actually occurred. A very comprehensive assessment of the benefits and costs of atmospheric pollution abatement required by the Clean Air Act between 1970 and 1990 found that the economic damages averted by air quality protection over these years were around 15 times as great as the environmental protection costs incurred (US EPA, 1996). Though there are substantial uncertainties surrounding these estimates of damages averted, since not all the identified benefits could be expressed in monetary values and the values that were estimated were subject to error, even the minimum estimate of benefits from air quality protection exceeded the estimated costs by a factor of 5. Nonetheless, despite this evidence that the emissions reduction achieved over this period reduced damages much more than they increased abatement costs, conventional productivity indicators used in all the studies cited above still imply that environmental protection reduced productivity growth. This strongly suggests that the current methodology does not provide a reliable indicator of environmental efficiency.

Although the majority of authors of those studies do recognize the methodological deficiency they still use the conventional productivity indicator. For example, Gollop and Roberts (1983, p. 672) note that 'any discussion of the desirability of environmental regulation must weigh . . . benefits against the costs of regulation'. Jorgenson and Wilcoxen (1990, p. 315) conclude that the results of their study 'cannot be taken to imply that pollution control is too burdensome or, for that matter, insufficiently restrictive'. Robinson (1995, p. 414) writes that 'reductions in productivity should be evaluated in light of the reductions in pollution achieved through regulation'.

Why then do government and academic economists continue to use the conventional productivity indicators? The reason is that waste products emitted to the environment, unlike saleable outputs, do not have market prices. As a cynic might say: 'The worst things in life are free'. Utilities sell their electricity but are perfectly willing to give you their particulate and sulphur emissions for nothing. Undeniably the fact that emissions lack market prices makes estimating their incremental cost to the economy difficult – but not impossible. Estimating the economic costs of pollution is the bread and butter of environmental economists; over the past 15 years the EPA has expended hundreds of millions of dollars to fund research into pollution damages. Though subject to uncertainties estimates do exist. Nor do these uncertainties validate the current approach, which in effect assigns a zero price to emissions in the index of outputs. Although quantifying the economic costs of emissions can only be done with a margin of error, 'zero' is usually not a good approximation.

Ironically the fundamental rationale for the existence of governments in the first place is to provide so-called 'public' goods and services that private parties cannot effectively provide for themselves through market activities: national defence, justice, enforcement of property rights, social insurance and environmental protection. Yet in attempting to measure improvements in economic productivity, governments have adopted a measure that ignores the extent to which the economy has been efficient in protecting the environment, an important public good.

Economists have tried to devise a productivity indicator that takes account of the costs of pollution. Russell Pittman (1983) proposed a 'multilateral productivity index' that included undesirable as well as desirable outputs, valuing emissions by shadow prices approximated by marginal abatement costs (rather than marginal damage costs). Though estimates of abatement costs are more readily obtainable, using them to value emissions will misrepresent efficiency gains from environmental protection unless firms are already controlling emissions optimally. Nonetheless, Pittman's conceptual approach represents an important step forward in productivity measurement, one that unfortunately has not been followed up even though his empirical results showed substantial differences from the conventional productivity measure.

The difficulty in estimating the costs of pollution prompted Fare et al. (1989) to use data from a sample of firms within an industry to estimate a 'socially feasible output set' that includes all possible combinations of desirable and undesirable outputs. From this they estimated the minimum level of emissions that firms can produce for different levels of marketable output. Then they define the efficiency of a single plant by its proximity to this frontier; that is, the extent to which its emissions exceed the minimum 'needed' to produce the same level of marketed output.

This analysis does get to grips with the fact that firms inevitably produce undesirable by-products along with their marketed outputs but it still exhibits two main limitations. First the estimated efficiency frontier represents at best what firms are currently achieving rather than what is potentially feasible. Second the 'distance' between any firm's actual performance and its most efficient performance can only properly be measured if the costs of emissions are known relative to the value of marketed outputs. Recognizing this, Fare et al. (1993) have attempted to derive plant-specific shadow prices for emissions equal to the cost of the desirable output that must be forgone to reduce the undesirable output by one unit. Nestor and Pasurka (1994) use this methodology to derive shadow prices for six air pollutants and then incorporate these prices into a measure of productivity for 20 manufacturing sectors. The adjusted total factor productivity measure is for all sectors but one (non-metallic minerals) higher than the conventional measure. However, these shadow prices still represent marginal abatement costs, simply derived differently than in Pittman's work. In Pittman's study the abatement costs are the monetary costs of reducing pollution through installation of abatement measures or equipment. In the Fare et al. (1993) study the abatement costs are estimated as the sacrifice in marketed output necessary to achieve a unit fall in emissions. In neither case do the shadow prices represent the economic costs of pollution damages.

A better approach

The conventional methodology used to derive the multifactor productivity index can be extended in a straightforward way to take account of environmental damage from emissions of industrial waste products. Emissions are simply considered to be joint outputs of the industrial process and are included in the output index with weights determined by their marginal damage costs (as opposed to marginal abatement costs).

The multifactor productivity indexes estimated by the Bureau of Labor Statistics are based on an assumed production function of the form:

$$Q(t) = A(t).f [K(t), M(t), L(t)] \tag{19.1}$$

where $Q(t)$ stands for real output in year t; $K(t)$, $M(t)$ and $L(t)$ represent capital, material and labour inputs respectively; and $A(t)$ is an index of neutral technological progress. From this function the rate of productivity index can be estimated as:

$$A'(t)/A = Q'(t)/Q - [s_k K'(t)/K + s_m M'(t)/M + s_l L'(t)/L] \tag{19.2}$$

where the primed quantities represent rates of change with respect to time. That is, the rate of productivity change is defined as the difference between the growth rate of the output index and the growth rate of the index of inputs. In turn the input index is derived by weighting each factor of production by the proportional change in output that results from a small change in that input alone (technically the output elasticity). If there is perfect competition in both the input factor markets and the output markets and there are constant returns to scale, these weights are equal to the shares of the individual factors in total costs and consequently sum to unity.

Environmental residuals can be incorporated into the framework by defining total output W as the aggregation of marketed output Q and emissions E. 'Total' output exhibits a rate of growth equal to:

$$W'(t)/W = s_q Q'(t)/Q + s_e E'(t)/E \tag{19.3}$$

That is to say, the rate of change of total output is equal to a weighted average of the growth of output and the growth of emissions. The weights are equal to the shares of output and emissions in the total value of output. Of course since emissions are damaging, they have a negative value rather than a benefit and so have negative shadow prices. Qualitatively their impact on productivity is the same as that of input costs.

If A^* is defined as the index of technological progress for the joint output function W, then the growth rate of A^* is:

$$A^{*\prime}(t)/A^* = s_q Q'(t)/Q + s_e E'(t)/E - [s_k K'(t)/K + s_m M'(t)/M + s_l L'(t)/L] \tag{19.4}$$

Comparing (19.2) with (19.4) gives:

$$A^{*\prime}(t)/A^* = A'(t)/A + s_e [E'(t)/E - Q'(t)/Q] \tag{19.5}$$

which shows how the two productivity indicators are related. Because s_e, the relative contribution of emissions to total output, is negative whenever emissions grow more slowly than output, the new productivity index will increase more rapidly than the conventional index. Furthermore, assuming that output increases or stays constant, *any* decline in emissions will lead to a faster rate of productivity growth than measured by the conventional index. Should emissions increase more rapidly than marketed outputs however, the conventional index will overstate the productivity growth rate.

In other words the revised methodology takes into account a source of productivity growth that the conventional methodology completely misses: a more rapid growth in the value of total output due to a shift toward highly valued marketable products and away from negatively valued waste products. Undoubtedly this is as valid and potentially as important an efficiency gain as any other. In some industries it has been the most important source of improvement.

For example, in the pulp and paper industry, fibre and chemicals can either end up in the paper product or they can end up in the river causing all manner of damages. Getting these materials out of the river and into the product raises productivity at the plant level not only (potentially) by raising output but also by reducing environmental damage. This latter source of improvement is not currently measured either at the plant or aggregate economy level, even though reduced pollution may reduce the firm's own potential liabilities as well as damages to the public.

Calculating the new productivity measure requires an estimate of s_e, the share of emissions in total output. In turn s_e is determined by both the quantity of emissions and its shadow price, which represents the total economic damages another unit of emissions would do. Damages can be of many different kinds: increased illness, reduced recreational opportunities, impairment of materials and ecological impacts. These damages are estimated by various techniques that have been the subject of extensive research and refinement over the last 20 years (Freeman, 1993). The US EPA has supported a great deal of research into damage estimation and has itself been required by Congress and the Office of Management and Budget to study the benefits and costs of the environmental regulations it has proposed. Estimates of pollution damages can draw on a large store of regulatory analyses, cost–benefit assessments and agency-sponsored research.

The wide confidence limits within which damage values are usually expressed arise partly from the complexity of the underlying physical and biological processes, each of which can be described only within some margin of error. Another source of variation stems from differences in valuation methodologies used in various studies. Moreover, damages from a unit of emissions will vary substantially depending on timing and location, the hydrological or meteorological conditions in the vicinity of the emissions source, the size of the population affected and other factors. Though damage studies have been carried out in many locations, extrapolating the results to other places or generalizing to larger areas creates a source of possible inaccuracy.

Despite their imprecision, the strongest justification for drawing on estimates of emissions damages is simply that pollution imposes real economic costs. In a study of the electric power industry (reported below) it was found that sulphur emitted from generating plant smokestacks has a higher cost per ton to the economy than elemental sulphur sold as an industrial raw material. To dispense with damage estimates because they are imprecise is implicitly to impute a zero value for emissions – a figure that is contradicted by the substantial evidence of pollution's adverse effects. A better conclusion is that more accurate estimates of the economic costs of pollution would be useful in several policy contexts and warrant further efforts to refine the available numbers.

The basic estimates of marginal damage costs per ton of conventional air pollutants that were used in calculating revised productivity estimates are given in Table 19.2. These estimates were selected from the available research literature as representing the most applicable and most reliable of available research results. The underlying sources are discussed in some detail in a previous publication (Repetto et al., 1996). These estimates refer only to a single year but actual damages probably varied from year to year influenced by two offsetting trends. On the one hand in earlier years emission rates were higher and air

Table 19.2   Marginal damages from air emissions

Pollutant	Marginal damage value (US$1987 per metric ton)
Nitrogen oxides (NO$_x$)	841
Sulphur oxides (SO$_x$)	964
Particulates	3192
Volatile organic compounds (VOCs)	1460
Carbon monoxide	1
Lead	1384

Source: Repetto et al., 1996.

quality was poorer, suggesting that marginal damage costs per ton were probably higher. On the other hand in earlier years population and the size of the economy were lower, implying fewer victims of pollution and lower damage costs. To take these contradictory forces into account, these base estimates were extrapolated to other years using two alternative assumptions: one time series was generated by assuming that costs per ton were constant in real terms; another series was based on the assumption that costs per ton increased in proportion to the rise in real GDP. The former assumption leads to higher rates of productivity growth once emissions trends are incorporated into the index since the early years – when emissions were relatively high – weighted more heavily.

Still marginal damage estimates are available where at all, only for some kinds of emissions. Few credible estimates have been made so far of the global damages from an additional ton of carbon dioxide emitted. Nor are there estimates of the environmental costs of most toxic substances listed on the Toxic Release Inventory. Consequently the pollution damages included in the revised figures are only a subset of the total. That this subset proves to be empirically significant reinforces the argument that the conventional productivity measurement methodology is misleading. Complete implementation of the proposed alternative would lead to an even starker reappraisal of the conventional conclusions.

Empirical results

Using these and other data, case studies of the electric power and pulp and paper industries have re-examined the record of productivity growth since the early 1970s in two environmentally sensitive US sectors. Both sectors were substantially affected starting in the 1970s by the Clean Air and Clean Water Acts as well as later amendments, which forced them to reduce 'conventional' air and water emissions substantially. These emissions included atmospheric emissions of particulates, sulphuric and nitrogen oxides, volatile organic compounds and carbon monoxide along with water emissions of suspended solids and organic materials (expressed as biochemical oxygen demand). Regulations essentially required large emissions sources to install pollution abatement equipment on smokestacks and water effluent discharge streams.

Over this period the electric power industry in the US has had high emissions per unit of output. Coal has been the primary energy source and producing 1 kilowatt-hour of electricity (worth about US$0.06) at average efficiency requires burning about a pound of coal, almost all of which then becomes a waste product. Moreover, most of the resulting

Table 19.3 Multifactor productivity in the electricity sector – conventional and revised estimates

Period	Multifactor Productivity (Average Annual Percentage Change)		
	Conventional MFP %	Revised MFP (constant damage values %)	Revised MFP (damage values proportional to GDP %)
1970–79	−0.75	1.02	0.38
1979–91	−0.05	0.43	0.37
1970–91	−0.35	0.68	0.38

emissions are dispersed into the atmosphere, giving rise to a variety of environmental impacts. The costs of the main atmospheric emissions other than carbon dioxides amounted to about 15 per cent of the value of total outputs from the sector, including good and bad outputs together, but these environmental damages costs might have been as much as 30 to 60 per cent of total output value back in 1970.

Somewhat less polluting has been the pulp and paper industry, a highly energy-intensive industry that uses large quantities of water in the manufacturing process. Both air and water emissions from this sector are significant. Environmental damage costs declined from over 7 per cent of the value of total output in 1970 to less than 2 per cent in 1990. These estimated damage costs do not include the effects of chlorinated organic compounds such as dioxins emitted from the bleaching process and suspected of causing significant adverse health and ecological effects. Starting in the 1980s the release of these organochlorine compounds has been reduced.

Consequently one would expect that over the past 20 years the conventional productivity indicators would most distort the record of efficiency gains in the electric power sector where the mix of 'good' outputs (kilowatt-hours of electricity) to 'bad' outputs (atmospheric emissions) has changed dramatically. This expectation is correct. It is particularly important to take account of environmental damage costs in the electricity sector because they have been large by any standard but have changed dramatically. Data from the EPA show that particulate emissions declined substantially after 1970 (US EPA, 1994). Power plant emissions of sulphur oxides, which almost doubled during the 1960s and by 1970 amounted to over half of the country's sulphur emissions, were stabilized and then moderately reduced in the succeeding two decades. The pace of reduction increased during the 1990s. By contrast, emissions of nitrogen oxides and volatile organic compounds increased in the 1970s through the early 1990s, though less rapidly than the growth of electricity output.

According to the conventional measurement, productivity apparently declined over the period 1970–91 at an average annual rate of −0.35 per cent per year (Table 19.3). Environmental regulations, as noted above, have been blamed for much of that decline. Using a more reasonable measure that takes account of the substantial increase of kilowatt-hour per ton of emissions, the record shows that productivity actually increased by an average rate of 0.38 to 0.68 per cent per year. The discrepancy between the conventional and revised estimates is even wider for the 1970s, when the industry came into compliance with Clean Air and Water Acts. The main source of productivity gain was not

Table 19.4   Multifactor productivity in the pulp and paper sector – conventional and revised estimates

Period	Multifactor Productivity (Annual Average Percentage Changes)		
	Conventional MFP %	Revised MFP (constant damage values %)	Revised MFP (damage values proportional to GDP %)
1970–79	0.36	0.88	0.72
1979–90	0	0.08	0.07
1970–90	0.16	0.44	0.36

improved output per unit of labour, capital or fuel input, rather it was improved output per unit of emissions.

Similar but smaller distortions were found for the pulp and paper sector (Table 19.4). The conventional methodology understates the annual average rise in productivity in the pulp and paper industry by a factor of 2 or 3. Revised estimates suggest that productivity rose by 0.36 to 0.44 per cent per year over the 1970s and 1980s but according to the conventional methodology the rate of increase was only 0.16 per cent annually. By neglecting the dramatic shift in industry outputs toward higher valued products (that is, paper that people might wish to buy rather than pollutants they would rather do without), the conventional measure understates substantial gains in economic efficiency.

Taken together the case study results provide some indication of the extent of the measurement bias. For pollution-intensive industries such as chemicals, metals, non-metallic minerals, mining, oil and gas and transportation, the conventional productivity measure probably seriously mis-states (and probably significantly understates) productivity gains during the last three decades. Moreover, as these industries continue their efforts at pollution reduction, the current productivity efforts will continue to give distorted results.

Conclusions and recommendations
The implications for government agencies concerned with productivity growth and with environmental protection are clear. It is important to introduce an unbiased measure of productivity that captures accurately the economic impacts of environmental protection, on which we now spend about 2 per cent of GDP. An unbiased measure would record the costs averted as well as the costs incurred throughout the economy as environmental quality is protected. It would also record more accurately economic progress in environmentally sensitive industries. This measure need not necessarily supplant, but should surely supplement, existing statistics.

Preparing a revised record of productivity growth depends on adequate information. The US EPA has developed environmental databases, using them for economic analysis and to meet community right-to-know requirements. The EPA should develop and publish consistent time series on emissions of regulated air and water emissions on an industry-by-industry and firm-by-firm basis. In the past EPA has classified these data by emissions source category (for example 'industrial boilers'). They should be cross-classified by industrial sector using the standard industry classifications. Having such time series records of emissions trends on an industry, firm, and plant basis would be useful not only for estimating productivity growth but also for other important purposes, such as community information, financial disclosure and monitoring industries' environmental commitments within alternative regulatory approaches.

The EPA should also continue to improve and publish estimates of marginal pollution damages, the other essential information needed to revise the productivity measures. Moreover, the Agency should continue to fund and carry out research to estimate the marginal damage costs of pollutants for which current knowledge is lacking or inadequate. Such information will be highly useful not only for productivity measurement but also for priority setting in environmental policy, regulatory analysis and other purposes. Such information should be accessible to researchers outside government as well.

However, insight need not wait on the results of further research. Sufficient information on these building blocks of productivity measurement is already available to permit the conventional estimates to be revised, at least for environmentally sensitive industrial sectors. The Bureau of Labor Statistics, which is the lead agency for productivity measurement in the US government, should undertake a joint study in cooperation with the EPA to develop a revised set of productivity growth estimates for pollution-intensive sectors using the basic methodology set out in this chapter. These revised estimates should cover a period starting in the 1970s and extending throughout the 1990s, to capture the true impact of environmental protection on the US productivity. Moreover, these estimates should be brought up to date every few years to maintain a current record and duly published.

Informed discussion of the true impacts of environmental protection on the national economy is highly desirable. In the past, discussion has tended to be rather one-sided, since the costs of controlling pollution can be quantified and estimated much more readily than the costs of *not* controlling pollution. To promote informed discussion, the Council of Economic Advisors, in its widely read annual *Economic Report of the President*, should include revised estimates of productivity growth in environmentally sensitive industries in its chapters on regulatory reform and economic efficiency. The Council on Environmental Quality should also take notice of these revised estimates in its annual report.

As noted earlier, individual companies are also keenly interested in their own productivity records and in environmentally sensitive industries companies are searching for performance metrics and indicators that can adequately reflect their individual progress toward 'eco-efficiency'. The methodology outlined above can readily be adapted as a superior performance metric for such companies. It would measure efficiency gains in the use of conventional inputs – capital and labour as well as raw materials and intermediates. In addition it would measure progress in reducing emissions and effluents and efficiency in the use of environmental assets. Extending the measurement of productivity at the level of the facility and firm has been beneficial to managers as the success of EVA (economic value added) measures demonstrates. Extending such measures further to environmental assets should also be beneficial.

Estimates of damage costs would have to be particularized to each company's own sites and the composition of its waste streams. However, doing so would provide environmental managers with information useful for priority setting. Environmentally progressive companies are urged to begin tracking their own productivity improvements using this basic methodology.

Notes

1. Of course an alternative is to recalculate the productivity measure by excluding both environmental costs and environmental benefits. In some respects this is the approach taken by Gray (1987) and Gray and Shadbegian (1993 and 1995). They attempt to pare down the productivity indicator to make it simply a measure of the efficiency of inputs devoted exclusively to the production of the marketed output. The real difficulty of this approach, as they recognize, is identifying which costs contribute only to pollution reduction.
2. Moreover, the notion of 'energy balance' (the second law of thermodynamics) embodies a similar principle for the conversion of fuels into noise and thermal pollution. The environmental effects of such converted energy are typically relatively small.

References

Baily, M.N. (1981), 'Productivity and the Services of Capital and Labour', *Brookings Papers on Economic Activity*, **1**, 1–50.
Baily, M.N. (1986), 'What Has Happened to Productivity Growth?', *Science*, **234**, 443–51.
Barbera, A.J. and McConnell, V.D. (1990), 'The Impact of Environmental Regulations on Industry Productivity: Direct and Indirect Effects', *Journal of Environmental Economics and Management*, **18**(1), 50–65.
Bureau of Labor Statistics (2005), 'News Release: Multifactor Productivity Trends', 2002, Washington, DC.
Christainsen, G.B. and Haveman, R.H. (1981), 'Public Regulations and the Slowdown in Productivity Growth', *American Economic Review*, **71**(2), 320–5.
Conrad, K. and Morrison, C.J. (1989), 'The Impact of Pollution Abatement Investment on Productivity Change: An Empirical Comparison of the US, Germany and Canada', *Southern Economic Journal*, **55**(3), 684–98.
Crandall, R.W. (1981), 'Pollution Controls and Productivity Growth in Basic Industries', in Cowing, T.J. and Stevenson, R.F. (ed.), *Productivity Measurement in Regulated Industries*, New York, NY: Academic Press Inc., pp. 347–68.
Denison, E.F. (1979), *Accounting for Slower Economic Growth: The US in the 1970s*, Washington, DC: Brookings Institution.
Fare, R., Grosskopf, S. and Pasurka, C. (1986), 'Effects on Relative Efficiency in Electric Power Generation Due to Environmental Controls', *Resources and Energy*, **8**, 167–84.
Fare, R., Grosskopf, S., Lovell, C.A.K. and Pasurka, C. (1989), 'Multilateral Productivity Comparisons When Some Outputs are Undesirable: A Non-parametric Approach', *Review of Economics and Statistics*, **71**(10), 90–98.
Fare, R., Shawna, G., Knox-Lovell, C.A., Yaisawarng, S. (1993), 'Derivation of Shadow Prices for Undesirable Outputs: A Distance Function Approach', *Review of Economics and Statistics*, **75**, 374–80.
Fischer, S. (1988), 'Symposium on the Slowdown in Productivity Growth', *Journal of Economic Perspectives*, **2**, 3–7.
Freeman, M.A. III (1993), *The Measurement of Environmental and Resource Values: Theory and Methods*, Washington DC: Resources for the Future.
Gollop, F.M. and Roberts, M.J. (1983), 'Environmental Regulations and Productivity Growth: The Case of Fossil-fuelled Electric Power Generation', *Journal of Political Economy*, **91**, 654–74.
Gordon, R.J. (1981), 'Productivity and the Services of Capital and Labour', *Brookings Papers on Economic Activity*, **1**, 51–8.
Gray, W.B. (1987), 'The Cost of Regulation: OSHA, EPA and the Productivity Slowdown', *American Economic Review*, **77**(5), 998–1006.
Gray, W.B. and Shadbegian, R.J. (1993), 'Environmental Regulation and Manufacturing Productivity at the Plant Level', Washington DC: US Department of Commerce, Center for Economic Studies, Discussion Paper 93–6.
Gray, W.B. and Shadbegian, R.J. (1995), 'Pollution Abatement Costs, Regulation, and Plant-level Productivity', Cambridge, MA: National Bureau of Economic Research, Working Paper W4994.
Jorgenson, D.W. and Wilcoxen, P.J. (1990), 'Environmental Regulation and US Economic Growth', *Rand Journal of Economics*, **21**, 314–40.
Joshi, S., Lave, L., Shih, J.S. and McMichael, F. (1997), *Impact of Environmental Regulations on the US Steel Industry*, Pittsburgh, PA: Carnegie Mellon University.
Munnell, A.H. (1990), 'Why Has Productivity Growth Declined? Productivity and Public Investment', *New England Economic Review*, January/February, 3–22.
Nestor, D.V. and Pasurka, C.A. (1993), *Productivity Effects of Environmental Regulations: A Cross-country Comparison*, Washington, DC: Economic Analysis and Research Branch, US Environmental Research Agency.
Nestor, D.V. and Pasurka, C.A. (1994), *Productivity Measurement with Undesirable Outputs: A Distance Function Approach*, Washington, DC: Economic Analysis and Research Branch, US Environmental Research Agency.
Norsworthy, J.R., Harper, M.J. and Kunze, K. (1979), 'The Slowdown in Productivity Growth: Analysis of Some Contributing Factors', *Brookings Papers on Economic Activity*, **2**, 387–421.
Oates, W., Palmer, K. and Portney, P. (1993), 'Thinking about the Porter Hypothesis', Washington DC: Resources for the Future Discussion Paper.

Pittman, R.W. (1983), 'Multilateral Productivity Comparisons with Undesirable Outputs', *Economic Journal*, **93**, 883–91.

Porter, M.E. (1990), *The Competitive Advantage of Nations*, New York: Free Press.

Porter, M.E. (1991), 'America's Green Strategy', *Scientific American*, April, 168.

Repetto, R., Faeth, P., Rothman, D. and Austin, D. (1996), *Has Environmental Protection Really Reduced Productivity Growth? We Need Unbiased Measures*, Washington, DC: World Resources Institute.

Robinson, J.C. (1995), 'The Impact of Environmental and Occupational Health Regulation on Productivity Growth in US Manufacturing', *The Yale Journal on Regulation*, **12**, 346–87.

Schmidheiny, S. with members of the Business Council for Sustainable Development (1992), *Changing Course: A Global Business Perspective on Development and the Environment*, Cambridge, MA: Massachusetts Institute of Technology.

US Environmental Protection Agency (EPA) (1994), *National Air Pollution Emission Trends, 1900–1993*, Washington, DC: Office of Air Quality Planning and Standards.

US Environmental Protection Agency (EPA) (1996), *The Benefits and Costs of the Clean Air Act, 1970–90*, Washington, DC: Government Printing Office.

20 Environmental accounting
Stefan Schaltegger and Roger Burritt

Introduction

Environmental and related issues are of increasing interest to managers of corporations for several reasons. First, demand for environmental accounting information from various interest groups is on the rise. The environmental impacts of corporate activities are increasingly being monitored by government agencies, local communities, non-governmental organizations, the finance sector and others (Schaltegger and Burritt, 2000). Second, the need to manage environmental liabilities has progressed in parallel with an increase in the amount of environmental legislation and the associated need for compliance. Third, the cost of technology associated with the generation of environmental information has been declining. Finally, growing recognition of the importance of links between business and its environmental impacts is generating a range of government incentives to encourage clean and green corporate activities that act in the interests of society at large.

Today a large number of companies in developed countries collect, use and distribute information related to the natural environment. This reflects a fundamental change compared with a decade ago (for example, Gray et al., 1996, p. 81; Schaltegger and Burritt, 2000, p. 30). Pressures from external stakeholders concerned about the impact of corporate activities on the environment have increased, the costs of environmental impacts have risen substantially (for example, through penalties established in new environmental legislation), and investments in environmentally benign processes and products have been encouraged by tighter environmental regulation and stakeholder pressure.

Such pressures together with the incorporation of environmental information in accounting systems have led to the emergence of a range of perceptions about the concept and practices of environmental accounting (for example, Burritt, 1997; EPA, 1995; Gray et al., 1993, 1996; Parker, 1999; Schaltegger, 1996; Schaltegger and Burritt, 2000; Schaltegger and Stinson, 1994; Schaltegger et al., 1996). An articulated framework for environmental accounting is thus essential for a pragmatic understanding of the linkages between these different approaches.

Framework and overview

Environmental accounting concerns an integrated set of systems that include monetary and non-monetary (physical) approaches to both internal and external corporate accounting (for example, Bennett and James, 1998; Burritt et al., 2002; ECOMAC, 1996; IFAC, 1998; Schaltegger and Burritt, 2000). As illustrated in Figure 20.1, these can be divided into two fundamental accounting systems: environmental management accounting (EMA) and external environmental accounting (EEA), which includes 'other' environmental accounting systems. Figure 20.1 provides an overview of the dimensions of these related systems.

It is generally agreed that there are two main groups of environmental impacts related to the activities of corporations (Schaltegger and Burritt, 2000, p. 58). The first group includes environmentally related impacts on corporate economic position and perform-

internal

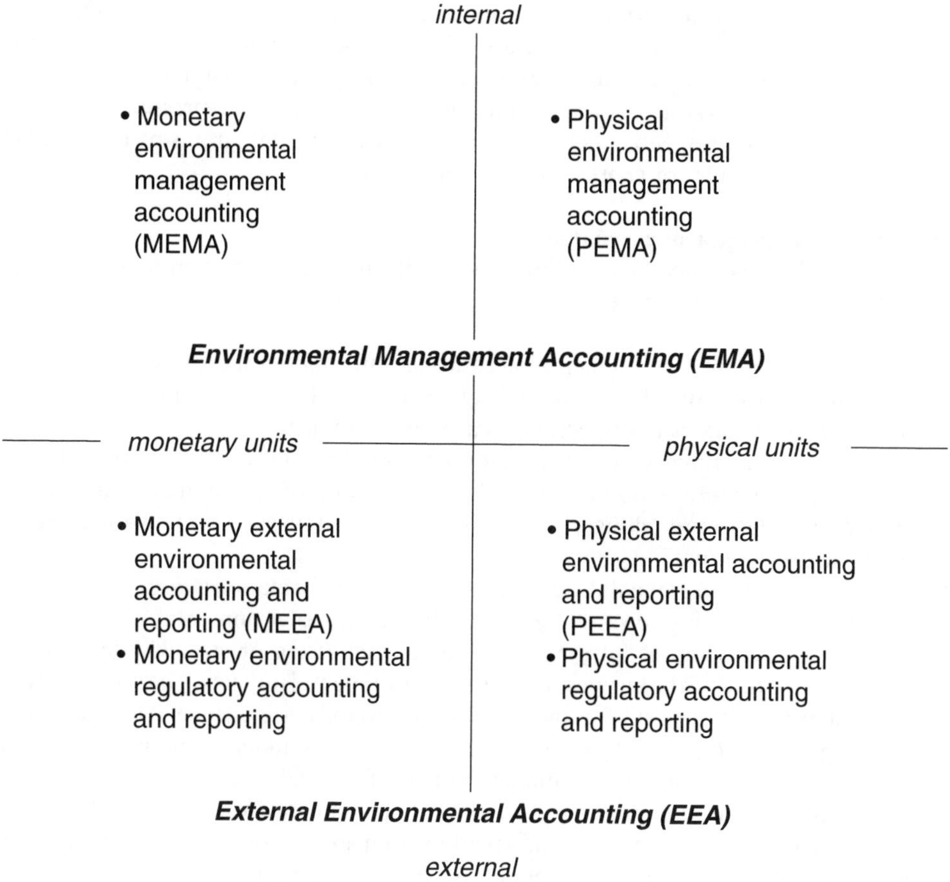

- Monetary
  environmental
  management
  accounting
  (MEMA)

- Physical
  environmental
  management
  accounting
  (PEMA)

Environmental Management Accounting (EMA)

——— monetary units ——————————— physical units ———

- Monetary external
  environmental
  accounting and
  reporting (MEEA)
- Monetary environmental
  regulatory accounting
  and reporting

- Physical external
  environmental accounting
  and reporting
  (PEEA)
- Physical environmental
  regulatory accounting
  and reporting

External Environmental Accounting (EEA)

external

Source: Modified from Bartolomeo et al., 2000, p. 33.

Figure 20.1   Environmental accounting systems

ance, whilst the second group includes corporate impacts on environmental (sometimes termed 'ecological') systems.

Environmentally related impacts on corporate economic systems are captured by and reflected through 'monetary environmental information'. Monetary environmental information addresses the effect of all corporate-related impacts on the past, present or expected future financial position and performance of the corporation (in other words, its financial stocks and flows), and is expressed in monetary units. Examples of monetary environmental information include measures expressed in terms of expenditure on cleaner production, the cost of fines for breaching environmental laws and monetary valuation of environmental assets and liabilities. Because they are based on the methods of conventional accounting, monetary environmental accounting systems can be thought of in terms of an extension of conventional accounting and may include developing, refining and expanding the scope of accounting in monetary units.

Impacts of corporate activities on environmental systems are reflected in 'physical environmental information'. At the corporate level, physical environmental information includes all past, present and future material and energy amounts that have an impact on ecological systems. Physical environmental information is always expressed in physical units – such as kilograms, cubic metres or joules (for example, kilograms of material per customer served or joules of energy used per unit of product).

Environmental management accounting
Environmental management accounting (EMA) information can be considered relevant to the needs of managers in three ways:

1. *Attention directing.* Environmental accounting information indicates critical environmental impacts and their potential significance to the organization. It draws the attention of managers to opportunities as well as problems.
2. *Problem-solving.* Environmental accounting provides information about potential alternative courses of action that include consideration of environmental impacts as part of the decision-making process. Such information is an important input to strategic decision-making.
3. *Scorekeeping.* Environmental accounting provides information about whether the corporation is moving towards targets established for environmental impact reduction through continuous improvement, value engineering or strategic decisions to change the nature of the business to reduce environmental impacts. The scorecard can be used internally as part of a 'balanced scorecard' of performance. Alternatively, it can be provided to external parties to help improve their understanding of corporate actions and impacts on the environment as part of the dialogue with stakeholders. In the absence of good communication about corporate environmental performance these parties need to draw their information from sources outside the organization. The importance of providing stakeholders with accurate information is their capacity to reduce or destroy the social legitimacy of a corporation.

Environmental management accounting can be defined as a generic term that includes both monetary environmental management accounting and physical environmental management accounting. Monetary environmental management accounting (MEMA) addresses the environmental aspects of corporate activities expressed in monetary units. It generates monetary information for internal management use (for example, payment of fines for breaking environmental laws and investment in capital projects that improve the environment). In terms of its methods, MEMA is an extension or adaptation of conventional management accounting to address the environmental aspects of corporate activities. This all-encompassing tool not only provides the basis for most internal management decisions but also addresses the issue of how to identify, track and treat costs and revenues incurred as a result of the corporation's impact on the environment (Schaltegger and Burritt, 2000, p. 59). Monetary environmental management accounting contributes to strategic and operational planning, acts as a control and accountability device and provides the main systematic source of information for decisions about how to achieve desired corporate goals or targets (Schaltegger and Burritt, 2000, p. 90).

Physical environmental management accounting (PEMA) also acts as an information tool for internal management decisions about corporate environmental impacts. However, in contrast to MEMA, it focuses on a company's impact on the natural environment and is expressed in terms of physical units, such as tonnes of carbon dioxide emissions (Schaltegger and Burritt, 2000, pp. 61–3). As an internal environmental accounting approach, PEMA has several functions (Schaltegger and Burritt, 2000, p. 261):

- as an analytical tool designed to detect ecological strengths and weaknesses;
- as an decision-support technique concerned with highlighting relative environmental quality;
- as a measurement tool that is an integral part of other environmental measures such as eco-efficiency;
- as a tool for direct and indirect control of environmental consequences;
- as an accountability tool providing a neutral and transparent base for internal and, indirectly, external communication;
- as a tool with a close and complementary fit to the set of tools being developed to help promote ecologically sustainable development.

Table 20.1 categorizes these environmental management accounting systems according to two dimensions: internal versus external information and monetary versus physical representation of that information.

Different types of managers rely on and have their performance assessed using physical information, monetary information or a combination of the two. For example, managers in the corporate environmental management department have various goals including:

- identifying environmental improvement opportunities;
- prioritizing environmental actions and measures;
- environmental differentiation in product pricing, mix and development decisions;
- transparency about environmentally relevant corporate activities;
- meeting the claims and information demands of critical environmental stakeholders to ensure resource provision and access;
- justifying environmental management division and environmental protection measures.

Physical environmental management accounting information can be used to guide and advise decisions in which these issues arise.

Typical styles of information used by different managers are provided in Table 20.2. All managers also need qualitative information and so this is not separately detailed.

Three additional dimensions of environmental management accounting tools are important, especially for managers of small- to medium-sized enterprises (see Table 20.1):

- *Time frame.* The chronological orientation of the tool (that is, past, present or future).
- *Duration.* The length of time addressed by the tool (that is, short term or long term).
- *Schedule.* The regularity by which the tool gathers information (that is, ad hoc versus routine).

Table 20.1   Illustrative environmental management accounting tools

	Environmental Management Accounting (EMA)			
	Monetary environmental management accounting (MEMA)		Physical environmental management accounting (PEMA)	
	Short term focus	Long term focus	Short term focus	Long term focus
1. Past/present orientated				
Routinely generated information	Environmental cost accounting (e.g. variable costing, absorption costing, and activity-based costing)	Environmentally induced capital expenditure and revenues	Material and energy flow accounting (short-term impacts on the environment, product, site, division and company levels)	Environmental (or natural) capital impact accounting
Ad hoc information	Ex post assessment of relevant environmental costing decisions	Environmental lifecycle (and target) costing Post investment assessment of individual projects	Ex post assessment of short-term environmental impacts (e.g. of a site or product)	Lifecycle inventories Post investment assessment of physical environmental investment appraisal
2. Future orientated				
Routinely generated information	Monetary environmental operational budgeting (flows) Monetary environmental capital budgeting (stocks)	Environmental long-term financial planning	Physical environmental budgeting (flows and stocks) (e.g. material and energy flow activity-based budgeting)	Long-term physical environmental planning
Ad hoc information	Relevant environmental costing (e.g. special orders, product mix with capacity constraint)	Monetary environmental project investment appraisal Environmental lifecycle budgeting and target pricing	Relevant environmental impacts (e.g. given short-run constraints on activities)	Physical environmental investment appraisal Lifecycle assessment of specific projects

Source: Based on Schaltegger and Burritt, 2000.

Different managers have different time frames and different demands for routine or ad hoc information. For example, top managers tend to have a longer time frame and need ad hoc information for critical decisions. A range of cost environmental accounting tools is available to different managers. Each tool is likely to be of use to different managers, given their particular level in the business and their own specialization. Only if managers have

Table 20.2   Generic aims and objectives of different types of managers

Relevant EMA users	Corporate Environmental Management Accounting System	
	Basic goals	Type of information desired
Top management	Long-term profitability and survival of company Securing legal compliance with minimal cost to the corporation Realization of all economically beneficial environmental protection measures Securing the provision of resources from the critical stakeholders	Highly aggregated financial and strategic (qualitative and quantitative) information on the business environment and the company's performance
Accounting and finance department	Identifying and realizing cost-saving potential Transparency about cost-relevant (environment-related) corporate activities Transparency about the impact of (environment-related) activities on the income statement and/or balance sheet Reduction of environmentally induced risks (Bennett and James, 1998, pp. 34ff) Compliance with accounting regulations Maximization of shareholder value	Financial measures about corporate activities (e.g. cost-, income- and balance sheet-related issues, risk assessments, investment decisions, mergers and acquisitions, and so on) Financial information on the value and economic performance of the enterprise
Environmental department	Identifying environmental improvement opportunities Prioritizing environmental actions and measures Environmental differentiation in product pricing, mix and development decisions Transparency about environmentally relevant corporate activities Meeting the claims and information demands of critical environmental stakeholders to ensure resource provision and access Justifying environmental management division and environmental protection measures (Bennett and James, 1998, pp. 34ff)	Physical measures on material and energy flows and stocks and related processes and products, and their impacts upon the environment
Health and safety department	Safeguarding the safety, health and welfare of employees at work from environmental accidents and disasters	Physical measures of health and safety
Quality department	Meeting the (environmental) product requirements of customers at the minimum cost for a given level of product quality	Information on cost of quality Physical measures of technical product requirements
Human resources department	Job-related (including environmental) concerns of employees Remuneration, including rewards for good environmental performance Physical jobs allocated and job conditions monitored	Information on financial rewards Physical information on turnover, satisfaction and morale
Legal department	Ensuring (environmental) legal compliance by the company's operations	Physical measures Qualitative compliance information

Table 20.2   (continued)

Relevant EMA users	Corporate Environmental Management Accounting System	
	Basic goals	Type of information desired

R&D and design department	Development and design of marketable products and services
	   Reducing (environmental) risks of investments
	   Development of improved production processes
		Strategic information about market demands
		   Financial information about costs of new products and services
		   Information on technical feasibility and environmental impacts of newly designed products and services

Corporate marketing and public relations department	Meeting external information demands of critical stakeholders
	   Meeting claims and information demands of shareholders, other economic stakeholders (including those interested in environmental reports)
	   Developing a green image of the company and its products
		Information about stakeholder claims
		   Physical and financial information on the company's environmental impacts and efforts for pollution reduction and prevention

Production management	Task control over operations
	   Optimizing energy and material consumption
	   Reduction of environmentally induced risks
		Information on material and energy flows and process records

Purchasing department	Efficient procurement of the inputs for corporate operations
	   Establishing and securing favourable relationships with suppliers
		Information on quality and environmental properties of the goods purchased
		   Financial information on prices

Logistics	Efficient organization, collection, storage and physical distribution of goods and products	Physical measures (e.g. on distribution means and storage facilities and related environmental impacts)

Marketing and sales department	Increasing sales and attracting and satisfying buyers
	   Provision of means by which buyers can purchase the product
	   Inducing customers to buy the enterprise's products through the tools of the marketing-mix (especially pricing, distribution and communication)
		Information on operational market conditions (e.g. pricing, competitor activities etc.)
		   Information on customer demands

Disposal and recycling	Efficient disposal and recycling of wasted or used material
	   Minimization of wastes to be treated, especially hazardous wastes
		Physical measures of the properties of disposable and recyclable goods
		   Technical information on treatment and recycling options

Source: Schaltegger et al., 2001.

a clear understanding of MEMA and PEMA tools will they be in a position to adopt the most appropriate environmental accounting tool for the particular decision-making or accountability setting. Environmental considerations relevant to corporate decision-making may include:

- adoption of 'green' options by product and production managers, when these are available;
- eco-efficiency improvements that reduce the volume of environmentally harmful materials used;
- use of environmental performance indicators in staff appraisal;
- government subsidies that are environmentally damaging and that may be removed in the future;
- corporate impacts of future environmental taxes and tightening regulations designed to bring corporations closer to tracking the full cost of their activities;
- impact of the corporation on 'environmental capital', including biodiversity and land, water and air quality;
- corporate impacts on, and contribution to, the goal of a sustainable society.

These settings span internal and external decisions and accountability of business and illustrate how EMA information interacts with internal and well as external uses.

External environmental accounting
The main audience for external environmental accounting (EEA) disclosures is external stakeholders with an interest in the corporation. External environmental accounting mirrors environmental management accounting inasmuch as it also records information in both monetary and physical terms.

Until recently monetary environmental impacts on business were considered to be addressed quite adequately under existing accounting and reporting standards and regulations. However, the increasing number of environmental issues has generated substantial financial consequences for companies. Therefore, external 'users' of financial statements have started to influence standard-setting bodies and regulators in order to induce them to alter existing, and to create new, reporting standards, regulations and guidelines. Three main groups (or 'providers') directly influence how corporate management addresses environmental issues in financial reports:

- *Regulatory bodies.* The US Securities and Exchange Commission (SEC) is one example of a regulatory body. The SEC, which was founded in 1934 in response to the Wall Street Crash, supervises the US securities exchanges. The SEC's mission is to administer federal securities laws and to issue rules and regulations that provide protection for investors and ensure that the securities markets are fair and honest. This is accomplished primarily by promoting adequate and effective disclosure of information to the investing public.
- *Standard-setting bodies.* Prominent standard-setting bodies include the International Accounting Standards Board (IASB) and the Financial Accounting Standards Board (FASB). The IASB was formed as a committee in 1973 to harmonize and improve financial reporting. It achieves this primarily by developing

and publishing International Accounting Standards (IAS). These standards are developed through an international process that involves national standard-setting bodies, the producers and users of financial statements and accountants all over the world (IASC, 1995, p. 7). The FASB is a private organization founded in 1973 (FASB, 1994) and has been responsible for the US generally accepted accounting principles (GAAP). The US government can influence the GAAP through the SEC. International accounting standards are currently being accepted by a broad range of countries looking for reliable standards in the wake of accounting problems associated with US standards, such as the Enron and WorldCom debacles.

- *Other stakeholders.* These may include professional accounting organizations, industry associations and international initiatives such as Canadian Institute of Chartered Accountants (CICA, 1992), and the Global Reporting Initiative (details of the Global Reporting Initiative are available online: http://www.globalreporting. org).

Regulators have the strongest direct influence on environmental disclosures as they create legally enforceable requirements. One example is the US Toxic Release Inventory, which requires a large range of specified industries to report releases of over 650 chemicals (Khanna et al., 1998).

In addition, a number of voluntary environmental reporting initiatives exist. Voluntary public environmental reporting initiatives enable businesses to self-regulate their external disclosures about environmental impacts. Some organizations have expanded the concept of public environmental reporting to incorporate the social and environmental aspects of operations, combined with traditional economic reporting. This has resulted in the production of 'triple bottom line' (TBL), or sustainability, reports (for example, Commonwealth of Australia, 2003).

Monetary external environmental accounting
A number of conventional financial accounting issues carry over to monetary external environmental accounting (MEEA) when the same conventions and principles are adhered to. These include decisions about, for example, whether to capitalize costs as assets, or how to treat indirect environmental costs (or environmental 'overheads').

Capitalize or expense monetary environmental costs   This is an ongoing and controversial issue driven by the choice made by accountants of 'cost' rather than 'market value' as the way to record transactions, transformations and events in the corporate accounts.

In practice, there is no definitive answer to the question of whether to capitalize environmental costs as assets, or whether to write them off against profits in the year they are incurred. From a strict economic perspective, capitalization of costs should only be allowed if these costs contribute to additional future economic benefits beyond those originally assessed. These are the incremental benefits that accrue if an alternative course of action is contemplated. However, in special cases, the costs of pollution clean-up or prevention may qualify as assets if they are absolutely necessary for the company to stay in business, even though they do not affect expected future cash flows. In this case expenditure is securing the value of future assets, a value that would fall, perhaps to a 'forced-sale' value, if the expenditure were not made.

A further issue relates to a movement from end-of-pipe improvement to precautionary investment in environmental improvement. If a firm is using old-style end-of-pipe technology, it is likely to be much easier to isolate the costs of environmental compliance. However, the more a firm adopts cleaner production approaches, the more difficult it becomes to identify its environmental compliance costs. If environmental management decisions are built into the whole production process and produce both environmental improvements and cost savings, it is not easy to separate environmental management costs from expenditure designed to return a commercial profit (A'Hearn, 1996).

From an environmental point of view, on the other hand, capitalization in the accounts should be favoured if pollution prevention creates future environmental benefits. Furthermore, capitalization facilitates its write off as an expense over a number of years, and thus enhances long-term thinking (Williams and Phillips, 1994).

The International Accounting Standards Committee has chosen the increased future benefits approach (IASC, 1995, IAS 16), while the Féderation des Expertes Comptables Européens and the Emerging Issues Task Force of the Financial Accounting Standards Board have adopted the additional cost of future benefits approach. This issue reveals to environmental managers just one type of problem that is associated with applying conventional accounting to environmental issues.

Environmental assets   Environmental expenditures on environmental capital are considered to represent a depreciable environmental asset under environmental accounting. Where there is no market the question arises of whether an asset can be recognized, and what measure can be used if an asset is to be recognized. Cost is suggested as one substitute. Net Present Value is another. The former is often out of date for the user of the information, whereas the latter relies too much on subjective assessments of future cash inflows and outflows, discount rates and the economic life of the asset being depreciated.

Environmental liabilities   In the past, environmental issues were not a high priority for management until they showed up as liabilities in the accounts. Yet some environmental liabilities have exceeded even the worst-case scenarios of management. In the US the Comprehensive Environmental Response, Compensation and Liability Act of 1980 aims at cleaning up abandoned waste sites (known as 'Superfund' sites). The liability is regarded as joint and retrospective for all costs incurred in the clean-up operation. All parties involved can be held liable for the total costs of remediating the landfill. The liability exists even if the activity that caused the environmental problem was legal and the Superfund legislation did not exist at the time. The EPA can require any person or company involved to bear the total of all remedial costs, no matter how much environmental damage each respective party has actually caused – termed 'joint and several environmental liability'. The Superfund example shows that environmental liabilities are one way of internalizing external costs. Even banks that have given mortgages or have managed properties secured for clean-up can be held liable as mortgagees in possession (Ernst and Young, 1992; Skellenger, 1992).

The main questions regarding the treatment of the environmental liabilities of a company are:

● What are (contingent) environmental liabilities?

- Should they be recognized? And if so, when?
- How can they be measured?
- If and when should they be disclosed?

Detailed discussion of these issues can be found in Schaltegger and Burritt (2000, p. 182). An 'environmental liability' is an obligation to pay future expenditures to remedy environmental damage that has occurred because of past events or transactions, or to compensate a third party that has suffered from the damage. A 'contingent environmental liability' is an obligation to remedy environmental damage dependent on the occurrence or non-occurrence of one or more uncertain future events, or to compensate a third party that would suffer from such damage. Examples of (contingent) environmental liabilities that can emerge from corporate activities include energy emissions (for example, heat, radioactive or electromagnetic emissions, noise), visual impact (because of buildings, for example), surface water contamination (such as those originating from point sources like industrial processes), air emissions (including fugitive emissions and transportation activities as well as sound, noise and light) and so on.

As a rule, environmental liabilities should be recognized in financial statements if they are material and if the liabilities or the events leading to the liabilities are probable and can be reliably measured (or reasonably estimated).

Management has a large discretionary latitude in deciding when to recognize a liability even if it is probably likely to occur. The main problem with environmental liabilities is the measurement, or estimation, of their amount. A liability must be measured or reliably estimated in order to qualify for recognition in the main body of a financial statement. Liabilities may be recognized even if they cannot be reliably measured, however. This is usually accomplished by making reserves, provisions or charges to income. Finally, disclosure of environmental liabilities is needed for an assessment of the monetary environmental risks facing the business.

Tradable emission allowances   In certain circumstances businesses are provided with government permission to pollute the environment. The total amount of pollution is strictly limited through the total number of pollution permits issued and the amounts of pollution permitted, usually calculated by natural scientists based on the perceived carrying capacity of environmental media. Often a process of 'grandfathering' is used to issue initial allowances; that is to say, existing polluters are provided with the permits first. From a monetary perspective the importance of emission permits is that they can often be sold in the market to other parties seeking the right to pollute. These permits will be reflected in the financial accounts, so that questions arise with respect to:

- whether emission allowances are assets or expenses;
- whether they should be recognized;
- how they should be measured;
- how they should be disclosed.

Before the process of putting monetary measures on environmental assets and liabilities can be undertaken, some physical assessment of the emissions or environmental media is required. Physical external environmental accounting addresses this issue.

Physical external environmental accounting

Environmental information has become part of marketing and public relations for many companies. Dubious claims and an improper use of information by some businesses in the past, especially in relation to the publication of positive environmental information and the neglect of negative environmental disclosure, have reduced the incentives for other companies to improve their own environmental record (Schaltegger, 1997). Introduction of a basic standard for physical external environmental accounting (PEEA) is needed in order to secure disclosure of a minimum standard of information quality.

The international form of environmental accounting and reporting has been led by the Féderation des Expertes Comptables Européens (see Adams et al., 1998), the United Nations Intergovernmental Working Group of Experts on International Standards of Accounting and Reporting, United Nations Division for Sustainable Development and Coalition of Environmentally Responsible Economies with its Global Reporting Initiative (GRI) on sustainability. Physical external environmental accounting standards are slowly being developed and harmonization is occurring. However, general reporting of PEEA information is only gaining strength in large companies, rather than in small and medium-sized enterprises. There is some confusion between macro PEEA indicators and business-level performance indicators as promoted by the GRI (2002). The recommendations of the GRI are based on a number of principles common to the preparation of accounts in business, such as the reporting entity, period and materiality (see Schaltegger et al., 2003, pp. 287ff).

Materiality in economic, environmental and social reporting is dependent on what is relevant either to reporting organizations or to external stakeholders. Those who believe a particular 'generally applicable' item is not applicable are asked to explain their reasoning. The reporting organization is asked to determine what to report on the basis of both applicable laws and the process of stakeholder dialogue and engagement. The economic, social and integrated indicators are presented for testing and experimentation by all reporting organizations. At this time, they are less developed than the environmental indicators.

The application of the 'materiality' concept to economic, environmental and social reporting is more complex than in financial reporting. In contrast to financial reporting, percentage-based or other precise quantitative materiality yardsticks will seldom be appropriate for determining materiality for GRI reporting purposes. Instead materiality is heavily dependent on the nature and circumstances of an item or event as well as its scale or magnitude. For example, in environmental terms, the carrying capacity of the receiving environment (such as a watershed or air shed) will be just one among several factors in the materiality of the release of one tonne or one kilogram of waste, air emissions or effluent. Similarly, health and safety information is likely to be of considerable interest to GRI reporting users despite its typical insignificance in traditional financial accounting terms. Different stakeholders may not agree on what is material. For the reporting organization the results of research into user needs as well as continuing interaction with stakeholders is necessary for determining materiality.

The GRI is not currently designed to tell the reader of a report what the actual situation is as far as the environment is concerned. The notion of the 'precautionary principle' is used to suggest that where scientific evidence about environmental impacts is unavailable, then preservation of the environment should dominate decision-making – erring on the side of caution as far as the environment is concerned. It would be far better for reasons of transparency to report matters as they are and then let policy-makers decide whether the environmental concerns outweigh economic considerations. To do otherwise is to invite manipulation of information, rather than 'tell it like it is'. Being cautious about upstream and downstream impacts and about lifecycle analysis would simply destroy the main purpose of these tools.

The principles guiding the form of environmental accounting and reporting are thus to be treated with great caution as they encourage manipulation of reported information. Debate about the principles of environmental accounting and reporting continues and is as yet unresolved.

The physical environmental accounting information reported externally has been much analysed and criticized because of the underlying fear that the environment is still being degraded, that natural resources are dwindling and that society is inching further away from sustainability all the time (for example, Maunders and Burritt, 1991). Many empirical studies have been undertaken recently, the results of which confirm that larger corporations are both engaging with environmental issues and looking to reduce their relative impact on the environment, even though growth in business means that absolute environmental impacts may get worse (for example, the increase in carbon dioxide emissions that are thought to lead to global warming). Potential advantages suggested for those businesses reporting environmental information include:

- improved public relations and corporate legitimacy;
- meeting the demands of shareholders;
- anticipation of tightening legal requirements for disclosure;
- gaining a competitive advantage;
- improved relationships with customers and environmental interest groups.

At the moment much improvement in the reporting practice and standardization is still necessary to create the 'true and fair view' in corporate environmental reports (e.g. Schaltegger and Burritt, 2000, p. 332).

Importance of environmental accounting and outlook
Environment-related issues are of increasing importance when considering the financial position of many companies, especially companies that operate in industries with a high potential to have a significant impact on the environment. Environmental accounting has developed to provide management and other stakeholders with an information base that helps make business impacts on the environment and environmental impacts on the business more transparent in decision-making and accountability processes.

To support the economic basis of decision-making by stakeholders, such as shareholders, creditors, suppliers and financiers, companies with environmentally sensitive businesses need to disclose monetary amounts induced by environmental impacts in their conventional financial reports. Likewise, to support improved accountability for corporate impacts on the environment, businesses need to disclose physical information about environmental impacts.

Management still has a large degree of discretion when deciding which environmental issues to recognize, how to measure these and what to disclose. During the 1990s, environmental accounting started to become an important issue in financial markets – hitherto a very conservative sector of the economy as far as environmental issues were concerned – and growth in the socially responsible investment sector added to this general trend. The importance of environmental accounting for other stakeholders cannot be played down either, as environmental accounting has the potential to be an enabler of stakeholder dialogue (Schaltegger et al., 2003, p. 144). Management, non-governmental organizations, shareholders, governments and other groups all are seeking information about the environmental risks and returns from business – information that environmental accounting systems are designed to capture, track and report.

For the future development of environmental accounting various aspects have to be considered for further improvement to support management effectively:

- Environmental accounting and reporting are but one element in the process of developing and institutionalizing a mindset that identifies environmental impacts and drives corporations towards improved environmental performance. On its own, the information provided by environmental accounting will not help managers to protect business from increasing environmental risk. It must be developed in combination with top-level support, corporate policies that integrate environmental policies, raise employee awareness, strategic environmental management and increase ownership of environmental problems (Hunt and Auster, 1990).
- The initial focus on environmental accounting as a support for improving business eco-efficiency needs to be extended to address issues relating to eco-effectiveness, where the aim is to achieve specific environmental outcomes established as targets, and eco-justice, where equity concerns between parties that have differential power bases need to be resolved (Burritt, 2004).
- Environmental accounting has a multi-functional emphasis. Information from environmental accounting systems can be used to motivate concern over the environmental impacts of business, to communicate environmental impacts to a range of parties and to aid in the strategic planning of business activities. Information to serve each of these functions is likely to differ and so it is important to identify the basis on which environmental accounting information is prepared. This is particularly important when predictive data are included in environmental accounting, as occurs in the budgeting process (Burritt and Schaltegger, 2001).
- Trade-offs between environmental accounting's focus on environmental improvement of business and broader accounting concerns related to movements towards sustainable society, or ecologically sustainable development, need to be recognized. In short, the role of environmental accounting in the context of sustainability accounting and reporting is starting to be addressed. Accounting for environmental risks and opportunities in the context of necessary trade-offs with social and economic impacts remains a key concern for all but very deep ecologists.

References

Adams, C.A., Roberts, C.B. and Hill, W.Y. (1998), 'Corporate Social Reporting Practices in Western Europe. Legitimating Corporate Behaviour?', *British Accounting Review*, **30** (1), 1–21.

A'Hearn, T. (1996), 'Environmental Management and Industry Competitiveness', Paper presented at the *Environmental Management and International Competitiveness* conference, Commonwealth of Australia, Canberra, Australia.

Bartolomeo, M., Bennett, M., Bouma, J.J., Heydkamp, P., James, P. and Wolters, T. (2000), 'Environmental Management Accounting in Europe: Current Practice and Future Potential', *The European Accounting Review*, **9**(1), 31–52.

Bennett, M. and James, P. (eds) (1998), *The Green Bottom Line. Environmental Accounting for Management. Current Practice and Future Trends*, Sheffield: Greenleaf Publishing.

Burritt, R.L. (1997), 'Environmental Disclosures in Annual Reports of Australian Gold and Copper Mining Companies with Activities in Papua New Guinea and/or Indonesia', Working Paper No. 1997/13, *Resource Management in Asia-Pacific Working Paper Series*, Canberra: Research School of Pacific and Asian Studies, Australian National University.

Burritt, R.L. (2004), 'Environmental Management Accounting. Roadblocks on the Way to the Green and Pleasant Land', *Business Strategy and the Environment*, **13**(1), 13–32.

Burritt, R.L. and Schaltegger, S. (2001), 'Eco-efficiency in Corporate Budgeting', *Environmental Management and Health*, **12**(2&3), 158–74.

Burritt, R.L, Hahn, T. and Schaltegger, S. (2002), 'Towards a Comprehensive Framework for Environmental Management Accounting. Links Between Business Actors and EMA Tools', *Australian Accounting Review*, **12**(2), 39–50.

Canadian Institute of Chartered Accountants (CICA) (1992), *Environmental Accounting and the Role of the Accounting Profession*, Toronto. CICA.

Commonwealth of Australia (2003), *Triple Bottom Line Reporting in Australia – A Guide to Reporting against Environmental Indicators*, Canberra: Environment Australia, Department of the Environment and Heritage, June, 68 pp.

ECOMAC (1996), *Synreport: Eco-management Accounting as a Tool of Environmental Management (The Ecomac Project)*, EIM Small Business Research and Consultancy, http://www.ukceed.org/ecomac.htm, accessed 25 November 2005.

Environmental Protection Agency, US (EPA) (1995), *Introduction to Environmental Accounting*, Washington DC, USA.

Ernst and Young (1992), 'Lender Liability for Contaminated Sites. Issues for Lenders and Investors', Ottawa: National Round Table on the Environment and the Economy, Working Paper No. 3.

Financial Accounting Standards Board (FASB) (1994), *Accounting Standards. Current Text*, Norwalk: FASB.

Global Reporting Initiative (GRI) (2002), *Sustainability Reporting Guidelines 2002*, Amsterdam: GRI.

Gray, R., Bebbington, J. and Walters, D. (1993), *Accounting for the Environment*, London: Chapman Publishing.

Gray, R., Owen, D. and Adams, C. (1996), *Accounting and Accountability. Changes and Challenges in Corporate Social and Environmental Reporting*, London: Prentice Hall Europe.

Hunt, C.B. and Auster, E.R. (1990), 'Proactive Environmental Management: Avoiding the Toxic Trap', *Sloan Management Review*, Winter, **31**(2), 7–18.

International Accounting Standards Committee (IASC) (1995), *International Accounting Standards 1995*, London: IASC.

International Federation of Accountants (IFAC) (1998), *Environmental Management in Organizations. The Role of Management Accounting, Financial and Management Accounting Committee*, IFAC, Study No. 6, New York, March.

Khanna, M., Quimio, W.R.H. and Bojilova, D. (1998), 'Toxics Release Information. A Policy Tool for Environmental Protection', *Journal of Environmental Economics and Management*, **36**, 243–66.

Maunders, K. and Burritt, R.L. (1991), 'Accounting and Ecological Crisis', *Accounting, Auditing & Accountability Journal*, **4**(3), 9–26.

Parker, L. (1999), *Environmental Costing. An Exploratory Examination*, Australian Society of Certified Practising Accountants, February.

Schaltegger, S. (1996), 'New Public Environmental Management', in *Information of the Basler Chamber of Commerce*, March, p. 3.

Schaltegger, S. (1997), 'Information Costs, Quality of Information and Stakeholder Involvement. The Necessity of International Standards of Ecological Accounting', *Eco-Management and Auditing*, **4**, November, 87–97.

Schaltegger, S. and Burritt, R.L. (2000), *Contemporary Environmental Accounting: Issues, Concepts and Practice*, Sheffield, UK: Greenleaf Publishing.

Schaltegger, S. and Stinson, C. (1994), 'Issues and Research Opportunities in Environmental Accounting', Basel: WWZ Discussion Paper No. 9124.

Schaltegger, S., Burritt, R.L. and Petersen, H. (2003), *An Introduction to Corporate Environmental Management: Striving for Sustainability*, Sheffield, UK: Greenleaf Publishing.

Schaltegger, S., Hahn, T. and Burritt, R.L. (2001), *EMA-Links. The Promotion of Environmental Management Accounting and the Role of Government*, Management and Stakeholders, Lüneburg, Germany: Center for Sustainability Management at the University of Lüneburg.

Schaltegger, S., Müller, K. and Hindrichsen, H. (1996), *Corporate Environmental Accounting*, London: John Wiley & Sons.

Skellenger, B. (1992), 'Limitation of Liability Clauses Gaining Popularity Among Environmental Consultants', *Environline*, **3**, Spring, 4–5.

Williams, G. and Phillips, T. (1994), 'Cleaning Up Our Act. Accounting for Environmental Liabilities. Current Financial Reporting Doesn't Do the Job', *Management Accounting*, 30–33.

21 Indicators for environmental innovation: what and how to measure

Anthony Arundel, René Kemp and Saeed Parto

Introduction

Environmental innovation is encouraged by many factors, including government regulation and subsidies, business opportunities for clean production, and the adoption by firms of an environmental ethic in response to public pressure. By the late 1980s environmental management had become an integral part of the corporate strategy of many firms in developed countries, diffusing increasingly also in developing countries. This brought into focus the potential economic benefits of environmental innovation from reduced material inputs and a shift away from end-of-pipe treatment solutions to preventive solutions.[1] However, despite the economic value of environmental innovation for industry and its potential societal benefits, there are only a few internationally comparable indicators[2] on the prevalence of environmental innovation, the factors that influence it and its economic effects. This shortage hampers the ability to design policies to encourage economically efficient environmental innovation and to track progress towards environmental goals.

The pollution abatement costs and expenditures (PACE) data are one example of the limitations of current indicators. PACE data on capital equipment spending by firms to reduce pollution have been collected on an annual basis in the United States, France and Germany and on an irregular basis in the Netherlands, the UK and Canada (Olewiler, 1994). The PACE estimates however, do not include investment in research and development (R&D) or other inventive activities, and are more likely to reflect end-of-pipe investments than investments in clean technology. The European Community Innovation Survey and similar surveys conducted in Australia and Canada are similarly lacking. Although these surveys include a few questions of relevance to environmental innovation, such as whether the firm's innovations have resulted in the reduction of per unit material and/or energy use, these questions only scratch the surface of environmental issues.

Our knowledge of environmental innovation largely comes from the extensive case study literature and from a few one-off surveys that focus on management and organizational responses to environmental issues. Although both are of value, the results of case studies may not apply to the majority of firms,[3] and only a few surveys specifically examine technical innovation.

Any attempt to collect an adequate set of environmental indicators for policy needs must address three main difficulties. First, impacts of environmental innovations are contingent upon the way the innovation is used and other factors. Second, environmental impacts, both positive and negative, occur during the entire lifecycle of a product. Third, many environmentally beneficial innovations are 'accidental' rather than developed intentionally to meet environmental goals. These include cost reduction measures and quality improvement techniques. The role and impact of these innovations on environmental protection need to be fully understood and accounted for.

In this chapter we develop a typology of environmental innovation and summarize the main questions for research for environmental innovation policy. An understanding of both is a necessary prerequisite for the design of indicators for environmental innovation. We then evaluate the advantages and disadvantages of several methods for obtaining environmental indicators. The main goal is to identify indicators that could assist public policy-makers and private firms to nurture the development, adoption and use of environmental innovation.

Typology of environmental innovation

Innovation is the commercialization of a new product, process or organizational system. According to the influential *Oslo Manual* (OECD, 1997), innovation does not require in-house investment in creative activities such as R&D. Firms can innovate by adopting technology developed by other firms or organizations.

Environmental innovation consists of new and modified processes, equipment, products, techniques and management systems that avoid or reduce harmful environmental impacts. A substantial fraction of environmental innovation is based on the simple *adoption* of new technology although firms may need to adapt the technology to their own production processes. A smaller fraction of environmental innovation is probably based on the firm's own *creative* activity.

In some cases reducing environmental impacts may be the sole purpose of an environmental innovation. In other cases the environmental benefit may be a fortuitous by-product of other innovation activities. *Intentional* environmental innovation is the product of an expressed goal to eliminate or reduce adverse environmental impacts. The use of flue-stack scrubbers to remove sulphur dioxide, for example, is relatively simple to identify as an environmental innovation. *Unintentional* environmental innovations are more difficult to identify but could be of even greater importance. An example is the photovoltaic energy cell for calculators. These permitted thinner calculators that never ran out of power but they also had the environmental benefit of reducing the use of batteries, most of which end up in landfills after use. Many innovative consumer products are also environmentally cleaner than older versions or alternatives.[4] These clean consumer products, which are often unintentional environmental innovations, should be differentiated from eco-products: goods that are marketed as ecologically sound. These constitute a small class of cleaner products.

Environmental innovation is 'technical' when it involves new equipment, products and production processes and 'organizational' when it involves structural change within the organization to institute new habits, routines, orientations and the use of tools such as Life Cycle Assessment and environmental programmes. Successful environmental innovation may often require both technical and organizational change.

Due to the complexity of environmental innovation, it is not enough simply to collect indicators on the percentage of firms that introduced an environmental innovation, or total expenditures on environmental innovation. This is because those approaches will miss the important category of innovations with environmental benefits that were not introduced for environmental reasons.

Typology of technical environmental innovation
Based on Skea (1995, p. 390), we propose a typology of six main types of technical environmental innovation: [5]

1. *Clean products.* Products designed to have minimal environmental impacts over their lifecycle.
2. *Cleaner production.* Process-integrated changes in the production system to reduce the amount of pollutants and waste materials generated during production.
3. *Pollution control.* Technology to prevent the direct release of environmentally hazardous emissions into air, surface water or soil.
4. *Recycling.* Identifying additional uses for certain production and post-consumer wastes to minimize waste generation.
5. *Waste management.* A formal system for handling, treatment and disposal of all waste.
6. *Clean-up.* A set of specific technologies to remediate contaminated environmental media (soil, water, air).

There are many other typologies for environmental innovation such as those based on a division between incremental and radical innovation, or between 'outcome-oriented' innovations such as end-of-pipe applications of technology and 'process-oriented' innovations to reduce inputs of energy and materials. We prefer the above typology for indicators because it has very little overlap between each category, full coverage of all possibilities within a short list of options, and clear and easy to understand definitions. Nevertheless, a better understanding of the factors that encourage the adoption of radical versus incremental environmental innovation would be of value to policy as discussed below.

Organizational environmental innovation
The literature on the 'greening' of industry emphasizes the need for firms to adopt an environmental consciousness or 'ethos', such that their business decisions and innovative activities automatically include environmental concerns. This requires changes to the firm's organizational practice and capability (Roome, 1994) and the inclusion of environmental goals in product design (Lenox and Ehrenfeld, 1997).

Relevant organizational innovation includes the adoption of environmental training and education programmes, incorporation of lifecycle analysis in design programmes, the introduction of programmes to make an existing plant or process more environmentally benign, and the establishment of inter-organizational networks and partnerships for achieving environmental goals. Firms may also need to learn continually about the environmental impact of their products and processes and to monitor regulatory, technical and social developments. Roome (1994) stresses the importance of such learning structures to integrate environmental knowledge into R&D processes.

Organizational changes can be driven by public pressure. An example is the chemical industry's 'Responsible Care' programme. Many large transnational corporations have voluntarily adopted environment management and auditing systems. Examples include ISO 14001 and the European Eco-Management and Audit Scheme (EMAS). Such systems help companies to identify and adopt environmentally beneficial solutions.

Environmental innovation as a firm strategy

Another area of research evaluates different firm strategies in response to green or environmental pressures. A common approach is to develop a normative model, where firms gradually internalize environmental goals into their R&D (Green and McMeekin, 1995; Miles and Green, 1996; Winn and Roome, 1993). Kolk and Mauser's (2002) comprehensive review of these models finds that most of them develop a linear classification system or a set of 'stages' for a firm's environmental strategy. Most, if not all, of these models are based on a small number of case studies and have not been fully tested empirically.

Figure 21.1 gives an example of one model developed by Steger (1993). The advantage of this model is that it links environmental strategies to innovation. For example, companies with a defensive attitude are unlikely to develop innovative solutions although they could adopt innovations developed outside of the firm. Furthermore, an environmental ethos is not required for a firm to adopt an innovative response. A firm could adopt an innovative strategy because it is in the market to develop environmental process innovations that are then sold to other firms. These characteristics make it possible to develop indicators to test the prevalence of each of these four strategies and the hypothetical relationship between environmental strategy and technology response empirically.

The purpose of environmental innovation indicators

Both firms and policy-makers would benefit from better indicators of environmental innovation. Firms could use these indicators to benchmark their own environmental strategies, while policy-makers could use them to develop policies to encourage firms to minimize their environmental impacts through innovation. In this section we examine four main policy-relevant areas that would benefit from better environmental indicators.

Types of environmental innovation

Although general indicators on the prevalence of any form of environmental innovation can be of value, basic indicators on specific types of environmental innovation by firms, including both technical and organizational innovation, are a prerequisite for developing policies to encourage firms to move along the innovation continuum (as in Figure 21.1). Another useful question to ask regards the firm's plans to phase out environmentally unfriendly products and processes. Details on the type of environmental innovation can also illuminate the cause–effect relationship between motivation and innovation. Many of these prevalence indicators could also be of value to firm managers, who could use them to benchmark their own strategic position.

Another question relates to the radical or incremental nature of innovation. Empirical research has shown that incremental improvements are an important source of cost reductions and product improvements. Although they often consist of mundane improvements that are not patented by the firm, incremental innovations can produce cumulatively significant improvements with environmental benefits. Yet a sole focus on incremental innovation may only result in limited environmental gains. In contrast, radical innovations – for example, replacing an entire production process or developing a new product – hold the potential to provide greater environmental benefits; these may require policy encouragement however. Therefore the problem is how to measure incremental versus radical innovation since these concepts vary by sector and the benefits of a 'radical' innovation might not be apparent for some time. One possible approach is to obtain information on

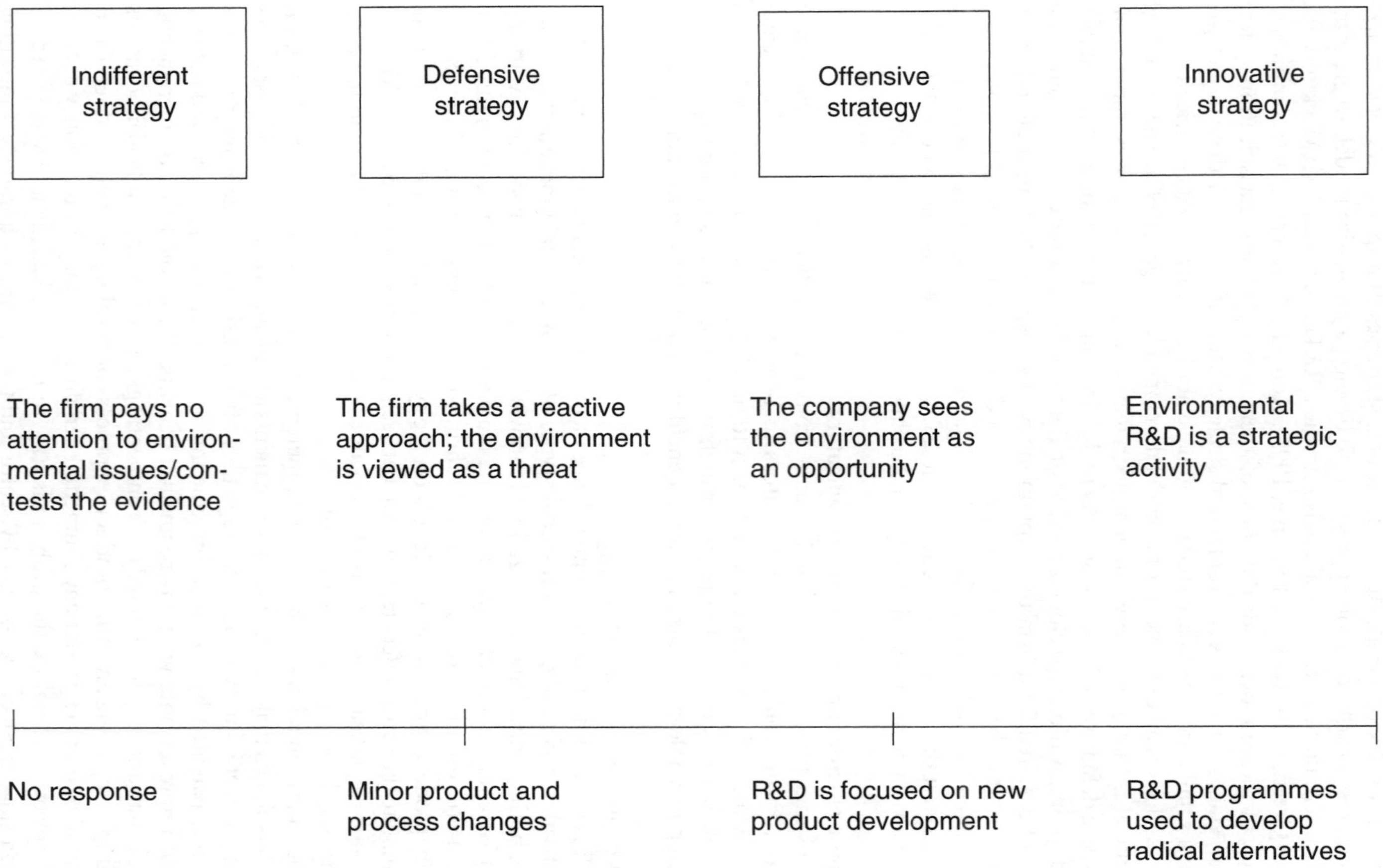

Indifferent strategy	Defensive strategy	Offensive strategy	Innovative strategy
The firm pays no attention to environ-mental issues/con-tests the evidence	The firm takes a reactive approach; the environment is viewed as a threat	The company sees the environment as an opportunity	Environmental R&D is a strategic activity
No response	Minor product and process changes	R&D is focused on new product development	R&D programmes used to develop radical alternatives

Figure 21.1   Technological responses to green pressures

the R&D cost of developing a specific innovation and the sources of information used to develop it. In addition it would be useful to improve our understanding of the factors that lead to radical responses to environmental challenges. These could include the role played by government technology development programmes, current and future regulation, and visions of environmental sustainability.

Indicators for organizational innovation could cover environmental mission statements and long-term goals for emission reductions, lifecycle analyses (LCAs), environmental audits, eco-design principles, collaboration with suppliers and users over environmental issues, contacts with public authorities, and the influence of environmentalists and citizens in the development of firm policy. Relevant questions include whether or not firms find these techniques useful for identifying and achieving environmental improvements and how firms determine which solution is best from an environmental point of view – do they use LCA or some other assessment method, make use of the company's own environmental knowledge base and information system, or use best available technology (BAT) guidance notes and lists for industry-specific processes?

Motivations and drivers

The economic incentive to invest in innovation is usually based on expectations of a higher profit level or at the minimum maintaining a satisfactory return on investment. In contrast, the motivation to undertake *environmental* innovation is considerably more complex and may be undertaken even when it is not profitable. Environmental innovation can be driven by such factors as accidents, regulation, vision, competition, and consumer pressure.[6]

Indicators of the motivations and drivers behind environmental innovation could help identify the conditions that favour environmental innovation and the circumstances in which environmental management operates as a self-propelling force. Questions on the firm's motivations for environmental innovation can be formulated as innovation goals. These include compliance with current regulations, pre-empting possible future regulations, cost savings, social responsibility and capture of a market for a new product. From a policy perspective, two important questions concern the main drivers for environmental innovation: (1) is it in response to regulation or in response to the firm's own innovation goals? And (2) are the beneficial environmental effects deliberate or fortuitous? The response to questions such as these would also be of interest to participating firms, which could use the data to compare their own behaviour to that of other companies in the same sector. These indicators would also be useful for testing strategic models, such as that given in Figure 21.1.

Government regulation and covenants can have a direct effect on environmental innovation strategies. Other possible public sector influences include R&D subsidies, technology adoption subsidies, technical assistance schemes or procurement policies (Foxon and Kemp, this volume; Hemmelskamp, 1996; Jaffe et al., 2001; Kemp, 1997; Kemp, 2000; Norberg-Böhm, 1999). Surprisingly we know very little about the incentive effect of these non-regulatory policies on environmental innovation.

The economic effects of environmental innovation

The economic effects of environmental policy are determined by the costs and benefits of the adopted solutions. It is unclear whether environmental policies impose a net cost on companies, and if they do, what the size of any such cost might be (Palmer et al., 1995; Welford and Starkey, 1996). A survey conducted by Jaffe et al. (1995) indicates that aggregate environmental investments do impose a cost on companies. Yet this may be true only for environmentally *motivated* investments and not true for all investments with environmental benefits. Surveys can provide data on the cost offsets of different types of environmental innovation among a large sample of firms active in different sectors. Such data are required to determine the usefulness of offsets as an incentive for environmental innovation.

There is growing evidence that environmental innovations can save costs and bring economic benefits. Hart and Ahuja (1996), using a sample of *Standard and Poor's* 500 firms in the United States, report a positive correlation between financial and environmental performance. The correlation was strongest for firms in high-pollution sectors. Porter and van der Linde (1995) have argued on the basis of case studies that offsets, combined with appropriate regulatory policies, could approximate or exceed the costs of pollution control, resulting in a 'win–win' scenario. However, it is possible that a win–win outcome may only occur under favourable conditions.

Knowledge sources and impediments
To innovate, companies draw on their in-house expertise and on a public knowledge pool. The source of knowledge varies by the type of innovation. Product innovation requires a good deal of internal knowledge. The development of effective public policies to stimulate environmental innovation requires data on the types of information sources that are most useful for different types of innovation. For instance, how useful are universities, public laboratories, government bodies, sector organizations and environmental transfer centres as a source of information for environmental innovation? What is the role of collaboration in environmental innovation? And what kind of knowledge is transferred: knowledge about environmental effects, requirements or solutions to environmental problems? One goal is to identify critical gaps in the knowledge infrastructure while another is to identify which knowledge sources are most useful.

Of interest also is the frequency with which process innovators keep their innovation secret or license it to other firms. The goal for policy is to encourage licensing and knowledge sharing.

Current indicators of environmental innovation
There are four main categories of existing indicators of environmental innovation: financial indicators collected by governments; patents; innovation surveys based on *The Oslo Manual* definition of innovation (includes both the development of innovations in-house and the adoption of new technology from external sources); and literature-based innovation output (LBIO) techniques. This section addresses the relevance of each of these methods to meet the four main goals of indicators as outlined above. These include indicators for: (1) the typology of environmental innovation, including shifts from end-of-pipe to cleaner production processes; (2) motivations and drivers; (3) the economic effects, which can be divided into costs (including cost savings), the impact on employment and the quality of employment, such as skill requirements; and (4) knowledge sources and impediments.

Financial indicators
A common measure of environmental innovation is the amount spent by firms on pollution abatement costs and expenditures (PACE). In the United States, the Commerce

Department survey asks respondents to estimate their *additional* capital expenditures due to environmental regulations (Jaffe et al., 1995). The question is partly hypothetical since these additional expenditures can only be estimated on the basis of no regulation. The PACE data have been useful for indirectly estimating the effect of government regulation on firm strategies as well as on economic factors, including output and competitiveness.

The US Department of Commerce has also collected data on cost offsets due to savings from investment in pollution abatement equipment. For example, investment in a pollution control system to capture heavy metals, such as cadmium, that were previously discharged into the environment can partly be recouped by selling or re-using the heavy metals. The ratio of the offset to the investment in pollution abatement could form a strong incentive to develop environmental innovations. This ratio has also been subject to strenuous debate, with Palmer et al. (1995) using the Department of Commerce data to show that the offset benefits amount to less than 2 per cent of US expenditures on pollution abatement. The main problem is that the cost offsets from environmental innovations that are not covered by PACE could be much larger, not to mention the financial advantages of unintentional environmental innovation.

The available financial indicators, based on PACE or offsets from PACE, provide some information on the economic effects of environmental innovation and indirectly on strategies and motivations. Their main disadvantage is that they have an extremely narrow focus, targeting only a limited range of environmental innovations, and provide no information on unintentional environmental innovation or on the locus of the innovation – in-house or externally sourced from outside the firm. Data on R&D expenditures for environmental innovation would be a very valuable indicator but this information is rarely collected. An exception is the 2000 R&D survey by Statistics Canada (2001), which asked respondents to 'estimate the percentage of total R&D expenditures (for 2000) attributable to prevention, treatment and re-use of pollutants and wastes, and reduction of material and energy use'. A second question asked if there were environmental benefits from R&D without an intentional environmental application. To date, the results of these questions have not been published.

Patents

Patent data can be used as a proxy for environmental innovation. Patents are not a direct measure of innovation because many patents are not commercialized and many innovations are not patented. Since it is difficult to link patent data to other measures of environmental innovation at the level of the firm, or to changes in competition and environmental pressures, most patent research correlates aggregate trends in patenting with either total innovation expenditures or PACE investments. An additional problem with using patent data is that there is no class of 'environmental patents'. Instead, the number of environmental patents must be laboriously identified using keywords or by analysing the patent description.

Jaffe and Palmer (1997) analysed the effects of PACE on total US patents and R&D and found no significant relation between PACE and total patents, but a positive relationship between PACE and total R&D. Lanjouw and Mody (1996), in an analysis limited to environmental patents in the US, Japan and Germany, found that patents followed PACE with a one- to two-year time lag. Marinova and McAleer (2003) identified US 'ecological' patents and reported a sharp increase in these patents after 1987, but the share of ecological patents remained close to 2 per cent of total patents between 1975 and 1997. Germany had the largest number of US ecological patents. Popp (2003) used patent data to study innovation in flue gas desulphurization under two policy regimes. He found that the rate of invention fell after the introduction of a market-based regime (the 1990 Clean Air Act, which established a market for sulphur allowances), but since 1990 the innovations were more oriented to improving removal efficiency. Brunnermeier and Cohen (2003) used data from the OTA (Office of Technology Assessment) on US environmental patents between 1983 and 1992 and found a positive association between PACE and environmental patents after controlling for changes in economic activity (total shipments) and government monitoring pressure (inspection visits to plants).

The patent literature strongly suggests that PACE and regulatory environments influence private incentives to invest in environmental innovation. These findings are of relevance to policy research on the motivations for and drivers of environmental innovation. Patent analyses would be more useful, however, if patent data were directly linked to firm-level variables on motivations, drivers, the type of environmental innovation and on knowledge sourcing (particularly licensing).

Survey indicators of environmental innovation
There are two basic sources of survey indicators. The first source consists of official, large-scale innovation surveys that sample thousands of firms and that are performed on a regular basis. The second source consists of smaller 'one-off' surveys by academics or government agencies. These usually focus on a limited region or set of sectors.

Large-scale innovation surveys in Europe, Canada and Australia include a few questions of relevance to environmental innovation. For example, the third Community Innovation Survey in Europe asks firms about the level of impact from innovation 'to reduce materials and energy inputs per unit of output' and the influence of innovation on 'improved environmental impact or health and safety aspects'. Unlike the patent and PACE data, these questions provide information on the prevalence of innovation with environmental benefits without limiting the results to intentional environmental innovation. Furthermore, a major advantage of these large-scale surveys is that the information on environmental innovation can be linked to other firm-level innovation strategies and characteristics. The main disadvantage of these surveys is that so far they have only collected data on a few aspects of environmental innovation.

Several smaller surveys, summarized in Table 21.1, have gone into environmental innovation in far greater depth.[7] Most of these surveys have not queried firms about their own in-house innovative activities but are included because they covered the adoption of environmental technology. For each survey, Table 21.1 describes the target population of firms, the number of responses and the response rate, and the types of questions asked in each survey of relevance to the four main policy issues. Many of the specialized environmental surveys suffer from low response rates and cannot match the response rates of official innovation surveys. Low response rates reduce confidence in the accuracy of prevalence rates (for example, the percentage of firms that report cost offsets). One option to address this problem is to follow up a sample of non-respondents and determine if they differ in any significant way from the respondent firms. To date, this technique has rarely been used in environmental innovation surveys.

Three surveys focus specifically on environmental innovation (Green et al., 1994;

Table 21.1 Environmental innovation surveys

Reference	Target Firms	Responses (response rate)	Results of Relevance to Main Policy Issues[a]			
			Type of innovation	Motivations and drivers	Economic effects	Knowledge sourcing/impediments
Steger, 1993	German manufacturing and service firms	592 (not given)	A	✓	C	
Green et al., 1994	UK firms interested in government support programmes	169 (21%)	A, CR	✓		
Arundel and Rose, 1999	Canadian firms in sectors with potential biotechnology applications	2010 (86%)	A, CR	✓	C	K, I
Blum-Kusterer and Hussain, 2001	German and UK pharmaceutical firms	32 (21%)	M	✓		I
Pfeiffer and Rennings, 2001[b]	German manufacturing firms	400 (45%)	A	✓	E, S	
Getzner, 2002	EMAS/ISO firms in Austria, Germany, Netherlands, Spain, Sweden	407 (16%)	A	✓	E, S	
Andrews et al., 2002	SMEs in Australia	145 (29%)	M, A		C	K
Lefebvre et al., 2003	SMEs in four industries in Canada	368 (quota sampling)	M, A, CR	✓		
Rennings and Zwick, 2003	Manufacturing and service firms in UK, Germany, Switzerland, Netherlands, Italy	1594 (not given for all countries)	A, CR	✓	C, E, S	
Zotter, 2004	Austrian firms introducing water conservation solution	87 (not given)	A			I
Zutshi and Sohal, 2004	ISO 14001 firms in Australia and New Zealand	143 (46%)	M	✓		K, I

Notes: [a] Type: M = management systems, A = technology adoption, CR = technology creation (innovation developed in firm, must be specified). Economic effects: C = costs, E = employment, S = skills; ✓ = motivations/drivers examined. Knowledge sourcing/impediments: K = knowledge sourcing, I = impediments to adoption.
[b] Includes results of Cleff and Rennings (1999) using the same data.

333

Lefebvre et al., 2003; Rennings and Zwick, 2003), while the fourth covers biotechnology but asks a large number of questions on environmental innovation (Arundel and Rose, 1999). These are the only four studies that differentiate between innovation as a creative and adoptive process. One of the first was the 1993 survey by Green et al. (1994), which was sent to a sample of 800 firms that had expressed an interest in the UK Department of Trade and Industry's Environmental Technology Innovation Scheme (ETIS). Under the scheme companies could receive a subsidy for industrial R&D that might improve environmental standards.

Most of these smaller surveys (Table 21.1) focus on the motivation and drivers for environmental innovation, followed by economic impacts on costs, employment or skills. All three studies on employment and skills (Getzner, 2002; Pfeiffer and Rennings, 2001; Rennings and Zwick, 2003) are from Europe and reflect the European policy interest in reducing high structural unemployment. None of these three studies obtain interval-level data on employment effects (such as percentage changes in job gains or losses) because of the difficulty for respondents to provide accurate estimates. Instead, the survey questions obtain either categorical data (employment increased, decreased or remained unchanged) or nominal-level data (yes or no). Consequently these results cannot be used to estimate changes in employment, skills or cost offsets. Nevertheless the results should still be of use to firms and to policy. As an example, Pfeiffer and Rennings (2001) report that between 84 per cent and 91 per cent of German firms (depending on the type of environmental innovation) found that the innovation had no effect on employment, while less than 5 per cent reported a decrease in employment.

Knowledge sourcing and impediments to environmental innovation have received the least attention in environmental innovation surveys. One exception is the survey by Andrews et al. (2002), which asked if firms shared their own knowledge and experience of cleaner production with other firms and with industry associations. This is a valuable area for future research if combined with data on licensing behaviour because the policy goal of encouraging knowledge sourcing could conflict with a strategic interest for the firm to keep cleaner production methods secret.

The Statistics Canada survey (Arundel and Rose, 1999) of biotechnology applications is the only study to cover all four policy areas. The respondents were asked if their firm currently used or planned to use one of five carefully defined environmental biotechnologies. Users of one or more of these technologies were then asked a series of questions on investment, their motivations for adopting the technology, difficulties with implementation, results from their use, and the principal internal and external sources of information to assist the adoption of environmental biotechnologies (Arundel and Rose, 1999).

An interesting avenue for future research on environmental innovation is to develop panel surveys that gather information from the same firms over time. An example is the Mannheim innovation panel of ZEW (Zentrum für Europäische Wirtschaftsforschung – Center for European Economic Research). The results of such surveys would permit sophisticated analysis of the effect of motivations and management systems on different types of environmental innovation.

Literature-based environmental innovation surveys
The literature-based innovation output (LBIO) method (Coombs et al., 1996; Kleinknecht et al., 1993) can collect objective data on environmental innovation that can

potentially cover all four main policy areas, particularly if the method is combined with a short survey to obtain additional data. Malaman (1996) uses this method to evaluate environmental innovations developed by Italian companies between 1970 and 1995. A database of 192 innovations from 168 companies was constructed from several information sources. Information on each innovation was obtained from published sources and from brief interviews with company representatives.

An advantage of this method is that it is easily amenable to identifying different types of environmental innovation, based on the description of each innovation in the literature. Malaman (1996) also used this method to assess the development stage for each innovation: R&D, prototype, first application, or mature technology. This type of information could be useful in addressing questions on the incremental versus radical nature of environmental innovations. Huber (2003) analysed environmental innovations from a German database of over 500 technological environmental innovations.[8] The database could be used to analyse the radical versus incremental nature of these innovations.

Depending on the use of follow-up surveys of the firms, the LBIO method can also provide data on information sources, the use of collaboration, cost offsets and many other topics. The main strength of the LBIO method is that it can obtain extensive information of relevance to all four policy areas on clearly defined innovations that can be classified into different typologies. The main disadvantage of the LBIO method concerns the identification of environmental innovations, which largely depends on steps taken by each firm to advertise their existence. This could bias the results towards the most successful environmental innovations and towards the most environmentally active firms. This is suggested by Malaman's (1996) results, which find that 41.8 per cent of the identified firms had adopted a proactive environmental strategy. This is a very high figure that is unlikely to represent general conditions in Italian industry.

The LBIO approach allows for longitudinal research over time. A good example is the study by Newell et al. (1999). The authors tested Hick's induced innovation hypothesis by estimating a product-characteristics model using a database on 735 room air-conditioner models offered for sale from 1958 through 1993; 275 central air-conditioner models for 1967 through 1988; and 415 gas water-heater models from 1962 through 1993. They found that the *direction* (but not the rate) of innovation was responsive to energy price changes and that the responsiveness increased substantially when energy efficiency labelling was required. This type of research can econometrically assess the effects of government policy and other factors on innovation.

The way forward
Patent data and the LBIO method provide an opportunity to develop a range of indicators of value to environmental policy and to firms that wish to benchmark their own environmental strategies. Nevertheless a full complement of environmental indicators that can address each of the four main policy areas will require survey data. The small-scale innovation surveys summarized in Table 21.1 point the way by developing appropriate questions on different aspects of environmental innovation. However, these surveys cannot be used to develop internationally comparable indicators because they are limited to a small number of economic sectors and almost always suffer from low response rates. Instead some of the questions developed by these surveys should be adopted by large-scale official surveys that cover all economic sectors and which have

the resources to obtain high response rates or use follow-up techniques to deal with non-response biases.

A large-scale survey on environmental innovation does not need to place a substantial burden on firm managers. Elsewhere we developed an example of an environmental innovation survey that could cover the four main policy issues in two pages and which almost entirely used ordinal and nominal question formats, which are the simplest types of questions for managers to answer (Kemp and Arundel, 1998). The questionnaire also included a third page of questions on one of the six environmental technologies summarized in the typology introduced earlier in this chapter. Different versions of the third page could be randomly assigned to the respondents so that the survey could obtain data on all six technology types. Many of the ideas developed for this survey were applied to the IMPRESS survey (Rennings and Zwick, 2003), an international research project that analyses the impact of clean production on employment in Europe. Alternatively the addition of only a few relevant questions to the Community Innovation Survey, such as whether they have innovated environmentally or undertaken R&D to improve the environmental performance of their products or processes, could substantially improve the availability of indicators for environmental innovation.

There is not one best type of indicator research for environmental innovation. All have a role to play to study the very heterogeneous issue of environmental innovation. To analyse the influence of environmental policy we also need better indicators for policy: stringency, enforcement, flexibility, orientation to prevention, combination of instruments used, long-term orientation and the multi-dimensional perspective employed. This serves to show that policy has several dimensions that are important for innovation (Jänicke et al., 2000). To make progress, studies should differentiate between different types of environmental innovation and should not measure policy through a single index such as pollution abatement costs. Further progress in better understanding the policy style–innovation relationship hinges on better indicators and on international comparative study, using micro-data for innovation and detailed sector data for environmental policy.

Notes

1. It is unclear what proportion of European environmental innovation is end-of-pipe. A European Commission (2002) white paper estimates that integrated solutions account for one-third of environmental investments, with end-of-pipe and clean-up accounting for two-thirds of such investments. Other research suggests that integrated solutions are more common than end-of-pipe technologies. Arundel et al. (2003) found that 71 per cent of Dutch firms in five sectors had introduced a production process change, compared with 52 per cent that had introduced an end-of-pipe solution. Frondel et al. (2006) found that shares of cleaner production technologies were higher than those for end-of-pipe technologies in the seven OECD countries being studied, with the highest share for Japan (86.5 per cent) and lowest share for Germany (57.5 per cent).
2. There is a difference between statistics and indicators. An environmental statistic is a data point, such as total spending on environmental innovation. Indicators place statistics in context, by giving the percentage of all private investment that is spent on environmental innovation, for example (Godin, 2003).
3. Examples of relevant case studies can be found in Clarke and Roome (1995); Fischer and Schot (1993); Groenewegen and Vergragt (1991); Gutowski et al. (2005); Lenox and Ehrenfeld (1997); Schmidheiny (1992); Shrivastava (1995) and Smith (1993).
4. It is estimated that 60 per cent of the innovations of the Dynamo Database in the Netherlands offer environmental benefits compared with existing technologies, and that 55 per cent of the innovations supported by a general innovation scheme for research cooperation offered environmental benefits. These two figures from the Netherlands suggest that the majority of technological innovations have environmental benefits.

5. Similar definitions can be found in Ashford (1993), Cleff and Rennings (1999) and OECD (1985). Technologies to monitor the condition of the environment, releases of pollutants and identification of pollutants are sometimes also included among environmental technologies (Skea, 1995).
6. Regulation is the most important factor for adopting end-of-pipe solutions but is possibly less important as a driver for integrated solutions.
7. Three very small surveys are excluded from Table 21.1 (Garrod and Chadwick, 1996; Pimenova and van der Vorst, 2004; Williams et al., 1993). Doyle (1992) only surveys environmental equipment manufacturers and is of less interest here.
8. Ecological consistency is juxtaposed to eco-efficiency in Huber (2003) but it is unclear exactly how it is defined.

References

Andrews, S.K.T., Stearne, J. and Orbell, J.D. (2002), 'Awareness and Adoption of Cleaner Production in Small to Medium-Sized Businesses in the Geelong Region, Victoria, Australia', *Journal of Cleaner Production*, **10**, 373–80.

Arundel, A. and Rose, A. (1999), 'The Diffusion of Environmental Biotechnology in Canada: Adoption Strategies and Cost Offsets', *Technovation*, **19**, 551–60.

Arundel, A., Chervenic-Poeth, L. and Kemp, R. (2003), 'The Dutch Survey', in Rennings, K. and Zwick, T. (eds), *Employment Impacts of Cleaner Production, ZEW Economic Studies 21*, New York: Physica Verlag Heidelberg, pp. 229–66.

Ashford, Nicholas (1993), 'Understanding Technological Responses of Industrial Firms to Environmental Problems: Implications for Government Policy', in Kurt Fischer and Johan Schot (eds), *Environmental Strategies for Industry: International Perspectives on Research Needs and Policy Implications*, Washington DC: Island Press.

Blum-Kusterer, M. and Hussain, S. (2001), 'Innovation and Corporate Sustainability: An Investigation into the Process of Change in the Pharmaceuticals Industry', *Business Strategy and the Environment*, **10**, 300–16.

Brunnermeier, S.B. and Cohen, M.A. (2003), 'Determinants of Environmental Innovation in US Manufacturing Industries', *Journal of Environmental Economics and Management*, **45**, 278–93.

Clarke, S.F. and Roome, N. (1995), 'Managing for Environmentally Sensitive Technology: Networks for Collaboration and Learning', *Technology Assessment and Strategic Management*, 7(2), 191–215.

Cleff, T. and Rennings, K. (1999), 'Determinants of Environmental Product and Process Innovation', *European Environment*, **9**, 191–201.

Coombs, R., Narandren, P. and Richards, A. (1996), 'A Literature-based Innovation Output Indicator', *Research Policy*, **25**, 403–13.

Doyle, D.J. (1992), *Building a Stronger Environmental Technology Exploitation Capability in Canada*, Report for Environment Canada and Industry, Science and Technology Canada, DSS contract KE144-1-2273/01-SS, Doyletech Corporation, Kanata.

European Commission (2002), *Environmental Technology for Sustainable Development*, COM(2002) 122, Brussels.

Fischer, K. and Schot, J. (eds) (1993), *Environmental Strategies for Industry: International Perspectives on Research Needs and Policy Implications*, Washington DC: Island Press.

Frondel, M., Horbach, J. and Rennings, K. (2006), 'End-of-pipe or Cleaner Production Machines? An Empirical Comparison of Abatement Decisions across OECD Countries', *Business Strategy and the Environment*, forthcoming.

Garrod, B. and Chadwick, P. (1996), 'Environmental Management and Business Strategy: Towards a New Strategic Paradigm', *Futures*, **28**(1), 37–50.

Getzner, M. (2002), 'The Quantitative and Qualitative Impacts of Clean Technologies on Employment', *Journal of Cleaner Production*, **10**, 305–19.

Godin, B. (2003), 'The Emergence of S&T Indicators: Why Did Governments Supplement Statistics with Indicators?', *Research Policy*, **32**, 679–91.

Green, K. and McMeekin, A. (1995), 'Excellent at What? Environmental Business and Technology Strategies', Working Paper 9505, Manchester: CROMTECH.

Green, K., McMeekin, A. and Irwin, A. (1994), 'Technological Trajectories and R&D for Environmental Innovation in UK Firms', *Futures*, **26**(10), 1047–59.

Groenewegen, P. and Vergragt, P. (1991), 'Environmental Issues as Threats and Opportunities for Technological Innovation', *Technology Analysis and Strategic Management*, **3**(1), 43–55.

Gutowski, T., Murphy, C., Allen, D., Bauer, D., Bras, B., Piwonka, T., Sheng, P., Sutherland, J., Thurston, D. and Wolff, E. (2005), 'Environmentally Benign Manufacturing: Observations from Japan, Europe and the United States', *Journal of Cleaner Production*, **13**(1), 1–17.

Hart, S. and Ahuja, G. (1996), 'Does it Pay to be Green? An Empirical Examination of the Relationship Between Emission Reduction and Firm Performance', *Business Strategy and the Environment*, **5**, 30–37.

Hemmelskamp, J. (1996), 'Environmental Policy Instruments and their Effects on Innovation', Discussion Paper 96–22, Zentrum für Europäische Wirtschaftsforschung GmbH.

Huber, J. (2003), 'Environmental Policy Shift through Technological Innovation', Paper for *Governance for Industrial Transformation* conference, Berlin.

Jaffe, A.B. and Palmer, K. (1997), 'Environmental Regulation and Innovation. A Panel Data Study', *Review of Economics and Statistics*, **79**, 610–19.

Jaffe, A., Newell, R. and Stavins, B. (2001), 'Technological Change and the Environment', in *Handbook of Environmental Economics*, North-Holland/Elsevier Science (see also Discussion Paper 00-47 REV, Resources for the Future, Washington DC).

Jaffe, A.B., Portney, P.R. and Stavins, R.N. (1995), 'Environmental Regulation and the Competitiveness of US Manufacturing: What Does the Evidence Tell us?', *Journal of Economic Literature*, **33**, 132–63.

Jänicke, M., Blazejczak, J., Edler, D. and Hemmelskamp, J. (2000), 'Environmental Policy and Innovation: An International Comparison of Policy Frameworks and Innovation Effects', in Hemmelskamp, J., Rennings, K. and Leone, F. (eds) (2000), *Innovation-oriented Environmental Regulations: Theoretical Approaches and Empirical Analysis, ZEW Economic Studies 10*, New York: Physica-Verlag, Heidelberg.

Kemp, R. (1997), *Environmental Policy and Technical Change. A Comparison of the Technological Impact of Policy Instruments*, Cheltenham, UK and Lyme, USA: Edward Elgar.

Kemp, R. (2000), 'Technology and Environmental Policy – Innovation Effects of Past Policies and Suggestions for Improvement', in *OECD Proceedings: Innovation and the Environment*, Paris: OECD, pp. 35–61.

Kemp, R. and Arundel, A. (1998), *Survey Indicators for Environmental Innovation*, Oslo: IDEA report, STEP Group.

Kleinknecht, A., Jeroen, O., Reijnen, N. and Smits, W. (1993), 'Collecting Literature-based Innovation Output Indicators: The Experience in the Netherlands', in Kleinknecht, A. and Bain, D. (eds), *New Concepts in Innovation Output Measurement*, New York: St Martin's Press, pp. 42–84.

Kolk, A. and Mauser, A. (2002), 'The Evolution of Environmental Management: From Stage Models to Performance Evaluation', *Business Strategy and the Environment*, **11**, 14–31.

Lanjouw, J.O. and Mody, A. (1996), 'Innovation and the International Diffusion of Environmentally Responsive Technology', *Research Policy*, **25**, 549–71.

Lefebvre, E., Lefebvre, L.A. and Talbot, S. (2003), 'Determinants and Impacts of Environmental Performance in SMEs', *R&D Management*, **33**(3), 263–83.

Lenox, M. and Ehrenfeld, J. (1997), 'Organizing for Effective Environmental Design', *Business Strategy and the Environment*, **6**, 187–96.

Malaman, R. (1996), 'Technological Innovation for Sustainable Development: Generation and Diffusion of Industrial Cleaner Technologies', *Nota di Lavoro*, 66.96, Fondazione Eni Enrico Mattei.

Marinova, D. and McAleer, M. (2003), 'Modelling Trends and Volatility in Ecological Patents in the USA', *Environmental Modelling and Software*, **18**, 195–203.

Miles, I. and Green, K. (1996), 'A Clean Break? From Corporate Research and Development to Sustainable Technological Regimes', in Welford, R. and Starkey, R. (eds), *The Earthscan Reader in Business and the Environment*, London: Earthscan, pp. 120–44.

Newell, R.G., Jaffe, A.B. and Stavins, R.N. (1999), 'The Induced Innovation Hypothesis and Energy-saving Technological Change', *The Quarterly Journal of Economics*, **114**, 941–75.

Norberg-Böhm, V. (1999), 'Stimulating "Green" Technological Innovation: An Analysis of Alternative Policy Mechanisms', *Policy Sciences*, **32**, 13–38.

Organisation for Economic Co-operation and Development (OECD) (1985), *Environmental Policy and Technical Change*, Paris: OECD.

Organisation for Economic Co-operation and Development (OECD) (1997), *The Measurement of Scientific and Technical Activities: Proposed Guidelines for Collecting and Interpreting Technical Innovation Data (The Oslo Manual)*, Paris: OECD.

Olewiler, N. (1994), 'The Impact of Environmental Regulation on Investment Decisions', in Benidickson, G. (ed.) *Getting the Green Light: Environmental Regulation and Investment in Canada*, Toronto: CD Howe Institute, Policy Study 22, pp. 53–113.

Palmer, K., Oates, W.E. and Portney, P.R. (1995), 'Tightening Environmental Standards: The Benefit–Cost or the No-cost Paradigm?', *Journal of Economic Perspectives*, **9**(4), 119–32.

Pfeiffer, F. and Rennings, K. (2001), 'Employment Impacts of Cleaner Production – Evidence from a German Study Using Case Studies and Surveys', *Business Strategy and the Environment*, **10**, 161–75.

Pimenova, P. and van der Vorst, R. (2004), 'The Role of Support Programmes and Policies in Improving SMEs' Environmental Performance in Developed and Transition Economies', *Journal of Cleaner Production*, **12**(6), 549–59.

Popp, D. (2003), 'Pollution Control Innovations and the Clean Air Act of 1990', *Journal of Policy Analysis and Management*, **93**, 390–409.

Porter, M.E. and van der Linde, C. (1995), 'Toward a New Conception of the Environment–Competitiveness Relationship', *Journal of Economic Perspectives*, **9**(4), 97–118.

Roome, N. (1994), 'Business Strategy, R&D Management and Environmental Imperatives', *R&D Management*, **24**(1), 65–82.

Rennings, K. and Zwick, T. (eds) (2003), *Employment Impacts of Cleaner Production, ZEW Economic Studies 21*, New York: Physica Verlag Heidelberg.

Schmidheiny, S. (1992), *Changing Course: A Global Business Perspective on Development and the Environment*, Cambridge, MA, MIT Press.

Shrivastava, P. (1995), 'Environmental Technologies and Competitive Advantage', *Strategic Management Journal*, **16**, 183–200.

Skea, J. (1995), 'Environmental Technology', in Folmer, H., Landis Gabel, H. and Opschoor, H. (ed.), *Principles of Environmental and Resource Economics. A Guide for Students and Business Makers*, Aldershot, UK and Brookfield, US: Edward Elgar, pp. 389–412.

Smith, D. (1993), *Business and the Environment. Implications of the New Environmentalism*, London: Paul Chapman Publ. Ltd.

Statistics Canada (2001), *Research and Development in Canadian Industry 2000*, Ottawa, Canada: Science, Innovation and Electronic Information Division, Statistics Canada.

Steger, U. (1993), 'The Greening of the Board Room: How German Companies Are Dealing with Environmental Issues', in Fischer, K. and Schot, J. (eds), *Environmental Strategies for Industry: International Perspectives on Research Needs and Policy Implications*, Washington DC: Island Press, pp. 147–66.

Welford, R. and Starkey, R. (eds) (1996), *The Earthscan Reader in Business and the Environment*, London: Earthscan.

Williams, H.E., Medhurst, J. and Drew, K. (1993), 'Corporate Strategies for a Sustainable Future', in Fischer, K. and Schot, J. (eds), *Environmental Strategies for Industry: International Perspectives on Research Needs and Policy Implications*, Washington, DC: Island Press, pp. 117–46.

Winn, S.F. and Roome, N. (1993), 'R&D Management Responses to the Environment. Current Theory and Implications for Practice and Research', *R&D Management*, **23**, 147–60.

Zotter, K.A. (2004), 'End-of-pipe versus Process Integrated Water Conservation Solutions: A Comparison of Planning, Implementation and Operating Phases', *Journal of Cleaner Production*, **12**(7), 685–95.

Zutshi, A. and Sohal, A. (2004), 'A Study of the Environmental Management System (EMS) Adoption Process Within Australasian Organisations – 2. Role of Stakeholders', *Technovation*, **24**(5), 371–86.

22 International trends in socially responsible investment: implications for corporate managers

Shanit Borsky, Diana Arbelaez-Ruiz,
Chris Cocklin and Doug Holmes

Introduction

The growth of socially responsible investment (SRI) over the past few years has drawn the attention of the financial community to the sustainability practices of companies globally. Socially responsible investments account for more than one out of every nine dollars under professional management in the United States today. The US$2.16 trillion managed by major investing institutions in the United States in 2003 (up from US$1.19 trillion in 1997) – including pension funds, mutual fund families, foundations, religious organizations and community development financial institutions – accounts for 11.3 per cent of the total US$19.2 trillion in investment assets under professional management (Social Investment Forum, 2003). In Europe, according to a recent survey of attitudes to SRI, 61 per cent of fund managers and financial analysts believe that interest in SRI has risen over the past two years. Total SRI assets in Europe now approximate €336 billion (CSR Europe, 2003).

With its antecedents in church and state value-based investment, SRI is still evolving. As a part of the broader move towards sustainable finance, SRI is a response to:

- the globalization of finance;
- the rising dominance of institutional investors;
- increasingly visible environmental and social impacts of corporate activity;
- changing community expectations about the limits of acceptable corporate impacts;
- new classes of company 'stakeholders' seeking information about corporate management; and
- a plethora of regulatory, industry and non-government initiatives calling for corporate accountability.

The rapid growth of the SRI sector since the mid-1990s can be further explained with reference to the erosion of public trust following spectacular transnational corporate failures such as Enron, WorldCom and countless examples from individual nations, prompting a convergence in the interests of socially responsible investors with principles of corporate governance. A more active investment community and a savvy and expectant civil society are now vocally expressing their discontent with, and increasing pressure on, the corporate sector.

Initiatives linking finance with sustainability outcomes have focused on the role of financial institutions in facilitating projects, underwriting corporate activity and creating economic opportunities. In each of these roles financial institutions factor environmental and social impacts into decisions about the activities they finance. This in turn increases the level of scrutiny exercised over financial investments, leading to a move beyond traditional financial measures and risk management approaches.

Socially responsible investment is a subset of sustainable finance in which investment is directed through a variety of investment styles and options products into enterprises that favour sustainability outcomes. Variously described as 'social investing', 'ethical investing' and 'socially aware investing', SRI is a process of identifying and investing in companies that meet certain standards of corporate social responsibility (CSR) or corporate sustainability. The UK Social Investment Forum (undated) describes SRI 'as investment which combines investors' financial objectives with their concerns about social, environmental and ethical issues'. The process is based on the premise that capital allocation decisions by the financial sector are founded on assessments of risk and opportunity, and that influencing these assessments to incorporate sustainability issues has the potential to leverage changes in business activity.

While the SRI sector is not new, the recent growth in the sector is characterized by the participation of large, mainstream financial institutions. These have a lower tolerance for risk but a significantly higher impact on corporate managers than traditional socially responsible investors such as state governments and individual socially and environmentally aware citizens. As institutional investors grapple with standards of corporate governance and re-examine systems of corporate monitoring and accountability, they have begun to influence the categories of information required to demonstrate effective internal controls.

The introduction of new approaches to incorporating environmental and social information into financial decision-making has created new challenges for corporate managers. Providing financial stakeholders with relevant and meaningful information in the context of the rising number of non-financial stakeholders and the multiple and often overlapping regulatory initiatives relating to corporate management and accountability has been a significant challenge, which many corporations have yet to resolve.

In this chapter we first outline the evolution of SRI. We begin with the more traditional concept of 'ethical investment' and then move on to discuss the more recent development of 'best-of-sector' approaches and the emergence of risk-based approaches. These three styles of SRI – which we refer to as the 'three waves' of SRI investment approaches – are an approximate chronology of how the sector has evolved. The discussion then turns to the influence of regulatory and industry-driven responses, which have served to promote the development and expansion of SRI. This leads into an analysis of the demands that the SRI sector imposes on business managers, demands that are in many respects different from those of traditional business management. The final section outlines how industry can respond constructively to the demands upon businesses in terms of environmental and social performance and how these responses can be communicated effectively to the SRI industry.

Trends in socially responsible investment

'Ethical' investment

The origins of SRI lie in the long-standing practices of religious and state institutions of selectively investing in projects or activities with preferred societal outcomes. Its modern roots were planted in the 1920s when the Methodist Church began investing in the stock market, avoiding the 'sin' companies involved in alcohol and gambling (Citizens Funds, 2005). This type of socially responsible investing emphasized the exclusion of activities deemed to be 'morally inappropriate' (such as alcohol, gambling and armaments) or 'environmentally damaging' (such as nuclear power, mining and forestry).

As a sector 'ethical investment' took a stronger foothold in the 1960s and 1970s. It is commonly accepted that the modern SRI movement began in the 1960s, a time when high levels of social activism gave SRI a popular underpinning. In 1971, during the Vietnam era, as conscientious investors sought to direct funds away from the industries support-ing the war effort, the Pax World Fund – widely considered to be the first SRI mutual fund – was launched.

In the 1980s interest in SRI grew substantially with the use of divestiture as a protest against apartheid in South Africa. Although foreign investment in South Africa contin-ued to grow, many American companies withdrew their funds, prompted in part by the activism of an increasingly strong and vocal SRI community (Domini Social Investments, undated). In the United Kingdom the Friends Provident launched the UK's first range of retail ethical funds, Stewardship, in 1984. Popular interest in SRI continued to increase through the 1990s with the heightened focus on tobacco-related litigation and inter-national campaigns against sweatshop labour targeting major multinational brands.

In the late 1990s the interest in sustainability led to a proliferation of ethical investment products. Drawing on the value-based approach, in which investment choices are person-alized to suit organizations' and individuals' preferences, investments were 'negatively' and 'positively' screened to eliminate specific commercial activities or industries from the investment universe and to enhance the portfolio with investments in environmentally or socially 'benign' industries. This first wave of SRI products was driven by personal values and was characterized by selective investments in industries. Economic sectors that created large social or environmental externalities were 'screened out' of the ethical funds. Environmental funds of this nature were often referred to as 'deep green funds', exclud-ing investment in activities that included high impact mining and uranium, and preferring instead innovative technology and renewable energy stocks. Domini Social Investment fund[1] and Jupiter's Ecology Fund[2] are prominent examples of such screened ethical funds.

These funds were appropriate for investors communicating strong environmental pref-erences, although a high tolerance for risk was also necessary as funds were less diversified than those invested across all industry sectors. The approach was therefore less appropri-ate for superannuation funds, which have a lower tolerance for risk and a need to protect their returns through diversified investment.

Following the first generation of ethical funds, subsequent portfolio management approaches have moved SRI closer to mainstream funds management (Figure 22.1).

Linking sustainability to financial performance: best-of-sector and constructive engagement
From the 1990s the growing interest in finance and sustainability and the rise in import-ance of institutional investors created the need for more sophisticated screening techniques for ethical investment products. Institutional investors, particularly superan-nuation, compensation and insurance funds, changed the way in which capital moved, creating a de facto shift from trading to ownership of investments. In 2001 institutional investors held more than 50 per cent of all listed corporate stock in the United States

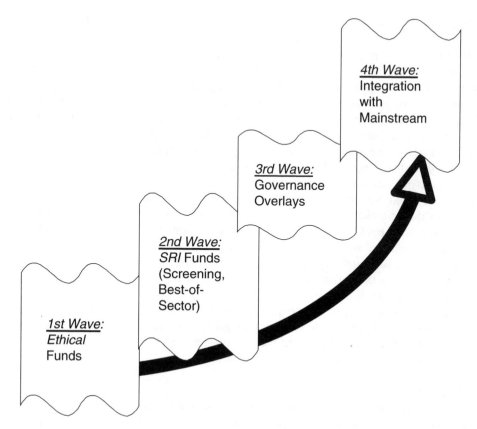

Figure 22.1   The evolution of socially responsible investment

(about 60 per cent in the largest 1000 corporations), with the largest 25 pension funds accounting for 42 per cent of the foreign equity held by all US investors (Goodman et al., 2002). This new class of investors was permanently invested in large parts of the equities markets. The increasing and long-term reliance on investment markets to provide retirement incomes led these investors to focus on principles of good governance and accountability. Interested in the sustainable outcomes of their investments, but constrained by a significantly lower tolerance for risk and strict obligations to meet fiduciary obligations to their funds' beneficiaries, these investors created incentives for the development of more sophisticated screening techniques based on the application of repeatable and rigorous criteria applied across the full industry spectrum of their investments.

The emerging screening techniques identified a range of performance improvement measures across a range of environmental and social criteria. The 'best-of-sector' approach emerged as a portfolio construction technique, enabling greater diversification – relative to negative screening – of investments across all industry sectors. To guide stock selection this approach draws on notions of 'best practice', defined by external standards of environmental and social performance.

The best-of-sector approach works by virtue of the incentive to companies of attaining preferred investment status if their environmental and social performance is exemplary.

Working in the other direction, the possibility of investment withdrawal if environmental and social performance is unsatisfactory acts as a disincentive to inferior practices, as only the top environmental and social performers in each sector are rewarded by investments and companies with lower sector rankings are encouraged to review their activities in pursuit of best practice. This investment philosophy is often accompanied by an engagement component that distinguishes it from most traditional 'ethical' investment funds. The emergence of best-of-sector approaches coincided with the popularization of the concept of 'triple bottom line' performance assessment (Elkington, 1997), increasing the legitimacy of the burgeoning SRI industry and providing investors with a platform to create a shared language with corporate managers.

With the objectives of achieving more mainstream application, best-of-sector funds draw on defined environmental and social criteria for investment and apply them to all industries. Market leaders in providing these criteria include the Dow Jones Sustainability Index Group[3] and the FTSE4Good family of social indexes.[4] Only the most contentious industries are excluded from investment and usually only by exception. By selecting the sustainability leaders in each market sector and across the triple bottom line, funds using a best-of-sector approach tend to be more diverse and are assumed to have a lower investment risk.

The best-of-sector approach represents a further shift toward the mainstreaming of SRI (Figure 22.1) by explicitly linking management of environmental and social risks with financial risk and performance: the best-of-sector SRI funds seek to optimize both socially responsible and financial performance.

Risk management
Long-term shareholders have an incentive to invest more in acquiring information about the effectiveness of management (Ayers and Cramton, 1994) in order to protect the long-term underlying assets of their investments. The type of information that investors use to determine corporate health and returns is changing, with information on management and other non-financial indicators becoming more important in determining investment decisions. International trends in 'value reporting' have developed to supplement traditional financial reporting, thereby helping companies provide a more detailed, transparent picture of their performance, including information on market opportunities, strategy, risks, intangible assets and other important non-financial value drivers. The use of environmental and social indicators to assist in protecting and enhancing long-term investment improves the underlying business value and minimizes the overall risk of the company by focusing on:

- improved workforce quality, productivity and retention;
- increased materials and energy efficiency; and
- effective reputation and brand management through active stakeholder engagement and supply chain management.

A number of approaches for analysing and assessing environmental and social risk for financial markets have been developed (see, for example, Repetto and Austin, 2000b). Methodologies that refine financial methods of evaluating environmental and social aspects can:

- uncover hidden liabilities or risks in acquisitions and mergers;
- measure the true value of a project, facility or company;
- capture the value of investment that would reduce environmental and social exposures; and
- benchmark a facility or a company against its competitors (Repetto and Austin, 2000a).

Assessing the exposure of companies to environmental and social risk through the development and use of indicators measuring governance responses as well as the quantification of environmental and social risks in financial terms, has provided the mainstream investment community with a lexicon to understand better – and therefore integrate – environmental and social practice into mainstream investment decisions. Funds of this sort are characterized by a 'governance overlay', an investment strategy that is superimposed over strictly financial investment decisions. This approach is well established in the UK, South Africa, Australia and the United States where some of the largest pension fund asset managers – such as Hermes,[5] ISIS Asset Management,[6] BT Financial Group[7] and CalPers[8] – all adopt a form of socially responsible and governance overlay to support their investment decisions.

As the tools for the quantification of financial impacts due to environmental and social risk become more focused and accessible, the likelihood that SRI will be taken up by the mainstream markets has increased. This will help to establish further the links between corporate environmental and social activity and shareholder value by defining the materiality of the risk or the scope of the opportunity through financial impacts or added business value.[9] At a minimum, concepts of corporate social responsibility are likely to extend to broader applications in the financial sector, including corporate lending, project finance and insurance underwriting.

Regulatory and industry responses affecting the SRI market
To understand better the impetus behind the SRI market, a number of regulatory and industry initiatives can be identified. The evolving tapestry of national legislation and industry self-regulation indicates the extent to which socially responsible investing is being integrated into the mainstream.

Elements of environmental and social management have long been a part of corporate risk management, aspects of which have been mandated by state-based authorities. For example, maximum levels of noise, air and water pollution are strictly controlled in most developed countries. A number of regulatory initiatives internationally, specifically addressing disclosure of non-financial information, are likely to provide new impetus to the SRI market. These have been complemented by a large number of voluntary industry initiatives and emerging reporting standards.

In the UK for example, the SRI Pensions Disclosure regulation was passed in July 2000. The regulation provided for all occupational pension funds to consider formally whether or not to develop policies on social, ethical and environmental issues. The UK Social Investment Forum survey published in October 2000 found that 59 per cent of the largest pension funds, representing over £230 billion of assets, had incorporated social responsibility issues into their investment strategies (Shepherd, 2001).

The Netherlands, Belgium, Switzerland, Sweden, Germany, France and Australia have followed suit with similar legislative initiatives (Table 22.1). The level of impact of these

Table 22.1   Global trends in SRI

	Common Screens/ Approaches	Legal/Regulatory Environment	Relative SRI Market Size (US dollars)
North America	Tobacco, alcohol, and gambling followed by environmental and human rights, are the most common screens; growing participation in shareholder advocacy and capitalization of community investing	Recent disclosure and transparency requirements by the Securities and Exchange Commission in US amid overall increased regulatory attention	**Canada**: $38.2 billion total in SRI (June 2002), with 53 retail mutual funds **US**: $2.16 trillion total in SRI (Dec. 2002), with 200 retail mutual funds
Europe	Environmental and labour screens are most popular; 'SRI' perceived as more than applying single exclusionary screen; shareholder engagement strategy commonly used	Leadership through new laws and SRI tax-free schemes by country and EU governments boost pro-SRI policy and participation growth	280 retail and institutional funds available as of 2001 $260 billion (including retail, pension, and shareholder advocacy)
Asia	Community investing active at local level; growing number of funds in Japan, Hong Kong and Australia, which provide opportunity for pension funds; environmental screens dominate	**Australia:** disclosure requirements on social and environmental issues in investment decisions **Rest of Asia:** no real SRI-related policy in place, but there is increasing attention especially among multinational fund managers	**Australia:** $14.3 billion (August 2003, with $2.2 billion in managed SRI funds) 74 funds total (31 per cent increase in total number of funds in one year) **Rest of Asia:** $2.5 billion total (December 2002)

Source: Adapted from Social Investment Forum (2003).

regulatory initiatives on a domestic SRI market will vary depending on a large number of factors.

Additionally, corporate law in some jurisdictions is beginning to recognize and integrate non-financial exposures as a core aspect of financial accountability. For example, the UK White Paper on UK Company Law Reform 2002 notes that: 'companies are increasingly reliant on intangible assets such as the skills and knowledge of their employees, their business relationships and their reputation. Information about future plans, opportunities, risks and strategies is just as important to users of financial reports as a historical review of performance'.[10]

The UK government has established an independent group of experts to help in the process of providing guidance on how directors can assess whether an item is material to their company and therefore must be reported. This approach is consistent with the approach taken by the Association of British Insurers (ABI), which holds one in five shares on the UK Stock Exchange. The ABI reporting guidelines[11] call on companies to disclose in their annual report that their boards have assessed the business and reputational risks arising from the way they manage social, environmental and ethical issues and that these risks are being properly managed. A review found that two-thirds of FTSE100 companies were providing either adequate or full disclosure.

There has also been an increase in the number of voluntary guidelines, codes and protocols directed at various environmental, social and other outcomes. These voluntary mechanisms sit outside the framework of formal legal rules but can become 'quasi-mandatory' through their explicit adoption by an organization or their general acceptance across an industry sector. Voluntary instruments have multiple sources, including national governments, non-governmental organizations (NGOs), the United Nations and its various humanitarian and environmental arms, business organizations and many others.

An example of a voluntary instrument is the United Nations Environment Programme Finance Initiative (UNEP FI), which was created to foster commitment to sustainable development and environmental management in a broad range of financial institutions.[12] Also of note is that steps toward the standardization of triple bottom line reporting have been given international status by the Global Reporting Initiative,[13] which was endorsed by the United Nation's Global Compact in 2002. Another international call to financial institutions came at the 2003 World Economic Forum in Davos where over 100 advocacy groups signed the so-called Colleveccio Declaration, calling on financial institutions to implement more socially and environmentally responsible lending policies. This was closely followed by the launch of the Equator Principles, which establish a framework for financial institutions to address environmental and social issues in their review of project proposals and require sponsor compliance with environmental and social policies based on World Bank guidelines and international finance corporate safeguard policies.

These international initiatives have had a significant effect on community views of the societal role of finance and financial institutions. There is now a large number of industry,[14] academic and commercial organizations that focus on defining, promoting and making the links between finance and sustainability.[15]

The demands on corporate managers

At the coalface corporate managers are under pressure to be on top of key sustainability developments, distil the relevant information and address only the most salient requirements crucial to protect and enhance their business. All this must be considered in the context of changing community expectations, new 'classes' of stakeholders with legitimate interests in corporate activity and a bewildering mix of regulatory, industry and non-governmental initiatives in which no clear management or reporting standard has yet emerged.

While the existence of non-financial company stakeholders is not new to corporations, the rise in the sheer number, impact and influence of stakeholder groups has vastly increased the workload required to demonstrate corporate accountability. An organization's stakeholders are those groups that affect or are affected by the organization and its activities. These may include, but are not limited to, owners, trustees, employees, trade unions, customers, members, business partners, suppliers, competitors, government and regulators, the electorate, NGOs, not-for-profit organizations and local and international communities.[16]

By and large corporate responses to stakeholder pressures have been ad hoc, often poorly understood by corporate managers (frequently coming out of the investor relations or communications office) and not integrated with company business management systems and corporate strategy. Systems for recording, handling and communicating the types of information required for deliberate and comprehensive accountability and transparency are vastly different from those targeting a more limited number of stakeholders. In an age of instant global communication and international markets, managing for corporate sustainability must be carefully considered and planned to address the key concerns of the company's most important stakeholder groups.

Amongst corporations leading the corporate sustainability debate, the management of environmental and social impacts is an integral part of business operations. Through transparency and accountability these companies have added environmental and social values to the core values reported to their stakeholders.

The growth and increasing sophistication of the SRI sector over the past several years has raised financial stakeholders' interest in standards of corporate social and environmental behaviour and disclosure. The SRI sector – and now increasingly other sectors in the financial community – understands sound social and environmental management and performance as a lead indicator of corporate health and a driver of future performance. Moreover, information on a company's approach to environmental and social responsibilities provides new insights into the quality of corporate management to the financial community (Commonwealth of Australia, 2003). Corporate social and environmental behaviour can therefore affect financial stakeholders' perceptions and decisions with regard to a specific organization including for example, equities and credit risk investment decisions.

The growing demand from the financial sector for sound social and environmental performance and disclosure represents a challenge for business. Traditionally business communication with the financial community has included reporting on some intangible assets such as the management of customer bases, marketing efforts, procurement processes, creation of new products and financial risk management. Management of social and environmental issues has tended to be a parallel dialogue between corporations and governments, pressure groups and local communities. The corporate sector as a whole has not developed expertise to communicate about social and environmental matters to financial stakeholders, a fact that is exacerbated by the lack of standards to guide communication on sustainability issues between business and the financial community. Appropriate and effective responses to the demands of the SRI sector – and emerging demands from mainstream financial markets – are crucial if corporations are to maximize the benefits of dialogue with the financial community.

Responding to SRI demands

While a number of stakeholder groups have an interest in the social and environmental conduct of corporations, our focus here is on how corporations can respond to the specific expectations of the SRI and broader financial communities on corporate behaviour and disclosure. Corporations can develop and demonstrate sound, value-adding sustainability behaviours to the SRI community and other financial sectors by:

- raising standards of social and environmental management;
- developing an understanding of SRI; and
- developing and implementing a strategy to communicate with the SRI community and financial markets on social and environmental management.

First, raising standards of social and environmental management will assist corporations in demonstrating sound business management as expectations within the financial community increase. High social and environmental management standards can demonstrate improvements in risk and opportunity identification and in delivering management initiatives that improve business performance, adding value to the companies and the communities in which they operate. Second, understanding the SRI market will facilitate the identification of specific information needs. Finally, developing and implementing a strategy to communicate with the SRI community is necessary in order to focus communication efforts to make the best use of resources, to ensure the relevance and completeness of information and to tailor communication to different segments of the SRI market. These three main dimensions of a proactive business response (Figure 22.2) are discussed in greater detail below.

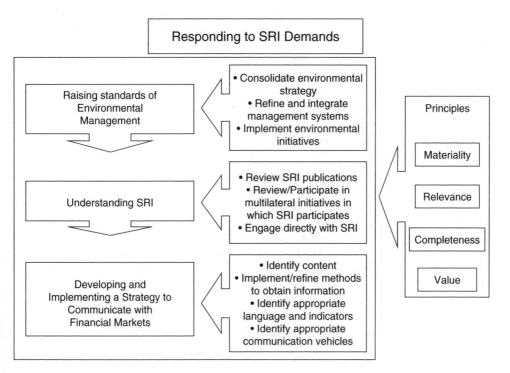

Figure 22.2   Responding to SRI demands

Raising standards of corporate social and environmental management
The first step towards demonstrating sound social and environmental management and performance to the SRI and broader financial sectors is to improve the organization's approach to its responsibilities and opportunities. The spectrum of responses ranges from basic legal compliance to full integration of sustainability principles into business strategy and management. Effort needs to be focused on ensuring a consistent approach to social and environmental management that can be communicated to the financial community in order to demonstrate how the company identifies and manages risks and opportunities, how it generates value through its initiatives and how cost-effective it is in delivering performance improvements.

In order to raise standards of corporate social and environmental behaviour it is necessary that corporations:

● consolidate their social and environmental strategies;
● implement, refine and integrate management systems for social and environmental issues; and
● implement initiatives that deliver social and environmental performance improvements.

A coherent approach at strategic, managerial and operational levels can illustrate sound management to the financial community.

Consolidating the firm's approach to social and environmental issues in a clear strategy serves the purpose of ensuring consistency of sustainability management efforts with business needs. Examples of companies that have elevated sustainability to the strategic level are more often found in the extractive industries, heavy industry and energy sectors. However, companies in sectors with a lower environmental footprint are increasingly adopting a similar strategic approach. For some of these companies, sustainability provides a framework to leverage other positive outcomes for the business, such as innovation and employee retention, whilst ensuring that social and environmental risks and opportunities are factored into day-to-day decision-making.[17]

Consolidating or formulating social and environmental strategy can be achieved by identifying the key risks and opportunities of the corporation. A joint analysis of these risks and opportunities, together with company values, objectives and strategy, will help determine possible courses of action to mitigate risks and realize opportunities. Identifying the tangible or intangible value impacts of actions will help prioritize them within the strategy. This is a key step that will assist meaningful communication with financial markets and can be approached with the use of a variety of business management and sustainability tools. The Sustainable Business Value Matrix (Figure 22.3) can assist as can the World Economic Forum's *Values and Value* report (2004), which provides an analysis of the value added through initiatives addressing the following risk or opportunity dimensions:

● protecting and enhancing reputation, brand equity and trust;
● attracting, motivating and retaining talent;
● managing and mitigating risk;
● improving operational and cost-efficiency;

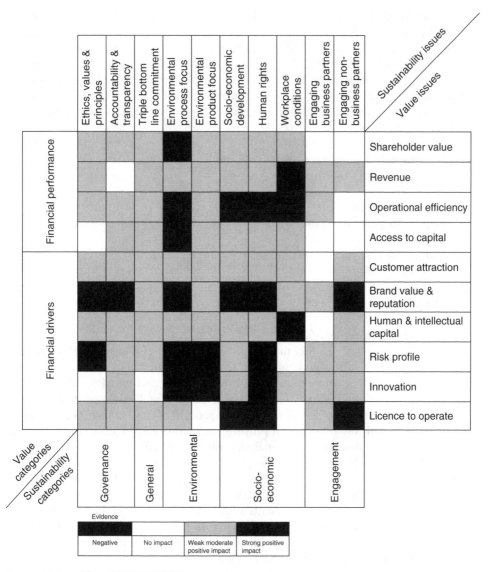

Source: Adapted from WBCSD (2002b).

Figure 22.3 The sustainable business value matrix

- ensuring licence to operate;
- developing new business opportunities, such as new products and services, new markets, new alliances, new business models; and
- creating a more secure and prosperous operating environment.

Once a clear strategic direction has been set to guide the company's social and environmental management efforts, procedures can be established or refined in line with the strategy, allowing the company to organize management efforts, monitor progress and implement corrective measures. The role of such systems is crucial to facilitate the collection of performance data to support communication with financial markets. Guidance on sustainability management procedures or systems can be sought from other chapters in this book and from internationally recognized standards such as ISO 14001. Industry initiatives such as Responsible Care, a voluntary industry code developed by the chemicals sector, may also provide useful frameworks to implement management systems and address particular sustainability risks.[18]

Finally, implementing initiatives that deliver social and environmental performance improvements will demonstrate operational implementation of strategic commitment. The focus of these initiatives might encompass:

- targeting the main social and environmental risks and opportunities;
- helping to achieve organizational goals or address key stakeholder concerns or interests; and
- adding value to the business and society.

These strategies serve the purpose of building stakeholder trust and illustrating organizational effectiveness in the eyes of the financial community.

Developing an understanding of SRI

A clear understanding of the SRI market will establish a useful basis for effective communication with the financial community. Research on business responses to the growing SRI sector indicates that communication between business and the SRI market remains a key challenge. For example, a 2003 survey of European fund managers, financial analysts and investor relations officers found that 56 per cent of fund managers and analysts were dissatisfied with the information provided by companies.[19] Hence corporate communication efforts in relation to sustainability can be more effective if business develops an understanding of SRI needs and appropriately responds to these demands. Of course, the SRI industry must also work to communicate its expectations clearly to businesses.

Corporations can improve their understanding of the SRI sector by reviewing publications, guidelines and position statements authored by members of the SRI community or SRI community associations,[20] by reviewing or participating in multi-stakeholder and/or multilateral initiatives in which the SRI sector is involved to leverage sustainability, and by engaging directly with the SRI sector.

Socially responsible investment publications can provide corporations with information on the sector's information needs. These publications also provide an SRI perspective on current corporate responses to SRI requirements as well as suggestions for improvement in social and environmental behaviour and disclosure. Some examples of publications by the SRI community and other financial stakeholders that may assist include:

- the 2003 survey of European fund managers, financial analysts and investor relations officers (CSR Europe, 2003);
- guidelines for disclosure provided by the Association of British Insurers;[21]
- publications from SRI national and international associations;

- position papers by SRI agencies, which are often available on websites and outline the agency's approach to particular environmental issues, such as those provided by Sustainable Asset Management;[22] and
- guidelines produced by SRI agencies, such as those now being designed by FTSE4Good addressing supply chain labour standards, which include an outline of reporting expectations.[23]

Direct engagement with the SRI community can also provide insights into how particular SRI organizations understand and analyse a specific company. Some SRI organizations may also provide specific suggestions for improving corporate social and environmental performance and comparative information to assist benchmarking against industry peers. Some channels for direct engagement with the SRI community are:

- responding to SRI questionnaires or requests for information. Some companies contract consulting services to analyse SRI information requests, including surveys, so that the company continuously prepares itself to meet information requirements;[24]
- discussing sustainability issues with SRI analysts and/or fund managers;
- organizing seminars to discuss sustainability issues with the financial community (for example, World Economic Forum);[25] and
- encouraging SRI actors and financial stakeholders to provide feedback on communication efforts, for example, using feedback forms in annual reports and corporate websites.

Developing and implementing a strategy to communicate with the SRI community
Having implemented efforts to enhance social and environmental performance and to understand the particular needs of the SRI sector, an organization is well placed to develop and implement a communications strategy. A strategy to communicate sustainability information to the SRI community provides the vehicle to realize benefits where sound social and environmental management can be demonstrated.

A component of the broader corporate sustainability communication strategy, the SRI-specific element will consolidate communication efforts targeting the SRI community under a common directive. This strategic direction will help the business focus its communication efforts, identify the appropriate content of corporate communications on environment and social issues, and determine the most appropriate language, indicators and communication vehicles to transmit the message to the SRI and broader financial community.

Where communications strategies are not clearly defined, corporate disclosure on sustainability often lacks focus and is typically characterized by broad descriptive information that is prospectively of limited value to the financial community. Disclosures to the financial sector should include discussion of material risks and opportunities and information on the business case for its key corporate social and environmental management initiatives.

Completeness and relevance are key principles to be applied when determining the scope of information to be disclosed to the SRI community (Institute for Social and

Ethical Accountability, 1999). To work towards completeness the organization needs to present information that is considered material, including:

- the key social and environmental risks and opportunities facing the business and their relative significance;
- the firm's strategic approach to social and environmental risks and opportunities;
- environmental and social risk management and reporting mechanisms in place;
- management initiatives addressing the key social and environmental risks and opportunities and their value impacts on the business (tangible business value such as improved financial performance by way of cost reduction or intangible business value such as increased intellectual capital after the creation of new environmental technologies); and
- performance information that illustrates the effectiveness of management efforts and their cost-efficiency.

The information provided to the SRI community should also be complete in geographical and chronological terms. Strategic directives, risk evaluation procedures, management efforts and performance need to be presented with clear reference to location, specifying their geographical scope. By way of example, a multinational corporation with sites in the developing world would need to clarify the extent to which management initiatives apply to these operations. With regard to chronology, sustainability performance data should be presented in the context of past performance, where possible. Omission of such information can negatively affect a company's assessment as it implies lack of consistency in initiatives and potential weaknesses in measuring and monitoring systems.

In the absence of a standard to disclose sustainability information to financial markets, corporations face the difficulty of ensuring completeness as it has been defined above. A focus on value and the concept of materiality, borrowed from traditional accounting, can facilitate the task of identifying information that is relevant to the SRI sector and other sectors of the financial community. In addition to the resources provided earlier in this chapter,[26] a number of frameworks to understand and define materiality have been developed. For example, AcountAbility (Zadek and Merme, 2003), presents a definition of materiality together with criteria to assess the materiality of specific sustainability issues. The criteria outlined by AccountAbility comprise: direct short-term financial impacts; policy-related performance (corporate, international, sector or national); peer-based business norms; stakeholder behaviours and concerns; and societal norms. This remains a contentious area however, and it may take years to resolve.

Benefits may also be realized by establishing mechanisms to analyse social and environmental data jointly with data from other business areas – such as finance, people management or business development – as this kind of analysis can help identify synergies between the sustainability and other business dimensions, which in turn provides valuable information to further support the business case for corporate sustainability management initiatives. This is an area that poses continuing challenges however, as it requires analysis of intangible values in which best practice is still evolving. Externally, corporations can benefit from obtaining information on industry's social and environmental performance, which will help the organization benchmark its efforts.

Selecting a suitable language and indicators to communicate sustainability efforts to the financial community will help firms maximize the benefits of disclosure efforts. Corporations can benefit from communicating sustainability management efforts in conventional business language as this helps to set sustainability in a broader business context, establishing the baseline to present the business value impacts of environmental and social initiatives to financial stakeholders. Indicators should provide a platform to clarify internal and external impacts of corporate social and environmental behaviour – including business value impacts and impacts on key stakeholder groups and ecosystems – as well as to communicate the effectiveness of management efforts. Eco-efficiency indicators for example, can condense information on product and service impacts and costs, and can be helpful in presenting the financial value impacts of social and environmental initiatives. Specific indicators of cost-efficiency are also of value. For example, when presenting a new waste management initiative, the financial value impact can be illustrated by comparing waste management costs prior to implementation with those realized with new waste management technologies. Guidelines such as those produced by the Global Reporting Initiative (GRI) can be of help for corporations exploring indicators to report on sustainability. The GRI provides specific guidelines for some sectors, which are the result of extensive, sector-specific stakeholder consultation.[27] However, indicators must be selected with a focus on relevance and value for the specific organization and its key stakeholders.

Communication with financial markets on sustainability issues will be more effective through mechanisms conventionally used by financial stakeholders or specifically tailored for this group. Conventional communication mechanisms that can be used by corporations to communicate sustainability-related information include annual reports, media releases, investor relations websites, investor newsletters, company announcements and analyst briefings. Corporations can also engage in active dialogue with the financial community on sustainability issues through specific sustainability workshops targeting this sector. The specific focus of such workshops will depend on the materiality of specific issues and the level of understanding within the financial community.

Conclusion

While the SRI sector is still in its infancy in many respects, its emergence has created a new set of demands on corporations. Rather than a short-term phenomenon, it appears that SRI is edging toward the mainstream in funds management (Figure 22.1), not only in investment applications but also in assessment of creditworthiness. This 'mainstreaming' is evidenced by the emergence of increasingly sophisticated investment products and applications, which integrate environmental and social performance considerations. Equally important are legislative initiatives designed to enhance disclosure and accountability in the finance sector in relation to social and environmental factors, as these will inevitably impact corporate disclosure. A direct consequence will be much greater scrutiny of measurable improvements in performance across relevant indicators and a greater focus on value impacts, in contrast to the simple negative screens used by the first generation of ethical funds. Moreover, the shift toward the mainstream implies that the assets represented by the SRI sector will continue to grow in weight, and hence in importance, as a driver of corporate strategy.

These developments pose both risks and opportunities for corporations. Understanding and anticipating the increased requirements for information and the changing nature of the information required from financial stakeholders, will provide a strong basis to develop internal information and reporting systems. Opportunities for differentiation from competitors in financial markets are emerging, but their realization will require careful attention to systems development and tailored communication to the financial sector.

Notes

1. http://www.jupiteronline.co.uk/.
2. James McRitchie, Editor, Corporate Governance, http://www.corpgov.net/ (accessed 23 December 2001). The majority of shares issued by quoted companies incorporated in Great Britain are held by institutional investors such as occupational pension funds, insurance companies, unit trusts, open-ended investment companies and investment trusts or, effectively, by the fund managers to whom they may delegate responsibility for governance and voting issues. (*Modernising Company Law: The Government's Policy*, UK White Paper on Company Law Reform, http://www.dti.gov.uk/companiesbill/part2.pdf, accessed 24 November 2005.)
3. http://www.sustainability-index.com.
4. http://www.ftse.com/ftse4good/.
5. http://www.hermes.co.uk/.
6. ISIS Asset Management is the new investment company formed by the merger of Friends Ivory and Sime and Royal and SunAlliance Investments (RandSAI), http://www.isisam.com.
7. BT Financial Group's Governance Advisory Service won the United Nations' 2003 Royal Award for Responsible Investment (http://www.btonline.com.au/downloads/press/20031021_UNRoyalAward_MR.pdf.)
8. California Public Employees' Retirement System, http://www.calpers.ca.gov/.
9. There are a large number of studies that have shown a positive link between improved corporate environmental and social performance and shareholder value. For example, Cooperative Insurance (2002), Sustainability Pays, www.forumforthefuture.org.uk/publications/Sustainabilitypays_page712.aspx; Butz (2004) and US EPA (2000).
10. http://www.dti.gov.uk/companiesbill/part2.pdf, accessed 24 November 2005.
11. *Investing in Social Responsibility – Risks and Opportunities, 2001* (www.abi.org.uk).
12. The Statement on Sustainability that financial institutions endorse is aspirational. The UNEP FI now requires signatories to report on their commitment to the Statement on Sustainability. There are 264 financial institutions in the initiative (down from 303 in early 2003).
13. Multi-stakeholder voluntary sustainability reporting guidelines are available at http://www.globalreporting.org/.
14. Information on regional SRI bodies is available for Asia (www.asria.org), Australia (www.eia.org.au), Canada (www.socialinvestment.ca), Europe (www.eurosif.org) and the United Kingdom (www.uksif.org).
15. See for example: Forum for the Future's Centre for Sustainable Investment (UK), Social Investment Forum (US), Ethical Investment Association (Australia).
16. See for example AccountAbility (1999).
17. See WBCSD (2002a). See also www.bp.com for a corporate example.
18. See for example, www.responsiblecare.com.au/.
19. CSR Europe (2003).
20. See websites of SRI associations: Asia, www.asria.org; Australia, www.eia.org.au; Canada, www.socialinvestment.ca; Europe, www.eurosif.org; and United Kingdom, www.uksif.org.
21. See www.abi.org.uk.
22. See www.sam-group.com.
23. See www.ftse.com/ftse4good.
24. Examples of these approaches can be found at www.stakeholderengage.com.
25. http://www.weforum.org/.
26. Resources to help identify the business value impacts of sustainability management initiatives were provided in the section entitled 'Raising standards of corporate social and environmental management'.
27. See www.globalreporting.org.

References

AccountAbility (Institute for Social and Ethical Accountability) (1999) *Accountability 1000 Standard*, November, www.accountability.org.uk/aa1000/default.asp.
Ayers, I. and Cramton, P. (1994), 'Relational Investing and Agency Theory', *Cardozo Law Review*, **15**, 1033.
Butz, C. (2004), 'Decomposing SRI Performance – Extracting Value Through Factor Analysis', *Pictet Quants*.

Citizens Funds (2005), *History of SRI*, http://www.citizensfunds.com, accessed April 2006.

Commonwealth of Australia (2003), *Corporate Sustainability – An Investor Perspective. The Mays Report*, Canberra: Department of the Environment and Heritage.

CSR Europe (2003), *Investing in Responsible Business: The 2003 Survey of European Fund Managers, Financial Analysts and Investor Relations Officers*, http://www.csreurope.org/pressroom/srisurvey_page4783.aspx, accessed 24 November 2005.

Domini Social Investments (undated), http://www.domini.com/index.htm, accessed 27 January 2004.

Elkington, J. (1997), *Cannibals with Forks: The Triple Bottom Line of 21st Century Business*, Oxford: Capstone.

Goodman, S.B, Kron, J. and Little, T. (2002), *The Environmental Fiduciary: The Case for Incorporating Environmental Factors into Investment Management Policies*, Oakland, California: The Rose Foundation for Communities & the Environment.

Institute for Social and Ethical Accountability (1999), *Accountability 1000 Standard*, www.accountability.org.uk/aa1000/default.asp, accessed 25 November 2005.

Repetto, R. and Austin, D. (2000a), 'Environmental Risk by Numbers', *Environmental Finance*, February, 2.

Repetto, R. and Austin, D. (2000b), *Pure Profit: The Financial Implications of Environmental Performance*, Washington, DC: World Resources Institute.

Shepherd, P. (2001), *A History of Ethical Investment, UK Social Investment Forum*, http://www.uksif.org/Z/Z/Z/lib/2001/05/art-ps-histupd/index.shtml, accessed 24 November 2005.

Social Investment Forum (2003), *Report on Socially Responsible Investing Trends in the United States*, SIF Industry Research Program, October.

US Environmental Protection Agency (EPA) (2000), *Green Dividends? The Relationship between Firms' Environmental Performance and Financial Performance*, http://www.epa.gov/ocem/nacept/green_dividends.pdf.

UK Social Investment Forum (undated), *Social Responsible Investment*, http://www.uksif.org/Z/Z/Z/sri/main/index.shtml, accessed 24 November 2005.

UNEP/SustainAbility (2001) *Buried Treasure*, www.sustainability.com/business-case/matrix.asp, accessed 25 November 2005.

World Business Council for Sustainable Development (WBCSD) (2002a), *Corporate Social Responsibility: Making Good Business Sense*, http://www.wbcsd.org/DocRoot/lunSPdlKvmYH5HjbN4XC/csr2000.pdf.

WBCSD (2002b), *Sustainable Development Reporting: Striking the Balance*, http://www.wbcsd.org, accessed 25 November 2005.

Zadek, S. and Merme, M. (2003), *Redefining Materiality*, UK: AccountAbility.

23 Comparison of international strengths in sustainable technological solutions
*Dora Marinova and Michael McAleer**

Introduction

The rapid development during the industrialization era not only created powerful economies and widespread urbanization, it also established the authority of modern technology. Technological advancement became, and continues to be, a major factor in efficient production, substituting for limited or expensive resources (including human power and natural resources), allowing enormous progress in our understanding of human health, prolonging life expectancies, and organizing and managing individual livelihoods. The technological race was the main arena of competition between the West and the former Eastern European economies, and now continues to play a major part in the development of the North as well as in the economic aspirations of the South.

Industrial development everywhere has been plagued by a lack of environmental and social sustainability. Deterioration in the natural environment, loss of biodiversity and climate change have occurred amid widespread acceptance of social disparities, division of the world and concentration of power. Many see science and technology as a panacea in solving the environmental problems created by centuries of industrialization, as well as in eliminating the gaps caused by a market-driven economic order, where the high consumption patterns of the elite dominate 'rational' behaviour and production. Irrespective of whether humans are capable of resolving the created problems with technology, it is clear that there is a need for new types of technologies that can support the currently emerging culture of sustainability, based on a holistic view of the economic, social and environmental aspects of human activities. These technologies must be economically viable, environmentally friendly and socially acceptable, and provide sustainable technological solutions.

Sustainable technologies

Some major changes in the last few decades gave rise to the understanding that sustainable development can only be achieved through the use of sustainable technologies. What are they and how can countries and companies position themselves against this trend?

The upsurge in economic globalization in the 1980s affected not only trade, labour and financial markets but also impacted profoundly on the nature and rate of technological change. Around the same time, environmental issues became a major concern across countries and industries. The opponents of globalization have ample evidence of the worldwide increase in environmental degradation and economic inequality (see for example, Borghesi and Vercelli, 2003). A major role has been played by technology and technological developments, such as the Green Revolution[1] or the large increase in greenhouse gas emissions from the fossil-fuel-based economies. According to Tisdell (2001, p. 185), the proponents of economic globalization view it 'as a positive force for environmental improvement and as a major factor increasing the likelihood of sustainable development through its likely boost to global investment'. Studies, such as that by the OECD (Johnstone, 1997) claim that there are positive environmental consequences from the interconnected world economy, from the opening of the domestic markets, introduction of environmental legislation and subjection of firms to international demand patterns (including green consumerism).

The development, implementation and use of technologies have been influenced largely by the imperatives of the day and the values embedded in the organizations holding the necessary resources. A significant component of this development has been new technologies that have less impact on the environment and/or help restore environmental health. The intent of these environmental technologies is to reduce the overall ecological impact by humans. Their advantages/benefits include a significant reduction in the environmental impacts of the activities of companies, agencies or people using them (Marinova and Altham, 2002). Companies with strong R&D capabilities and the capacity to bring innovative products and processes realized that a good environmental performance can enhance market performance (Skea, 1994), and started to generate ranges of clean technologies.

The sustainability agenda of the 1990s and 2000s re-emphasized concerns raised earlier during the appropriate technology movement about social responsibility in technology development (see for example, Schumacher, 1973). Socially responsive and environmentally sound (benign or restorative) technological solutions are now expected to become more mainstream on commercial markets. Their number, availability and coverage of sectors and industries are continually increasing. The renewable energy sector is seen as having enormous potential to deliver sustainable technologies. Another example of a large sector that offers opportunities for application of sustainable technologies is the housing industry. The roles of sustainable technologies in the food market, land use and water management are other extremely important and contested areas (for example, Government of Western Australia, 2003). There is also substantial hope associated with the new emerging frontier technologies such as nanotechnologies, which are perceived as inherently green and with great potential to influence current practices (for example, Marinova and McAleer, 2003a).

Sustainable technologies or sustainable technological solutions need to simultaneously and synergistically address market profitability, environmental considerations and social accountability. By balancing these three aspects, they allow for sustainable practices. Both the market and society expect sustainable technology ventures to deliver economic, environmental and social benefits in a synergistic way. These innovations need to be socially acceptable and contribute to appropriate natural resource management.

The sustainable technologies concept has enormous growth potential for each country and globally, and also within each separate industry or across industries. Although the prevailing business attitude may still be to maintain the status quo that demonstrates almost unilateral preference for economic values, the changes towards a more sustainable way of doing things are accelerating. The pressure is coming from all directions, including civil society, government and non-governmental organizations.

There is an emerging consensus that the sustainability imperatives require new approaches and new ways of thinking to respond to the current environmental and social concerns. Innovative companies and research organizations are investing at the forefront of sustainable technology development. This has resulted in intensive investment in research and development, and consequently in numerous new technologies that have the potential to be sustainable. This development is what constitutes the focus of interest for this chapter. It can be expected that in the future all technological development will be sustainable. However, for the time being many traditional technologies have started to incorporate environmental features that make them safer to use. There are also certain new groups of technologies that are viewed as encouraging sustainability more than others.

Patent data have traditionally been widely and intensively used to analyse trends in emerging technologies. Some patent trends in anti-pollution, renewable energy, environmental and nanotechnologies are discussed in the following section. They reveal some interesting, and frequently not so positive, tendencies. Patents however, capture only a limited section of the reality of technological change when it comes to sustainable development.

The deficiencies of patents as indicators for sustainable technological solutions relate broadly to the following issues:

- By its nature, patenting is an extremely prohibitive exercise that favours those countries, organizations and individuals who have the resources, including finances and skills, to register and maintain a patent, be it domestically or internationally. For example, Marinova (2000) observed that in the case of Eastern European countries, there are large variations between firms and national economies as to what costs they can afford (such as patenting fees) to protect their inventions or to purchase the rights of usage of patents that have originated elsewhere. Sustainability requires all countries to act on issues related to climate change and natural resource management, irrespective of their economic resources.
- Moreover, the issuing of a patent is only the beginning of the commercialization process. There is significant evidence that a large number of patented inventions do not become innovations. A study of Australian medical patents in the USA for example, shows that only 50 per cent have been subsequently transformed into commercialized products (Mattes et al., 2006). Only technologies that are implemented and used in real life can provide technological solutions for sustainability.
- According to Griliches (1990, p. 1669), '[n]ot all inventions are patentable, not all inventions are patented, and the inventions that are patented differ greatly in quality'. There are large differences between sectors and the use of know-how, trade secrets, tariffs and regulations and marketing tactics are well-known alternatives to patenting. The criteria used for evaluating an invention in a patent application do not incorporate any sustainability assessment. Moreover, the goals pursued in finding sustainable technological solutions can be in some cases contradictory to the well-entrenched patenting practices.
- Although the concept of sustainability is an expression of global concerns for current and future generations, its implementation requires local technological solutions, that is, technologies and processes that will best suit a particular locality according to its specific bio-regional, geographic, social, cultural, economic, legislative and regulatory specifics. Consequently not all technologies developed are potentially suitable for patenting in a foreign environment. Also, many environmental companies do not actively seek expansion into markets outside their domestic sphere and do not pursue patenting as a strategic goal.

- Many sustainable technological solutions originate from traditional wisdom and knowledge, such as, for example, indigenous peoples' practices of looking after the land, fauna and flora. The patent system is not suited to providing adequate protection and adequately rewarding the communities that have developed and maintained these knowledge and capabilities (Marinova, 2005c). The building up of partnerships between traditional communities, government and private industry for the purpose of achieving sustainability goals can be a much better alternative to the privately driven intellectual property protection mechanisms.
- The transition to a more sustainable way of living and operating requires innovative approaches that go beyond the rigidity and imperatives of the current market system and its rationality. The use of voluntary standards, such as ISO 14001 or the EMAS, the sharing of profits with the communities where some of the technological knowledge originated, the use of the Internet and other information and communication technology, are only a few examples of how things can be done differently.

The remainder of this chapter is as follows. The next section analyses patenting trends in selected sustainable technologies based on information from the US Patent and Trademark Office (PTO). According to Sclove (1995), patents are a source of ideology. Once committed to particular (patented) technologies, states and large corporations could develop an interest in suppressing alternatives in order to keep their market status quo. If they have invested heavily in a particular technology, they are likely to attempt to appropriate the maximum benefits from this. Hence an increasing presence of sustainable technologies in the patenting activities in the USA would potentially indicate some positive trends.

The remaining two sections examine alternative sources of information about sustainable technological solutions, namely, country trends in ISO 14001/EMAS registrations and information in the Internet-based *Green Pages*. The conclusion outlines major differences in comparing countries' performances.

Patent trends in selected technologies

Four representative technological groups have been chosen for analysis to shed light on the quantitative trends in the development of sustainable technologies. These are the anti-pollution, environmental, renewable energy and nanotechnologies. The approach used in this analysis is based on previous studies of patented technologies in the USA (see for example, Marinova and McAleer, 2003a, 2003b, 2006). The data are derived directly from the US Patent and Trademark Office (PTO). It should be noted that, as the assigning of patents to a particular group of technologies is on the basis of keywords, rather than through a thorough analysis of the patents' descriptions and specifications, the actual numbers are essentially of informative value. They are however, sufficiently robust to outline trends and changes that have occurred over a 27-year period, namely between 1975 and 2002.[2]

The first group of patents, namely anti-pollution technologies, is traditionally perceived as a reaction to fix the damage or restore environmental health. Figures 23.1 and 23.2 show that, although the overall number of anti-pollution technologies has been increasing since the mid-1980s (Figure 23.1), their share of all patented technologies has fallen (see Figure 23.2). The most obvious explanation for this is that pollution prevention has become a mainstream requirement for technology development rather than an end-of-the pipe solution.

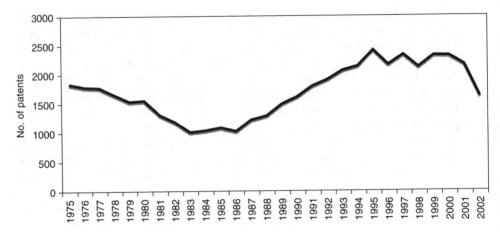

Source: Compiled from US PTO, accessed 8 April 2005.

Figure 23.1   Annual US anti-pollution patents, 1975–2002

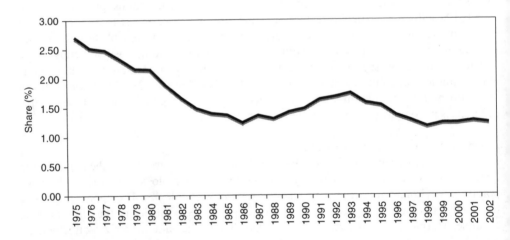

Source: Compiled from US PTO, accessed 8 April 2005.

Figure 23.2   Share of anti-pollution to total US patents (%), 1975–2002

The second group of patented technologies (see Figures 23.3 and 23.4) covers broadly technologies that are designed to benefit or restore the ecology. The absolute and relative
numbers of these environmental technologies have been increasing steadily until the mid-
1990s and then have levelled off. Some concern can be raised in relation to the issue that their
relative share in the overall patented technologies in the USA has remained low at below 2.5
per cent, and has dropped to around 1.8 per cent in recent years. If a significant shift towards
sustainability is expected to occur, then more prominent technology development activities

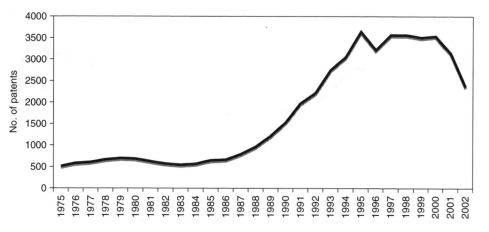

Source: Compiled from US PTO, accessed 8 April 2005.

Figure 23.3   Annual US environmental patents, 1975–2002

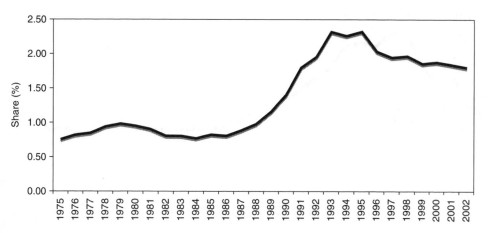

Source: Compiled from US PTO, accessed 8 April 2005.

Figure 23.4   Share of environmental to total US patents (%), 1975–2002

should be observed. However, a possible explanation for the relatively low levels of patenting of environmental technologies can be found in the fact that sustainability often requires technological solutions that are locally grounded. Their application is restricted to the specific conditions of the locality and linked to geo, climate, biodiversity and cultural factors and are not easily transferable; hence the redundancy for patenting.

The group of renewable energy technologies comprises the wide range of solar, wind, wave, tide, geothermal, hydro and biogas patented energy solutions. During the 27-year period, their absolute numbers have been increasing slightly since the late 1980s

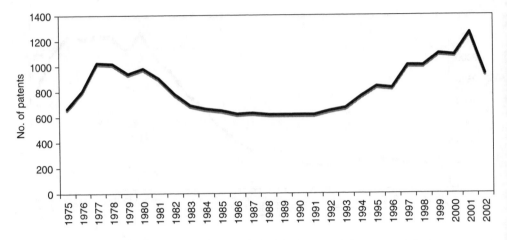

Source: Compiled from US PTO, accessed 8 April 2005.

Figure 23.5   Annual US renewable energy patents, 1975–2002

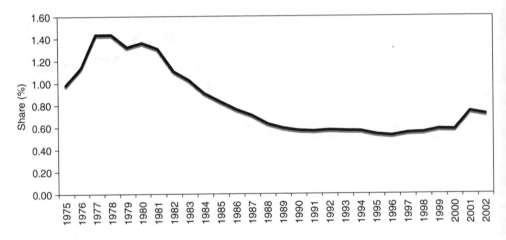

Source: Compiled from US PTO, accessed 8 April 2005.

Figure 23.6   Share of renewable energy to total US patents (%), 1975–2002

(Figure 23.5). However, their relative share has experienced a significant fall from 1.4 per cent to below 0.6 per cent, and only recently has increased slightly to a low 0.7 per cent (Figure 23.6).

The last group of technologies analysed, namely nanotechnologies, are very different in nature as far as sustainability is concerned. These technologies are perceived to be inherently ecological and to have a great future potential in many sectors such as medicine, agriculture, manufacturing, construction, transport and communications, among others. They typically use few resources and, for example, can process all types of waste or can be used

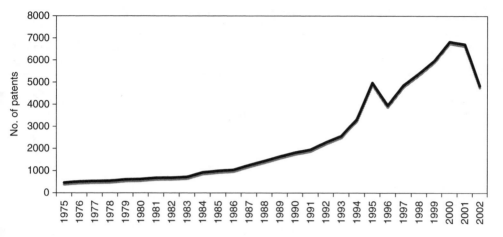

Source: Compiled from US PTO, accessed 8 April 2005.

Figure 23.7   Annual US nanotechnology patents, 1975–2002

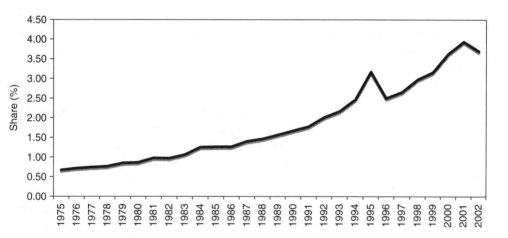

Source: Compiled from US PTO, accessed 8 April 2005.

Figure 23.8   Share of nanotechnology to total US patents (%), 1975–2002

in preventive medicine (Marinova, 2005a). Nanotechnology is also representative of the challenges in interdisciplinary and transdisciplinary research (Nicolau et al., 2000) that permeate sustainability. The absolute number and relative shares of these technologies have dramatically increased in recent years (Figures 23.7 and 23.8). Currently their share in new patented technology is just below 4 per cent. Whether the use of nanotechnologies by society, companies and individuals will lead towards a more sustainable way of living is something yet to be seen and will depend on how deeply the sustainability culture has been adopted in the mainstream.

Table 23.1   Country rankings in technology development, 1975–2002

Country	Anti-pollution Technologies	Environmental Technologies	Renewable Energy	Nano-technologies
Australia	9	3	3	5
Canada	6	5	6	5
France	2	10	1	1
Germany	3	1	3	3
Italy	3	8	9	10
Japan	1	3	2	2
Korea	5	10	10	8
Netherlands	7	2	11	4
Sweden	10	8	8	12
Switzerland	10	6	5	9
Taiwan	7	12	12	10
UK	12	7	7	7

Source: Marinova (2005b).

It is also interesting to examine how different countries fare in the development of these new technologies, the application of which can lead potentially to more sustainable economies. Table 23.1 presents comparative data on a national level for the four groups of technologies for the leading 12 foreign countries patenting in the USA. The USA itself is not included due to disparities between comparing domestic and international patenting. It is estimated that around 50 per cent of all patents registered in the USA originate from foreign countries (Griliches, 1990). The US performance would certainly be the best in all categories but the USA also has the advantage of performing in a domestic market. The ranking of the countries is based on: their share of US patents – as a measure of their global presence; technological specialization index – as a measure of the national importance of the technologies; and proximity to the market – as a measure of commercialization (for further details see Marinova, 2000).

During the 1975–2002 period examined, the leading 12 foreign countries covered 45 per cent of all anti-pollution technologies, 29 per cent of nanotechnology, 27 per cent of environmental technologies and 26 per cent of renewable energy. This is also indicative of the relative importance that foreign countries assign to the implementation of the sustainability concept in the US market. Japan ranks first in anti-pollution technologies, second in renewable energy and nanotechnologies and third in environmental technologies. France is ranked first in nanotechnologies and renewable energy technologies, while Germany is ranked first in environmental technologies.

Overall, Japan appears to be the best performing country, followed closely by Germany, while France is ranked relatively highly in three of the four technological groups. Australia is ranked fourth on average, followed by Canada, the Netherlands, Italy, Switzerland, South Korea, UK, Sweden and Taiwan. The presence of the newly industrializing countries of South Korea and Taiwan is mainly the result of a concerted effort in research and technology development in more recent years. As indicated previously, notwithstanding the bias of using domestic patenting, the US presence in all four groups of technologies is highly significant. Nevertheless the relatively low shares of these patents indicate that the dominant business interests and practices remain still distant from sustainable technological solutions.

Against this background, it is essential to look for changes in other practices within society and the business community. This is examined below.

Green businesses

In the current networked world, the Internet is a powerful tool in providing and finding information about sustainable technological solutions. The *Green Pages* on www.eco-web.com has been a useful database for environmental technologies since 1994. Based in Switzerland and currently in its 45th issue, the *Green Pages* lists 7000 technological solutions from 146 countries claiming to be the online gateway to environmental products and services that 'facilitates cross-border direct trade by bringing together suppliers and buyers. It aims to deepen and broaden consumers' ability to make a positive impact through their purchasing, and promotes a shift of public and private financial flows into projects and programs that encourage and contribute to environmentally and socially sustainable development' (*Green Pages*, 2005).

The database covers nine major groups of environmental technologies, namely (*Green Pages*, 2005):

- waste water treatment, including domestic waste water, industrial effluent recycling, sewer pipe systems, physical treatment, bio-augmentation, chemical treatment, sludge handling and general waste water services;
- water supply and purification, including water analysis and instruments, water sources, water distribution, filtration and conditioning, membrane processes, demineralization and general water purification;
- air pollution control, including air pollution monitoring, climate change mitigation, dust cyclones, dust and gas scrubbers, electrostatic precipitators, air filtration, vapour recovery processing and general air purification;
- waste management, including refuse collection and transportation, refuse containing, sorting, shredding, compacting, incineration and disposal and waste management services;
- recycling, including material recycling, material transformations, resource recovery, composting processes and general recycling services;
- soil preservation, including soil diagnostics, decontamination and rehabilitation, forestry, agriculture, ecotourism, ecosystem management and general geotechnical engineering;
- noise protection, including noise analysis and instruments, noise abatement facilities and general noise protection engineering;
- power generation, including photovoltaic solar cells, solar electric systems, wind power systems, hydro power engineering, bio reactors, fuel cells, geothermal power generation, natural gas co-generation and general power engineering; and
- energy efficiency, including burners and boilers, heat exchangers and heat pumps, cooling and ventilation, lighting devices, insulation technology, housing development, batteries and storage devices, new engine and fuel technology, transportation systems, energy-saving consumer products and general services.

Table 23.2   Countries with the largest presence in the Green Pages

Country	Number of Technologies	% of Total Technologies in the Database	Intensity per 1 Million Population
USA	1371	19.59	4.64
UK	643	9.19	10.64
India	480	6.86	0.44
Canada	453	6.47	13.81
Germany	337	4.81	4.09
Australia	212	3.03	10.55
Netherlands	210	3.00	12.80
Italy	189	2.70	3.25
China	172	2.46	0.13
Turkey	148	2.11	2.12
Spain	147	2.10	3.64
France	136	1.94	2.24
Switzerland	100	1.43	13.35
Total selected countries	4598	65.69	1.47
Total	7000	100.00	1.08

Source: Calculated from www.eco-web.com, http://www.census.gov/ipc/www/idbprint.html, accessed 7 November 2005.

The popularity of the web-based database has been growing consistently with the number of hits per day increasing from just above 2500 in 1997 to about 43 500 in 2004. The range of technologies and representative companies is vast, as is the country presence. The USA continues to be the leading country with 1371 technologies registered. However, the list of the next 12 leading countries is quite different than the top ranking patenting countries. In addition to nine developed countries, namely Australia, Canada, France, Germany, Italy, the Netherlands, Spain, Switzerland and UK, it includes three developing countries, namely Turkey, China and India (the latter two having the world's largest populations).

Table 23.2 lists these 13 countries according to their contribution to providing sustainable technological solutions. The most technology-intensive country from the list is Canada, followed by Switzerland, the Netherlands, UK and Australia. India and China are ranked lowly according to technological intensity, but India is the third largest contributor to this range of technologies, behind the USA and the UK. It is also interesting to note that Japan, which consistently provides the largest number of foreign patents in the USA, does not appear among the leading 13 *Green Pages* countries. The latter observation also holds for Korea and Taiwan.

Table 23.3 shows for each country the environmental technology group with the highest value of their technology specialization index.[3] Again the diversity between countries is significant, with all but two (namely water supply and purification, and waste management) technology groups appearing as an area of strength for at least one country. Overall energy efficiency technologies appear to be the most important and are also areas of

Table 23.3   The highest technology specialization index by country

Country	Technology	Technology Specialization Index
USA	Energy efficiency	1.16
UK	Recycling	1.20
India	Air pollution control	1.22
Canada	Energy efficiency	1.43
Germany	Power generation	1.38
Australia	Noise protection	1.21
Netherlands	Soil preservation	1.23
Italy	Recycling	1.26
China	Energy efficiency	1.98
Turkey	Wastewater treatment	1.64
Spain	Noise protection	2.37
France	Power generation	1.67
Switzerland	Power generation	1.69
Total selected countries	Energy efficiency	1.11

Source: Calculated from www.eco-web.com, accessed 7 November 2005.

specialization for the USA, Canada and China. This is followed by power generation, which is of great interest to Germany, France and Switzerland.

Environmental Management Systems

The increasing popularity of the uptake of the voluntary standards of ISO 14001 world-wide and the EMAS (Eco-Management and Audit Scheme) in Europe is presented on Figures 23.9 and 23.10. The number of ISO 14001 certified sites has increased four-fold, from 20 842 in January 2001 to 88 800 in April 2005 (Figure 23.9a). Most importantly the number of countries where companies and organizations have ISO certified sites has increased globally by around 50 per cent, from 97 in 2001 to 140 in 2005 (Figure 23.9b). This is now similar to the web-based presence in the *Green Pages*.

The uptake of EMAS certification in Europe is also very impressive (Figure 23.10). The number of organizations grew from 1269 in December 1997 to 3797 in 2002, after which it appeared to have levelled off. However, in 2004 there was a change in the way statistics on the EMAS were collected in order to reflect more accurately the implementation of environmental management, which is site-specific. Instead of reporting only the number of organizations, the counting of actual sites is now more representative of the changes occurring within companies. This also makes EMAS data more compatible with ISO 14001 statistics. As an indication of the significant changes in certification, in the nine months between December 2004 and September 2005 alone, the number of EMAS sites increased by 4 per cent from 4093 to 4253 (Figure 23.10).

The countries with the largest number of ISO 14001 certified sites are shown in Table 23.4. It is interesting to note that Japan heads the list with 18 104 sites, which is more than 20 per cent of all sites certified globally. This is followed by China and Spain. The USA is sixth, but on a per capita basis, the world's largest economy is ranked even lower, with only 16 sites

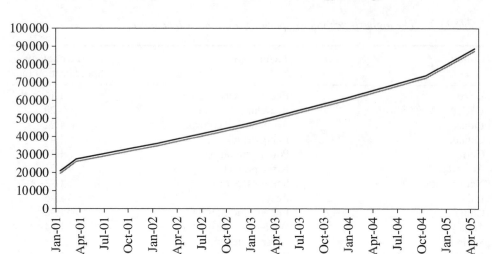

a. ISO 14001 sites

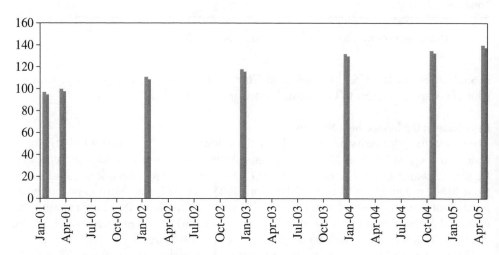

b. Number of countries with ISO 14001 certified sites

Source: Compiled from http://www.ecology.or.jp/isoworld/english/analy14k.htm.

Figure 23.9   ISO 14001 certified sites globally, 2001–05

per 1 million people (Table 23.4). Sweden leads this group of countries by ISO 14001 site intensity, with 413 sites per 1 million of population, or more than 20 times the average of the group. Spain and Japan are also ranked very highly according to site intensity.

The presence of China and India in the group of countries with a large number of ISO 14001 certified sites is very encouraging as it is a sign of the positive changes in relation to the management of technologies and care for the natural environment. Another significant observation that this approach to comparing countries reveals is the development in Latin America, with Brazil appearing in the group of the leading countries (Table 23.4).

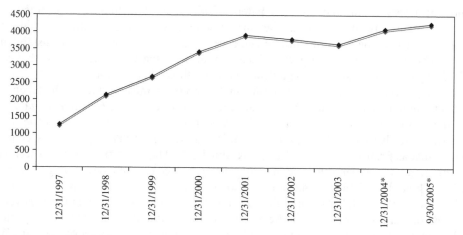

Note: Oganizations/sites*.

Source: Compiled from http://europa.eu.int/comm/environment/emas/, accessed 7 November 2005.

Figure 23.10   *EMAS certified organizations (1997–2003) and sites (2004–05)*

Table 23.4   *ISO 14001 certified sites by country, 2005*

Country	Certified Sites April 2005	% of Total ISO 14001 Certified Sites	ISO 14001 Sites per 1 Million
Japan	18 104	20.39	142
China	8 863	9.98	7
Spain	6 523	7.35	162
UK	6 223	7.01	103
Italy	5 304	5.97	91
USA	4 671	5.26	16
Germany	4 440	5.00	54
Sweden	3 716	4.18	413
Korea, South	2 610	2.94	54
France	2 607	2.94	43
Brazil	1 800	2.03	10
Canada	1 706	1.92	52
India	1 500	1.69	1
Average	*5 236*	*5.90*	*20*

Source: Compiled from http://www.ecology.or.jp/isoworld/english/analy14k.htm, http://www.census.gov/ipc/www/idbprint.html, accessed 7 November 2005.

Conclusion

A comparison of countries reveals that, although the concept of sustainability has been in existence for decades now, the practical adoption of the concept of sustainable technological solutions has only just started. Therefore, it is very important to rely on information that goes beyond the official patenting data. Sustainable technologies still represent a very small share of new patents. The types of technologies patented are however, highly informative about the current state and priorities of the economically strong countries. There is only limited evidence of changes towards a more sustainable world. This also raises very important issues about the disparity between those that can afford to pay for more sustainable technologies and those that cannot.

Alternative ways of accessing environmental technologies and assuring that better environmental practices are gaining momentum can be witnessed by the *Green Pages* gateway and the ISO 14001/EMAS certifications. There is also a limited number of countries that perform relatively well in all three ways of making technological changes, namely six countries: Canada, France, Germany, Italy, UK and USA. Their contributions and technological interests however, differ significantly. The presence of some developing countries such as China, India, Turkey and Brazil, in some of the groups is extremely encouraging.

The efforts of the global community to combat the problems of social justice, poverty and environmental degradation in the pursuit of a better quality of life will be supported and dramatically enhanced by smart and breakthrough technological developments. We are currently witnessing the emergence of the new sustainable technologies that will contribute to the development of the ethics, values and practices of the new sustainability culture. However, it will be up to the global community to make the best use of them for the sustainability of all people on the planet Earth.

Notes

* The authors wish to acknowledge the financial support of the Australian Research Council.
1. Shiva et al., 1994 describe the detrimental effect of the Green Revolution on village life in India and in other developing countries.
2. The patent data are by date of patent application, not by date of issue, which tends to distort technological trends. It takes at least two years after an application is lodged for a patent to be issued. In some cases, this delay can take as much as ten years. The data were extracted from the US PTO database on 8 April 2005. The patent numbers for recent years are likely to increase slightly as successful delayed applications are gradually approved.
3. The technology specialization index compares a country's performance in a technology with the overall presence of this technology in the database (for further details see Paci et al., 1997). An index of 1 indicates a performance that is similar to the average; a value less than 1 indicates lower interest than the average, and a value greater than 1 indicates specialization. The higher the value, the more important is this technology for the country.

References

Borghesi, S., Vercelli, A. (2003), 'Sustainable Globalization', *Ecological Economics*, **44**(1), 77–89.
Government of Western Australia (2003), *Hope for the Future: The Western Australian State Sustainability Strategy*, Perth: Department of the Premier and Cabinet, Western Australian Government.
Green Pages (2005), Preface, http://www.eco-web.com/, accessed 6 November 2005.
Griliches, Z. (1990), 'Patent Statistics as Economic Indicator: A Survey', *Journal of Economic Literature*, **28**(4), 1661–707.
Johnstone, N. (1997), 'Globalization, Technology and Environment', in *Globalization and Environment: Preliminary Perspectives*, Paris: OECD, pp. 227–52.
Marinova, D. (2000), 'Eastern European Patenting Activities in the USA', *Technovation*, **21**(9), 571–84.
Marinova, D. (2005a), 'Development of Sustainable Technologies', in Banse, G., Hronszky, I. and Nelson, G. (eds), *Rationality in an Uncertain World*, Berlin: Verlag Edition Sigma, pp. 211–20.
Marinova, D. (2005b), 'Trends in Developing Sustainable Technologies', in *Proceedings of the 3rd Dubrovnik Conference on Sustainable Development of Energy, Water and Environment Systems*, Dubrovnik, Croatia: CD ROM.
Marinova, D. (2005c), 'Protecting Indigenous Knowledge', in Zerger, A. and Argent, R. (eds), *MODSIM 2005 International Congress on Modelling and Simulation*, Melbourne, Australia: MSSANZ.

Marinova, D. and Altham, W. (2002), 'Environmental Management Systems and Adoption of New Technology', in Jamison, A. and Rohracher, H. (eds), *Technology Studies and Sustainable Development*, München: Profil, pp. 199–220.

Marinova, D. and McAleer, M. (2003a), 'Nanotechnology Strength Indicators: International Rankings Based on US Patents', *Nanotechnology*, **14**(1), R1–R7.

Marinova, D. and McAleer, M. (2003b), 'Modelling Trends and Volatility in Ecological Patents in the USA', *Environmental Modelling and Software*, **18**(3), 195–203.

Marinova, D. and McAleer, M. (2006), 'Anti-pollution Technology Strengths Indicators: International Rankings', *Environmental Modelling and Software* (in press, available online 28 September 2005).

Mattes, E., Stacey, M. and Marinova, D. (2006), 'Predicting Commercial Success for Australian Medical Inventions Patented in the United States: A Cross-sectional Survey of Australian Inventors', *Medical Journal of Australia*, **184**(1), 33–8.

Nicolau, D.E., Phillimore, J., Cross, R. and Nicolau, D.V. (2000), 'Nanotechnology at the Crossroads: The Hard or the Soft Way?', *Microelectronics Journal*, **31**(7), 611–16.

Paci, R., Sassu, A. and Usai, S. (1997), 'International Patenting and National Technological Specialization', *Technovation*, **17**(1), 25–38.

Schumacher, E.F. (1973), *Small is Beautiful: A Study of Economics as if People Mattered*, London: Vintage.

Sclove, R.E. (1995), *Democracy and Technology*, New York: The Guilford Press.

Shiva, V., Noeberg-Hodge, H., Goldsmith, E. and Khor, M. (1994), *The Future of Progress. Reflections on Environment and Development*, Dehra Dun, India: Natraj Publishers.

Skea, J. (1994), 'Environmental Issues and Innovation', in Dodgson, M. and Rothwell, R. (eds), *The Handbook of Industrial Innovation*, Aldershot, UK and Brookheld, US: Edward Elgar, pp. 421–31.

Tisdell, C. (2001), 'Globalization and Sustainability: Environmental Kuznets Curve and the WTO', *Ecological Economics*, **39**(2), 185–96.

24 Sustainability assessment
Peter Newman

Introduction
There are three types of sustainability assessment emerging:

1. sustainability assessment of complex and strategic projects;
2. sustainability assessment of policies, programmes and plans; and
3. sustainability assessment of buildings and developments.

This chapter will examine all three approaches within a global context and within the local context of recent developments in Australia, especially the State Sustainability Strategy in Western Australia (Government of Western Australia, 2003), which is the first sub-national strategy of its kind, and the Metropolitan Strategy in New South Wales (Government of NSW, 2004).

Sustainability assessment of complex and strategic projects
The approach to sustainability assessment adopted in Western Australia (WA) was created as part of the Sustainability Strategy developed over the period 2001 to 2003. The newly elected Gallop Government had a commitment to develop a Sustainability Strategy across all sections of government, and following much public discussion the definition of sustainability adopted was that of 'meeting the needs of current and future generations through an integration of environmental protection, social advancement and economic prosperity' (Government of Western Australia, 2003, p. 2). In the framing of this definition, the Strategy sought to address what sustainability could mean for assessment. The approach taken in the Strategy was to develop a series of background papers on contentious issues, and eventually two papers dealing with sustainability assessment were placed on the Government's website.[1] The second of these papers covers the results of a working party, which developed ideas from industry and the community. These ideas included the detailed approach adopted by Hammersley Iron (Rio Tinto Company) for sustainability assessment of projects, which was developed by the company for integrating social, environmental and economic factors into their internal project development framework. The final approach adopted by the Sustainability Strategy is relevant to sustainability assessment of all projects, whether they be large resource projects in remote areas or complex infrastructure projects in the city.

A parallel government process initiated to evaluate the major projects approvals process (Government of Western Australia, 2002 or the Keating Review) also generated a commitment to sustainability assessment in WA. This commitment was based on the potential of sustainability assessment to speed up and integrate decision-making through its capacity to consider economic and social factors in parallel to the existing environmental assessment process. In the course of the Keating Review, a commitment was made to create a demonstration project, rather than shy away from controversy, it was decided that the Gorgon gas development would be subject to a sustainability assessment.[2]

The sustainability assessment of the Gorgon project involved a steep learning curve for government agencies and consultants. It has been criticized on the grounds that it did not include enough options for assessment, but rather aimed to develop a detailed and integrated approach to the one option favoured by the company (Pope, 2003 and 2004). Thus in the review of the sustainability assessment process within government, it was decided that if a sustainability approach was going to work, much greater emphasis had to be placed on the 'scoping stage' of project assessment, as well as on the final integration of economic, social and environmental factors. This realization echoes a lot of the international discussion on sustainability assessment (for example, Gibson, 2005), which has been examined in some detail by Pope et al. (2004) in a paper that uses the WA sustainability assessment experience to further conceptualize the ideas.

Essentially what was proposed in the Western Australian Sustainability Strategy was an approach to policy based on 11 sustainability principles (Table 24.1).

These principles were then applied to 42 areas of government, one of which was sustainability assessment. The Strategy proposed that assessment should move from minimizing impacts to promoting positive outcomes in an integrated way. It therefore suggested a set of criteria for sustainability assessment based on the sustainability principles, which contrasts the traditional 'impact assessment' approach with that of 'sustainability assessment' (Table 24.2).

This framework for decision-making is a challenge for any government; it certainly has not been a straightforward task of simply adopting it in WA. That will require institutional changes to create expertise in the economic and social 'bottom line' areas, as well as to integrate these with the normal environmental assessments done through the Environmental Protection Authority. Surprisingly there is very little economic analysis done of projects at present, and all social assessment has been stripped from the bureaucracy. Moreover, there is a recognition of the need to create an integrative function within the Department of the Premier and Cabinet, which would conduct its work before options are taken to Cabinet. These institutional ideas are still being worked through, though in general it is considered that sustainability assessment will emerge as the major way that complex and strategic projects are evaluated in WA.

The next phase of sustainability assessment in WA is to apply the approach to a major new port development in the outer harbour of Fremantle and to a major new water supply project. The rationale behind the adoption of a sustainability assessment process, rather than an ordinary environmental assessment process, is that the State Government wants to see how it can address complex and strategic projects in a way that enables trade-offs to be minimized. Politicians are accustomed to addressing the full triple bottom line on projects, even where the majority of the work has traditionally been done on environmental impacts rather than socio-economic considerations. Particularly in urban systems the distinctions between these areas are blurring, and the need to provide more detailed analysis of options for politicians has become increasingly apparent. Other large industries seeking to expand to take advantage of the resources boom currently under way in WA are also asking for sustainability assessments. Sustainability assessment of complex and strategic projects is thus the next stage upon which sustainability assessment will be trialled and developed in WA.

The political imperative for sustainability assessment is now quite large in a state that

*Table 24.1   Sustainability principles as developed for the Western Australian State
Sustainability Strategy*

	Foundation Principles
Long-term economic health	Sustainability recognizes the needs of current and future generations for long-term economic health, innovation, diversity and productivity of the earth
Equity and human rights	Sustainability recognizes that an environment needs to be created where all people can express their full potential and lead productive lives and that significant gaps in sufficiency, safety and opportunity endanger the earth
Biodiversity and ecological integrity	Sustainability recognizes that all life has intrinsic value, is interconnected and that biodiversity and ecological integrity are part of the irreplaceable life-support systems upon which the earth depends
Settlement efficiency and quality of life	Sustainability recognizes that settlements need to reduce their ecological footprint (that is, less material and energy demands and reduction in waste), while they simultaneously improve their quality of life (health, housing, employment, community. . .)
Community, regions, 'sense of place' and heritage	Sustainability recognizes the significance and diversity of community and regions for the management of the earth and the critical importance of 'sense of place' and heritage (buildings, townscapes, landscapes and culture) in any plans for the future
Net benefit from development	Sustainability means that all development, and particularly development involving extraction of non-renewable resources, should strive to provide net environmental, social and economic benefit for future generations
Common good from planning	Sustainability recognizes that planning for the common good requires equitable distribution of public resources (like air, water and open space) so that ecosystem functions are maintained and so that a shared resource is available to all
	Process Principles
Integration of the triple bottom line	Sustainability requires that economic, social and environmental factors be integrated by simultaneous application of these principles, seeking mutually supportive benefits with minimal trade-offs
Accountability, transparency and engagement	Sustainability recognizes that people should have access to information on sustainability issues, that institutions should have triple bottom line accountability, that regular sustainability audits of programmes and policies should be conducted, and that public engagement lies at the heart of all sustainability principles

Table 24.1   (continued)

Precaution	Sustainability requires caution, avoiding poorly understood risks of serious or irreversible damage to environmental, economic or social capital, designing for surprise and managing for adaptation
Hope, vision, symbolic and iterative change	Sustainability recognizes that applying these principles as part of a broad strategic vision for the earth can generate hope in the future, and thus it will involve symbolic change that is part of many successive steps over generations

Table 24.2   Criteria for sustainability assessment in the WA Sustainability Strategy

Managing the Negative (Impact Assessment)	Promoting the Positive (Sustainability Assessment)
Provides short-term economic gain but long term is uncertain	Provides both short- and long-term economic gain
Minimizes impacts on access, equity and human rights in the provision of material security and effective choices	Increases access, equity and human rights in the provision of material security and effective choices
Avoids damage to biodiversity, ecological integrity and life-support systems	Improves biodiversity and ecological integrity and builds life-support systems
Minimizes the increase in ecological footprint while improving the quality of life	Reduces the ecological footprint while improving the quality of life
Minimizes impacts on community and regions, 'sense of place' and heritage protection	Builds up community and regions, 'sense of place' and heritage protection
Minimizes conservation loss and social impact while providing economic benefit	Provides conservation benefit and net social-economic benefit
Minimizes the reduction of 'common good' resources	Increases 'common good' resources
Minimizes the risks that are not understood	Ensures there are acceptable levels of risk with adaptation processes for the worst scenarios
Brings change without hope for the future as it is not part of a broader strategic vision	Brings change and a sense of hope for the future as it is linked to a broader strategic vision

has a history of environmental and social awareness and that is under considerable pressure for economic expansion (8 per cent Gross State Product growth per annum in 2004). However, the language and the institutions for providing the integrative approaches required are lagging well behind this political imperative.

This process of examining the institutional framework whilst trialling the process reflects the methodological questions that are being asked internationally in academia (Biegelbauer and Borrás, 2003; Lakoff, 2004; Pretty, 2003; Schön and Rein, 1994; Scott and Gough, 2003). These questions include:

- What are the best ways to enable the scoping phase to include all the critical factors?
- How do you integrate these to enable pre-feasibility studies to occur on the best options?
- How do you incorporate agency consideration and public participation into this phase of assessment so that a sense of the critical issues can be determined?
- What models do you use to provide integrated advice that evaluates options for decisions based on sustainability criteria that can then be evaluated by Cabinet?
- How do you bring social issues into the process in a meaningful way?

These issues will be examined in the next two areas of sustainability, which are indicating some solutions because of the scale at which they are being applied.

Sustainability assessment of policies, programmes and plans
The application of sustainability to the evaluation of policies, programmes and plans is a major thrust in international literature as the importance of strategic environmental assessment emerges. This process, which has evolved from the project-based environmental impact assessment process introduced in the previous section, has raised questions about why the other elements of the triple bottom line should not be included in any strategic analysis of the future (Verheem, 2002). The Dutch and the Canadians are doing a lot of work in this area, as is the UK Government.[3]

Strategic analysis is critical for any urban system, so that the planning system has developed to ensure that cities are able to integrate any factor into their future. Thus planning developed in the late nineteenth century as a way to integrate the health factors associated with waste and water into the design of the city and its neighbourhoods, as well as to address other social and environmental issues such as parks and transport. Planning has since become a complex system of legal and bureaucratic processes that is reviewed every decade or so to check for any changes in strategic direction. It is in the context of this process that sustainability assessment is now emerging in a different way to the assessment of projects discussed above.

Strategic planning is based on an assessment of the underlying values guiding the long-term planning of a city as well as how the city can best accommodate the expected number of houses and businesses in an economically, socially and environmentally acceptable way. Thus it is not unexpected that strategic planning exercises on cities in the past decade have increasingly looked to sustainability as a guiding concept. In Australia sustainability is at the heart of the Melbourne Metropolitan 2030 Strategy, the Sydney Metropolitan Strategy, the Perth 'Network City' Plan, the Tasmania Together Plan, the Brisbane SEQ 2010 Plan and the Adelaide Metropolitan Strategy.[4]

The UK has probably gone as far as any place in directing its planning agencies to incorporate sustainability into the planning system (Benson and Jordan, 2004; Owens and Cowell, 2002; Selman, 1996). They call it 'sustainability appraisal' and it is well established in law and in the culture of managing change in the UK.

The sustainability assessment process, as it is being applied to the strategic planning process, will be discussed in terms of the role of politics in this process and the roles of indicators and stories, as well as regional planning. These aspects of the process will be examined as a way of trying to see how the required language and institutions for sustainability assessment can begin to catch up to the political need for action in sustainability.

Politics and sustainability assessment in planning
Owens and Cowell (2002) stress the importance of seeing that planning for sustainability cannot be divorced from politics. Any process that suggests that policies, programmes and plans can be subject to a sustainability audit, integrated with a whole new set of assumptions and then given a discursive participation process, cannot expect this to be other than a highly political exercise. Technical inputs will always be needed, but in the end the changes demanded for sustainability – reduced resource consumption, less impact and greater benefit to the common good – cannot occur without some pain and some resentment from those who do not wish the change to happen. This resistance can be from financial interests, environmental interests and social interests alike. The importance of the underlying principles that guide the process is therefore highlighted, as is the realization that sustainability assessment cannot ever become some monolithic machine that takes over from the need for elected politicians to make hard decisions. It is hoped however, that it can help politicians to make those decisions.

Nothing more clearly supports Owens and Cowell's proposition than the process of strategic transport planning in cities; especially issues related to reducing car dependence, which has been the main subject of my research (see Newman and Kenworthy, 1999). No city that addresses its future can any longer neglect to see the impact of the car on energy, greenhouse gases, air quality, land-take from rural areas and bushland, noise and physical separation due to large roads, community and health impacts from excessive car dependence and loss of economic competitiveness. Yet when issues to do with car dependence are raised in strategic planning reviews there is an immediate rush to defend political interests from within and outside government over land development and priorities in infrastructure provision, as well as the so-called, in Australia, love affair with the car and the suburb.

Given that political realities cannot be by-passed what can sustainability assessment do to take them in hand? The main way of addressing these realities is to ensure first that good sustainability analysis of all the urban options for the future are conducted, showing data on all the variables outlined, and second that a large-scale public process is conducted to help provide an evaluation of the options available. There is substantial evidence that public servants and planning bureaucracies interpret options to overcome car dependence in much less radical ways than the public. This was demonstrated in a 2003 'Dialogue for the City' conducted in WA's capital city, Perth. This process involved 1100 people in a sophisticated planning process over a weekend.[5] Members of the public involved in the process concluded that environmental issues were much more significant, and the need for public transport a much greater priority, than had been interpreted by prevailing administrative processes. The main issues chosen by the public to be the basis for how they wished the city to be re-planned are set out in Table 24.3.

The development of future strategies to deliver this vision will always remain contested and contentious, but it is possible to translate the priorities into transport and land use plans with a degree more certainty after such a process. Strategic planning is of course a

Table 24.3   Vision priorities from the 'Dialogue for the City' process in Perth, 2003

1. Strong local communities (city of villages)
2. Clean, green city
3. Urban growth boundary
4. Connected, multi-centred city
5. Reduced car dependence – better public transport, especially more rail, better local bike/walk and integrated transport/land use
6. Housing diversity (more options)
7. Access to city services for all

two-way process, and the educational value of delving into such issues with a large cross-section of the community cannot be discounted as a means of driving change. The other thing that such processes do is to help develop a set of sustainability indicators for the city.

Indicators and stories
Indicators have become part of the tool box of sustainability assessment in all three areas of sustainability assessment, but most of all when it comes to policies, programmes and plans. The United Nations has developed a set of guidelines, and sustainability indicators are gathered by most nations, with Canada and the UK having perhaps the most developed examples.[6]

These international and national approaches are now being supplemented by states conducting their own sustainability strategies, which are often much more urban-oriented as they are closer to their cities.[7] Few of these are as well developed for cities, however, as reported in *Quality of Life in New Zealand's Eight Largest Cities*.[8] In addition, sustainability indicators are now being extended to regions and to local governments within cities.[9]

Sustainable Seattle was probably the first to show how indicators developed in a sustainability context could be used for assessing how decisions are made across a city. As a result of that pioneering work, Alan AtKisson has become one of the most sought after consultants on sustainability indicators in cities (AtKisson, 1999).[10] Developing the indicators requires a public process of engagement similar to the Dialogue process described above, and is to all intents and purposes a way of doing a sustainability assessment of a city's plan for the future.

The problem that arises quite quickly when indicators are seen as part of the tool box of sustainability assessment is that they rapidly grow into a totally unmanageable list when the task of managing a city is subjected to its triple bottom line, and even its refined list of integrated, sustainability indicators. The key to moving from a broad set of sustainability indicators to effectively using them in assessment is to choose what are the critical indicators that are going to change a system – for better or for worse. The vast lists of sustainability indicators will all be of interest to particular people but they often will be only marginal in the change process that is needed for sustainability. Choosing the critical indicators, or as some say the 'killer indicators', requires an understanding of the system being evaluated. In urban systems this will invariably include an emphasis on transport infrastructure and urban design as these factors shape the way people live in cities (Newman and Kenworthy, 1999).

Planning is one of the key ways that social issues can be considered by government in how it attempts to see the future unfolding. Yet the social arena remains the area least well integrated methodologically into sustainability assessment of policies, programmes and plans, especially ones that reduce social issues down to a set of measurable sustainability indicators. This is because the social cannot always be reduced to quantifiable data. Social issues such as equity, housing, health, education and so on can be quantified, but there are other issues such as identity, sense of place, heritage and belonging that are not measurable as they are about values and worldviews. The dilemma of how these may be integrated into sustainability assessment has been discussed by Bradbury and Raynor (2002) as 'reconciling the irreconcilable'. They see the 'descriptive approach' of the sciences and social sciences to have fundamental disciplinary differences to the 'interpretive approach' of the humanities but suggest that there are emerging techniques that allow the two to be reconciled in sustainability assessment.

The main approach I have found to be of value for reconciling the interpretive approach of the humanities with the descriptive approach of the sciences in sustainability assessment is to highlight and prioritize stories as well as statistics. Stories are an emerging technique in the social sciences for integrating issues and enabling their values and political scope to be mainstreamed. As Sandercock says in her book *Cosmopolis II* (which is a plea for how the social dimension can be better integrated into city planning):

> For the longest time, 'story' was thought of in the social sciences as 'soft', inferior, lacking in rigour, or, worst insult of all, as a 'woman/native/other' way of knowing. . ... But as Alasdair MacIntyre put it: 'I can only answer the question "What can I do?" if I can answer the prior question, . . . 'of what story or stories do I find myself a part?'. (Sandercock, 2003a, p. 182)

Eckstein and Throgmorton (2003) have provided an edited collection of papers that seek to establish the links between story and sustainability. These papers are a coherent collection, rich in theory and real stories about the way sustainability is being approached in American cities. Although the US is the focus of the book, the conceptual approach is such that it can be applied to any city or to any region.

The main value of the book is that it offers a way to reinvigorate democracy at the scale of the community, city and region. The global economy is making nation-state democracies impotent as it moves more and more to being a series of competing global cities; but as the editors say in the Introduction, 'sustainability, story and democracy mutually construct one another' (Eckstein and Throgmorton, 2003, p. 4). The main way this happens is by giving back to the social sciences a sense of values and ethics. The triumph of the descriptive, the quantitative, has meant that not only is economic capital and natural capital understood solely through measurement, social capital has now been accounted for in this way. Thus sustainability could be seen in such a model as the integration of these three forms of capital. 'Story' makes this triumph meaningful by emphasizing the importance of interpretation, of making sense of these forms of capital, and most of all, giving them policy direction.

This is not an easy exercise. Bringing together the descriptive and the interpretive is the most important challenge in the policy arena opened up by the sustainability agenda. However, the approaches to doing this are rare and the simple model of 'storytelling' usually doesn't appear on the policy radar. Eckstein and Throgmorton's book is therefore of great significance as it fills a need that is being felt by academic and policy-maker alike who are seeking the holy grail of integration.

The emerging area of sustainability challenges all disciplines and professions to think more holistically, more globally and more long term. However, this can still be an expert's game, one involving the collection of data to describe a problem and the development of technical options for solutions. Important as this is, the issues of sustainability in cities and regions go broader and deeper than such analysis. Only through stories can the will to change be generated in such matters as racially segregated cities, car dependence, consumerism, declining community, the loss of habitat and climate change. The power of the story is in its empowerment of ordinary people, the setting of boundaries around 'place' and the ability to 'imagine communities' thus creating a 'shared sense of moral purpose at a regional scale' (Eckstein and Throgmorton, 2003, p. 5).

Leonie Sandercock (2003b) tells a number of stories in her chapter of the Eckstein and Throgmorton book and expands on them in her later (Sandercock, 2003a) publication. She tells of a social planner, Wendy Sarkissian, going to a new suburb in Australia where families were struggling. After collecting statistics she felt nothing in her report truly reflected their situation and instead told the story of a typical family, their hopes and their pain as the place did not fulfil their dreams. When she took the story back to the family they said finally someone had understood them. However, that was the beginning of a process to try to change their future and redeem some of their lost dreams. This was the real power of the story.

Regional planning and sustainability
Applying the idea of 'story' to a methodology for sustainability assessment seems to make most sense when it is applied to a region. Thus each major bioregion, which has natural resource boundaries that help define its natural capital and economic plans and markets that help define its economic capital, can now have a process that will help to define its social capital through telling the core story of the place. This was the conclusion of the Western Australian Sustainability Strategy on regional sustainability. It now is being applied initially in a Regional Sustainability Strategy for the Pilbara region of that state, where around 20 stories are being written around fictional characters that are used in a community visioning process called 'Dialogue for the Pilbara'. It is aimed to make this the kind of strategic planning process that can form a basis for any future project assessment work. It is believed that this can go beyond the social impact assessment work that tends to just list problems.

Most regions are made of a city set within a bioregion. The kind of principles needed to make regional sustainability into a valuable addition to the planning system require a coherent set of sustainability principles to guide the integration process. One of the newly emerging set of principles is the Melbourne Principles, which were developed through the United Nations Environment Programme.[11] These Melbourne Principles, named after the 2002 Workshop held in Melbourne, are set out in Table 24.4 and were adopted by local government at the World Summit on Sustainable Development (WSSD) conference in Johannesburg. They are part of an approach called 'Cities as Sustainable Ecosystems'; a book explaining the principles has been put together by Newman and Jennings (2004).

The regional scale of sustainability assessment and its links to the planning system is the process adopted by the New South Wales (NSW) government for its Metropolitan Strategy. The exercise is to create a visionary planning strategy that is being evaluated by three NSW Sustainability Commissioners as it is being developed. This enables more input on sustainability at the 'scoping stage' as suggested earlier.

Table 24.4   The Melbourne principles for cities as sustainable ecosystems

1.	Vision	Provide a long-term vision for cities based on intergenerational, social, economic and political equity; and their individuality
2.	Economy and Society	Achieve long-term economic and social security
3.	Biodiversity	Recognize the intrinsic value of biodiversity and natural ecosystems, and protect and restore them
4.	Ecological footprint	Enable communities to minimize their ecological footprint
5.	Model cities on ecosystems	Build on the characteristics of ecosystems in the development and nurturing of healthy and sustainable cities
6.	Sense of place	Recognize and build on the distinctive characteristics of cities, including their human and cultural values, history and natural systems
7.	Empowerment	Empower people and foster participation
8.	Partnerships	Expand and enable cooperative networks to work towards a common, sustainable future
9.	Technology	Promote sustainable production and consumption, through appropriate use of environmentally sound technologies and effective demand management
10.	Governance and hope	Enable continual improvement, based on accountability, transparency and good governance

The model of a regional plan that sees the city in its bioregion and seeks to minimize its ecological footprint while simultaneously improving its quality of life, is at the heart of this approach to sustainability assessment. This 'extended metabolism model' of cities in their regions was first adopted in *Sustainability and Cities* (Newman and Kenworthy, 1999) and has been applied in the past two *Australian State of the Environment Reports* (Department of the Environment and Heritage, 2001) for the chapter on settlements. The approach adopted was to gather as much data as possible on cities so that best practice could be found, as well as to outline case studies (stories) to help see how cities can change. The combination is the power of an integrated approach.

Perhaps the most advanced sustainability assessment methodology along these lines has been developed by British Columbia (BC) in Canada. A review of its settlements, mostly in the Greater Vancouver Region, is given in the *BC Sprawl Report: Economic and Livable Communities, 2004*.[12] The approach has been to examine a range of indicators in each of the 24 communities and then to combine them into three composite indicators – the Urban Form Index, the Livability Index and the Economic Vitality Index – and one overall index, the Overall Smart Growth Index. The power of the report is that it also tells a short story about each place first so that its indicators have policy meaning. The story and indicators are easily understood as the overall composite index is shown diagrammatically in a spider web of the 27 indicators, so that it is immediately clear which indicators the place does well on, and which it does not. Thus policy responses are drawn out of the sustainability assessment very directly. As data on the indicators have now been collected between 2001 and 2004 it is possible to get a sense of direction for each area and for the region. The report concludes: 'communities that are developing smarter and with less dependency on automobiles, also tend to be associated with a higher quality of life, and seem to be more adept at attracting the leading edge sectors of the economy'.[13]

Such a result would suggest that sustainability assessment of strategic policy directions in British Columbia has been very worthwhile. Some of the language and the institutions appear to be in place, perhaps because from the UN Habitat Conference in 1968 on they have been grappling with these issues as long as any city. The questions that arise from strategic planning and sustainability include:

- How do you make stories and statistics fit into a coherent, integrated sustainability strategy?
- How do you validate the sustainability assessment priorities chosen given the political consequences of any such choices?
- What are the best institutional arrangements for providing integrated advice from across government?
- How do you translate this sustainability assessment into a land use plan?
- Can sustainability assessment be regulated or must it always be facilitated only?

Sustainability assessment in buildings and developments

Buildings and groups of buildings (developments) are heavily regulated in the planning system. The process of seeking approval for even the simplest addition to a family dwelling has been a revelation to most people. These regulations have come from experience over many years with health, safety, environmental and social issues and are a collection of national building by-laws, state planning requirements and local town plans. The system together is called the statutory planning system as opposed to the strategic planning system described above.

The statutory control system is the latest target of sustainability assessment. Following the 1992 UNCED conference in Rio, innovative local governments signed up to a commitment to 'Local Agenda 21' or 'Cities for Climate Protection' and began seeking ways to apply sustainability. Their main power is through the statutory planning system and hence it should be no surprise that in the absence of national and state sustainability guidelines for development that local governments would seek to define their own.

Across Australia and the world, local authorities have created their sustainability assessment systems for development control decisions. Often they concentrate on how to achieve green buildings, the approach taken by the City of Scotsdale in the US. Sometimes these schemes have become rather arbitrary, which was the case in NSW when one local authority would not approve any development unless it had a worm farm, and in another case when a mayor was elected on the platform that all developments had to use photovoltaic lighting in their streets (one of the least cost-effective ways to reduce greenhouse gases).

Much of the debate about these systems is similar to the debate about the statutory control system in general:

- Are these regulations really needed?
- Do they not stifle good design and in fact sometimes work against good sustainability outcomes?
- How can a system of control be more outcome- or performance-oriented?

National approaches to green buildings have been implemented mostly on a voluntary basis, with accreditation being provided to any innovative builder. The US system is called Leadership in Energy and Environmental Design (LEED) but there are many others (Beyer, 2002).

Because of the chaotic approach to these issues in Australia there has been increased interest by state governments in how they could create a state-based 'Sustainability Scorecard' for all developments. This has the potential to help industry have greater certainty, communities to have a better way to reduce their ecological footprint and government to have a coherent way of achieving their sustainability objectives. This kind of partnership, which the World Business Council for Sustainable Development calls 'Jazz', is in contrast to leaving the market to find its own way, or imposing heavily from above.[14]

The system that has been adopted in NSW to provide a Sustainability Scorecard for residential development is called BASIX, or the Building Sustainability Index.[15] There are other models that have been developed, like First Rate and NatHers (Government of South Australia, 2004) but BASIX is the first system with the following characteristics:

- It is a tool for developers and councils (regulated and now in operation since 1 July 2005 in Sydney).
- It is web-based (this removes the huge need for documentation on most sustainability issues but allows information to be found for any particular problem through the links and the tool kit provided).
- It measures potential performance against sustainability indices (this is via a step-wise process, and although it has only been applied to water and energy so far, it will be developed for other important areas such as construction materials, waste in construction, site ecology, universal design for disability access and so on).
- Applicable to all residential dwellings (this applies to new buildings and renovations).

Local Councils deliver the BASIX certification once it is clear that a development can meet the requirements of 40 per cent less water and 25 per cent less energy (measured as greenhouse gases) compared with the average Sydney home. These are heroic goals in the sustainability arena and few other places in the world could claim such a system, yet it has happened in just a few years and with a partnership between key stakeholders. A benefit–cost analysis shows that it has a positive outcome for the economy. Some considerable concern is now being expressed by elements of the housing industry who hadn't quite seen that they would indeed have to change from the normal project home. It has been applied to other areas of NSW from 1 July 2005. As far as current housing goes it is hoped that BASIX certification will be seen by homeowners as a way to upgrade their homes before selling, or just as a way to make their contribution to sustainability. It could be regulated in future on all homes that are to be re-sold.

Across Australia there is considerable interest in the BASIX system. National seminars have been held and the Western Australian Minister for Planning has announced that WA will be the second state to adopt BASIX. Others are likely to follow, although state government bureaucracies are often taking the position that they would prefer their particular voluntary rating tools to be kept. Some scientific work to validate the BASIX model for different areas is needed. This does not need to prevent its application in trials as the approach to assessing the outcomes of different designs is still relevant, only the level of savings will be uncertain.

The idea of BASIX as a tool for sustainability assessment in buildings could be extended to commercial construction. The Australian Green Building Council has a well developed rating scheme and those developments (for example, '60 L' or '30 The Bond') have shown that five-star rating does not damage commercial outcomes and creates much better working arrangements.[16] The question remains as to whether this standard should be regulated more.

The next phase of sustainability assessment is at the sub-division or neighbourhood scale where urban design issues can be dealt with, such as water-sensitive urban design, solar orientation of streets, transit orientation, walkability and permeability of streets, level of mix in terms of housing diversity and commercial/services and other community-oriented issues. In NSW a system is being developed called METRIX.

There are a number of experiments in this area, indeed local governments and NGOs are developing them all across Australia. State governments are beginning to create these subdivision scale models, such as WA's Liveable Neighbourhoods Design Code and a new form of Sustainability Accreditation based on web processes by the Armadale Redevelopment Authority.[17] In addition, the Total Environment Centre in NSW has developed a set of criteria for subdivisions (Alexandra and Associates, 1998) and the Australian Housing and Urban Research Institute have a new project to assess affordability and sustainability in traditional developments compared with master planned communities (Blair et al., 2004). This latter study concluded that:

- The methodology of using sustainability assessment via a set of indicators does work, though they need to be reduced in number and integrated more.
- Designed communities are better off in sustainability terms than traditional development (car dependent with mostly monocultural project homes).
- The most significant measures for achieving affordability and sustainability are increasing development densities and starting a trend to smaller houses.
- The more radical notions of having more cooperative systems for house construction, using unconventional materials and introducing full cost pricing on housing developments, are also raised as ways to assist urban design in achieving sustainability outcomes.

The language and institutional processes for dealing with sustainability assessment down at the building and neighbourhood scale are starting to appear as quite manageable. This is mostly because local government has had a much longer history in dealing with these issues. Whether this can be achieved at the right scale for the larger, more bioregional level of sustainability issues remains to be seen. City-wide and bioregional groupings of local government will almost certainly be needed.

Monitoring remains an issue for all aspects of urban sustainability. Sustainability assessment can deliver ways of designing better suburbs and better houses, even mandating better appliances and new efficient and renewable infrastructure technologies. However, the operational aspects of all this depend on household behaviour as well as management systems that can monitor and report on progress. Often such monitoring has to be done by utilities and agencies whose main task is to sell more rather than to reduce consumption of resources, whether it be water, energy, or urban land.

Conclusions

1. The value of sustainability assessment is so inherently obvious it is bound to develop as a methodology and as a priority for government, business and the community. However, the disciplinary and professional understanding of how to do this is lagging behind the political will to implement it, especially in cities.
2. Sustainability assessment needs to progress at all levels from the assessment of complex and strategic projects, to the strategic planning process associated with policies, programmes and plans, to the statutory planning process associated with buildings and developments.
3. Demonstrations are still needed in all three areas. As state governments are the main authorities responsible for managing cities and their bioregions, it is necessary for strong leadership to be shown at this level of government. Only in this way can an integrated, partnership approach be developed.
4. Sustainability assessment needs to be seen as an aid in the politics of more sustainable decision-making, rather than as a monolithic process that somehow will avoid politics. Change will still demand hard decisions, though the goal of sustainability assessment is still to enable politicians to have better options to consider.
5. Regulating for sustainability assessment should be seen as a goal to be implemented when the various demonstration projects have been evaluated and it is clear that benefit can be derived overall.
6. Institutional processes are not proceeding quickly enough to cope with the integrative processes required within government to enable sustainability assessment to occur. The importance of local government and regions of local government in sustainability assessment will grow.
7. The language for sustainability assessment continues to lag behind the need. The importance of finding a balance of statistics and stories to adequately express the insights of all disciplines cannot be underestimated. Disciplinary and professional change may have to be led by institutional change or the political opportunity for sustainability assessment may be lost.
8. Monitoring of the results of sustainability assessment should be instituted to ensure that indicators and stories of sustainability can be evaluated and communicated.

Notes

1. These documents and others can be found at www.sustainability.dpc.wa.gov.au.
2. For details of the assessment, see www.doir.wa.gov.au.
3. See for the Netherlands: http://www.kit.nl/development/, Canada: http://www.ec.gc.ca and UK: http://www.odpm.gov.uk/settlement/.
4. See http://www.portphillip.vic.gov.au/melbourne_2030.html; http://www.planning.nsw.gov.au/; http://tasmaniatogether.tas.gov.au/ and http://www.planning.sa.gov.au/planstrat/documents.htm.
5. See http://www.dpi.wa.gov.au/dialogue/.
6. The UN guidelines are available at www.un.org/esa/sustdev/natinfo/indicators/isd. National approaches to sustainability indicators are also available for Canada (http://www.nrtee-trnee.ca/eng/programs/

currentprogrammes/SDIndicators), the UK (http://www.sd-commission.org.uk/pubs) and Australia (www.deh.gov.au/esd/national/indicators/report/).
7. States conducting sustainability assessments include in the US Oregon and Minnesota (Dernbach, 2005 and http://egov.oregon.gov/DCBS/), and in Australia the ACT (http://www.sustainability.act.gov.au/), Victoria (www.dpc.vic.gov.au/) and Tasmania (http://www.tasmaniatogether.tas.gov.au/).
8. The report is available online at http://www.bigcities.govt.nz.
9. Especially well developed are the local government approach taken by the City of Melbourne (http://www.melbourne.vic.gov.au/) and the regional approach taken in Western Australia (www.dlgrd.wa.gov.au).
10. See also http://www.atkisson.com.
11. See www.unep.or.jp/ietc/.
12. See http://www.smartgrowth.bc.ca.
13. The report is available at http://www.smartgrowth.bc.ca.
14. See http://www.wbcsd.ch.
15. See www.basix.nsw.gov.au/.
16. See www.gbcaus.org.
17. WA's Liveable Neighbourhoods Design Code is available at http://www.wapc.wa.gov.au/Initiatives/Liveable+neighbourhoods/default.aspx and the Armadale Redevelopment Authority's Sustainability Accreditation is available at http://www.ara.wa.gov.au/.

References

Alexandra and Associates (1998), *The ESD of Subdivision: An Assessment Framework for Urban Development*, Sydney: Total Environment Centre.
AtKisson, A. (1999), *Believing Cassandra: An Optimist Looks at a Pessimist's World*, Vermont: Chelsea Green Publishing Co.
Benson, D. and Jordan, A. (2004), 'Sustainability Appraisal in Local Land Use Planning: Patterns of Current Performance', *Journal of Environmental Planning & Management*, **47**(2), 269–86.
Beyer, D. (2002), 'Sustainable Building, Design and Construction Sector in Western Australia', Honours thesis, Perth, Western Australia: Murdoch University, http://www.sustainability.dpc.wa.gov.au/BGPapers/DavidBeyerSustainableBuildingAndConstruction.pdf, accessed 12 November 2005.
Biegelbauer, P. and Borrás, S. (2003), 'Conclusion: Policy Changes, Actors, Institutions and Learning', in Biegelbauer, P. and Borrás, S. (eds), *Innovation Policies in Europe and the US. The New Agenda*, Aldershot, UK: Ashgate, pp. 285–311.
Blair, J., Prasad, D., Judd, B., Zehner, R., Soebarto, V. and Hyde, R. (2004), *Affordability and Sustainability Outcomes: A Triple Bottom Line Assessment of Traditional Development and Master Planned Communities, Vol 1*, Melbourne: Australian Housing and Urban Research Institute.
Bradbury, J. and Rayner, S. (2002), 'Reconciling the Irreconcilable', in Abaza, H. and Baranzini, A. (eds) *Implementing Sustainable Development: Integrating Assessment and Participatory Decision-making Processes*, Cheltenham, UK and Northhampton, MA, USA: Edward Elgar, pp. 15–31.
Department of the Environment and Heritage (2001), *Australia State of the Environment 2001 Report*, Canberra: State of the Environment Australia, http://www.deh.gov.au/soe/2001/, accessed 12 November 2005.
Dernbach, J. (2005), *Sustainability at State and Local Levels in the United States*, Washington DC: Widener University Law School.
Eckstein, B. and Throgmorton. J.A. (2003), *Story and Sustainability: Planning, Practice and Possibility for American Cities*, Cambridge, MA: MIT Press.
Gibson, R.B. (2005), *Sustainability Assessment*, London: Earthscan.
Government of New South Wales (2004), 'Metropolitan Strategy', Discussion Paper, Sydney: Department of Planning, Infrastructure and Natural Resources.
Government of South Australia (2004), *Fact Sheet: Make a House Energy Efficient – Rate it with Stars!*, Adelaide, Australia.
Government of Western Australia (2002), *Review of the Project Development Approvals System*, Perth: Independent Review Committee.
Government of Western Australia (2003), *Hope for the Future: the State Sustainability Strategy*, Perth: Department of the Premier and Cabinet.
Lakoff, G. (2004), *Don't Think of an Elephant: Know your Values and Frame the Debate. The Essential Guide for Progressives*, White River Junction, VT: Chelsea Green Publishing.
Newman, P. and Jennings, I. (2004), *Cities As Sustainable Ecosystems*, Osaka: United Nations Environment Programme, International Environmental Technology Centre.
Newman, P. and Kenworthy, J. (1999), *Sustainability and Cities: Overcoming Automobile Dependence*, Washington DC: Island Press.

Owens, S. and Cowell, R. (2002), *Land and Limits: Interpreting Sustainability in the Planning Process*, London: Routledge.

Pope, J. (2003), 'Integrated Strategic Assessment of the Proposed Gorgon Gas Development in Western Australia', Paper presented to the 23rd Annual Meeting of the International Association for Impact Assessment, Marakech, Morocco, June.

Pope, J. (2004), 'Conceptualizing Sustainability Assessment – Three Models and a Case Study', Paper presented to the 24th Annual Meeting of the International Association for Impact Assessment, Vancouver, Canada, April.

Pope, J., Annandale, D. and Morrison-Saunders, A. (2004), 'Conceptualizing Sustainability Assessment', *Environmental Impact Assessment Review*, **24**, 595–616.

Pretty, J. (2003), 'Social Capital and the Collective Management of Resources', *Science*, **302**, 1912–15.

Sandercock, L. (2003a), *Cosmopolis II: Mongrel Cities of the 21st Century*, London and New York: Continuum.

Sandercock, L. (2003b), 'Dreaming the Sustainable City: Organizing Hope, Negotiating Fear, Mediating Memory', in Eckstein, B. and Throgmorton, J.A. (eds), *Story and Sustainability: Planning, Practice, and Possibility for American Cities*, Cambridge, MA: MIT Press, pp. 143–66.

Schön, D. and Rein, M. (1994), *Frame Reflection: Toward the Resolution of Intractable Policy Controversies*, New York: Basic Books.

Scott, W. and Gough, S. (2003), *Sustainable Development and Learning: Framing the Issues*, London: Routledge.

Selman, P. (1996), *Local Sustainability: Managing and Planning Ecologically Sound Places*, London: Paul Chapman Publishing.

Verheem, R. (2002), *Recommendations for Sustainability Assessment in the Netherlands*, Utrecht, the Netherlands: Commission for Environmental Impact Assessment.

25 Trends and volatility of ecological and anti-pollution technology patents in the USA

*Felix Chan, Dora Marinova and Michael McAleer**

Introduction

Technology has dominated the economic, social and environmental fabric of society since the Industrial Revolution in the eighteenth century. Ironically, while industrial development has positive impacts on economic growth, it has also been associated with pollution from its earliest days. For example, smoke, fumes and fog-laden air can be seen clearly in the beautiful city landscapes painted by impressionists who depicted images directly from nature. The general tones and vivid colours in paintings such as Claude Monet's *The Tames at Westminster* (1871), Georges Seurat's neo-impressionist *Bathing at Asnières* (1983–84) and *Entrance to the Port of Honfleur* (1886), or Camille Pissaro's *Pont Neuf – A Winter Morning* (painted near his death in 1903), show that pollution has long left its ominous mark on the ecological environment. Only in recent decades however, has the serious environmental damage caused by pollution been recognized as a problem that requires attention.

Dunn and Peterson (2001) reported that billions of tons of carbon and other harmful elements produced through the burning of fossil fuels are added to the Earth's atmosphere each year. The World Bank (2001) also revealed that the emission of CO_2 alone totalled more than 22.6 billion in 1996 and despite the attempts of the Kyoto Protocol this figure has only been on the increase. The global community is now experiencing the impact of this enormous pollution through the many ecological, social and economic problems caused by global warming, ozone layer depletion, land erosion, depletion of natural resources and acid rain. These problems have drawn the attention of politicians, business people and researchers to the challenge of sustainable development, an important aspect of which is the development and use of technologies.

In the current globalized world economy, albeit quite controversial, patents have become an important tool, conducive to economic growth. For example, the efforts of dealing with ecological problems should result in more ecological and/or anti-pollution innovations being patented. Thus, the greater the number of such patents, the more likely will market economies adopt a course of sustainability. A number of studies have confirmed that patenting activities cause immediate and subsequent market changes (see for example, Ernst, 1997; Griliches et al., 1991; Soete, 1987). Patents are also used to trace technological strengths and national priorities (see for example, Marinova and McAleer, 2003b). Is the importance of the natural environment increasing and is change occurring?

The best way to attempt to answer these questions is by examining the patenting activities within the USA because of its large and technologically advanced economy that has always been highly favourable to companies and individuals interested in protecting their intellectual rights. In absolute numbers, the US Patent and Trademark Office (PTO) receives by far the largest number of foreign applications (see Archibugi, 1992) and overall is the largest source of information on technological developments. Of interest to this chapter are technologies related to the ecological environment as well as those related to the development of anti-pollution technology.

This chapter will also analyse the volatility (or conditional variance) in ecological and anti-pollution patents. In a typical time series model, the conditional mean can provide insightful information about the dynamic of the series under investigation, with the conditional variance assumed to be constant. This assumption is often found to be unrealistic, therefore modelling the conditional variance can provide a more complete picture for understanding the dynamics of the time series. In the present case, the conditional variance of patent registrations can provide insight into the variability of research activities that lead to technological innovation for ecological and anti-pollution technologies. This information allows formal investigation to be conducted in establishing a relationship between R&D expenditure and technology innovation, and subsequently provide helpful guidance for policies related to environmental issues.

There have been a number of studies recently analysing the volatility of patents (see Chan et al., 2003, 2004b, 2005a), including ecological (Chan et al., 2005a; Marinova and McAleer, 2003a) and anti-pollution (Chan et al., 2004a) patents. They all use time series of patents according to the application date rather than the approval date in order to avoid artificial distortions of the data caused by organizational delays in the process of granting patents. The only drawback of this approach is that the historical data will no longer be static until the decisions on every application have been made. Therefore, historical data on patents will need to be updated on a yearly basis as the process of granting a patent could take more than ten years in some cases. For this reason, some of the analysis in Marinova and McAleer (2003a) and Chan et al. (2004a) is updated in this chapter using monthly data from the PTO as at 11 November 2005. The two classes of patents examined are: (1) patents related to the development of more ecologically friendly technologies, or technologies that assist in abating existing ecological problems; (2) patents related to anti-pollution technologies.

Data
Empirical information on patents data is collected from the US PTO through its online search engine (http://164.195.100.11/netahtml/search-adv.htm). The time series data used consist of monthly observations for the number of ecological patents with application dates between January 1975 and December 2003 (the data were extracted on 11 November 2005). The current US patent classification system does not provide special categories that cover ecological or anti-pollution patents. Consequently, extracting empirical information is a challenging exercise that requires the development of working rules and definitions (see Marinova and McAleer, 2003a for detailed analysis of ecological technology). The approach given in Marinova and McAleer (2003a) was used to identify such patents: a patent is considered to be related to the ecological environment if its abstract or full text contains words such as 'ecology', 'ecological', 'ecologically' (or any other word beginning with 'eco') or 'environmentally'.[1] It is highly unlikely that a patent related to ecologically sustainable technology would not include one or more of these definitional words. In addition, the same approach was used consistently across the time series, which makes it possible for trends and patterns in the data to be analysed. Similarly, a registered patent is defined to fall within the category of 'anti-pollution patents' if its abstract,

claims or specifications include the word 'pollution'. It is important to note that the anti-pollution patents data is not a subset of the ecological patents data and vice versa, although a substantial overlap is expected between the two data sets given their definitions.[2]

Trends in ecological and anti-pollution technology patents
Figure 25.1 shows the trends in ecological and anti-pollution technology patents in the USA, based on monthly data from January (1) 1975 to December (12) 2003. It is clear that the trend for ecological patents is upward sloping, in general, with the 1990s being a period of intensive patenting of technologies that are related positively to the ecological environment. In fact, the total number of approved ecological patents reached 3577 in 1997. However, the momentum for ecological technology development seems to have slowed down after 2000 as the number of ecological patents appears to be decreasing.

The trends in anti-pollution technology patents reveal a very similar story. The major difference between the two time series is that the graph of anti-pollution technology patents exhibits two trends, namely, downward sloping from the mid-1970s to the mid-1980s, and upward sloping from the mid-1980s to the mid-1990s, followed by stabilization in the late 1990s. It would seem that the interest in anti-pollution patents diminished in the mid-1980s, but has been resurrected slightly in the 1990s.

Interestingly, there seems to have been greater effort into developing anti-pollution technology before 1990 as the number of anti-pollution technology patents exceeds the number of ecological patents. However, the efforts in developing ecological technology seem to have increased after 1990 resulting in higher numbers of patents being registered than for anti-pollution technology. This would seem to suggest more research effort has gone to improving and preserving our environment in general rather than just minimizing and preventing pollution.

It is important to note that the decrease in ecological and anti-pollution technology patents in the late 1990s and early 2000s could be artificial as the process of granting patents can be time-consuming. It takes an average of two years for a patent application to be approved. However, in many cases it can take much longer, delays of seven to eight years are not uncommon and delays of over ten years are not unknown. It is likely that the number of approved applications in more recent years will have increased.

The total number of patents registered in the USA during the same period has also been increasing steadily (Figure 25.2 shows the monthly data), reaching a peak of 183 581 approved patents from applications lodged in 1997. Note that the number of total patents has also decreased dramatically in the late 1990s and 2000s. This provides evidence to support the case that the decrease in patents during this period is artificial and is due to delays in the process of granting patents. For these reasons, the patent shares, namely the ratio of ecological patents to total patents (henceforth ecological patent share) as shown in Figure 25.3 and the ratio of anti-pollution technology patents to total patents (henceforth anti-pollution patent share) as shown Figure 25.4 may be more useful in analysing the trends in the development of ecological and anti-pollution technology patents.

As shown in Figures 25.3 and 25.4, the trends in ecological patent share and anti-pollution patent share are quite different between the 1970s to mid-1980s. In particular, anti-pollution patent share is generally decreasing between 1975 and the mid-1980s as shown in Figure 25.4. However, they both seem to have experienced a structural change

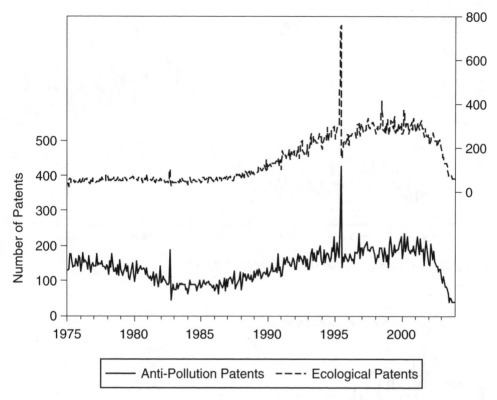

Figure 25.1   *US ecological and anti-pollution technology patents by date of application,*
              1975(1)–2003(12)

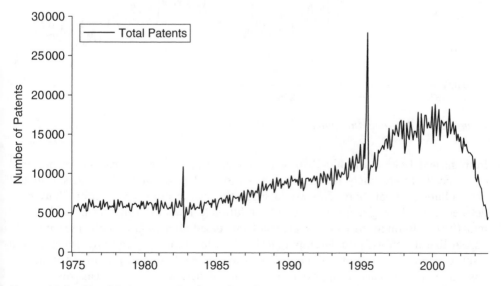

Figure 25.2   *Total US patents by date of application, 1975(1)–2003(12)*

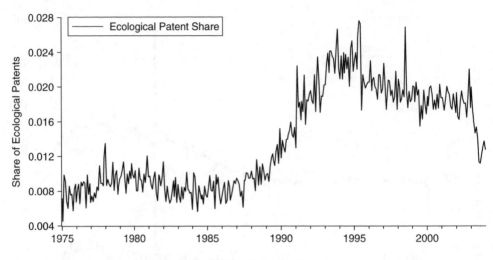

Figure 25.3   Ecological patent share, 1975(1)–2003(12)

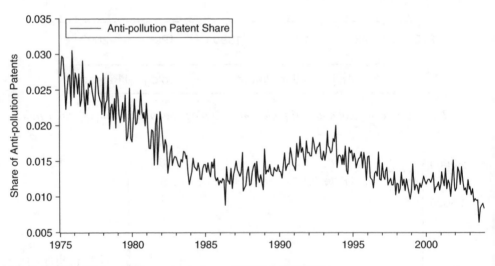

Figure 25.4   Anti-pollution patent share, 1975(1)–2003(12)

from the mid-1980s to the mid-1990s where the patent shares in both cases have increased dramatically. Interestingly, the rate of increase for ecological patent share exceeds that of anti-pollution patent share. This would suggest that intensive research efforts have been devoted to developing ecological and anti-pollution technology during the mid-1980s and mid-1990s. Although the patent shares have decreased after 1995 in both cases, it would appear that the efforts into developing technology for improving and preserving natural resources and environment have continued where research for anti-pollution technology has slowed down. This is supported by the rate of decrease in ecological patent share, which is lower than that of anti-pollution patent share.

To further analyse the trends and relationships between ecological and anti-pollution patent shares, the sample has been divided into three periods, namely January 1975 to December 1985, January 1986 to December 1994 and January 1995 to December 2003. The correlations between ecological and anti-pollution patent shares for the three periods are 0.175, 0.681 and 0.677, respectively. Given the correlation between the two patent shares has increased dramatically after the mid-1980s, it would appear that there is a more integrated approach into the research and development of ecological and anti-pollution patents.

Volatility in ecological and anti-pollution patent share

The sample volatility, v_t, of a time series is often calculated as

$$v_t = \varepsilon_t^2 = (y_t - E(y_t|\Im_{t-1}))^2,$$

where y_t denotes the time series of interest and $E(y_t|\Im_{t-1})$ denotes the conditional expectation of y_t given the information set, $\Im_{t-1}$, which contains all information up to $t-1$. In other words, the sample volatility is the square of the residuals, which is obtained by taking the difference between the patent share and its trends (conditional mean).

The volatility in the ecological and anti-pollution patent share can be found in Figures 25.5 and 25.6, respectively. Although the volatility is relatively low in the early period for the ecological patent share, persistence in volatility is evident throughout the entire sample. Furthermore, extreme observations and outliers are not uncommon in this sample, particularly from the early to the late 1990s. Similar to ecological patent share, persistence of volatility is also evident throughout the entire sample for the anti-pollution patent share as shown in Figure 25.6. Moreover, the mid-1970s to the mid-1980s seems to represent a period of high volatility and volatility clustering seems to appear throughout the entire sample.

These observations seem to suggest the dynamics of volatilities in ecological and anti-pollution patent shares can be captured by models such as the generalized autoregressive conditional heteroskedasticity (GARCH)-type models. In fact, the LM (Lagrange multiplier) tests of no ARCH (autoregressive conditional heteroskedasticity) effects are highly significant in both cases. The test statistics for the ecological and anti-pollution patent shares are 18.515 and 12.481, respectively, rejecting the null hypothesis of no ARCH errors. This suggests that the conditional variances for these series are not constant and GARCH-type models might be appropriate in modelling the dynamic of volatility for both ecological and anti-pollution patent shares.

Models specifications

The remainder of the chapter focuses on modelling volatility in ecological and anti-pollution patent shares. This approach is based on Engle's (1982) idea of capturing time-varying volatility (or uncertainty) using the autoregressive conditional heteroskedasticity (ARCH) model, and subsequent developments forming the ARCH family of models. Of these models, the most popular has been the generalized ARCH (GARCH) model of Bollerslev (1986), especially for the analysis of financial and economic data. In order to accommodate asymmetric behaviour between negative and positive shocks (or

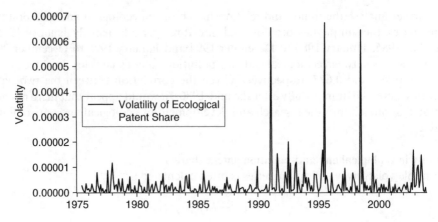

Figure 25.5   Volatility of ecological patent share, 1975(1)–2003(12)

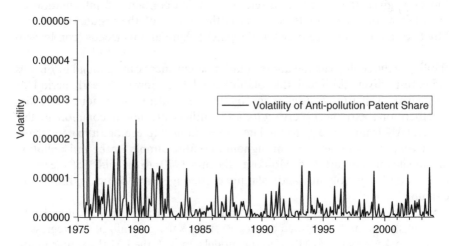

Figure 25.6   Volatility of anti-pollution patent share, 1975(1)–2003(12)

movements in the time series), Glosten et al. (1992) proposed the GJR model, and Nelson (1991) proposed the exponential GARCH (EGARCH) model.

Consider the stationary AR(1)-GARCH(1,1) model for the patent share, y_t:

$$y_t = \phi_1 + \phi_2 y_{t-1} + \varepsilon_t, \quad |\phi_2| < 1 \tag{25.1}$$

where the shocks to the patent share, ε_t, are given by:

$$\varepsilon_t = \eta_t \sqrt{h_t}, \quad \eta_t \sim iid(0,1)$$

$$h_t = \omega + \alpha \varepsilon_{t-1}^2 + \beta h_{t-1}, \tag{25.2}$$

and $\omega > 0$, $\alpha \geq 0$, $\beta \geq 0$ are sufficient conditions to ensure that the conditional variance $h_t > 0$. The ARCH (or α) effect indicates the short-run persistence of shocks, while the

GARCH (or β) effect indicates the contribution of shocks to long-run persistence (namely, $\alpha + \beta$). In equations (25.1) and (25.2), the parameters are typically estimated by the maximum likelihood method to obtain quasi-maximum likelihood estimators (QMLEs) in the absence of normality of η_t.

Ling and McAleer (2002a) established the necessary and sufficient moment conditions for the univariate GARCH(p,q) model. Ling and McAleer (2003) showed that the QMLE for GARCH(p,q) is consistent if the second moment is finite, that is, $E(\varepsilon_t^2) < \infty$, Ling and Li (1997) showed that the local QMLE is asymptotically normal if the fourth moment is finite, that is, $E(\varepsilon_t^4) < \infty$, and Ling and McAleer (2003) showed that the global QMLE is asymptotically normal if the sixth moment is finite, that is, $E(\varepsilon_t^6) < \infty$. The necessary and sufficient condition for the existence of the second moment of ε_t for GARCH(1,1) is $\alpha + \beta < 1$ and, under normality of η_t, the necessary and sufficient condition for the existence of the fourth moment is $(\alpha + \beta)^2 + 2\alpha^2 < 1$.

Using a weaker condition, Elie and Jeantheau (1995) and Jeantheau (1998) showed that the log-moment condition is sufficient for consistency of the QMLE for the univariate GARCH(p,q) model, and Boussama (2000) showed that it is also sufficient for asymptotic normality. Based on these theoretical developments, the sufficient log-moment condition for the QMLE of GARCH(1,1) to be consistent and asymptotically normal is given by:

$$E(\ln(\alpha\eta_t^2 + \beta)) < 0. \tag{25.3}$$

However, this log-moment condition is not straightforward to check in practice as it involves a function of an unknown random variable and unknown parameters. Although the sufficient moment conditions for consistency and asymptotic normality of the QMLE for the GARCH(p,q) model are stronger than their log-moment counterpart, the moment conditions are far more straightforward to check in practice.

The effects of positive shocks (or upward movements in the patent share) on the conditional variance, h_t, are assumed to be the same as the negative shocks (or downward movements in the patent share) in the symmetric GARCH model. Asymmetric behaviour is accommodated in the GJR model, for which GJR(1,1) is defined as follows:

$$h_t = \omega + (\alpha + \gamma I(\eta_{t-1}))\varepsilon_{t-1}^2 + \beta h_{t-1}, \tag{25.4}$$

where $\omega > 0$, $\alpha + \gamma \geq 0$, $\beta \geq 0$ are sufficient conditions for $h_t > 0$, and $I(\eta_t)$ is an indicator variable defined by:

$$I(\eta_t) = \begin{cases} 1, & \varepsilon_t < 0 \\ 0, & \varepsilon_t \geq 0 \end{cases}$$

as η_t has the same sign as ε_t. Such an indicator variable differentiates between positive and negative shocks, so that asymmetric effects in the data are captured by the coefficient γ, with $\gamma > 0$. In the GJR model, the asymmetric effect, γ, measures the contribution of shocks to both short-run persistence, $\alpha + \gamma/2$, and long-run persistence,

$$\alpha + \beta + \frac{\gamma}{2}.$$

Ling and McAleer (2002b) showed that the regularity condition for the existence of the second moment of GJR(1,1) under symmetry of η_t is $\alpha + \beta + 1/2\gamma < 1$, and the condition for the existence of the fourth moment under normality of η_t is $\beta^2 + 2\alpha\beta + 3\alpha^2 + \beta\gamma + 3\alpha\gamma + 3/2\gamma^2 < 1$. Using a weak novel condition, McAleer et al. (2002) showed that the log-moment condition for GJR(1,1), namely

$$E(\ln[(\alpha + \gamma I(\eta_t))\eta_t^2 + \beta]) < 0, \tag{25.5}$$

is sufficient for consistency and asymptotic normality of the QMLE.

An alternative model to capture asymmetric behaviour in the conditional variance is the exponential GARCH (EGARCH(1,1)) model of Nelson (1991), namely:

$$\log h_t = \omega + \alpha|\eta_{t-1}| + \gamma\eta_{t-1} + \beta\log h_{t-1}, \quad |\beta| < 1. \tag{25.6}$$

The distinct differences between EGARCH, on the one hand, and GARCH and GJR, on the other, include the following: (1) EGARCH is a model of the logarithm of the conditional variance, which implies that no restrictions on the parameters are required to ensure $h_t > 0$; (2) Nelson (1991) showed that $|\beta| < 1$ ensures stationarity and ergodicity for EGARCH(1,1); (3) Shephard (1996) noted that $|\beta| < 1$ is likely to be a sufficient condition for consistency of QMLE for EGARCH(1,1); (4) as the conditional (or standardized) shocks appear in equation (25.4), McAleer et al. (2002) observed that $|\beta| < 1$ is likely to be a sufficient condition for the existence of moments; (5) in addition to being a sufficient condition for consistency, $|\beta| < 1$ is also likely to be sufficient for asymptotic normality of the QMLE of EGARCH(1,1).

As GARCH is nested within GJR, based on the theoretical results in McAleer et al. (2002), an asymptotic t-test of $H_0: \gamma = 0$ can be used to test GARCH against GJR. However, as EGARCH is non-nested with regard to both GARCH and GJR, non-nested procedures are required to test EGARCH versus GARCH and EGARCH versus GJR. A simple non-nested procedure was proposed by Ling and McAleer (2000) to test GARCH versus EGARCH, in which the test statistic is asymptotically N(0,1) under the null hypothesis. Following a similar approach, McAleer et al. (2002) derived a non-nested procedure for testing EGARCH versus GJR, in which the test statistic is asymptotically N(0,1) under the null hypothesis.

In the next section, the volatilities of the ecological and anti-pollution patent share will be modelled by estimating GARCH, GJR and EGARCH using the full sample, and through the use of rolling windows with size 240. The dynamic paths of the estimates, second moments and log-moments will provide insightful information regarding the short- and long-run persistence over time, as well as the stability and performance of each model.

Empirical results
All the models in this chapter were estimated using EViews 4 with the Brendt-Hall-Hall-Hausman algorithm. Estimates, second and log-moment conditions for GARCH(1,1),

Table 25.1 *Estimates and t-ratios of GARCH(1,1), GJR(1,1) and EGARCH(1,1) for ecological patent share*

	ω	α	γ	β	Log-moment	Second moment
GARCH(1,1)	1.668E-06	0.188		0.245	−0.991	0.433
	2.393	2.622		0.890		
	1.601	1.627		0.715		
GJR(1,1)	1.663E-06	0.191	−0.014	0.250	−0.984	0.434
	1.960	1.605	−0.113	0.736		
	1.530	1.291	−0.066	0.679		
EGARCH(1,1)	−9.409	0.308	−0.041	0.282		
	−1.957	2.849	−0.525	0.756		
	−1.098	1.950	−0.352	0.421		

Note: The three entries are the estimates, asymptotic t-ratios and Bollerslev-Wooldridge robust t-ratios, respectively.

Table 25.2 *Estimates and t-ratios of GARCH(1,1), GJR(1,1) and EGARCH(1,1) for anti-pollution patent share*

	ω	α	γ	β	Log-moment	Second moment
GARCH(1,1)	3.409E-08	0.032		0.952	−0.016	0.984
	0.868	1.607		30.079		
	0.466	1.794		38.708		
GJR(1,1)	4.671E-08	0.031	0.018	0.940	−0.036	0.981
	0.929	1.283	0.557	23.616		
	0.527	1.305	0.333	29.686		
EGARCH(1,1)	−1.831	0.218	−0.127	0.871		
	−2.093	2.182	−2.268	13.420		
	−1.900	2.526	−1.960	11.989		

Note: The three entries are the estimates, asymptotic t-ratios and Bollerslev-Wooldridge robust t-ratios, respectively.

Table 25.3 *Non-nested tests for ecological and anti-pollution patent shares*

	H_0 : GARCH H_A : EGARCH	H_0 : EGARCH H_A : GARCH	H_0 : GJR H_A : EGARCH	H_0: EGARCH H_A : GJR
Ecological patent share	3.362	1.492	3.153	1.687
Anti-pollution patent share	4.398	0.827	5.032	1.272

GJR(1,1) and EGARCH(1,1) using the full sample can be found in Tables 25.1 and 25.2 for ecological and anti-pollution patent shares, respectively. The results of the non-nested tests for choosing between GARCH and EGARCH and between GJR and EGARCH can be found in Table 25.3.

Full sample estimation

Before modelling the volatility of ecological and anti-pollution patent share, it is important to ensure that the patent shares do not follow non-stationary conditional means. However, the conventional augmented Dicky-Fuller (ADF) test is not robust against non-normal and GARCH errors. For this reason, the Phillips-Perron (PP) test is conducted with truncated lags of order 5. The test statistics were calculated to be -4.336 and -6.428 for ecological patent share and anti-pollution patent share, respectively. The two statistics are clearly less than the 5 per cent critical value of -3.424, and hence, rejected the null hypothesis of non-stationarity.

The full sample estimates, second and log-moment conditions for GARCH(1,1), GJR(1,1) and EGARCH(1,1) can be found in Tables 25.1 and 25.2 for ecological and anti-pollution patent shares respectively. As shown in Table 25.1, the log-moment and second moment conditions are satisfied for both GARCH and GJR. This suggests that the estimates are consistent and asymptotically normal. In addition, the β estimate for EGARCH is also less than 1. However, the γ estimates for both GJR and EGARCH are not statistically significant. This implies positive shocks have the same impacts on the volatility of ecological patent share as negative shocks.

Similar to the previous case, the log- and second moment conditions for GARCH and GJR are satisfied in the case of anti-pollution patent share as shown in Table 25.2. The γ estimate is not statistically significant in GJR but it is significant in EGARCH. In addition, the α estimates are not statistically significant in both GARCH and GJR but it is significant in EGARCH. This provides evidence to suggest that EGARCH is superior to GARCH and GJR to model the volatility in anti-pollution patent share. Moreover, the γ estimate in EGARCH is negative; this implies a positive shock induces higher volatility than a negative shock. One plausible explanation is that a period of high patent share (positive shock) could encourage further subsequent research in the same area. This would increase uncertainty in the patent share depending on the outcome of the research. However, a period of low patent share (negative shocks) might be an indication of low research activities or negative research outcome and therefore, the uncertainty in patent share for the next period is being reduced.

Table 25.3 contains the results for two non-nested tests, namely the Ling and McAleer (2000) test for selecting between GARCH and EGARCH and the McAleer et al. (2002) test for selecting between GJR and EGARCH. Given both test statistics are asymptotically normal, the null hypotheses of GARCH and GJR are rejected in both ecological and anti-pollution patent shares at 5 per cent significant level as shown in Table 25.3. However, the null hypothesis of EGARCH cannot be rejected in either case. This suggests EGARCH is the preferred volatility model for both ecological and anti-pollution patent shares.

Concluding remarks

This chapter analysed the trends and volatility in ecological and anti-pollution patent shares using monthly PTO data from January 1975 to December 2003. The results show that the research and development in ecological and anti-pollution patents might have become more integrated after the mid-1980s. Moreover, research into ecologically-related technology seemed to have resulted in more registered patents than anti-pollution technology. In terms of the volatility of the patent shares, EGARCH seemed to be the preferred model for capturing the time-varying volatility for both ecological and anti-pollution patent shares. In addition, asymmetric behaviour was detected for anti-pollution patent share but not for the ecological patent share. This would seem to suggest that negative shocks discourage research efforts into anti-pollution technology as it reduced the uncertainty in its patent shares.

Notes

* The authors wish to acknowledge the financial support of the Australian Research Council.
1. The word 'eco' was excluded because it generated only patents referring to the so-called Eco enzyme, which is somewhat outside the area of this chapter. It was impossible to incorporate in the definition of ecological patents a keyword search using 'environment' or 'environmental' because of their widespread use outside the area of ecological environment, such as in the digital, physical or economic environments. Individual reading and checking of each of the thousands of American patents containing 'environment' or 'environmental' would have been an excessively time- and labour-intensive exercise.
2. It is possible to find examples in which anti-pollution technology patents do not have terms such as 'ecology', 'ecological', 'ecologically' (or any other word beginning with 'eco') or 'environmentally' in the abstracts, claims or specifications, and vice versa.

References

Archibugi, D. (1992), 'Patenting as an Indicator of Technological Innovation: A Review', *Science and Public Policy*, **19**(6), 357–68.
Bollerslev, T. (1986), 'Generalized Autoregressive Conditional Heteroskedasticity', *Journal of Econometrics*, **31**, 307–27.
Boussama, F. (2000), 'Asymptotic Normality for the Quasi-maximum Likelihood Estimator of a GARCH Model', *Comptes Rendus de l'Académie des Sciences*, Série I, **331**, 81–4 (in French).
Chan, F., Marinova, D. and McAleer, M. (2003), 'Trends and Volatilities in Foreign Patents Registered in the USA', *Applied Economics*, **36**, 585–92.
Chan, F., Marinova, D. and McAleer, M. (2004a), 'Modelling the Asymmetric Volatility of Anti-pollution Technology Patents Registered in the USA', *Scientometrics*, **59**(2), 179–97.
Chan, F., Marinova, D. and McAleer, M. (2004b), 'Modelling the Asymmetric Volatility of Electronics Patents in the USA', *Mathematics and Computers in Simulation*, **64**, 169–84.
Chan, F., Marinova, D. and McAleer, M. (2005a), 'Modelling Thresholds and Volatility in US Ecological Patents', *Environmental Modelling and Software*, **20**(11), 1369–78.
Chan, F., Marinova, D. and McAleer, M. (2005b), 'Rolling Regressions and Conditional Correlations of Foreign Patents in the USA', *Environmental Modelling and Software*, **20**(11), 1413–22.
Dunn, S. and Peterson, J.A. (eds) (2001), *Hydrogen Futures: Towards a Sustainable Energy System*, Washington, DC: Worldwatch Institute.
Elie, L. and Jeantheau, T. (1995), 'Consistency in Heteroskedastic Models', *Comptes Rendus de l'Académie des Sciences*, Série I, **320**, 1255–8 (in French).
Engle, R.F. (1982), 'Autoregressive Conditional Heteroskedasticity with Estimates of the Variance of United Kingdom Inflation', *Econometrica*, **50**, 987–1007.
Ernst, H. (1997), 'The Use of Patent Data for Technological Forecasting: The Diffusion of CNC-technology in the Machine Tool Industry', *Small Business Economics*, **9**(4), 361–81.
Glosten, L., Jagannathan, R. and Runkle, D. (1992), 'On the Relation Between the Expected Value and Volatility of Nominal Excess Returns on Stocks', *Journal of Finance*, **46**, 1779–801.
Griliches, Z., Hall, B.H. and Hausman, J.A. (1991), 'R&D, Patents and Market Value Revisited: Is There a Second (Technological Opportunity) Factor?', *Economics of Innovation and New Technology*, **1**, 183–201.
Jeantheau, T. (1998), 'Strong Consistency of Estimators for Multivariate ARCH Models', *Econometric Theory*, **14**, 70–86.
Ling, S. and Li, W.K. (1997), 'On Fractionally Integrated Autoregressive Moving-average Models with Conditional Heteroskedasticity', *Journal of the American Statistical Association*, **92**, 1184–94.
Ling, S. and McAleer, M. (2000), 'Testing GARCH versus E-GARCH', in Chan, W.-S., Li ,W.K. and Tong, H. (eds), *Statistics and Finance: An Interface*, London: Imperial College Press, pp. 226–42.
Ling, S. and McAleer, M. (2002a), 'Necessary and Sufficient Moment Conditions for the GARCH(r,s) and Asymmetric Power GARCH(r,s) Models', *Econometric Theory*, **18**, 722–9.
Ling, S. and McAleer, M. (2002b), 'Stationarity and the Existence of Moments of a Family of GARCH Processes', *Journal of Econometrics*, **106**, 109–17.

Ling, S. and McAleer, M. (2003), 'Asymptotic Theory for a Vector ARMA-GARCH Model', *Econometric Theory*, **19**, 278–308.

Marinova, D. and McAleer, M. (2003a), 'Modelling Trends and Volatility in Ecological Patents in the USA', *Environmental Modelling and Software*, **18**, 195–203.

Marinova, D. and McAleer, M. (2003b), 'Nanotechnology Strength Indicators: International Rankings Based on US Patents', *Nanotechnology*, **14**, R1–R7.

McAleer, M., Chan, F. and Marinova, D. (2002) 'An Econometric Analysis of Asymmetric Volatility: Theory and Application to Patents', Paper presented to the Australasian Meeting of the Econometric Society, Brisbane, July, to appear in *Journal of Econometrics*, 2007.

Nelson, D.B. (1991), 'Conditional Heteroscedasticity in Asset Returns: A New Approach', *Econometrica*, **59**, 347–70.

Shephard, N. (1996), 'Statistical Aspects of ARCH and Stochastic Volatility', in Barndorff-Nielsen, O.E., Cox, D.R. and Hinkley, D.V. (eds), *Statistical Models in Econometrics, Finance and Other Fields*, London: Chapman & Hall, pp. 1–67.

Soete, L. (1987), 'The Impact of Technological Innovation on International Trade Patterns: The Evidence Considered', in Freeman, C. (ed.), *Output Measurement in Science and Technology*, Amsterdam: North-Holland, pp. 47–76.

World Bank (2001), *Making Sustainable Commitments: An Environment Strategy for the World Bank*, Washington, DC: World Bank.

PART V

CASE STUDIES IN NEW TECHNOLOGIES FOR THE ENVIRONMENT

26 Integrated waste management
Robert Hughes, Martin Anda, Goen Ho and Kuruvilla Mathew

Introduction

The concept of integrated waste management (IWM) has been developed to provide a means towards sustainability. Population growth coupled with increasing consumption have increased the amount of waste generated across the world while also facilitating the use of IWM to divert waste from landfill toward more long-term sustainable options such as reuse and recycling programmes – programmes that also maintain the longevity of a product's life and reduce pressure on natural resources (Figure 26.1) (Aini et al., 2002). The other major factor influencing the formation of IWM strategies is social pressure (Huang et al., 2001). Social pressure has been found to cause authorities to implement IWM strategies even where landfill space is available (Barth, 2000) and has originated with increased education levels among consumers about current unsustainable and sustainable waste practices (Clarke et al., 1999; Huang et al., 2001). The success of IWM strategies is largely attributed to the degree of social acceptance, such that the landfill problem associated with high population density areas has been found to stem from a low social acceptance of reuse and recycling programmes, rather than their absence (Aini et al., 2002; Bradshaw and Ozores-Hampton, 2002; Mohee, 2002; Poon et al., 2001). An assessment of the social acceptance of IWM strategies is now a precursor for most new programmes and the technologies chosen for IWM may be dependent on social factors (Kwawe, 2002).

The low social acceptance of IWM strategies has been due to low consumer awareness and knowledge about waste management and how this affects human quality of life (Aini et al., 2002). A number of authors have noted that the most crucial aspect of an IWM strategy is the initial sorting of the materials into associated waste streams (such as organic, recyclable and non-recyclable), which is where consumer awareness and knowledge as well as enthusiasm is vital, given that there is no economic incentive for consumers at this stage (Beccali et al., 2001; Maltbaek, 1999; Palacios et al., 2002; Poon et al., 2001; Woodard et al., 2001). Hazardous material production is also seen as a problematic factor with IWM strategies because it reduces the lifecycle of the product to one use and may reduce the value of heterogeneous wastes (Palacios et al., 2002). Efficient primary waste collection systems, such as different coloured 'wheelie bins', have been shown to reduce problems with heterogeneous waste contamination because they offer the user a simple method of waste separation (Woodard et al., 2001) and is one of the main factors in IWM strategy success (Maltbaek, 1999).

Lifecycle assessment (LCA) is a decision-making tool used in conjunction with IWM as a practice for finding the most acceptable and efficient reuse pathways for waste (Clift et al., 2000). The LCA strategy is also known as the 'cradle-to-grave' approach because it covers each aspect of a product's lifecycle and will include an assessment of a product's extraction and manufacture, use and reuse and final disposal (Dennison et al., 1998). Dennison et al. (1998) noted that LCA consists of four main stages: goal definition and

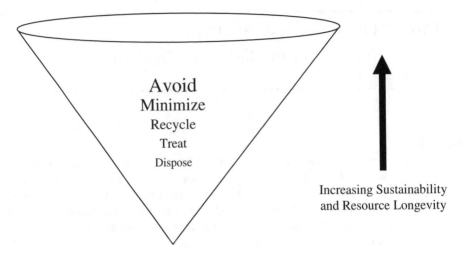

Avoid
Minimize
Recycle
Treat
Dispose

Increasing Sustainability
and Resource Longevity

Figure 26.1   The waste management hierarchy aims to promote sustainable practices and reduce single use of products

scoping, inventory analysis, impact assessment and improvement assessment. The goal definition and scoping stage relate to the purpose of the study, the system boundaries and data quality assurance (Clift et al., 2000). Lifecycle inventory analysis quantifies the environmental inputs and outputs and assesses the impacts of the processes involved in the use and manufacture of a product (Beavis and Lundie, 2003; Seo et al., 2004). This includes an inventory of the energy and water consumption during the production and life of the product, the environmental impact during its life and upon disposal (greenhouse gases, ecotoxicity, remediation cost) and the social impacts associated with the product (Costi et al., 2004; Mellor et al., 2002; Rodriguez-Iglesias et al., 2003). This form of inventory uses comparable functional units that allow an impact assessment of the different pathways for reuse and/or disposal (Ashley et al., 1999). The final state, improvement assessment, is used to compare pathways and find new – or change existing – pathways towards more sustainable approaches (Beccali et al., 2001; Clarke et al., 1999; Clift et al., 2000; McDougall and Hruska, 2000; Wang, 2001). As a part of the inventory, the future costs of each pathway are also quantified until the product is deemed to be adequately safe for re-introduction into the environment either through disposal and/or reuse, which allows long-term sustainable options to be established that are economically, environmentally and socially acceptable (Huang et al., 2001).

The use of LCA is a vital tool for reaching policy guidelines with IWM (for example, the future zero waste policies set down by many continents, nations, states and municipalities) and can show how each pathway for waste management can or cannot achieve these guideline values (Palacios et al., 2002). However, LCA is a decision-support tool and should be used appropriately. Factors like community support and involvement are the most cost-effective IWM strategies but these also involve many uncertainties, such as different social beliefs about waste within a population (Beccali et al., 2001).

The use of lifecycle assessment for IWM is addressed within this chapter through an examination of appropriate resource recovery techniques that provide suitable pathways towards sustainability. The use of IWM will be assessed in the context of three major waste streams:

- organic waste;
- construction and demolition waste; and
- urban and municipal wastewater.

In each waste stream the pathways chosen are those currently employed across the world in different sustainability scenarios that have been identified as practical measures towards global sustainability for use in decision-support tools (such as LCA) as well as by local, national and international authorities.

Integrated management of organic waste

Organic waste is generated in large quantities and is one of the most common forms of solid waste from established urban and rural areas (Felder et al., 2001). Organic waste comes from many sources, including parks and gardens, offices and workplaces, agriculture and households. This type of waste includes green wastes such as tree prunings, grass clippings and vegetable scraps, agricultural wastes such as manures, wastewater treatment by-products such as sewage sludge, and non-recyclable paper and cardboard wastes. Organic waste is often the biodegradable portion of municipal waste and is the most appropriate form of waste for recycling because it can be easily converted to fertilizers, vermicompost, compost, soil remediation mediums and soil conditioners – all of which offer a highly valuable resource, especially in peri-urban and rural areas that rely on solid organic waste to boost crop yields (Eaton and Hilhorst, 2003). Due to the high value of solid organic waste in a recycled form and its ability to bolster crops, it is often the first form of waste resource recovery employed, in some cases illegally against authority consent (Eaton and Hilhorst, 2003; Shehata et al., 2004).

The present practice of landfilling organic waste has created unseen environmental impacts. The open anaerobic digestion of organic waste that is facilitated by landfilling produces high quantities of methane (a greenhouse gas that is 21 times more damaging than carbon dioxide) and leachate contaminating groundwater with nutrients, metals and organics (Liu et al., 2002). The burning of methane emissions from landfills to convert them to carbon dioxide has been proposed as a means of reducing their impact on the greenhouse effect (Ayalon and Shechter, 2001). Landfill bioreactor technologies coupled with appropriate landfill leachate control have been put forth to capture biogas from landfills (Clarke, 2000). The extraction of waste products from landfills, lining of landfills and the treatment of landfill leachate via wastewater treatment methods have also been proposed (Lin and Chang, 2000). These technological approaches may reduce current landfill problems but they do not offer a method to facilitate organic waste recycling and natural resource longevity, nor do they offer the best energy yields from organic waste (Ayalon and Shechter, 2001; Liu et al., 2002). The first waste hierarchy developed in Europe (Figure 26.1) has specifically emphasized reuse and recycling programmes for organic waste through composting and anaerobic digestion technologies near the source of production because in the LCA of organic waste they offer the best method for resource longevity and the highest energy and resource yields (Avery, 2000).

Lifecycle analysis of integrated organic waste management has placed an importance on reducing the contamination of organic waste, as the value of the final recovered resource (for example, compost) is highly dependent on the level of contamination (Barth, 2000; During and Gath, 2002; Gajdos, 1998a; Palacios et al., 2002). As During and Gath (2002) explain, the reuse of sewage sludge in particular – because it is usually the most contaminated organic waste (including heavy metals, nonylphenol ethoxylates and linear alkybenzoates) – has to be controlled to avoid soil pollution and subsequent groundwater and aquatic pollution. They further note that the level of contamination can reduce stabilized sewage sludge (biosolid) application rates and annual application. Contamination of municipal organic waste with hazardous objects such as syringes and batteries has also been found to increase secondary sorting costs at material reclamation facilities (MRFs) and to reduce the effectiveness of recycling and compost programmes (Maltbaek, 1999).

Organic waste technologies for resources recovery
The biological treatment of organic waste through aerobic composting and vermicomposting, and anaerobic digestion, have been shown to be effective treatment methodologies (Schober et al., 1999). Composting and vermicomposting are principally employed as a simple means to produce a humus-rich compost suitable for land application (Kwawe, 2002) whereas anaerobic digestion is principally employed to produce energy but also has low yields of stabilized organic waste (Liu et al., 2002; Schober et al., 1999). Analysis of the different methods shows they are acceptable in different situations and the choice of technology depends on the desired inputs (capital cost), available technical expertise and desired outputs (energy and/or compost) (Ayalon and Shechter, 2001).

Composting is the most common method for resource recovery from organic solid wastes and can be employed on a number of regional and economic levels (Kwawe, 2002). Huang et al. (2001) and Mohee (2002) all found from LCA that composting of green wastes such as grass clippings and food scraps with manures (such as fowl manures) at the household level in urban areas reduces the overall cost of IWM programmes because the waste is recycled at the source before it is transferred at a cost to MRFs. In rural areas recycling of organic waste through composting at the farm level has been found to play an important role in increasing agricultural production but involves larger-scale composting techniques than those found in urban backyards (Shehata et al., 2004; Smith and Hughes, 2004). In developing nations, composting or direct recycling of organic waste plays a major role in crop production and has been used as a way to recycle organic wastes from urban areas to rural areas where much of the organic matter originated (for example, food crops), which closes the nutrient carbon cycle on a bioregional scale (Eaton and Hilhorst, 2003). There are a number of techniques for composting of organic waste, some involving passive aeration and others involving forced aeration. Windrows have been found to be the most effective passive method but as these involve the use of machinery (front-end loaders and windrow turners), they are not applicable to the household scale that use compost piles or containers (Adediran et al., 2003).

The use of passive windrow composting techniques is generally employed on agricultural, low socio-economic and peri-urban areas due to its cost-effectiveness and requirement of space (Huang et al., 2001; Maniadakis et al., 2004). The windrow technique uses a mechanical device, such as a front-end loader, to combine the organic wastes together in long rows – often two to three metres high, and three to six metres wide. The rows are left to go through thermophilic (55°C) and mesophilic (35°C) stages, and are turned regularly with a windrow turner or front-end loader to ensure thorough composting of all the organic waste. Once the organic matter is stabilized into compost it is then screened through a mesh grill to increase uniformity in particle size, and the final stabilized compost is reused or sold (Cooperband, 2000).

In urban areas a static windrow technique can be employed that involves aeration pipes below the windrow, forcing air through the organic waste and reducing the need for turning. The forced aeration technology has been applied to a number of compost designs and, unlike passive techniques, may work as well in compost plants that apply water through sprinklers and humidifiers and that are able to filter the emissions (NO_x) that occur during composting (Bari and Koenig, 2001; Clemens and Cuhls, 2003). Wet composting processes have also been facilitated using forced aeration with in-vessel facilities and are able to work with higher nitrogen ratios and produce higher thermophilic temperatures, which increase pathogen and seed destruction (Malmen et al., 2003).

The heat produced during the thermophilic (55°C) and mesophilic (35°C) stages of composting has been found to produce a number of beneficial effects by increasing degradation of the organic waste and some contaminants (for example, pesticides) and inactivating pathogens and seeds (Adediran et al., 2003). To ensure proper composting a number of elements need to be involved, such as a carbon to nitrogen ratio between 15 to 1 and 30 to 1, adequate aeration, sufficient time and appropriate micro-organisms (Smith and Hughes, 2004). The higher nitrogen ratios will speed up the composting process but they involve the use of high nitrogen content wastes such as manures, which are often less available than high carbon content wastes such as tree prunings and leaf litter (Adediran et al., 2003). Co-composting of high nitrogen content industrial or municipal wastes (such as sewage sludge) with low carbon content wastes (such as wood chips from tree maintenance) is often facilitated to ensure proper compost stabilization and offers a means to reuse common urban wastes in a sustainable manner (Stelmachowski et al., 2003).

Vermiculture is another common aerobic method for producing compost or, more accurately, vermicompost (Sinha et al., 2002). Vermicompost is generally more fine and uniform in character than compost because the use of worms ensures thorough stabilization (Benitez et al., 1999). The process uses specific worm species that are typical of the top layers of forest floors, which contain high quantities of organic matter (leaves, decaying branches and so on) and relies on the worms to grind down the organic waste into fine particles (Eastman et al., 2001). The reduction in the size of the waste increases the aeration of the waste and is a precursor to biodegradation from micro-organisms. The final vermicompost, which is composed of vermicasts, is concentrated in nutrients and has a fine humus texture better suited to use in potting mixes and home gardens (Sinha et al., 2002). Vermicomposting used for the biodegradation of highly contaminated organic waste (such as sewage sludge) has been found to reduce pathogens to an acceptable level and also to reduce heavy metal and organic compound contamination (Bajsa et al., 2004; Benitez et al., 1999; Maboeta and van Rensburg, 2003a; Matos and Arruda, 2003).

Anaerobic digestion is a newer development than traditional windrow and vermicomposting technologies and has been specifically used to reduce greenhouse gas emissions and produce energy in the form of biogas; that is, 60 per cent methane and 40 per cent carbon dioxide (Mata-Alvarez et al., 2000). Liu et al. (2002) found from a lifecycle analysis that steam pressure disruption during anaerobic digestion of organic wastes in a closed reactor, which uses the resultant methane as biogas, was the most effective means to reduce greenhouse gas emissions and increase energy yields from organic wastes. The anaerobic digestion of organic waste can also recover most of the nutrients, such as nitrogen, which are lost through ammonia volatilization in aerobic composting (Ayalon and Shechter, 2001). The degradation of common contaminants in anaerobic digestion has also been found to occur but this requires high temperatures in the thermophilic (> 50°C) range (Hartmann and Ahring, 2003).

Anaerobic digestion technologies are particularly suited to fast biodegradation processing of high nitrogen wastes such as agricultural manures (Hammad et al., 1999; Misi and Forster, 2001). However, they have also been applied effectively to co-digest a range of organic wastes such as sewage sludge and municipal organic wastes (Kubler et al., 2000). The systems employed can involve a thermophilic and mesophilic stage in one reactor, or two separate thermophilic and mesophilic reactors (Mata-Alvarez et al., 2000). The use of two-stage digestion through anaerobic mesophilic and thermophilic chambers has been found to be the best technology for sludge and manure digestion (Oles et al., 2000). The pre-treatment of organic waste by pulping, for example, as a means to reduce particle size, has been found to increase waste digestibility (Palmowski and Muller, 2000).

The combination of anaerobic and aerobic systems has been proposed for maximum yields of both energy and compost (Gajdos, 1998b). The combination of the two systems may reduce problems that have been found to occur with the anaerobic digestion of volatile fatty acids and aerobic composting gas emissions (Clemens and Cuhls, 2003). In vermicomposting systems pre-composting is often facilitated because the worms cannot break down larger organic molecules and cannot cope with higher temperatures, including those associated with both thermophilic and mesophilic compost stages (Alauzet et al., 2002).

Integrated management of construction and demolition waste
The largest component of the waste stream by weight (sometimes up to 50 per cent) is construction and demolition (CD) waste and is commonly comprised of building materials and products such as concrete, asphalt, flooring, furniture, gypsum wallboard, roofing, insulation, doors, windows and frames, sand, soil, bricks and rocks (Brown and Buranakarn, 2003). Poon et al. (2001) noted that CD waste – though often viewed as inert waste – is comprised of both inert (bricks, sand and soil) and non-inert waste (plastics, paper and wood). The origin of CD materials is from renewable resources such as wood from sustainable forestry as well as from non-renewable resources such as metal, crushed stone, sand, gravel, rocks and gypsum, which are mined or quarried. Brown and Buranakarn (2003) noted that the recycling of CD waste materials with high production costs (for example, refining costs in the case of aluminium) is more readily applied due to their value in the marketplace. In contrast, the low prices of some products (gravel, sand and crushed stone, for example) result in major environmental impacts (such as destruction of riverine habitats) but the cost is only indirectly related to construction prices and is borne in large part by the public and the environment (Hsiao et al., 2002). The LCA of construction and demolition wastes shows that the longevity of resources by either sale or reuse onsite can reduce redevelopment costs by reducing transport, environmental impacts, landfill and resource purchase costs (Schachermayer et al., 2000). The argument to increase landfill prices to encourage reuse of low-cost materials has been used

(Fleming, 2002) but this approach also encourages illegal disposal of wastes, especially if cost-effective reusable methodologies are not available (McGrath, 2001).

More sustainable approaches to the integrated management of CD wastes are available and ensure more thorough reuse. A study by Poon et al. (2001) in Hong Kong found that the best methodology for IWM success is on-site sorting of CD wastes, coupled with specified development contracts that provide documentation on reuse pathways. Poon (1997) from earlier accounts found that contract incentives allow developers to view the reuse of CD waste as a business opportunity because without these incentives developers view reuse as a constraint to the demolition and construction industry and prefer landfill options. The impetus behind the use of contractual agreements has been facilitated due to a lack of landfill space (CD waste constitutes up to 50 per cent of landfill globally by mass), which has evidently increased landfill prices as well as inappropriate disposal of hazardous substances (Fatta et al., 2003; Hsiao et al., 2002; Poon et al., 2001; Schachermayer et al., 2000; Seo and Hwang, 1999; Winter and Henderson, 2003). The facilitation of reuse programmes are more than pertinent in some regions of the world such as Taiwan and Hong Kong, which due to increased economic status and ageing buildings, are currently moving through redevelopment phases in high-rise areas and have little available space for landfill (Hsiao et al., 2002; Poon, 1997).

The IWM programmes for CD waste have identified a number of key steps that are essential to increase the efficiency of reuse programmes. Pre-planning has been noted as a major step to increase the speed of reuse programmes and is used in the Athens Olympic Games development programme (Fatta et al., 2003). Pre-planning includes the use of decision-support tools such as LCA to find acceptable reuse pathways such as the tax-deductible sale of reusable products to charities, and to identify hazards such as leachate of contaminants from landfill (Jang and Townsend, 2003; Lin and Chang, 2000; Townsend et al., 1999). Once pathways for reuse are identified, the demolition process is the next important step and has been appropriately renamed the deconstruction process because reuse (sale and donation) of products such as windows, flooring, tiles, bricks, plumbing is reliant on their quality (Schachermayer et al., 2000). In this sense a relevant contract noting the value of the reusable products either in the construction process or elsewhere is vital and to ensure that the products are actually reused (Poon et al., 2001).

Pathways for reuse of CD waste
Throughout the literature a number of different pathways have been identified for CD wastes, including wood, metal, aggregate and concrete wastes. Unlike organic wastes, CD wastes are often dealt with using different pathways because the waste includes biodegradable fractions (such as wood), non-biodegradable reusable fractions (such as aggregates) and non-biodegradable recyclable fractions (glass, aluminium, plumbing, and so on). These wastes can in some cases be reused on-site in a redevelopment stage but are more commonly reused off-site after sorting on-site, and moved in different pathways after sorting. As with all other types of waste, the reuse of CD waste depends largely on the contamination level and there is an environmental risk and economic disadvantage involved with all contaminated fractions (such as crushed concretes), making some methods of reuse or recycling inappropriate for environmental and economic reasoning (Krook et al., 2004). Each waste stream, once divided, also has its own specific sorting techniques to increase the reuse potential (Huang et al., 2002).

The major component of the biodegradable fraction of CD waste is wood. The reuse of wood wastes has been studied extensively, finding that they are often used as a means to boost biomass for combustion in biofuel boilers or incinerators (Blassino et al., 2002; Chang et al., 2001; Hiramatsu et al., 2002; Krook et al., 2004; Tolaymat et al., 2000; Weber et al., 2002). Chang et al. (2001) found a unique method for reuse of wood waste, through combustion in a redundant brick kiln, which produces hot water for secondary uses. However, the problem encountered with such approaches is the emission of volatile organic compounds into the air stream. In controlled situations with air filtration, such as with some incinerator designs, there have also been problems with the residual ash from chromated copper arsenate (CCA) wood. Solo-Gabriele et al. (2002) found that a small ratio of contaminated wood waste to non-contaminated wood waste causes the residual ash after combustion to have leachate properties of hazardous substances. Krook et al. (2004) found that most residual ash from the combustion of CCA wood is unusable as a filler material and potentially hazardous to human and environmental health. The major constituents that cause the hazardous nature of CCA wood waste are elevated levels of copper, chromium and arsenic, all of which are toxic metals (especially arsenic) that are also incombustible and are concentrated during combustion (Tolaymat et al., 2000).

The leachate of CCA ash in landfills is not purely attributed to combusted CCA wood and has been found to occur from uncombusted CCA wood (Weber et al., 2002). Krook et al. (2004) explained that for recovered wood waste from CD to contribute to sustainability, cleaner waste wood flows are essential. More prerequisites for efficient waste wood flows have also been identified (Alderman and Smith, 2000; Alderman et al., 2003). Alderman et al. (2003) in an American study found that spent wood waste was not recovered for two reasons, one being the lack of recovery programmes and the other the lack of facilities. The facilities for sorting waste wood components, such as removing CCA-contaminated wood, have been identified and provide mechanisms to reuse more wood waste. The use of on-line sorting of wood wastes using laser-induced breakdown spectroscopy (LIBS) has been shown to be 92 to 100 per cent accurate in identifying contaminated wood because it can identify the chromium emission signal in CCA wood (Moskal and Hahn, 2002). Another method that uses an x-ray fluorescence spectrometer has been found to have an accuracy between 95 to 97 per cent and shows potential for on-line work using arsenic as the indicator of CCA wood (Blassino et al., 2002). The efficient sorting of wood waste should allow more feasible means for reuse.

The use of simple composting or vermicomposting techniques to recover CD wood waste is often more economically viable than incinerators (Ayalon and Shechter, 2001). The co-composting of wood chips with organic wastes such as sewage sludge has been used as a means to 'value-add' the wood wastes whilst also stabilizing and reusing sewage sludge (Maboeta and van Rensburg, 2003b). The recycling of wood waste through diverse utilization has been found to be an appropriate methodology as well (Hiramatsu et al., 2002). Hiramatsu et al. (2002) discovered a technique that separates wood into woodchips, fibre bundles and strands by use of vapour pressure. The authors noted that the wood waste can then be easily transformed into the right size fraction for processing and reuse as a valuable product such as paper.

Concrete and brick are the major constituents of all CD waste from highly urbanized areas and are a non-biodegradable inert waste fraction (Hsiao et al., 2002; Seo and Hwang, 1999). They are often removed for reuse using hand tools, vehicle-mounted equipment, explosive blasting, demolition agents, mechanical splitters, thermal tools and hydro demolition (Lamond et al., 2002). Landfill, or in some cases land reclamation, is often viewed as a reuse methodology because of the largely inert nature of this waste (Poon et al., 2001). This involves risks however, as the landfilling of CD wastes has been found to cause leachate problems such as acidity from sulphate associated with gypsum drywalls (Townsend et al., 1999). The sulphate may in some cases form hydrogen sulphide (rotten egg gas) and would pose a health risk if untreated (Jang and Townsend, 2001a).

Quality brick waste can be salvaged and reused (Klang et al., 2003) but concrete and broken bricks are often mechanically crushed into a number of size fractions (Huang et al., 2002). The crushed waste is sorted according to specific qualities, including size using screening and aeration (Huang et al., 2002) and chemical composition using wet and dry techniques (Schachermayer et al., 2000). The aggregates from the crushed concrete and brick can be reused in construction (low-grade concrete aggregate, for example) but have to be assessed to ensure their properties are suitable (water absorption, plasticity, relative density and so on). Compared with natural aggregates, recycled aggregate generally has coarser surface texture, lower density, higher water absorption and lower load-bearing capacity (Poon, 1997). The crushed brick and concrete aggregates are most commonly reused as road bases because of their properties (Wahlstrom et al., 2000). Leachate from CD waste used in road bases has been found to be acceptable because the contaminants are usually in an insoluble form, which further enhances the use of recycled aggregates in road bases (Wahlstrom et al., 2000).

The reuse of the smaller portion of the aggregates, the crushed waste fines, has been shown to possess some reuse hazards because trace elements such as heavy metals are often concentrated in the fines (Schachermayer et al., 2000). Soil fines have also been found to have low concentrations of volatile organic compounds (Jang and Townsend, 2001b). Soil fines have a number of applications, such as in the production of aluminium substituted tobermorite (Klimesch et al., 2002), but are often used as a soil substitute (Jang and Townsend, 2001a; Jang and Townsend, 2001b). The reuse of waste fines has been found to cause leachate problems because they are used in a similar method to landfilling (Jang and Townsend, 2001a). Unknown qualities of concrete fines have been found such as their ability to re-cement over time, which should be taken into consideration before they are reused (Arm, 2001).

Integrated management of domestic and municipal wastewater
Domestic and municipal wastewater is derived from a variety of sources, including households, industry and businesses. The water can be categorized into black, yellow and grey waters. These waters contain numerous contaminants (more preferably termed elements), which include heavy and trace metals, nutrients, macronutrients and some organic compounds such as surfactants (Esperanza et al., 2004; Gutierrez et al., 2002; Ren and Frymier, 2003). The elements of the wastewater can be harmful, however, their toxicity or environmental consequence (for example, eutrophication) is dependent on their concentration and load. Wastewater also contains viruses such as waterborne hepatitis and pathogens such as *E. coli* (Bajsa et al., 2004). Effective wastewater treatment to reduce some elements and to inactivate viruses and pathogens is thus essential to reuse or disposal and has been practised for over a century (Panikkar et al., 2003). Water and wastewater quality guidelines (for example, ANZECC and ARMCANZ 2000) are used to

regulate both the discharge of wastewaters into treatment facilities and the disposal or reuse of water from those facilities (Patterson, 2001).

There are a number of different methods to treat wastewater (such as activated sludge and trickling filter) that were principally developed to comply with water guidelines and to reduce wastewater treatment operational costs (Clauson-Kaas et al., 2001). Wastewater treatment methods differ in respect to their sustainability and are commonly assessed with current decision-making tools such as LCA to determine their environmental impacts (Balkema et al., 2001). Balkema et al. (2001) explained that the indicators chosen for wastewater treatment sustainability include environmental efficiency aspects such as energy balances, resource use and emissions to soil, air and water. Aspects such as sewage sludge disposal or reuse through appropriate incineration, digestion or composting are often included within the boundaries of these assessments because the methods chosen for wastewater treatment (such as activated sludge) often produce high quantities of sewage sludge (Mels et al., 1999; Nakano et al., 2003; Seo et al., 2004) and the resulting sludge possesses environmental health risks (Beavis and Lundie, 2003; Dennison et al., 1998).

These assessments of wastewater treatment sustainability have also focused on more encompassing global sustainability aspects by attempting to assess differences with water and wastewater integration (Jeppson and Hellstrom, 2002) and by quantifying the nutrient budgets to ensure nutrients are used in the least detrimental way (Mulder, 2003). In particular there has been an emphasis on nutrient removal efficiencies during wastewater treatment because the use of fertilizers in agriculture to produce foodstuffs has increased nutrient loadings to sewers. These nutrients in the agricultural fertilizers are made from atmospheric inputs, which during wastewater treatment, need to be emitted back into the atmosphere (for example, through nitrification – denitrification) to ensure the nutrient budget (that is the nutrient cycle) is maintained (Mulder, 2003). The environmental impacts from nutrient disposal into lakes, streams and oceans from wastewater treatment has some important implications such as eutrophication and pollution of water resources and contamination of seafoods such as shellfish (Maurer et al., 2003; Vidal et al., 2002). Maurer et al. (2003) noted that nutrient removal through wastewater treatment has high energy consumption, and that removal of nutrients from wastewater and reuse before treatment would be more energetically viable. An ideal integrated approach to this problem where nutrients are reused and energy is recovered during the treatment process, is shown in Figure 26.2.

Appropriate design for nutrient recovery is an aspect of integrated urban water management, where nutrients are reused on-site and for nearby agriculture and horticulture. The use of decentralized treatment systems to ensure the efficiency of such a system has been proposed (Ho, 2003) and found to be cost-effective whilst posing similar health risks to centralized approaches (Fane et al., 2002). The use of urine separation (yellow water) on-site for agricultural purposes has been found to be an important step to ensure adequate nutrient recovery because urine has been found to be a complete nutrient fertilizer with low heavy metal concentrations (Jonsson, 2002). Urine also contains most of the household nutrients and so reduces the need for costly nutrient-removal methodologies. The transportation of urine waste, as with treated wastewater, reduces the cost-effectiveness of reuse programmes and can increase health risks, hence decentralized approaches that reuse the wastewaters nearby are more cost-effective (Jonsson, 2002).

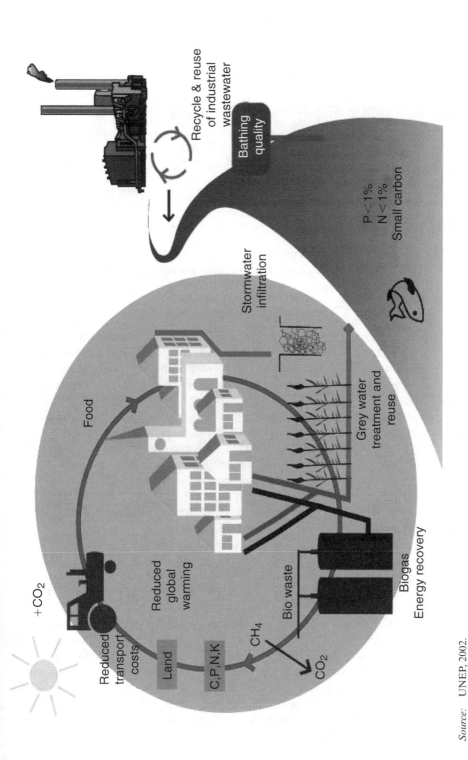

Source: UNEP, 2002.

Figure 26.2   An integrated wastewater treatment approach uses all available resources within the system and closes the nutrient cycle which subsequently protects natural ecosystems

415

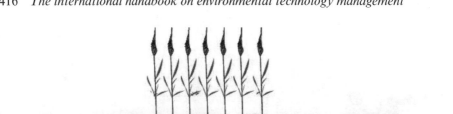

Source: UNEP, 2002.

Figure 26.3   *The use of natural processes such as the wetland to treat wastewater offers a cost-effective approach whilst also adding aesthetic and ecological qualities to the environment*

Proper urban development with available spaces for urban water reuse are essential for the effectiveness of nutrient recycling programmes and are components of integrated urban water management. These integrated urban water management programmes are not only beneficial for urine separation but can be used with appropriate technology to reuse black waters, which can be treated with appropriate on-site or cluster technologies (Fane et al., 2002; Hammes et al., 2000; Wilderer and Schreff, 2000). Hammes et al. (2000) further explained that decentralized methods such as anaerobic digestion of household wastewater and some organic wastes can close the carbon cycle and produce energy (Figure 26.2). The use of soft technologies such as constructed wetlands may offer viable alternatives for reuse programmes because they efficiently clean grey and yellow water, and ensure the nutrients are transferred back into the natural bio-geo-chemical cycle without detrimental impacts (Figures 26.2 and 26.3) (Steer et al., 2003). Wetlands and natural treatment methods offer functions within nature as well because they are a wildlife habitat and hold aesthetic values for the public (Brix, 1999).

The use of centralized systems along with decentralized systems is the most likely future adaptation to municipal and urban wastewater programmes because soft technological approaches and decentralized systems may not always be appropriate for wastewater treatment. For example, in many locations that receive high rainfalls in winter the treated wastewater may not be able to be reused on common ground because they exceed the capacity of the soil to absorb water and of plants to uptake the nutrients. In these cases centralized technologies may offer a way to integrate the approach because they can help buffer the impacts during these periods. The treated water from centralized technologies (Patterson, 2001) may then be reused in less sensitive urban areas to recharge aquifers for summer reuse. It has been suggested however, that transporting wastewater over large distances for reuse would be highly unsustainable (Ho, 2003) and that areas for aquifer recharge or wastewater reuse from centralized treatment may have to be developed in conjunction with urban water-sensitive design.

Conclusions

The IWM is an approach towards sustainability through appropriate reuse pathways that ensure natural resource longevity and low environmental impacts. In this chapter the use of LCA and IWM has been found to provide suitable pathways for waste reuse in three major waste streams.

In the organic waste stream suitable pathways through efficient sorting, composting, vermicomposting and anaerobic digestion were shown to reduce the environmental impacts of landfilling organic waste. On-site composting at the urban level was emphasized in IWM programmes to reduce operation costs. The general composting approaches used were found to include backyard methods, passive windrows and forced aeration, which could all yield suitable compost for reuse. The use of vermicomposting was also suitable and produced finer grades of compost. Anaerobic digestion of wastes was a method to produce biogas from organic wastes and has the added advantage of reducing environmental emissions.

The second waste stream addressed in this chapter was CD waste. It was found that CD waste reuse was best facilitated by contractual incentives and on-site sorting, allowing developers to view IWM as a business opportunity. The pathways for CD waste reuse were highlighted for wood waste and brick and concrete waste. Construction and demolition wood waste was found to be commonly incinerated but this poses waste management problems because the CCA wood waste leaves hazardous ash residues. A number of efficient sorting techniques are available however, which should ensure wood waste is reused properly without such a problem. The brick and concrete waste was also found to pose problems but these are mainly experienced in landfills where the waste can produce sulphide gases. Therefore, these wastes have been best used in road bases but there is scope for their application to other areas of construction.

Urban and municipal wastewater was the final waste stream addressed in this chapter. These wastewaters were found to be best reused in a decentralized approach with the inclusion of urine separation and soft technologies to reduce the energy used in nutrient removal and fertilizer production. Integration of decentralized technologies with proper urban planning would ensure all nutrients are recycled on-site without pollution of ecosystems. In some cases, where urban planning has not accounted for seasonal variation in rainfall and evaporation, the use of centralized systems would be beneficial with decentralized systems because centralized systems offer a buffer when the limit of natural ecological water and nutrient uptake is exceeded.

References

Adediran, J.A., Taiwo, L.B. and Sobulo, R.A. (2003), 'Effect of Organic Wastes and Method of Composting on Compost Maturity, Nutrient Composition of Compost and Yields of Two Vegetable Crops', *Journal of Sustainable Agriculture*, **22**(4), 95–109.

Aini, M., Fakhrul-Razi, S.A., Lau, S.M. and Hashim, A.H. (2002), 'Practices, Attitudes and Motives for Domestic Waste Recycling', *International Journal of Sustainable Development and World Ecology*, **9**(3), 232–8.

Alauzet, N., Garreau, H., Bouche, M. and Vert, M. (2002), 'Earthworms and the Degradation of Lactic Acid-base Stereocopolymers', *Journal of Polymers and the Environment*, **10**(1/2), 53–8.

Alderman, D.R. and Smith, R.L. (2000), 'Solid Wood Received and Marketed by Virginia's Landfill Facilities', *Forest Products Journal*, **50**(6), 39–44.

Alderman, D., Smith, R. and Araman, P. (2003), 'A Profile of CCA-treated Lumber from Service in the Southeastern United States', *Forest Products Journal*, **53**(1), 38–45.

Arm, M. (2001), 'Self-cemeting Properties of Crushed Demolished Concrete in Unbound Layers: Results from Triaxial Tests and Field Tests', *Waste Management*, **21**(3), 235–9.

Ashley, R., Souter, M.N., Butler, D., Davies, J., Dunkerley, J. and Hendry, S. (1999), 'Assessment of the Sustainability of Alternatives for the Disposal of Domestic Sanitary Waste', *Water Science and Technology*, **39**(5), 251–8.

Avery, I. (2000), 'Composting', *Proceedings of the Institution of Civil Engineering-Municipal Engineer*, **139**(3), 159–65.

Ayalon, O.A.Y. and Shechter, M. (2001), 'Solid Waste Treatment as a High-priority and Low-cost Alternative for Greenhouse Gas Mitigation', *Environmental Management*, **27**(5), 697–704.

Bajsa, O., Nair, J., Mathew, K. and Ho, G. (2004), 'Pathogen Die Off in Vermicomposting Process', *6th Specialist Conference on Small Water and Wastewater Systems and 1st International Conference on On-site Wastewater Treatment and Recycling*, Perth, Australia, p. 104.

Balkema, A.J., Preisig, H.A., Otterpohl, R., Lambert, A.J.D. and Weijers, S.R. (2001), 'Developing a Model Based Decision Support Tool for the Identification of Sustainable Treatment Options for Domestic Wastewater', *Water Science and Technology*, **43**(7), 265–9.

Bari, Q.H. and Koenig, A. (2001), 'Effect of Air Recirculation and Reuse on Composting of Organic Solid Waste', *Resources Conservation and Recycling*, **33**(2), 93–111.

Barth, J. (2000) 'Compost Quality, Quality Assurance and Use – The Basis for Sustainable Organic Waste Management in Europe', *Proceedings of the International Composting Symposium (ICS'99)*, pp. 14–32.

Beavis, P. and Lundie, S. (2003), 'Integrated Environmental Assessment of Tertiary and Residuals Treatment – LCA in the Wastewater Industry', *Water Science and Technology*, **47**(7–8), 109–16.

Beccali, G., Cellura, M. and Mistretta, M. (2001), 'Managing Municipal Solid Waste – Energetic and Environmental Comparison among Different Management Options', *International Journal of Life Cycle Assessment*, **6**(4), 243–9.

Benitez, E., Nogales, R., Elvira, C., Masciandaro, G. and Ceccanti, B. (1999), 'Enzyme and Earthworm Activities during Vermicomposting of Carbyl-treated Sewage Sludge', *Journal of Environmental Quality*, **28**(4), 1099.

Blassino, M., Solo-Gabriele, H. and Townsend, T. (2002), 'Pilot Scale Evaluation of Sorting Technologies for CCA Treated Wood Waste', *Waste Management and Research*, **20**(3), 290–301.

Bradshaw, J. and Ozores-Hampton, M. (2002), 'Recycle Florida Today and Florida Organics Recyclers Association: Then, Now, and Beyond', *Hort-echnology*, **12**(3), 328–31.

Brix, H. (1999), 'How "Green" are Aquaculture, Constructed Wetlands and Conventional Wastewater Treatment Systems?', *Water Science and Technology*, **40**(3), 45–50.

Brown, M.T. and Buranakarn, V. (2003), 'Energy Indices and Ratios for Sustainable Material Cycles and Recycle Options', *Resources Conservation and Recycling*, **38**(1), 1–22.

Chang, N.B., Lin, K.S., Sun, Y.P. and Wang, H.P. (2001), 'An Engineering Assessment of the Burning of the Combustible Fraction of Construction and Demolition Waste in a Redundant Brick Kiln', *Environmental Technology*, **22**(12), 1405–18.

Clarke, M.J., Read, A.D. and Phillips, P.S. (1999), 'Integrated Waste Management Planning and Decision-making in New York City', *Resources Conservation and Recycling*, **26**(2), 125–41.

Clarke, W.P. (2000), 'Cost–benefit Analysis of Introducing Technology to Rapidly Degrade Municipal Solid Waste', *Waste Management and Research*, **18**(6), 510–24.

Clauson-Kaas, J., Poulsen, T.S., Jacobsen, B.N., Guildal, T. and Wenzel, H. (2001), 'Environmental Accounting – A Decision Support Tool in WWTP Operation and Management', *Water Science and Technology*, **44**(2–3), 25–30.

Clemens, J. and Cuhls, C. (2003), 'Greenhouse Gas Emissions from Mechanical and Biological Waste Treatment of Municipal Waste', *Environmental Technology*, **24**(6), 745–54.

Clift, R., Doig, A. and Finnveden, G. (2000), 'The Application of Life Cycle Assessment to Integrated Solid Waste Management – Part 1 – Methodology', *Process Safety and Environmental Protection*, **78**(B4), 279–87.

Cooperband, L.R. (2000), 'Composting: Art and Science of Organic Waste Conversion to a Valuable Soil Resource', *Laboratory Medicine*, **31**(5), 283–9.

Costi, P., Miniciardi, R., Robba, M., Rovatti, M. and Sacile, R. (2004), 'An Environmentally Sustainable Decision Model for Urban Solid Waste Management', *Waste Management*, **24**(3), 277–95.

Dennison, F., Azapagic, J.A., Clift, R. and Colbourne, J.S. (1998), 'Assessing Management Options for Wastewater Treatment Works in the Context of Life Cycle Assessment', *Water Science and Technology*, **38**(11), 23–30.

During, R.-A. and Gath, S. (2002), 'Utilisation of Municipal Organic Wastes in Agriculture: Where Do We Stand, Where Will We Go?', *Journal of Plant Nutrition and Soil Science*, **165**, 544–56.

Eastman, B.R., Kane, P.N., Edwards, C.A., Trytek, L., Gunadi, B., Stermer, A.L. and Mobley, J.R. (2001), 'The Effectiveness of Vermiculture in Human Pathogen Reduction for USEPA Biosolids Stabilization', *Compost Science and Utilization*, **9**(1), 38–41.

Eaton, D. and Hilhorst, T. (2003), 'Opportunities for Managing Solid Waste Flows in the Peri-urban Interface of Bamako and Ouagadougou', *Environment & Urbanization*, **15**(1), 53–63.

Esperanza, M., Suidan, M.T., Nishimura, F., Wang, Z.M., Soria, L.G.A., Zaffiro, A., McCauley, P., Brenner, R. and Sayles, G. (2004), 'Determination of Sex Hormones and Nonylphenol Ethoxylates in the Aqueous Matrixes of Two Pilot-scale Municipal Wastewater Treatment Plants', *Environmental Science and Technology*, **38**, 3028–35.

Fane, S.A., Asholt, N.J. and White, S.B. (2002), 'Decentralized Urban Water Reuse: The Implications of System Scale for Cost and Pathogen Risk', *Water Science and Technology*, **46**(6–7), 281–8.

Fatta, D., Papadopoulos, A., Avramikos, E., Sgourou, E., Moustakas, K., Kourmoussis, F., Mentzis, A. and Loizidou, M. (2003), 'Generation and Management of Construction and Demolition Waste in Greece – An Existing Challenge', *Resources Conservation and Recycling*, **40**(1), 81–91.

Felder, M.A., Petrell, J.R.J. and Duff, S.J.B. (2001), 'A Solid Waste Audit and Directions for Waste Reduction at the University of British Columbia, Canada', *Waste Management and Research*, **19**(4), 354–65.

Fleming, G. (2002), 'Waste Management Removing the Barriers', *Proceedings of the Institution of Civil Engineering–Civil Engineering*, **150**(1), 4.

Gajdos, R. (1998a), 'Bioconversion of Organic Waste by the Year 2010: To Recycle Elements and Save Energy', *Resources Conservation and Recycling*, **23**(1–2), 67–86.

Gajdos, R. (1998b), 'Efficient Bioconversion of Solid and Liquid Waste – Composting and Anaerobic Digestion in Novel Systems', *International Symposium on Composting and use of Composted Materials for Horticulture*, **469**, 149–56.

Gutierrez, M., Etxebarria, J. and de las Fuentes, L. (2002), 'Evaluation of Wastewater Toxicity: Comparative Study Between Microtox (R) and Activated Sludge Oxygen Uptake Inhibition', *Water Research*, **36**(4), 919–24.

Hammad, M., Badarneh, D. and Tahboub, K. (1999), 'Evaluating Variable Organic Waste to Produce Methane', *Energy Consumption and Management*, **40**(13), 1463–75.

Hammes, F., Kalogo, Y. and Verstraete, W. (2000), 'Anaerobic Digestion Technologies for Closing the Domestic Water, Carbon and Nutrient Cycles', *Water Science and Technology*, **41**(3), 203–11.

Hartmann, H. and Ahring, B.K. (2003), 'Phthalic Acid Esters Found in Municipal Organic Waste: Enhanced Anaerobic Degradation Under Hyper-thermophilic Conditions', *Water Science and Technology*, **48**(4), 175–83.

Hiramatsu, Y., Ysunetsugu, Y., Karube, M., Tonosaki, M. and Fujii, T. (2002), 'Present State of Wood Waste Recycling and a New Process for Converting Wood Waste into Reusable Materials', *Materials Transactions*, **43** (3), 332–9.

Ho, G. (2003), 'Small Water and Wastewater Systems: Pathways to Sustainable Development?', *Water Science and Technology*, **48**(11/12), 7–14.

Hsiao, T.Y., Huang, Y.T., Yu, Y.H. and Wernick, I.K. (2002), 'Modeling Materials Flow of Waste Concrete from Construction and Demolition Wastes in Taiwan', *Resources Policy*, **28**(1–2), 39–47.

Huang, G., Sae-Lin, H.N., Chen, Z. and Liu, L. (2001), 'Long-term Planning of Waste Management System in the City of Regina – An Integrated Inexact Optimization Approach', *Environmental Modeling and Assessment*, **6**, 285–96.

Huang, W.L., Lin, D.H., Chang, N.B. and Lin, K.S. (2002), 'Recycling of Construction and Demolition Waste via a Mechanical Sorting Process', *Resources Conservation and Recycling*, **37**(1), 23–32.

Jang, Y.C. and Townsend, T. (2001a), 'Sulfate Leaching from Recovered Construction and Demolition Debris Fines', *Advances in Environmental Research*, **5** (3), 203–17.

Jang, Y.C. and Townsend, T.G. (2001b), 'Occurrence of Organic Pollutants in Recovered Soil Fines from Construction and Demolition Waste', *Waste Management*, **21**(8), 703–15.

Jang, Y.C. and Townsend, T. (2003), 'Effect of Waste Depth on Leachate Quality from Laboratory Construction and Demolition Debris', *Environmental Engineering Science*, **20**(3), 183–96.

Jeppson, U. and Hellstrom, D. (2002), 'Systems Analysis for Environmental Assessment of Urban Water and Wastewater Systems', *Water Science and Technology*, **46**(6–7), 121–9.

Jonsson, H. (2002), 'Urine Separating Sewage System – Environmental Effects and Resource Usage', *Water Science and Technology*, **46**(6–7), 333–40.

Klang, A., Vikman, P.A. and Brattebo, H. (2003), 'Sustainable Management of Demolition Waste – An Integrated Model for the Evaluation of Environmental, Economic and Social Aspects', *Resources Conservation and Recycling*, **38**(4), 317–34.

Klimesch, D., Ray, S.A. and Guerbois, J.P. (2002), 'Differential Scanning Calorimetry Evaluation of Autoclaved Cement-based Building Materials Made with Construction and Demolition Waste', *Thermochimica Acta*, **389**(1–2), 195–8.

Krook, J., Martensson, A. and Eklund, M. (2004), 'Metal Contamination in Recovered Waste Wood Used as Energy Source in Sweden', *Resources Conservation and Recycling*, **41**(1), 1–14.

Kubler, H., Hoppenheidt, P., Hirsch, K., Kottmair, A., Nimmrichter, R., Nordsieck, H., Mucke, W. and Swerev, M. (2000), 'Full-scale Co-digestion of Organic Waste', *Water Science and Technology*, **41**(3), 195–202.

Kwawe, D.B. (2002), 'Refuse Treatment Options: A Case Study', *Journal of Environmental Management*, **66**, 345–59.

Lamond, J.F., Campbell, R.L., Giraldi, A., Jenkins, N.J.T., Campbell, T.R. and Halczak, W. (2002), 'Removal and Reuse of Hardened Concrete', *ACI Materials Journal*, **99**(3), 300–25.

Lin, S.H. and Chang, C.C. (2000), 'Treatment of Landfill Leachate by Combined Electro-Fenton Oxidation and Sequencing Batch Reactor Method', *Water Research*, **34**(17), 4243–9.

Liu, H.W., Walter, H.K., Vogt, H.S. and Holbein, B.E. (2002), 'Steam Pressure Disruption of Municipal Solid Waste Enhances Anaerobic Digestion Kinetics and Biogas Yield', *Biotechnology and Bioengineering*, **77**(2), 121–30.

Maboeta, M.S. and van Rensburg, L. (2003a), 'Vermicomposting of Industrially Produced Woodchips and Sewage Sludge Utilizing *Eisenia Fetida*', *Ecotoxicology and Environmental Safety*, **56**, 265–70.

Maboeta, M.S. and van Rensburg, L. (2003b), 'Bioconversion of Sewage Sludge and Industrially Produced Woodchips', *Water, Air and Soil Pollution*, **150**, 219–33.

Malmen, L., Palm, O. and Norin, E. (2003), 'A Collection and Treatment System for Organic Waste and Wastewater in a Sensitive Rural Area', *Water Science and Technology*, **11–12**, 77–83.

Maltbaek, C.S. (1999), 'MRFs and Municipal Waste – Bold Success or Heroic Failure', *Proceedings of the Institution of Civil Engineering-Municipal Engineer*, **133**(1), 19–24.

Maniadakis, K., Lasaridi, K., Manios, Y., Kyriacou, M. and Manios, T. (2004), 'Integrated Waste Management through Producers and Consumers Education: Composting of Vegetable Crop Residues for Reuse in Cultivation', *Journal of Environmental Science and Health Part B – Pesticides Food Contamination and Agricultural Wastes*, **39**(1), 169–83.

Mata-Alvarez, J., Mace, S. and Llabres, P. (2000), 'Anaerobic Digestion of Organic Solid Wastes. An Overview of Research Achievements and Perspectives', *Bioresource Technology*, **74**(1), 3–16.

Matos, G.D. and Arruda, M.A.Z. (2003), 'Vermicompost as Natural Adsorbent for Removing Metal Ions from Laboratory Effluents', *Process Biochemistry*, **39**, 81–8.

Maurer, M., Schwegler, P. and Larsen, T.A. (2003), 'Nutrients in Urine: Energetic Aspects of Removal and Recovery', *Water Science and Technology*, **48**(1), 37–46.

McDougall, F.R. and Hruska, J.P. (2000), 'Report: The Use of Life Cycle Inventory Tools to Support an Integrated Approach to Solid Waste Management', *Waste Management and Research*, **18**(6), 590–94.

McGrath, C. (2001), 'Waste Minimization in Practice', *Resources Conservation and Recycling*, **32**(3–4), 227–38.

Mellor, W., Wright, E., Clift, R., Azapagic, A. and Stevens, G. (2002), 'A Mathematical Model and Decision-support Framework for Material Recovery, Recycling and Cascaded Use', *Chemical Engineering Science*, **57** (22–23), 4697–713.

Mels, A.R., van Nieuwenhuijzen, A.F., van der Graaf, J.H.J.M., Klapwijk, B., de Koning, J. and Rulkens, W.H. (1999), 'Sustainability Criteria as a Tool in the Development of New Sewage Treatment Methods', *Water Science and Technology*, **39**(5), 243–50.

Misi, S.N. and Forster, C.F. (2001), 'Batch Co-digestion of Two-component Mixtures of Agro-wastes', *Process Safety and Environmental Protection* (B6), 365–71.

Mohee, R. (2002), 'Assessing the Recovery Potential of Solid Waste in Mauritius', *Resources Conservation and Recycling*, **36**(1), 33–43.

Moskal, T.M. and Hahn, D.W. (2002), 'On-line Sorting of Wood Treated with Chromated Copper Arsenate Using Laser-induced Breakdown Spectroscopy', *Applied Spectroscopy*, **56**(10), 1337–44.

Mulder, A. (2003), 'The Quest for Sustainable Nitrogen Removal Technologies', *Water Science and Technology*, **48**(1), 67–75.

Nakano, K., Minami, W. and Kim, H. (2003), 'Life Cycle Inventory Analysis of a Sewage Treatment System Using Statistics', *Kagaku Kogaku Ronbunshu*, **29**(5), 640–45.

Oles, J., Dichtl, N. and Niehoff, H.-H. (2000), 'Full Scale Experience of Two Stage Thermophilic/Mesophilic Sludge Digestion', *Water Science and Technology*, **36**, 449–56.

Palacios, J.M., de Apodaca, G.A.R., Rebollo, C. and Azcarate, J. (2002), 'European Policy on Biodegradable Waste: A Management Perspective', *Water Science and Technology*, **46**(10), 311–18.

Palmowski, L.M. and Muller, J.A. (2000), 'Influence of the Size Reduction of Organic Waste on their Anaerobic Digestion', *Water Science and Technology*, **41**(3), 155–62.

Panikkar, A., Shrestha, P., Hackney, P. and Riley, S. (2003), 'A Residential Blackwater and Municipal Waste Treatment System – Safety Issues and Risk Management', *ORBIT 2003: Organic Recovery and Biological Treatment; Proceedings of the Fourth International Conference on Biological Processing of Organics: Advances for a Sustainable Society*, Perth, Australia, pp. 118–24.

Patterson, R.A. (2001), 'Water Quality Relationships with Reuse Options', *Water Science and Technology*, **43**(10), 147–54.

Poon, C.S. (1997), 'Management and Recycling of Demolition Waste in Hong Kong', *Waste Management and Research*, **15**, 561–72.

Poon, C.S.A., Yu, T.W. and Ng, L.H. (2001), 'On-site Sorting of Construction and Demolition Waste in Hong Kong', *Resources Conservation and Recycling*, **32**(2), 157–72.

Ren, S. and Frymier, P.D. (2003), 'The Use of a Genetically Engineered Pseudomonas Species (SHK1) as a Bioluminescent Reporter for Heavy Metal Toxicity Screening in Wastewater Treatment Plant Influent', *Water Environmental Research*, **75**(1), 21–30.

Rodriguez-Iglesias, J., Maranon, E., Castrillon, L., Riestra, P. and Sastre, H. (2003), 'Life Cycle Analysis of Municipal Solid Waste Management Possibilities in Asturias, Spain', *Waste Management and Research*, **21**(6), 535–48.

Schachermayer, E., Lahner, T. and Brunner, P.H. (2000), 'Assessment of Two Different Separation Techniques for Building Wastes', *Waste Management and Research*, **18**(1), 16–24.

Schober, G., Schafer, J., Schmid-Staiger, U. and Trosch, W. (1999), 'One and Two-stage Digestion of Solid Organic Waste', *Water Research*, **33**(3), 854–60.

Seo, S. and Hwang, Y. (1999), 'An Estimation of Construction and Demolition Debris in Seoul, Korea: Waste Amount, Type, and Estimating Model', *Journal of the Air and Water Management Association*, **49**(8), 980–85.

Seo, S., Aramaki, T., Hwang, Y.W. and Hanaki, K. (2004), 'Environmental Impact of Solid Waste Treatment Methods in Korea', *Journal of Environmental Engineering-ASCE*, **130**(1), 81–9.

Shehata, S.M., El Shimi, S.A., Elkattan, M.H., Ali, B.E., El-Housseini, M. and El Sayad, S.A. (2004), 'Integrated Waste Management for Rural Development', *Journal of Environmental Science and Health Part A – Toxic/Hazardous Substances and Environmental Engineering*, **39**(2), 341–9.

Sinha, R.K., Heart, S., Agarwal, S., Asadi, R. and Carretero, E. (2002), 'Vermiculture and Waste Management: Study of Action of Earthworms *Eisinia foetida*, *Eudrilus euginae* and *Perionyx excavatus* on Biodegradation of Some Community Wastes in India and Australia', *The Environmentalist*, **22**, 261–8.

Smith, D.C. and Hughes, J.C. (2004), 'Changes in Maturity Indicators during the Degradation of Organic Wastes Subjected to Simple Composting Procedures', *Biology and Fertility of Soils*, **39**, 280–86.

Solo-Gabriele, H.M., Townsend, T.G., Messick, B. and Calitu, V. (2002), 'Characteristics of Chromated Copper Arsenate-treated Wood Ash', *Journal of Hazardous Materials*, **89**(2–3), 213–32.

Steer, D., Aseltyne, T. and Fraser, L. (2003), 'Life-cycle Economic Model of Small Treatment Wetlands for Domestic Wastewater Disposal', *Ecological Economics*, **44**(2–3), 359–69.

Stelmachowski, M., Jstrzebska, M. and Zarzycki, R. (2003), 'In-vessel Composting for Utilizing of Municipal Sewage-sludge', *Applied Energy*, **75**(3–4), 249–56.

Tolaymat, T.M., Townsend, T.G. and Solo-Gabriele, H. (2000), 'Chromated Copper Arsenate-treated Wood in Recovered Wood', *Environmental Engineering Science*, **17**(1), 19–28.

Townsend, T.G., Jang, Y. and Thurn, L.G. (1999), 'Simulation of Construction and Demolition Waste Leachate', *Journal of Environmental Engineering-ASCE*, **125**(11), 1071–81.

United Nations Environmental Programme (UNEP) (2002), *Environmentally Sound Technologies for Wastewater and Stormwater Management–An International Source Book*, London: IWA Publishing.

Vidal, N., Poch, M., Marti, E. and Rodriguez-Roda, I. (2002), 'Evaluation of the Environmental Implications to Include Structural Changes in a Wastewater Treatment Plant', *Journal of Chemical Technology and Biotechnology*, **77**(11), 1206–11.

Wahlstrom, M., Laine-Ylijoki, J., Maatanen, A., Luotojarvi, T. and Kivekas, L. (2000), 'Environmental Quality Assurance System for Use of Crushed Mineral Demolition Waste in Road Constructions', *Waste Management*, **20**(2–3), 225–32.

Wang, F.S. (2001), 'Deterministic and Stochastic Simulations for Solid Waste Collection Systems – A SWIM Approach', *Environmental Modeling and Assessment*, **6**, 249–60.

Weber, W.J., Jang, Y.C., Townsend, T.G. and Laux, S. (2002), 'Leachate from Land Disposal Residential Construction Waste', *Journal of Environmental Engineering-ASCE*, **128**(3), 237–45.

Wilderer, P.A. and Schreff, D. (2000), 'Decentralized and Centralized Wastewater Management: A Challenge for Technology Developers', *Water Science and Technology*, **41**(1), 1–8.

Winter, M.G. and Henderson, C. (2003), 'Estimates of the Quantities of Recycled Aggregates in Scotland', *Engineering Geology*, **70**(3–4), 205–15.

Woodard, R., Harder, M.K., Bench, M. and Philip, M. (2001), 'Evaluating the Performance of a Fortnightly Collection of Household Waste Separation into Compostables, Recyclates and Refuse in the South of England', *Resources Conservation and Recycling*, **31**(3), 265–84.

27 Renewable energy technologies: key to sustainable futures

Ian Lowe

Introduction

The case for increased use of renewable energy rests on three foundations: the depletion of petroleum reserves, global climate change and the need for a secure, equitable world. Together these make an irresistible case for using renewable energy. There are however, some serious issues that we have to confront. I will first outline the case for making greater use of renewable energy technologies, then discuss their limitations and possible negative consequences.

Resources

About 60 years ago, King Hubbert (1949) used statistical data on US oil discoveries and associated production to predict that US oil output would peak about 1970. It did, leading to a change in the relationship between petroleum-producing nations (OPEC) and petroleum-consuming nations. The 1970s' oil 'shocks' dispelled the myth of infinite resources, causing significant policy changes in many northern hemisphere countries. Hubbert's techniques were being used in the 1970s to estimate that world oil production would peak about 2010, plus or minus ten years. That is still the best estimate: there are optimists who think the peak might be as far away as 2020 while there are pessimists who think it happened in the year 2000! Whoever is right, there can be no escaping the fundamental geological truth that we are using petroleum much faster than it was produced naturally, so it will not be plentiful for much longer (Fleay, 1997). Some analysts think the struggle for the remaining oil is already under way. Professor Gretchen Daily of Stanford University made the obvious point when she asked a 2003 forum in Sydney to reflect on how differently the Bush administration would have regarded Iraq if it only had 10 per cent of the world's broccoli.

Military and other geo-political issues aside, most decision-makers are still in denial about the approach of the peak in world oil production and the consequent need to change our transport systems, which are still based on the presumption that fuel will continue to be plentiful and cheap. While fuel prices in Australia are much higher than in North America, they are much lower than in Europe. In fact, Australians pay more per litre for beer, cask wine, milk, orange juice or even bottled water than they do for motor spirit! Cheap fuel leads to profligate use, with many urban commuters driving long distances as the sole occupant of large and inefficient cars. The situation is being worsened by the current fad of buying large four-wheel-drive vehicles (SUVs) to cope with the uneven terrain of suburban streets. A recent survey in Australia found that the dominant reason for buying these urban assault vehicles is that the increasing numbers mean that drivers no longer feel safe on the road in sedans! So we are witnessing an urban arms race leading to increasing use of vehicles that are heavy and extremely inefficient.

Alternatives to petroleum

There are alternatives to oil as a transport fuel but most of them pose problems. In the short term it is relatively easy to envisage natural gas being used. Liquid petroleum gas (LPG) is already the fuel for most taxis in Australian cities, while Brisbane City Council decided a few years ago to move its bus fleet from diesel to compressed natural gas. The problem is that gas is also a finite resource. While the known reserves of gas are much greater than the reserves of oil, using gas for all transport purposes would roughly quadruple demand, so the change would only buy a decade or two of breathing space. Many see gas as a bridge to a sustainable future taking advantage of that breathing space. That possibility leads to a questioning of the current Australian enthusiasm to export gas. If all the large projects currently under discussion were to materialize, Australian gas exports would increase from the present seven million tonnes a year to about 40 million tonnes per year by 2020. The sale of the gas would produce short-term economic benefits but it may be doing future generations a disservice by selling a resource they might need.

From time to time other hydrocarbon resources such as oil shale and tar sands are promoted as the solution to the problem of transport fuels. There is no doubt that it is technically possible to produce synthetic fuels from these sources. The problem with oil from shale and tar sands is the economics but the economic issue is actually a manifestation of a deeper and more fundamental obstacle. Energy analysis by Leach (1979) showed that the energy used to process typical shale deposits is about the same as the energy content of the product. So in energy terms the process is effectively paying one lot of men to dig a hole and another lot to fill it up. However high the price of oil goes, the economics will always be dubious. Many shale deposits are very large, with huge amounts of potential hydrocarbons, but the low grade of the resource makes the economics questionable. The low grade of the resource also means there would be serious environmental problems if we were ever to process shale on the scale needed to make an impact on fuel needs.

There is a group of alternative fuels that can be produced sustainably from plant material – in other words renewable energy technologies. Australia has produced ethanol from sugar since the 1930s while Brazil and the USA also produce large quantities from sugar and maize respectively. There are three problems with ethanol. The collection and processing of crops like sugar requires significant amounts of transport fuel, so the energy benefits are small – or, in the view of some analysts, possibly negative. The second problem is that growing sugar leads to other environmental problems; in Australia pollution of the waters around the Great Barrier Reef is a serious consequence. The third problem is the ethical dilemma of whether it is appropriate to use food-growing land to produce transport fuel in a world where millions go hungry. The scale of the potential contribution is also limited; as one example, converting Australia's total sugar production to ethanol would meet about 10 per cent of transport fuel needs. In Australia's case, Commonwealth Scientific and Industrial Research Organisation (CSIRO) studies have found that pyrolysis of rapidly growing trees could produce enough methanol to supply the country's transport fuel but only if it were possible to use an area about the same as that used for all agriculture (Foran and Crane, 2000). So plant-based alcohols might be a useful supplement to petroleum fuels but they are unlikely to be produced on a scale sufficient to be a replacement, while scaling up production would also increase the environmental impacts associated with the growing and processing of the plant materials.

Finally it is possible to envisage using hydrogen as a transport fuel. The cheapest way to produce the hydrogen in the short term would be to use natural gas as the feedstock but the only potentially sustainable path would be based on renewable electricity to split water.

Global climate change and other environmental issues

The second reason for supporting renewable energy is to reduce the environmental impact of energy use. Petroleum fuels are largely responsible for poor air quality in urban areas, with the emission of particulate matter and the contribution to 'photochemical smog' having tangible direct health effects. The burning of petroleum fuels in transport vehicles and the use of coal to produce electricity are the major contributors to global climate change. While the natural range of carbon dioxide levels in the atmosphere have varied between about 180 and 280 parts per million for the last half million years, the figure is now about 380 and increasing rapidly (Lowe, 2005). This is a direct result of burning huge amounts of coal, oil and gas since the Industrial Revolution. The Earth is now about 0.6 degrees warmer than it was 100 years ago with consequent changes to rainfall patterns, plant growth, distribution of animal species, sea levels and the frequency of severe events like storms, floods and droughts. Some Australian cities such as Perth are facing severe water problems as a direct result of climate change.

All the projections suggest the situation will get much worse. The UN's advisory body, the Inter-governmental Panel on Climate Change, gave in its *Third Assessment Report* a range of possible future outcomes depending on the pattern of future fuel use and taking account of uncertainties in the science (IPCC, 2001). The worrying 'bottom line' is that the most optimistic future based on a rapid phasing out of fossil fuels and the best interpretation of the scientific uncertainty still involves a further 1.5 degrees increase in average global temperature with associated changes in other outcomes influenced by temperature. That is why the governments of most industrial nations have agreed to begin a process of reducing emissions of carbon dioxide. The first step was the Kyoto Protocol, still unfortunately being obstructed by short-sighted politicians in the USA and Australia, but now ratified and in legal force. Some nations are looking well beyond the first timid steps in the Kyoto agreement. For example, the UK recently adopted a target of reducing its carbon dioxide emissions by 60 per cent by the year 2050. Even if reserves of fossil fuels were unlimited, climate change would be requiring us to look seriously at ways of reducing its use for unnecessary purposes.

The International Geosphere-Biosphere Programme recently released a major report on its decade-long scientific study, *Global Change and the Earth System, A Planet Under Pressure* (Steffen et al., 2004). It points out that global change is more than just changes in the climate. It is real, is happening now and in many ways it is accelerating as the multiple interacting effects of human activity cascade through the natural systems of the Earth. In addition to carbon dioxide levels, several other parameters of the Earth's systems are now well outside the natural variation observed over the last half million years or more. We also know that the dynamics of the natural systems of the Earth are characterized by critical thresholds and abrupt changes when those thresholds are exceeded. So, the report warns, it is entirely feasible that human activity 'could inadvertently trigger changes with catastrophic consequences' (Steffen et al., 2004, p. 4). The analysis leads to an obvious conclusion: 'Dramatic increases in energy efficiency, decarbonisation and the development and utilisation of new sustainable energy technologies, such as a hydrogen-based energy system, are needed' (Steffen et al., 2004, p. 38). A 2005 conference in the UK city of Exeter (DEFRA, 2005) considered dangerous implications of climate change and warned that the ocean chemistry is changing and the West Antarctic ice sheet thinning. These sorts of major changes to natural systems underline the urgency of a concerted response to the issue of global climate.

So the second leg of the case for renewable energy technologies is the need to move away from the energy sources that are changing the global climate and threatening our future.

Social issues

The third leg is the need to move toward a more equitable world if we want a sustainable future. I don't believe that entrepreneurs in the OECD countries can be secure doing property deals on mobile telephones as they speed around in large cars, in a world where the majority of humans have never ridden in a car, never made a phone call and never owned property. The division between the 'haves' and the 'have-nots' is widening all the time. In 1980 the richest 20 per cent of the world had 70 per cent of the wealth and the poorest 20 per cent had 2.3 per cent: a ratio of about 30:1. By 1995 the ratio was 60:1 and today it is 75:1. That trend cannot lead to a secure future. The dominance of US media has also made many people in the poorest countries acutely aware of the relative material comfort of the wealthy nations. An increasingly inequitable world is very likely to be an increasingly insecure world, with literally millions of people from poor countries risking their lives to make their way to North America, Western Europe or the relatively affluent countries of the Pacific Rim: Japan, Australia and New Zealand. It is possible to imagine social and political solutions to basic needs. For example, the UN Development Programme has estimated that the entire developing world could be given adequate nutrition, clean drinking water, reasonable shelter, basic education and health care for about 5 per cent of the global military budget! However, there is no prospect even in principle of extending to the developing world the sort of access to energy taken for granted in OECD countries. So the only prospect of more equitable access to energy, arguably a prerequisite for a secure and peaceful future, involves the development of energy technologies based on plentiful resources. That makes a compelling case for promotion of renewable energy technologies.

Renewables: the main options

Most people see renewables as solar energy, used either directly or indirectly in the form of wind energy. The most common form of renewable energy, biomass, is essentially stored solar energy. Hydro-electricity is also a form of stored solar energy since it is the Sun's heat that evaporates water and lifts it onto the land surfaces that feed the rivers used to produce the power. In the medium term geothermal energy is likely to become increasing important. Historic use has concentrated on the sites where steam is produced naturally. These produce significant amounts of electricity and process heat in New Zealand and Italy. The available heat in hot dry rocks constitutes a much larger resource. Feasibility studies were being done as this chapter was finalized on the possible use of geothermal energy from hot dry rocks five kilometres below the surface of the Cooper Basin.

Significant amounts of power are generated in many countries by hydro-electric schemes, which were once seen as the epitome of clean energy. In the 1980s the environmental movement became aware of the impact of reservoirs on riverine ecosystems. Since the changes to river flow have wide-ranging effects on plant and animal species as well as the cultural values of the streams, there is now concerted opposition to any attempt to dam rivers for power generation. This is a good example of a general problem to which I return in the next section: harnessing renewable energy inevitably has environmental effects. While these may be less serious than the environmental impacts of using fossil fuels, they are by no means negligible.

Wind energy is the obvious 'success story' in the recent history of renewables. At the time of writing there was about 40 000 megawatts of installed wind power around the world, almost all installed in the last ten years; the 1992 figure was 2500. The installed wind capacity grew at about 30 per cent per year from 2000. There are many reasons for the success of wind power. It is relatively cheap, it uses relatively little land area and it can co-exist with other land uses such as grazing or cropping. It is also a relatively mature technology, with significant cost and quality improvements since its serious expansion began about 20 years ago. Manufacture of wind turbines has become a major industry in Denmark.

Solar energy is a huge resource, in total about 10 000 times total human energy use. Put another way, the solar energy that hits Australia in one summer day alone is about half the total global energy use for an entire year! It can be used directly, mainly as a source of heat. The heat can also be used to generate electricity or sunlight can be turned directly into electricity by photovoltaic cells. The most widespread use of the Sun's heat is in solar hot water systems, now accounting for about 5 per cent of the Australian domestic market. Social and economic factors lead to much higher uses in particular regions, with solar used in about 25 per cent of Western Australian homes and about 45 per cent of homes in Australia's Northern Territory. This is often perceived as being mainly due to economic differences, but a 1984 study (Lowe et al., 1984) found the economic benefits of solar hot water were greater in Queensland, where it has only 7 per cent of the market, than in Western Australia. Since the pay-back time in most parts of mainland Australia is less than the guarantee period of modern solar hot water systems, there is a solid case for mandating solar hot water in new houses. Hot water is estimated to account for about 35 per cent of domestic energy use in Australia, so solar water heating makes a significant difference.

Solar energy can also be focused to produce enough heat to raise steam. A large-scale solar thermal power station was built in Central Australia in 2003. Ten large dishes, each 14 metres across, capture solar energy. The overall array supplies about 170 kilowatts to Umuwa and three nearby communities at a lower cost than diesel generators, the alternative in such remote areas. Since it is quite technically demanding to use solar energy to raise steam, there have been feasibility studies of a scheme to use solar energy to pre-heat water for a coal-fired power station, thus reducing the amount of coal-fired energy needed to produce steam. There are also serious plans to build a solar chimney one kilometre high in rural Australia; it would generate 200 megawatts of electricity. This design is based on a successful smaller version in Spain. It uses the Sun's heat to produce an updraft that operates a wind turbine, making power available around the clock.

The energy in sunlight can be turned directly into electricity using the photovoltaic effect: when light strikes a suitably prepared semiconductor material such as silicon, electrons are released and an electric current can be produced. The process requires extremely pure materials to ensure the energy is not re-absorbed, so solar cells have been quite expensive. Roof-integrated systems are now being installed, usually feeding power to the grid during the day; owners of such systems often sell more electricity during the day than they buy at night and receive a cheque rather than paying a power bill! Japan, the USA and the European Community all have targets for huge numbers of roof-integrated systems: 1.5 million, one million and half a million respectively. The Prime Minister of Spain has installed solar cells on the roof of his official residence to demonstrate his support for renewable energy. In South Australia the government is building what it calls 'the North Terrace solar power station' by installing solar cells on the main public buildings along that street: Parliament House, Government House, the museum and the art gallery.

Globally by far the most common form of renewable energy is biomass, mainly in the form of wood. Several other forms are in use today. Ethanol has been produced from various plant materials since the 1930s. It can be used as a 10 per cent supplement to petrol without any modification of standard engines. Methanol, a simpler alcohol, can be produced from any vegetable material. A range of vegetable oils can be used as bio-diesel: peanut oil, sunflower oil, canola oil and many others are all suitable and are likely to be used increasingly by farmers as commercial diesel becomes more expensive. When organic matter decays in the absence of air, it produces methane, which is the major component of natural gas. Several domestic refuse tips are now being used as methane generators, taking advantage of the process that has been going on for many years under the covering of soil.

There are several other forms of renewable energy that are being developed or are already in use in other parts of the world. A tidal generator at La Rance in France extracts energy from the daily flow of the tides. Prototype wave generators have been developed in Europe, designed to harness the huge resource of wave energy; in the North Atlantic the energy is about 100 kilowatts per metre of wavefront! The engineering problem is to build devices that will harness that energy without being destroyed by the awesome power of storm waves, which are strong enough to dismantle reinforced concrete barriers. Experimental devices in the Pacific islands have used the temperature difference between the sea surface and deep ocean water to run a heat engine. One remote Australian town, Birdsville, took advantage of the fact that artesian bore water comes to the surface at a temperature of about 70 degrees to run a small heat engine, turning the heat energy of the water into electricity for the town.

The basic point is that there are many forms of usable renewable energy. While it would be unwise to rely on any one form, feeding electricity from the various technologies into an integrated grid is a sensible strategy. A 1992 Australian government report (Stevens, 1992) estimated that a mix of renewables could supply 30 per cent of Australia's electricity by 2020 at an extra cost of about 10 per cent, or could provide all of Australia's electricity by 2030 at a cost about 50 per cent greater than the present system, which gets almost all its power from coal. These are robust conclusions because several different combinations of renewable energy can be used to achieve the targets. Those price supplements are arguably less than the extra cost imposed by climate change resulting from the use of coal.

Problems of widespread use of renewables
The wise US journalist H.L. Mencken was quoted as saying that every complex question always has a simple answer but it is always wrong! It is never possible to change only one thing in a complex system because other changes always follow. Though there is a compelling case for making greater use of renewables, it is also important to be aware of the outstanding issues.

There are some technical problems still to be solved before we can be confident of having reliable energy systems based on renewables. In general we can be reasonably confident that problems that are *purely technical* can be solved if sufficient resources are devoted to the task.

The technical issues overlap with the economic questions. Both technically and economically as one example, the most attractive way to produce large amounts of hydrogen for transport would be to use natural gas as the feedstock. Analysis however, shows that this approach does not reduce the greenhouse emissions from transport. If we want the hydrogen economy to reduce our impact on the atmosphere, we will need to produce hydrogen by using renewable electricity to split water. Turning solar or wind energy into hydrogen is a way of storing those intermittent sources, so it has obvious appeal. The problem is that the present technologies make hydrogen from renewable electricity economically unattractive. So the prospect of hydrogen fuel cell futures eventually being based on clean renewable energy hinges on the development of less expensive ways of harnessing those energy forms. While transport energy is almost certain to become more expensive as increasing demand in the USA and China chases a fixed or declining supply, in the real world of politics there are few leaders willing to risk a voter backlash by deliberately increasing fuel costs.

The third issue is the environmental impact of renewable energy technologies. The most significant problem with wind energy arises from it being extremely site-specific. Since the power output of a wind turbine increases as the cube of the wind velocity, there is a great incentive to find sites that are especially windy, like the tops of mountains or ocean headlands. These are sites of scenic importance, so there are now often objections to proposals to install wind generators. There have also been claims that turbines would kill birds but modern machines have large blades that turn relatively slowly and pose very small risks to birds. Finally it should be noted that manufacture of wind turbines requires conventional resources: metals, alloys, composites and so on, requiring fuel energy for extraction and processing. So the production of energy is not environmentally benign but it is much less damaging than extracting and burning fossil fuels.

This is an example of the general truth that there are environmental issues if any technology is scaled up to meet the demands of a modern industrial society. Producing solar cells, or wind turbines, or any other renewable energy devices on the scale needed to supply energy on the scale we use will have environmental impacts. Specific technologies have their particular problems. As an extreme example, a study by a group of researchers at CalTech, published in the journal *Science*, calculated that a future global hydrogen economy could result in 60 to 120 million tonnes of hydrogen being lost each year into the global atmosphere (Tromp et al., 2003). We just don't know what the impact would be. An optimistic view would hope that it would just react with oxygen to form water and return to the oceans, or that it would be absorbed in ways that don't cause unforeseen problems, but we cannot be certain. So we should be doing the research now. Nobel Laureate Paul Crutzen has pointed out that we were just lucky that CFCs were based on chlorine rather than bromine and so thinned the ozone layer rather than destroying it entirely! Some researchers have also pointed out that the lightness of hydrogen makes it likely that significant amounts would drift into the stratosphere and possibly even be lost from the Earth system. They argue that taking water from the oceans to transfer hydrogen into the atmosphere might in time lead to reductions in sea level. This could be desirable in the short term, given the problems of rising sea level caused by global warming, but we would obviously have to be very confident about our sums if we were to consider this as a conscious strategy! The serious point once again is that this is an issue that should be studied. Before we embrace the hydrogen economy we need to have done enough research to be confident we are not getting out of the climate change frying pan into some unknown environmental fire.

The fourth issue is the social and political impact of working toward a renewables future. The time, intellectual effort and other resources expended on one area of research and development inevitably reduces the capacity to work in other fields. Some well-intentioned people see the nub of the transport problem not as oil and the internal combustion engine but as the inefficiency of the average vehicle. This is a point that cannot easily be dismissed. Imagine giving a group of engineering students the task of designing a transport vehicle to carry a fragile payload, typically between 50 and 100 kilograms. If they produced a design that weighed more than a tonne, you would almost certainly recommend they review their career options, possibly steering them toward a future for which numeracy would not be important. It is easy to imagine a vehicle like the Amory Lovins hyper-car, weighing about 250 kilograms instead of five times as much and as a direct result using only one-fifth the fuel (Lovins, 1977). Making such vehicles the norm would make the limited oil reserves last five times as long and dramatically reduce emissions. Others go back one step further and see mobility as a response to poor urban design. Again, these people have a point. The reason we move into cities is to access the wider range of services that are available there. We don't move to cities so we can spend hours travelling to access those services. So a fundamental priority should be improved urban design, making the services people want easily accessible, rather than accepting the inadequacy of current design and flattening increasing fractions of our cities to allow people to drive by themselves for hours in pursuit of the services that are now inaccessible. Finally, it could be argued that the technical task of expanding hydrogen fuel cell transport to the entire human population is so huge that embracing this approach will do nothing to bridge the gap between the mobile and the immobile.

I concede the relevance of these criticisms. We should of course encourage improved urban design and smaller, more efficient vehicles. These measures however, have long time lines. Much of the structure of the cities of 2030 is already in place. Most of the vehicle fleet of 2015 is already on the roads. So we need to be pursuing cleaner energy systems as well as trying to develop better vehicles and putting greater effort into urban planning. While the prospect of extending transport to the entire human population by using renewable energy to produce hydrogen seems daunting, it is in principle possible if there is the political will, whereas there is no prospect even in principle of universal access to oil-based transport. There are also, of course, resource issues. Before we could embark on a programme to produce millions of hydrogen fuel cell vehicles, we would need to be confident that the materials needed are actually available! Like the other problems I have discussed, these issues should not deter us from working toward futures based on renewable energy technologies; they simply provide the wider context within which we should be doing that work.

Conclusion: the case for sustainable futures

I have no doubt that we should be working toward sustainable futures. That is a moral imperative. It is indefensible to be developing futures that we know cannot be sustained, producing inevitable problems for future generations. As the second report in the UNEP *Global Environmental Outlook* series said, our present approach is not sustainable, so doing nothing is no longer an option (UNEP, 1999). A sustainable future will be one in which we are not depleting the resources future generations will need, are not doing serious damage to natural systems, and are moving toward an equitable and secure world. So market-led wealth generation and government-guided technological change has to be supplemented by a values shift toward a new global vision marked by equity and marked by durability.

I believe we should be looking at strategic goals, like stabilizing the population and eliminating hunger. That does not require technical advances, it simply requires a more equitable distribution of the two kilograms of food per person per day we now produce, rather than a market approach in which those who cannot afford food go hungry, while land that formerly grew cheap food for subsistence living in Africa is now increasingly being used to grow flowers to be air-freighted to rich consumers in the developed world. We should also be aiming at a dematerialization of society. Some European nations have now adopted the goals suggested by the Wuppertal Institute of reducing energy use to a quarter of the present level and reducing material use to 10 per cent of the present level; they see those as realistic targets (Yencken and Wilkinson, 2000). But above all else, we need a values shift, perhaps away from *Homo sapiens*, which is gendered and a link back to our past, towards what my partner Patricia Kelly, based on Pentti Malaska's idea, calls *Globo sapiens*. The idea of being wise citizens of the planet recognizes that we share it with all other species and that we hold it in trust for all future generations. That means that we need to see the economy as a *means* to service human needs rather than an end in itself, and that we should be committed to *genuine* globalization rather than the current fad of simply reducing the constraints on corporations. We also need to improve our social institutions and processes for making difficult decisions. It will only be possible to take difficult decisions if there is an open and transparent process, involving the community and allowing time to work through the costs and benefits of alternatives. Changing one thing in a complex system always produces other changes, so no change is ever universally beneficial; there are always losers as well as winners. In a fair world those who lose out from a change that benefits the community as a whole should be compensated by the rest of the community.

The fundamental problem is still that most decision-makers are operating under what could be called the pig-headed model of the world, in which the world is seen like the head of a pig, with the economy a large shape like the face, while society and environment are minor protuberances like the ears (SoEAC, 1996). For those who still have that primitive worldview, it actually makes sense to say that the economy is supreme and the minor problems of society and environment can be handled as long as the economy is thriving. This approach is not working; economic progress since 1990 has been accompanied by serious social problems and environmental degradation. The only intellectually defensible model is one that accepts that *the economy is a part of society*; it is a very important part, to be sure, but only a part, because we expect from our society a range of services that are not part of the economy, like a sense of place, our cultural identity, security, companionship and love. Our societies are totally enclosed within natural ecological systems on which we depend for breathable air, drinkable water, adequate nutrition, a sense of cultural identity, spiritual sustenance and so on. We tend to behave as though we are not part of natural systems. We should each remember that every molecule of our bodies was once part of the natural systems of this planet and every one of those molecules will, in time, once again be part of the natural systems of this planet.

We need to accept that our social and economic planning should be within an ecological framework, that we do need planning and conscious decision-making, rather than trusting the magic of the market that cannot, even in principle, represent the interests of other species and future generations. So we need new social institutions, we need new technologies to meet our needs, but above all else we need values for a sustainable future based on the principle of *Globo sapiens* and continuous adaptive management based on social learning.

References

Department for Environment, Food and Rural Affairs (DEFRA) (2005), *Avoiding Dangerous Climate Change*, http://www.stabilisation2005.com/, accessed 29 November 2005.

Fleay, B. (1997), *The End of the Age of Oil*, Sydney: Pluto Press.

Foran, B. and Crane, D. (2000), *Modelling the Transition to a Biofuel Economy in Australia*, http://www.cse.csiro.au/publications/2000/biofuel-00-10.pdf, accessed 29 November 2005.

Inter-governmental Panel on Climate Change (IPCC) (2001), *Third Assessment Report*, Cambridge: Cambridge University Press.

Kelly, P. (2006), 'Letter from the Oasis: Helping Engineering Students to Become Sustainability Professionals', *Futures*, **38**, 696–707.

King Hubbert, M. (1949), 'Energy from Fossil Fuels', *Science*, **109**(2823), 103–9.

Leach, G. (1979), *The Energy Question*, Harmondsworth: Penguin.

Lovins, A. (1977), *Soft Energy Paths*, Harmondsworth: Penguin.

Lowe, I. (2005), *Living in the Hothouse*, Newham: Scribe Books.

Lowe, I., Backhouse, D. and Sheumack, M. (1984), 'The Experience of Solar Hot Water Systems', *Search*, **15**, 165–7.

State of the Environment Advisory Council (SoEAC) (1996), *State of the Environment Australia 1996*, Collingwood, Australia: CSIRO Publishing.

Steffen, W., Jager, J., Matson, P., Moore, B., Oldfield, F., Richardson, K., Sanderson, A., Schnellnhuber, J., Turner, B.L., Tyson, P. and Wasson, R. (2004), *Global Change and the Earth System: A Planet Under Pressure, Executive Summary*, Berlin: Springer-Verlag.

Stevens, M. (1992), 'Renewable Electricity for Australia', National Energy Research Development and Demonstration Council (NERDDC) Discussion Paper 2, Canberra: Australian Government Publishing Service.

Tromp, T.K., Shia, R.-L., Allen, M., Eiler, J.M. and Yung, Y.L. (2003), 'Potential Environmental Impact of a Hydrogen Economy on the Stratosphere', *Science*, **300**, 1740–42.

United Nations Environment Programme (1999), *Global Environmental Outlook 2000*, London: Earthscan.

Yencken, D. and Wilkinson, D. (2000), *Resetting the Compass*, Collingwood, Australia: CSIRO Publishing.

28 Internet tools for environmental technology management learning
*Dorit Maor and Dora Marinova**

Introduction

Environmental technology management (ETM) is still an emerging area of education and learning for the tertiary sector. It is extremely encouraging that some universities have started recently to offer qualifications in this area. Examples are Pennsylvania College of Technology with a Bachelor of Science (BSc) in ETM (since 2005), Arizona State University with a BSc in Industrial Technology: Environmental Technology Management and Masters of Science and Technology with concentration in ETM (including online), California State University Bakersfield with a BSc in ETM (online), Utah Valley State College with a technical area in ETM as part of a BSc degree in Technology Management. The Texas Technology Institute offers a Masters of Science degree in ETM as distance education with an emphasis on engineering solutions, and this has been part of its strategy to address the challenges of the state to maintain a healthy economy in the next decade (Heintze and Hays, 2003). In Europe the University of Applied Sciences in Augsburg, Germany offers a Masters of Engineering in ETM and the Liverpool John Moores University has a BSc in ETM. Other places, such as Cuyamaca College or Palomar College in California, USA tend to combine Environmental Technology with Environmental Management. Outside the western educational system, University of Mobile, Latin-American campus in Nicaragua offers a BSc in ETM and Universities Putra (formerly Pertanian) Malaysia offer a Masters in ETM.

This short list of teaching institutions would be very close to being exhaustive unless we consider the immense power of continuous learning through the corporate, government and non-government sectors including the Internet teaching and learning resources, which are of particular interest to this chapter.

Since the 1990s, resources available online through the worldwide web have become a major part of the ongoing learning process across all ages and institutions. The Internet can potentially be very influential in promoting and facilitating changes in attitudes towards the use of technologies, and particularly environmentally sound technologies. The sustainability agenda requires fast cultural changes at all levels in society and while the universities are being relatively slow in reshaping their degrees to respond to this demand, the Internet has the ability to be a strong tool in this direction. The aim of this chapter is to describe and critically analyse the progress made in reaching the global virtual community and contribute to the greening of learning, with a particular emphasis on environmental technology management. The approach we have taken is to analyse from a pedagogical point of view the most popular and valued Internet sites that deal with issues of sustainability. The potential of online learning is enormous as it also contributes to the establishment of network communities of learners and learning organizations that can significantly influence the development, adoption and management of environmentally (sound, sensitive, friendly, healthy) technologies. Such networks expand synergistically the resources available within one particular organization and allow participants to draw on global knowledge, expertise and experience as to how best to contribute to the sustainability agenda.

According to Ashford (n.d.), the transition to sustainability requires the corporate and government sectors to address the following questions related to technology management:

- What are the current technologies (for example, in manufacturing, extraction, transportation, services, agriculture, energy generation, waste management) that present significant sustainability problems?
- What features and characteristics will the technologies have that will replace the current technologies?
- Do such technologies exist, can they be adapted or do they need to be developed?
- Which firms or organizations are in the best position to provide such technologies?
- What policies need to be put into place to encourage the transition to the desired technologies?

The expectations are that all new technologies should be environmentally sound as well as economically viable and socially acceptable. The issues of ETM clearly belong to the sustainable development area. However, more recently they have also been linked to terrorism, disaster management and emergency interventions because the technological choices we make and how we manage technology affects the flexibility and robustness of the lifestyles we create. For example, a centralized energy system built around fossil fuels technology (as most grid systems in the developed world are) is not only unsustainable from an environmental and often economic point of view but can be an easy terrorist target and can pose difficult disaster recovery challenges.

Addressing these issues on the Internet learning sites would be a significant step towards building knowledge about environmental technology management. Is this happening? How useful are the online teaching resources?

Pedagogical requirements for online learning websites

Before examining the particular selected websites whose purpose is to spread the message about sustainable development and the use of technology, it is important to have a clear picture of what constitutes an effective learning environment. The learning resources in an educational setting are concerned with the learning task that will cause the learner to engage with the material in a meaningful way and achieve a learning outcome. Oliver (1999) presents a framework (triangle) that identifies and distinguishes between three critical elements of the learning setting and they equally apply in the case of online learning, namely:

- the learning tasks;
- the learning resources; and
- the learning support.

The lack of one of these three crucial components would impede the efficiency of the learning and would delay the achievement of the educational/learning objectives. For example, it is not enough only to have the resources available without encouraging the learners to undertake specific tasks and provide the needed advice and support.

Alternatively, providing only tasks without the necessary resources or support is equally inefficient from a pedagogical point of view.

In relation specifically to online learning, Oliver and Herrington (2001, p. 114) provide a checklist of items that can be used to assess the quality of resources, tasks and support available on websites. These include:

1. accessibility that examines whether the information is organized in ways that make it easily accessed and located;
2. currency that examines whether the age of the materials is appropriate to the subject matter;
3. richness that examines whether resources, tasks and support reflect a rich variety of perspectives;
4. purposeful use of various media that examines whether the electronic media is suitable for the intended purpose of the learning process;
5. inclusivity that examines whether materials demonstrate social, cultural and gender inclusivity.

The above criteria correspond completely with the sustainability concept that argues for: (1) transparency and accessibility to information (item 1 from the above list); (2) appropriateness of solutions to the local environment; (3) acknowledgement and respect for diversity; (4) good communication of the issues; and (5) holistic approach. It is this framework that we are going to adopt in analysing the websites offering opportunities for online learning in ETM.

Websites visited
The amount of web resources directly aimed at encouraging environmental technology management learning is very limited; in effect the majority of them refer to the limited provision by the higher education sector. There is however, a myriad and growing number of websites that address the issues of sustainability. A 2004 GlobeScan survey of web pages (WBCSD, 2005) shows that the sites most valued by sustainability experts from a business and organizational point of view are the ones provided by the World Business Council for Sustainable Development (http://www.wbcsd.ch) valued by 40 per cent of the experts, the International Institute for Sustainable Development (http://www.iisd.org) valued by 28 per cent, various United Nations sites (www.unep.org, www.un.org/esa/desa.htm, www.unglobalcompact.org) at 27 per cent, the World Resources Institute (www.wri.org) and SustainAbility (www.sustainability.com) both at 10 per cent, GreenBiz (www.greenbiz.com) at 8 per cent and Business for Social Responsibility (www.bsr.org/) at 5 per cent (see Figure 28.1). The popularity of these websites among the experts across 40 countries in Europe, Asia, the Americas, Africa and the Middle East, should also be acknowledged taking into consideration the fact that the survey did not prompt any answers and experts were allowed to indicate multiple choices. How do these sites cater for ETM learning?

World Business Council for Sustainable Development
(http://www.wbcsd.ch, accessed 22 April 2005)
A keyword search for 'environmental technology management' on the website of the World Business Council for Sustainable Development (WBCSD) returned no entries for

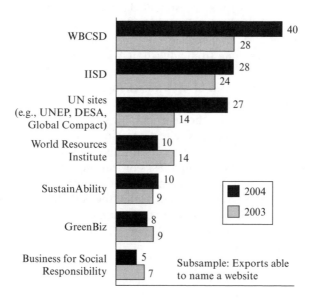

Best Websites for Information on
Sustainable Development

Unprompted Up to Three Mentions, 2003–2004

Source: WBCSD, 2005.

Figure 28.1   Best websites survey results

both documents available through the site and the site itself. 'Environmental technology' however, returned 22 entries, 'technology management' eight and 'environmental management' 167, indicating that these areas are still perceived as unrelated. As expected, the cross-cutting theme of innovation/technology claims that technology is a key tool for a sustainable future and that business clearly needs to play a stellar role in meeting this challenge. The news, case studies, publications and speeches are information resources with limited use as they are targeted mainly as a public relations exercise without helping those who are looking for real solutions (for example, what a database of environmental solutions could provide) or those who still have not been converted or are recent converts to the sustainability concept. Nevertheless, the popularity of this website seems to have grown by a significant 12 per cent since a similar survey in 2003 (see Figure 28.1).

The site of the WBCSD has an animated learning tool called 'An Interactive Sustainable Development Learning Map', which claims to present a graphical (cutting down on words) description of 'the critical links between the natural, the economic and the social dimensions of our world'. The goal of the learning tool is to present a holistic picture of the complex and interrelated layers of reality and help the user understand 'how humanity is hooked on growth, how values or social tensions drive technology/innovation, how production not only creates environmental degradation but also the social capacity for solutions'. The learner immediately gets a clear picture of the intricacies in the topic as well as a sense of intrigue to continue to push the button. The presentation of the sustainability concepts is organized around six levels, which add new dimensions to the basic issues of population (presented through financial security, standard of living, employment and unemployment, desire for more growth), consumption (surplus wealth), production, material input and capital investment. This leads to environmental degradation and the 'life saving' devices are eco-efficiency and mitigation investment that we all need to undertake to succeed. A level higher, the power of creative knowledge offers new approaches to solutions ranging from education, which delivers technology and innovation, to the development of social capital and redistribution. The final outcome is new values and awareness for the need of systemic change, which leads to new allocation of investments and institutional improvements.

Aside from the criticism for the choice of graphic symbols (everything is presented by fish) and the difficult to read colour/size text combination, from a pedagogical point of view the learning tool as well as the website itself satisfy only the condition to present the learning resources. There are no learning tasks or learning support provided. Also there is no proactive encouragement for the learning process to happen.

As the same GlobeScan survey also predicts that the WBCSD will be one of the most influential international institutions in terms of advancing sustainable development in the next five years, only behind the European Union (Chennelle, 2005), its educational role leaves ample space for improvement.

International Institute for Sustainable Development
(http://www.iisd.org/, accessed 26 March 2005)
The website of the International Institute for Sustainable Development (IISD) also has a sustained and increasing popularity (up by 4 per cent since 2003). Its website has a lot to offer in relation to electronic resources. However, pedagogies and delivery strategies are not clearly identified or transparent within the website. The International Institute for Sustainable Development sees its vision as a 'better living for all – sustainably' and its mission to 'champion innovation, enabling society to live sustainably'. It includes a lot of innovative ideas but does not take them to a level of facilitating the learning process. Examining the website from the perspective of the learning environment provided by the web-based technology, there are a few concerns about its effectiveness.

The IISD provides only one side of Oliver's triangle (Oliver, 1999), namely the learning resources. From an educational focus the 'learner' is replaced by the 'user' and the goal is to engage with the information and to be able to use it critically. The quality of the website therefore, is determined by the richness of the information, the authenticity, the updated nature of it and the relevancy to the user. It is also very important for the website to allow the learner to engage with the material in a meaningful way. Firstly, the website contains functions such as Calendar, Media Room, Employment, Funders and Contact, which help to organize and provide the virtual community with functional services, upon which the users can participate in making decisions and choices such as conference participation, view employment opportunities and so on.

The Institute is clearly interdisciplinary and provides knowledge from business, climate change, communities and livelihoods, economic policy, energy, investment, knowledge network, leadership development, measurement and assessment, natural resources, security and trade. For each topic an easy to navigate webpage contains description, information and weblinks to extend the topic to other relevant sites. It is comprehensive and enables the downloading of PDF files on a variety of topics. It also covers current features and links to the latest news and research from IISD and other international events (for example, 'Integrating Economic and Social Policies to Achieve the United Nations Developmental Agenda' forum held 14–15 March, 2005, at the United Nations headquarters in New York). The website also provides access to electronic commentary, publications, an internship research library and more that can be used by scholars, naive users or students in cross-disciplinary areas. The research tools offer database and catalogues facilities such as 'links to libraries with sustainable development focus that have online searchable databases'.

The use of index pages enables easy navigation to all areas of the website. It uses hyperlinks to provide ready access to related information quickly and effectively (for example, the link to another environmental education site such as PEMBINA Institute). It includes navigation markers and links within the page, graphics to improve the look and feel of the page and colour images where necessary. There is clear layout, different sections with titles, navigation is easy and there are organizational cues to aid orientation and scooping, but there is no variety of media format such as video, audio or animation. Only text is being used.

Although the knowledge network is extensive, and provides a good digital repository, the part that is missing is interactivity with the user. A simulation on a particular topic can provide the user with a sense of interactivity, which can increase the quality of learning in using the electronic resources. The virtual experience can stay in the level of information exchange or can be taken to a higher level of knowledge construction by building in interactivity or simulation, which will require additional input from the user.

The keyword search for 'environmental technology management' again did not return any results, with 'environmental technology' returning nine, 'technology management' two and 'environmental management' 55 entries. It is clear that ETM is still an emerging area of professional education and research but the extensive website suggests that there is a growing interest in the sustainability area. However, the main questions still remain: do the resources promote the awareness for and facilitation of sustainable technologies and how can the learning task and the learning support be incorporated to increase the learning opportunities?

UN websites (www.unep.org, accessed 24 April 2005, www.un.org/esa/desa.htm, accessed 25 April 2005, www.unglobalcompact.org, accessed 25 April 2005)
A characteristic feature of the UN websites valued by the sustainability experts is again the richness of resources. Nevertheless, it would have been better if there were an integrated starting webpoint rather than the current fragmentation between environmental, economic, social and human rights issues and hence the need to look at all these websites in combination.

The first website is of the United Nations Environment Programme (UNEP) and it offers resources for governments, scientists, journalists, business persons, civil society, children and youth. A keyword search on 'environmental technology management' on the home webpage generated zero hits, 'environmental technology' returned 426 entries, 'technology management' 30, but 'environmental management' returned results in the tens of thousands (23 100) indicating the importance of the latter as well as the disconnection between these areas.

The UNEPNet portal available through the resources for scientists is an extremely valuable connection to the major thematic environmental issues, namely climate change, freshwater, mountains, socio-economic, urban environments (the access to the latter two was not available at the time of access) as well as to the Global Environmental Outlook (GEO) database. A keyword search on 'environmental technology management' and 'technology management' generated zero results in UNEPNet and the GEO database; 'environmental technology' produced only five for UNEPNet and again zero for GEO but 'environmental management' generated 40 for UNEPNet and six for GEO. However, the fuzzy word search of the GEO database (a search engine that uses fuzzy logic) is a highly recommended brilliant feature, which returned 972 entries to 'environmental technology management' covering the range of issues similar to the ones included in this book. A beautifully planned feature of the UNEPNet website are the regional portals of which only the Arctic was available at the time of writing.

The resources available to journalists are of paramount importance because of the power of media in influencing public opinion and behaviour, and the UNEP website has a good coverage of this through information about particular initiatives, such as (at the time of writing) the 35th Earth Day 'Protect Our Children & Our Future' and the 'World Environment Day'. The business resources available under the UNEP's Department of Technology, Industry and Economics point to centres of global importance such as the International Environmental Technology Centre in Osaka, the Production and Consumption Branch in Paris and the Chemicals Branch in Geneva. A range of useful publications is available. None of these resources, however, encourages active learning and problem-solving. This is also true about the rich information for civil society.

The educational functions of the UNEP's website are targeted at children and youth. The use of graphics, animation and various media is excellent. The interaction encourages active learning through tasks listed under 'What can you do!', game playing and open competitions for children, and Speaker's Corner, E-Dialogue, Live Chat, Share your Story for the youth. Although the targeted issues are predominantly environmental, there is potential in the designed interactive activities to develop and apply the holistic approach required within a sustainability framework and potentially to introduce a focus on technology management. The 'Interact with UNEP' option for youth contains an excellent training kit on sustainable consumption entitled YouthXchange (produced by the UNEP-UNESCO Partnership on Youth & Sustainable Lifestyles) as well as other action learning tasks such as a photo competition, prompt to the GEO database and its educational products, and information on environmental campaigns. A lot of the youth educational material would have been very appropriate also for the business community and there is potential for adjusting the learning tasks towards the way business operates.

An interesting feature of this website is its positioning within events that help raise the awareness of the importance of sustainability and sustainable lifestyles, such as the Asian Tsunami Disaster (from the main page) and sports, including the Olympics (from the Youth page). The 15 photo images also allow for a quick link to the important issues covered with resources on the website. Although online learning about ETM is not an explicit focus on this website, it contains very good resources and action learning tasks.

The second UN website is that of its Department of Economic and Social Affairs (DESA). Again, the richness of information and resources is very good but there is no straight indication for online learning options. The opening page states the mission, recent highlights, materials from conferences, sessions and meetings, and sub-programmes, including sustainable development, social development, financing for development, population, advancement of women and forests. The page also has statistics, development trends (including policies), many publications and reports as well as links to the various divisions and offices.

The search on 'environmental technology management' generated zero results, while 'environmental technology', 'environmental management' and 'technology management' produced respectively 163, 2110 and 8150 entries. Technical cooperation is a significant part of the Division for Sustainable Development's webpage with two wide areas of issues: energy, transport and atmosphere, and water and natural resources. Overall, the new developments in sustainability are well covered on the web pages of DESA through information on sustainability assessment, sustainability indicators (for example, energy indicators for sustainable development) and national sustainable development strategies.

The UN Commission for Sustainable Development (CSD) has a learning centre for each of its meetings (the most recent one being CSD-13 in April 2005) whose sessions are open to a mix of delegates, NGO representatives and participating experts from developed and developing countries. Only very brief summaries of the presented papers or case studies are accessible through the DESA site. Overall, both DESA and its Division for Sustainable Development do not have any online learning tools or programs. Hence, from the three pillars of e-learning only resources are available with no coverage of tasks and support.

The Global Compact initiative launched in 2000, which is the third UN website, is aimed explicitly at the business community and attempts to bring companies and firms together with UN agencies, NGOs and civil society to support the UN ten principles in the areas of human rights, labour standards, the environment and anti-corruption. The 9th principle specifically relates to environmental technologies, namely: 'encourage the development and diffusion of environmentally friendly technologies'. The importance of management of these technologies is clearly stated in the first sentence describing this principle: 'Encouraging the development and diffusion of environmentally friendly technology is a longer-term challenge for a company that will draw on both the management and research capabilities of the organization.'

The homepage of the Global Impact site has quick links to the most important functions of the initiative, including how companies can participate, search for partners from any country, view communications on progress and frequently asked questions, as well as full information on the initiative itself, related news and events, and issues and policy dialogue. A distinctive feature of this website is the learning tool, named Learning Forum. This so far is the only specifically designed learning environment that promotes ETM and it is extremely good.

The goal of the Learning Forum is 'to establish a rich and useful repository of both corporate practice and fundamental research, a platform of knowledge that integrates the views of all relevant stakeholders, while simultaneously increasing the transparency of companies' activities'. It allows for accessing real-life experiences from organizations around the world through short, detailed project descriptions. It enables a search as well as the submission of examples. The Learning Forum specifically targets academia and university business schools in order to promote social corporate responsibility, which is the main focus of this initiative. The search engine of the Learning Forum found 103 examples and reports (of a total of 242) for the 9th principle, 24 case studies (of a total of 74) and 12 projects (of 31); although the keywords 'ETM', 'environmental technology' and 'technology management' did not return any entries and 'environmental management' returned only one example, one case study and no projects. Overall, the Global Impact Learning Forum promotes engagement with the issues of ETM by sharing best practices and looking for partnerships and projects. This is a very good alternative to learning tasks but the initiative is left to the e-learner. As far as support from the UN's website is concerned, there is no direct contact person named and not even an option of whom to contact from the e-learning environment. The available resources again are of a high standard.

World Resources Institute (www.wri.org, accessed 26 April 2005)
The website of the World Resources Institute (WRI) covers most areas of sustainability through its research topics, namely agriculture and food, biodiversity and protected areas, business and economics, climate change, energy and transportation, coastal and marine ecosystems, forests, grasslands and drylands, governance and institutions, population, health and human well-being, resource and materials use, and water resources and freshwater ecosystems. Each area connects to publications, the EarthTrends Database (a comprehensive free online database, which uses renowned and reliable data sources to present the information in multiple formats), overview of key issues and WRI's work as well as staff contacts. Technology does not appear to be a major explicit focus as keyword searches on 'ETM', 'environmental technology' and 'technology management' generated no results and 'environmental management' produced only four entries. Technology issues however, are covered within the various research areas. There is a newsroom, an invitation to take action and several of WRI's projects cover areas of learning and leading by action (for example, in CO_2 reduction and 'green' office space) but no specific e-learning environment is provided. From a pedagogical point of view apart from availability of resources, there is little effort put into encouraging the learning for sustainable technologies and practices.

SustainAbility (www.sustainability.com, accessed 26 April 2005)
SustainAbility, whose website has ranked high with around 10 per cent of the experts' votes, is 'an independent think tank and strategy consultancy' company, which employs 'experts in corporate responsibility and sustainable development' and provides advice 'on market risks and opportunities'. E-learning is not one of the areas covered by the company. However, it provides useful free resources to those who join the Sustainability Compass, which creates a homepage allowing the display of information matching the user's chosen preferences as well as full access to the SustainAbility public resource archives. There are also useful showcases and stories from business leaders and good opportunities for e-networking, which potentially can contribute to ETM learning. The keyword search for 'environmental technology management', 'environmental technology' and 'technology management' did not generate any results, and 'environmental management' returned only five results.

GreenBiz (www.greenbiz.com, accessed 26 April 2005)
The website of GreenBiz, which is a non-profit, non-partisan company, aspires to use 'the power of technology to bring environmental information, resources, and tools to the mainstream business community'. The provision of learning opportunities is part of the company's mission together with access to clear and accurate information and

resources helping to integrate environmental responsibility with profitable business practices. The amount of information and advice is impressive, but most importantly the website represents a very efficient e-learning environment through the Business Toolbox. It targets the private sector as well as NGOs and government organizations. It is extremely well designed and a keyword search on 'ETM' generated one entry (the University of Arizona Masters as described above), 'technology management' had five hits, 'environmental technology' returned 264 matches and 'environmental management' 700.

The GreenBiz's Business Toolbox covers a wide range of topics, namely best practices, climate change, compliance, corporate reporting, Environmental Management Systems (EMSs), energy efficiency, green building, green careers, green design, green printing, green purchasing, partnerships, pollution prevention, renewable energy, small business, supply chain, sustainable development, waste reduction and water conservation. There are no learning tasks but the materials are organized in a way that is extremely useful and easy to use for anybody who is looking to make a change in the way business is conducted. There are GreenBiz essentials, how-to, reports and tools. The tools for example, are a collection of resources that allow a company to assess itself in terms of environmental impact or to locate useful software. There are also links to a bookstore, available jobs and reference collection. Although there is not a direct learning support for the website, the reference option provides information on the Mentor Center, which is a link to hundreds of organizations offering free or low-cost assistance to help companies address sustainability issues, 'from regulatory compliance to cutting-edge initiatives'.

Overall, this website comes at the top according to its impact on and potential for e-learning about environmental technology management and sustainability.

Business for Social Responsibility (http://www.bsr.org/, accessed 26 April 2005)
As the name states, the website of Business for Social Responsibility does not focus on technology. The site's search engine returned no entries for either 'ETM' and 'environmental technology', one for 'technology management' and only 14 for 'environmental management'. Business for Social Responsibility (BSR) is a non-profit organization with memberships by companies from around the globe that offers them information, tools, training and advisory services to help them 'achieve success in ways that respect ethical values, people, communities and the environment'. Although learning is part of BSR's raison d'être, the website does not provide any learning tools. It includes information about BSR services by areas (such as consumer goods, information and communication technology, extractives, pharmaceuticals and biotechnology, food and agriculture, and transportation) and type of services (such as corporate social responsibility reporting, strategy, structure, assessment and policy development, implementation, stakeholder engagement, working groups and supply chain management). There is a wide range of free resources available as well as information on conferences, links to the BSR store and invitation to join.

As with some of the other visited websites discussed earlier, this one also contributes to the e-learning environment only through the richness of materials and ideas.

Sustainable technologies websites
In addition to the websites cited in the GlobeScan survey, there is also a myriad of sustainable technologies websites (see Appendix to this chapter), which are organized around

particular technologies, for example, solar medical equipment and solar hospitals, or issues, for example, eliminating hunger or disaster help. They contribute to the technical learning of the sustainability practitioners but tend to service a particular purpose rather than provide a holistic approach. We are certain that they make a significant contribution to e-learning in their respective fields and we leave it to the readers to assess their specific features.

Designing a learning online environment for ETM

Table 28.1 presents a summary assessment of the websites of the seven most popular organizations according to the sustainability learning checklist. The notional ranking system used is on a five-level scale, namely, not available, limited, good, very good and outstanding. The striking resemblance between all these sites is their relatively high ranking in relation to the availability of resources and relatively low ranking when it comes to learning tasks. Only one of the examined organizations, namely the website of GreenBiz, passes the assessment criteria and is a truly valuable resource for ETM and sustainability practitioners. Although the combined UN websites pass most of the checklist criteria, their major weakness is the lack of a holistic approach, which is essential for any sustainability learning tools. This leaves a lot of space for improving the learning potential of the available Internet tools.

The following are some suggestions to improve the design of the e-learning environment:

- The importance of learning tasks should be explicitly recognized and learners should be encouraged to engage in a meaningful way with practical problems related to the management of environmental technologies at organizational level.
- There should be an explicit role and transparency for the learning support. There is potential for an innovative approach based around facilitating the establishment of communities of learners and interactive networks (for example, through chat rooms) rather than relying only on contact name(s) from the organization.
- There should be a wider range of learning tools used, which, for example, encourage interactivity, involve simulation, provide virtual experience and go beyond the traditional information exchange.
- The learning tools should be designed to encourage and maintain long-lasting interest that translates into practice and contributes to the development of a sustainability culture across as wide an Internet community as possible. They need to cater for the newcomers as well as for the people who have already adopted and are practising sustainability values.

Conclusion

The sustainability agenda is not an issue that is likely to go away and the new approaches to the way we live and do business need to be learned. This emerging field of formal education needs the strong support of the Internet community and the power of online learning to deliver a broad band of knowledge, interactivity (Maor, 2003), creativity and connectivity. The Internet can potentially be very influential in promoting and facilitating changes in attitudes towards the use of technologies (Garrison and Anderson, 2003; Harasim et al., 1995), and particularly pedagogical sound technologies (Reeves and Harman, 1994). We are fully aware that things on the Internet are dynamic and what we

Table 28.1   Websites assessed according to sustainability learning checklist

Organization and Website(s)	Transparency and Accessibility	Appropriateness	Richness and Diversity	Communication	Holistic Approach	Overall
WBCSD (*http://www.wbcsd.ch*)						
– tasks	limited	limited	limited	limited	limited	good
– resources	outstanding	outstanding	good	good	very good	
– support	limited	good	good	good	limited	
IISD (*http://www.iisd.org*)						limited
– tasks	not available	not available	not available	not available	not available	
– resources	very good	very good	very good	very good	good	
– support	limited	good	limited	limited	limited	
UN (*http://www.unep.org, http://www.un.org/esa/desa.htm, http://www.unglobalcompact.org*)						good
– tasks	very good	very good	very good	good	limited	
– resources	outstanding	outstanding	very good	very good	limited	
– support	good	very good	good	very good	limited	
WRI (*http://www.wri.org*)						good
– tasks	limited	limited	limited	limited	limited	
– resources	outstanding	very good	very good	very good	good	
– support	good	good	limited	limited	limited	
SustainAbility (*http://www.sustainability.com*)						limited
– tasks	limited	not available	limited	limited	not available	
– resources	very good	very good	good	good	good	
– support	not available	not available	not available	not available	not available	
GreenBiz (*http://www.greenbiz.com*)						very good
– tasks	outstanding	outstanding	outstanding	very good	outstanding	
– resources	outstanding	outstanding	very good	very good	very good	
– support	very good	very good	very good	very good	good	
Business for Social Responsibility (*http://www.bsr.org*)						limited
– tasks	not available	not available	not available	not available	not available	
– resources	very good	outstanding	very good	very good	good	
– support	limited	very good	good	limited	very good	

have analysed is only a snapshot, nevertheless, it is representative of the implementation of pedagogical thinking. Although there are some positive examples, the Internet's learning potential for encouraging the implementation and management of environmentally sound technologies is yet to be fully explored.

The quality of the websites is determined by the richness of the information, the authenticity, the updated nature and the relevancy to the user. All examined websites pass these assessment criteria and are truly valuable resources for ETM and sustainability practitioners. For effective learning, however, it is also very important for the websites to allow the learner to engage with the material in a meaningful way and not to be drawn in to the sea of information. Learning support and learning tasks should be an integral part of the website design in order to create real impact and change perceptions and practices. Most websites have yet to make this contribution.

Note

* The authors want to acknowledge the financial support of the Australian Research Council.

References

Ashford, N. (n.d.), 'The Corporate Role in Protecting Health Safety and the Environment', *Biopolitics*, http://www.biopolitics.gr/HTML/PUBS/EVEA/english/ash2_usa.htm (accessed 22 April 2005).

Chennelle, J. (2005), *WBCSD Predicted to Play Major Role Advancing SD Next Five Years*, World Business Council for Sustainable Development, http://www.wbcsd.org/plugins/DocSearch/details.asp?type=DocDet&ObjectId=13416, accessed 22 April 2005.

Garrison, D.R. and Anderson, T. (2003), *E-learning in the 21st Century: A Framework for Research and Practice*, London: Routledge Falmer.

Harasim, L., Hiltz, S.R. and Turoff, M. (1995), *Learning Networks: A Field Guide to Teaching and Learning Online*, Cambridge, MA: MIT Press.

Heintze, M.R. and Hays, S.J. (2003), 'Closing the Gaps in Texas: The Texas Tech University Pathway Programme', *Journal for Higher Education Strategists*, **1**(2), 167–90.

Maor, D. (2003), 'Teachers' and Students' Perspectives on Online Learning in Social Constructivist Learning Environment', *Technology, Pedagogy and Education*, **12**(2), 201–18.

Oliver, R. (1999), 'Exploring Strategies for Online Teaching and Learning', *Distance Education*, **20**(2), 240–54.

Oliver, R. and Herrington, J. (2001), *Teaching and Learning Online: A Beginner's Guide to E-learning and E-teaching in Higher Education*, Perth, Western Australia: Edith Cowan University Publications.

Reeves, T.C. and Harmon, S.W. (1994), 'Systematic Evaluation Procedures for Interactive Multimedia for Education and Training', in S. Reisman (ed.), *Multimedia Computing: Preparing for the 21st Century*, Harrisburg, PA: Idea Group Publishing, pp. 472–505.

World Business Council for Sustainable Development (2005), www.wbcsd.ch, accessed 16 March 2005.

Appendix: Sample of websites on sustainable technology compiled by Independent Media Centre, South Africa

- *Dangerous pollution from biogas cookers*:
  http://www.bhopal.net, http://www.aprovecho.net
- *Disaster help and economic development*:
  http://www.southafrica.indymedia.org/news/2005/04/8040.php
  translate: http://www.babelfish.altavista.com
- *Electric bicycles and vehicles*:
  http://www.electroauto.com, http://www.solarsailor.nl, http://www.etraction.com,
  http://www.greenspeed.com.au, http://www.solarmobil.net,
  http://www.solarwaterworld.de, http://www.feelgoodcars.com,
  http://www.rqriley.com, http://www.solomontechnologies.com,
  http://www.sourceguides.com/energy, http://www.oxygenworld.net,
  http://www.zapworld.com, http://www.powabyke.com,
  http://www.sunswift.com,
  http://www.sunversy.de, http://www.eddos.com/solarcycling,
  http://www.electrichorse.co.uk, http://www.elbil.no, http://www.heliev.gr,
  http://www.apve.pt, http://www.evthai.com, http://www.egovehicles.com,
  http://www.e2v.co.uk, http://www.itiselectric.com, http://www.biketec.ch,
  http://www.ihpva.org, http://www.elweb.info, http://www.a-e.ch,
  http://www.kta-ev.com, http://www.electroportal.com, http://www.nevco.com,
  http://www.revaindia.com, http://www.avere.org, http://www.acev.fi,
  http://www.twike.ch, http://www.elcat.fi, http://www.evparts.com,
  http://www.eveez.com, http://www.idbike.com,
  http://www.ecopowertechnology.com, http://www.aae.it, http://www.alel.biz,
  http://www.solarcycle.net, http://www.evt-scooter.de, http://www.seiko.net.tw,
  http://www.kewet.com, http://www.haveblue.com
- *Eliminating hunger*:
  http://www.carbon.org, http://www.growingsolutions.com,
  http://www.treesforlife. org, http://www.emnz.com, http://www.acresusa.com,
  http://www.thrive.org.uk, http://www.growbiointensive.org, http://www.tilth.org,
  http://www.permaculture.org.uk, http://cityfarmer.org,
  http://www.communitygarden.org, http://www.echonet.org,
  http://www.biodynamic.org.uk, http://www.livingnutrition.com,
  http://www.healthmasters.com,
  http://www.southafrica.indymedia.org/news/2005/04/ 7985. php,
  http://www.southafrica.indymedia.org/news/2005/04/8037.php
- *Free journal about solar energy, sustainable technologies*:
  http://www.homepower.com, http://www.sses.ch
- *Houses from the Earth*:
  http://www.calearth.org, http://www.rammedearthworks.com,
  http://www.cobcottage.com
- *Indestructible and energy efficient buildings*:
  http://www.monolithicdome.com, http://www.kalwall.com,
  http://www.solatube.com, http://www.sloanled.com

- *Micro-hydro energy*:
  http://www.microhydropower.com, http://www.hydroscreen.com,
  http://www.microhydropower.net, http://www.sourceguides.com/energy,
  http://www.harris hydro.com, http://www.ampair.com,
  www.canyonindustriesinc.com
- *Ocean energy and river energy*:
  http://www.uekus.com, http://www.marineturbines.com,
  http://www.inverpower.com,
  http://www.hydroscreen.com, http://www.itpower.co.uk,
  http://www.bluenergy. com, http://www.poemsinc.org
- *Organic vegetable diesel fuel*:
  http://www.biofuels.ca, http://www.greasecar.com, http://www.dieselveg.com,
  http://www.vegetableoildiesel.co.uk, http://www.greaseworks.org,
  http://www.veggievan.org, http://www.biodiesel.org.au,
  http://www.worldenergy.net, http://vegburner.co.uk,
  http://www.biodieselnow.com, http://www.fmso.de,
  http://www.boulderbiodiesel.com, http://www.journeytoforever.org
- *Rainwater harvesting*:
  http://www.fogquest.org, http://www.rainwaterharvesting.org,
  http://www.about rainwaterharvesting.com, http://www.rainCentre.org,
  http://www.rainharvesting. co.uk
- *Solar cookers and solar ovens*:
  http://www.solarcooking.org, http://www.sungravity.com,
  http://www.konacooker.com, http://www.sunoven.com,
  http://www.cleardomesolar.com, http://www.solarbakeovens.com,
  http://www.eg-solar.de,http://www.cuisinesolaire.com,
  http://www.sunspot.org.uk, http://solar-ovens.com, http://www.enolar.com
- *Solar energy*:
  http://www.powerlight.com, http://www.beaconpower.com,
  http://www.sourceguides.com/energy, http://www.solar.sharpusa.com/solar,
  http://www.solargenix.com, http://www.firstsolar.com,
  http://www.solarsystems.com.au (http://www.waterfurnace.com and
  http://www.twapanels.co.uk), http://www.solarfoerderung.de
- *Solar homes and building tours*:
  http://www.ases.org
- *Solar medical equipment and solar hospitals*:
  http://www.sunutility.com/html_pg/healthcare.html
- *Solar refrigerators for vaccines and medicines*:
  http://www.sunfrost.com, http://www.staber.com
- *Solar water desalination*:
  http://www.sunutility.com/html_pg/clean_water.html, http://www.ette.no,
  http://www.sourceguides.com/energy
- *Solar water heaters*:
  http://www.rpc.com.au, http://www.heliosenergies.com,
  http://www.heliodyne.com, http://www.thermotechs.com,
  http://www.sourceguides.com/energy, http://www.sundasolar.com

- *Solar water pumps*:
  http://www.riferam.com, http://www.kyocerasolar.com,
  http://www.solarenergy.com
- *Solar water purification*:
  http://www.sunutility.com/html_pg/thermos_bottle.html,
  http://www.ecological-engineering.com, http://www.oceanarks.org,
  http://www.wolvertonenvironmental.com, http://www.solartoilet.com,
  http://www.dharmalivingsystems.com, http://www.oasisdesign.net,
  http://www.sourceguides.com/energy
- *Sustainable technologies – contacts and links*:
  http://www.southafrica.indymedia.org/news/2004/08/6503.php,
  http://www.southafrica.indymedia.org/news/2005/04/8037.php,
  http://www.southafrica.indymedia.org/news/2005/04/8038.php
- *Training for people in developing countries and emergency response teams*:
  Email: <ian@solarenergy.org>, http://www.solarenergy.org,
  http://www.scoraigwind.com, http://www.watermotor.net,
  http://www.greenandcarter.com
- *Waste not, want not; a dollar saved is a dollar earned*:
  http://www.wormwoman.com, http://www.zeri.org, http://www.emtech.org
- *Wind energy*:
  http://www.bergey.com, http://www.bwea.org, http://www.activepower.com,
  http://www.provenenergy.com, http://www.vestas.com,
  http://www.sourceguides.com/energy, http://www.ropatec.com,

Please share this info with other people. Thank you.

29 Russian greenhouse gas emissions: options for sustainable reduction

Malcolm Hill

Introduction

This chapter focuses on one particular aspect of environmental pollution in the Russian Federation, namely, atmospheric emissions of greenhouse gases. This group of gases was selected in view of the global warming effects attributed to greenhouse gas emissions and the existence of the Kyoto Protocol to achieve their reduction. Within the basket of six greenhouse gases included within the Kyoto Protocol, carbon dioxide and methane were selected for particular study in this chapter in view of their major contribution to global warming. Russia was a signatory to the Protocol in 1997 and also completed the process of ratification in November 2004 to enable the Protocol to come into operation in February 2005.

The first section of the chapter provides summaries of published estimates of Russian emissions of CO_2 and CH_4, focusing primarily on the fuels and power-generation sectors because of the high levels of emissions from those industries. Reference is also made to other emitting industries, such as metals production, but emissions from road, rail and air transport are not discussed as the focus of this chapter has been reductions of pollution from 'point sources' rather than 'mobile sources'. Furthermore, carbon dioxide sequestration through land utilization and forestry has not been discussed in this chapter, as the focus is on the reduction of the emissions themselves.

The next section of the chapter provides information on technological options for the reduction of carbon dioxide and methane emissions, within a framework of available fuels and combustion processes utilized in power generation and the fuels industries within the Russian Federation. The chapter concludes with a discussion of the Russian economic, political and commercial factors that will influence the practical implementation of those technologies in that region, paying particular attention to the potential for the development of indigenous Russian technologies and the assimilation of western know-how.

Sources of data

The information presented in this chapter has been compiled from several sources of data, including specialized technical and industrial journals and environmental reports, but particular use has been made of six official publications, namely the *First*, *Second* and *Third National Communications* of the Russian Federation to the United Nations Framework Convention on Climate Change (UNFCCC) published in 1995, 1998 and 2002 (Mezhvedomstvennaya . . ., 1998; Mezhvedomstvennaya . . ., 2002; Russian Federation, 1995); and the draft (2000), approved (2001) and revised (2003) energy strategy documents developed by the Ministry of Energy of the Russian Federation for 2000–20 (Minenergo, 2001; Mintopenergo, 2000; Proekt . . ., 2003).

The authors of these documents have made various assumptions, as explained in Hill (2002a) and Hill (2002b) for the publications between 1995 and 2002, but the consequent effects of these assumptions on forecasts for greenhouse gas emissions do not appear to be widely different. The 2003 revised document however, constructed three energy scenarios compared with two in the 2001 approved document. These three scenarios for 2001–20 were based on ranges for the following factors: growth in GDP, increase in physical investment, world oil and gas prices, GDP per capita and the percentage of the fuel and energy industries in Russian industrial output. In addition, assumptions were made concerning proportions of energy to be saved in relation to energy-saving costs: 20 per cent at less than $20 per tonne of fuel equivalent (one tonne of fuel equivalent or 'tonna uslovnogo topliva' is defined as 7000 kilocalories of heat content in Goskomstat SSSR, 1990, p. 377), some two-thirds at $20–$50 per tonne of fuel-equivalent and the remaining 15 per cent at more than $50 per tonne of fuel-equivalent. Assumptions were also made concerning the proportions of savings available from technical and organizational changes and economic restructuring (37 per cent and 63 per cent respectively in 2010). Although these estimates of fuel savings and proportions are different from those in the 2001 approved document, they do not appear to make major differences to the estimates of future greenhouse gas emissions.

Russian greenhouse gas emissions – the size of the problem
Carbon dioxide emissions
Total emissions (for base year 1990) Emissions of CO_2 from the Russian Federation were some 2360 Mt CO_2 in 1990 (selected as the base year by the UNFCCC at its 1997 conference in Kyoto), accounting for some 77 per cent of total Russian greenhouse gas emissions (3050 Mt CO_2-equivalent) (Mezhvedomstvennaya . . ., 2002, p. 8). Approximately 70 per cent (1670 Mt) of the total CO_2 emissions from the Federation emanated from the area defined as 'European Russia' (that is, west of 60°E longitude) (Economic Commission for Europe, 1995, Table 5).

Compared with the other signatories of the Kyoto Protocol, Russian total emissions of CO_2 were surpassed by only one other Annex 1 country in 1990, namely the United States (4400 Mt), and collectively by member states of the European Union (2765 Mt). In addition, Russian greenhouse emissions in 1990 approximated to those of China, the non-Annex 1 country with the highest emissions (Grubb et al., 1999, pp. 82, 168). Russian energy consumption per unit of GDP and per capita were higher than for Western Europe and Japan however (Saneev et al., 2003), suggesting that Russian power, heating and industrial processes were less energy-efficient than in those OECD countries, although it is important to remember that the Russian climate is far harsher. Compared with the other western industrialized countries however, CO_2 emissions from the Russian Federation reduced significantly between 1990 and 1995, namely by more than 30 per cent (Grubb et al., 1999, p. 170; Mezhvedomstvennaya . . ., 2002, p. 8). These reductions occurred partly as a consequence of continued fuel switching to natural gas, but mainly as a consequence of post-1991 deindustrialization following the fragmentation of the former USSR. Investment in contemporary technologies needs to be considered therefore, to enable Russian output to recover without an associated increase in emissions. There are commercial constraints to such investments to obtain sustainable emissions reductions however, as discussed in the final section of this chapter.

Industrial emissions   The *Third National Communication* quoted figures of 2320 Mt CO_2 for emissions from fossil fuels in 1990 and 1470 Mt CO_2 in 1999, including products from their processing. The majority of these 1999 emissions (1452 Mt CO_2) were from fossil fuel combustion for the production of thermal, mechanical and electrical energy, and the remainder (18 Mt CO_2) comprised losses and leakages. Carbon dioxide emissions from industrial processes were quoted as 42 Mt CO_2 in 1990, of which 41 Mt CO_2 were from the cement industry, but from 1997 data were also included for emissions from the production of lime, soda, ammonia, carbides and ferrous alloys. In 1999 the total CO_2 emissions from industrial processes were some 39 Mt CO_2, of which cement accounted for some 14 Mt CO_2 (Mezhvedomstvennaya . . ., 2002, pp. 9, 10).

Estimates of CO_2 from the electricity generation industry were quoted by Eremin (1999) as 708.5 Mt CO_2 in 1990, reducing to 517.9 Mt CO_2 in 1995 and to 485.3 Mt CO_2 in 1998, indicating a reduction of more than 30 per cent between 1990 and 1998. The data provided in the *Third National Communication* are very similar to those provided by Eremin, namely 708 Mt CO_2 in 1990, 517 Mt CO_2 in 1995, 490 Mt CO_2 in 1998, reducing to 475 Mt CO_2 in 1999 (Mezhvedomstvennaya . . ., 2002, p. 44), indicating that fossil-fuelled power stations accounted for some 30 per cent of emissions from fossil fuel combustion. A more recent estimate (Saneev et al., 2003) provides a higher figure of 830 Mt CO_2 for emissions from fossil-fuel-generating facilities in 1990, reducing to 560 Mt CO_2 in 2000. This higher estimate from Saneev et al. (2003) however, could be due to the inclusion of emissions from fossil-fuel-generating facilities located in industries other than electricity generation (for example, combined heat and power facilities in chemical plants).

The reductions in carbon dioxide emissions from the power-generation sector were achieved mainly as a result of the post-1990 severe fall in industrial output following the fragmentation of the former USSR and consequent lower demand for electricity and associated reduction in fossil fuel combustion. Total sustainable reductions of some 5 per cent (or 85 Mt CO_2) per year were claimed to have been achieved between 1990 and 1994 however, by the increasing use of natural gas in the energy balance (from 42.6 per cent to 48.9 per cent) in place of other fossil fuels with a higher hydrocarbon ratio, and a significant proportion of these savings would be in the electricity-generating sector. In addition, a growing proportion of electricity was generated at hydro and nuclear stations (Mezhvedomstvennaya . . ., 1998, pp. 5, 6), which emit no carbon dioxide from the electricity generation process although contributing to other environmental problems.

The remaining 70 per cent of carbon dioxide emissions were presumably accounted for by other stationary sources such as ferrous and non-ferrous metals facilities and domestic heating (although combined heat and power – CHP – electricity-generating stations also accounted for some 35 per cent of the USSR's low temperature supply in 1990 according to Hill, 1997, p. 129) and non-stationary sources such as road transport. It is probable therefore that CO_2 emissions from Russia were divided fairly equally between power generation, co-generation and district heating, industrial combustion and processes, and road transport, with some 60–70 per cent emanating from 'point sources'. The focus of this chapter will consequently be one of reducing carbon dioxide emissions from electricity generation as the single largest collection of point sources of those emissions, although reference will be made to the reduction of emissions from gas pipeline compressor stations as those installations are also definable as point sources, accounting for almost 90 Mt CO_2 emissions in the mid-1990s. In addition, compressor stations share a core technology with electricity generation, namely gas turbines.

Future CO_2 emissions According to the approved energy strategy for the Russian Federation, emissions from the energy sector are expected to increase to between 1750 Mt CO_2 and 1870 Mt CO_2 in 2010, and to continue to increase to 2015 (1800–2000 Mt CO_2) and 2020 (1840–2200 Mt CO_2) as a consequence of economic growth (Minenergo, 2001, p. 75). The updated draft energy document (Proekt . . ., 2003, p. 24) assumes that greenhouse gas emissions will reach some 75–80 per cent of their 1990 levels by 2010, suggesting an emissions level of 1770–1888 Mt CO_2 (assuming that CO_2 emissions increase at the same rate as greenhouse gas emissions in total) and that by 2020 emissions would return to their 1990 levels of 2360 Mt CO_2. The *Third National Communication* (Mezhvedomstvennaya . . ., 2002, p. 18) produced an estimate of 75 per cent to almost 90 per cent of 1990 emissions in 2010 (that is, 1770 Mt CO_2 – 2100 Mt CO_2), increasing to some 81 per cent to 113 per cent of 1990 emissions in 2020 (that is, 1912 Mt CO_2 –2643 Mt CO_2). Saneev et al. (2003) quoted an estimate of some 1933 Mt CO_2 for emissions in 2010, which is almost identical to the average for the range of emissions predicted in Mezhvedomstvennaya . . . (2002, p. 18).

Assuming that as in the past, emissions from the electricity sector continue to account for approximately 30 per cent of these estimates of emissions from the energy sector, then the power generation industry could be expected to emit some 510–630 Mt CO_2 in 2010, suggesting that some 78–198 Mt CO_2 could be available for emissions trading (see below), although Saneev et al. (2003) quoted a figure of 680 Mt CO_2 suggesting a possible 150 Mt CO_2 available for trading compared with 1990. As mentioned previously however, Saneev et al.'s data may include emissions from generating plant not included in the electricity generating industry. If there is some substitution of coal for natural gas in electricity generation (see below) then levels of CO_2 emissions from the industry may be higher and the quantities available for trading would consequently be lowered by some 16–23 Mt CO_2 (Hill, 2002b), although this increase in emissions from fuel switching may be compensated for by some further substitution of nuclear, hydro power (presumably assuming no increase in methane emissions from reservoir vegetation) and additional use of renewable sources.

Methane emissions
Total emissions The emissions of methane from the Russian Federation in 1990 were reported as 26.1 Mt CH_4 or some 548 Mt CO_2-equivalent (using the IPPC ratio of 21) (Mezhvedomstvennaya . . ., 2002, pp. 9, 11), which amounted to some 18 per cent of total Russian greenhouse gas emissions (3050 Mt CO_2-equivalent) (Mezhvedomstvennaya . . ., 2002, p. 8). Carbon dioxide and methane therefore accounted for some 95 per cent of greenhouse gas emissions from the Russian Federation in 1990 and the remaining 4 per cent was comprised of emissions of N_2O, HFC, PFC and SF_6 (Mezhvedomstvennaya . . ., 1998, p. 6).

The 2002 *Third Russian National Communication* also provided data on greenhouse gas emissions from the Russian Federation for 1994, 1997, 1998 and 1999 (Mezhvedomstvennaya . . ., 2002, p. 11). These data showed that methane emissions had decreased by 26 per cent to 19.3 Mt CH_4 in 1994, 14.4 Mt CH_4 in 1997, 14.7 Mt CH_4 in 1998, and 13.8 Mt CH_4 in 1999.

Industrial emissions   The 2002 *National Communication* also provided information on the sources of methane emissions within the Russian Federation. For 1990 from the total of 26.1 Mt CH_4 emitted, 19.1 Mt CH_4 (or 73 per cent) were from the energy sector, including 16 Mt CH_4 (or 61 per cent) from the processing and leakage of oil and natural gas and 2.9 Mt CH_4 (or 11 per cent) from the mining of solid fuels. The remaining 5.0 Mt CH_4 (or 19 per cent), 1.9 Mt CH_4 (or 7 per cent) and 0.1 Mt CH_4 were from agriculture, waste disposal and forest fires respectively (Mezhvedomstvennaya . . ., 2002, p. 11).

By 1999 the levels of methane emissions from every category described in the previous paragraph had decreased apart from waste disposal, and the proportions had changed slightly. Some 9.4 Mt CH_4 were emitted from the energy sector (or some 68 per cent of the total of 13.8 Mt) including 7.9 Mt CH_4 (or 57 per cent) from oil and gas extraction, transport and distribution and 1.4 Mt CH_4 (or 9 per cent) from the mining of solid fuels, 2.4 Mt CH_4 (or 17 per cent) from agriculture and 1.8 Mt CH_4 (or 10 per cent) from waste disposal (Mezhvedomstvennaya . . ., 2002, p. 11).

From these data it is apparent that the oil and gas industries have been the major emitters of methane in the Russian Federation accounting for some 60 per cent of total Russian methane emissions in both 1990 and 1999. This figure includes emissions from liquid and gaseous fuels through the complete supply chain from extraction to delivery including process emissions, procedural and faulty leakages and fumes from oil and oil products. According to data from Gazprom quoted in the *Second National Communication*, emissions from the processing of gas and liquid fuels accounted for some 1.45 Mt CH_4 in 1994 and losses and leakages accounted for some 6.59 Mt CH_4 (Mezhvedomstvennaya . . ., 1998, p. 38). As total emissions from oil and gas processing and leakage were some 11.5 Mt CH_4 (Mezhvedomstvennaya . . ., 1998, Table 7A) it is likely that process emissions and fumes from the oil industry accounted for some 3.5 Mt CH_4, assuming that the data from Gazprom referred to liquefied and natural gas from the gas industry alone. This chapter has therefore focused on losses and leakages as the single highest source of methane emissions.

The tonnage of 6.59 Mt CH_4 quoted for Gazprom's losses and leakages in 1994 is equivalent to some 9.41 bcm (billion cubic metres), assuming that the density of gas is some $700g/m^3$ at atmospheric pressure between temperatures of 0°C and 20°C (UNFCCC, 1997), which accounts for some 1.55 per cent of gas output during that year (607 bcm quoted in Mezhvedomstvennaya . . ., 1998, p. 19). That estimate of leakage is of the same order as the 2 per cent provided for Soviet gas pipeline losses by Gazprom (2003), Makarov and Bashmakov (1990) and Stern (1993) (that is, emissions of 11 bcm of methane in 1990 according to Gazprom, 2003, p. 8 from production of 637 bcm according to Stern, 1993, pp. 14, 15) and 1.3 per cent quoted for losses within Gazprom's network by the head of the company's department responsible for ecology (Sedikh, 1995). Moe and Tangen (2000, pp. 82–4) however, use a variety of sources to arrive at a figure of 8 bcm for leakages in 1998 from production, high pressure trunk pipelines and compressor stations equivalent to some 114 Mt CO_2, suggesting a weaker gas than pure methane and a further 5 bcm (equivalent to some 66 Mt CO_2, again suggesting a weaker gas than pure methane) from leakages during distribution.

The CH_4 leakages from coal mines were some 2.75 Mt in 1990 and the majority (2.18 Mt CH_4) was emitted during coal extraction from underground mines. Total emissions from Russian coal mines had reduced by 36 per cent to 1.75 Mt CH_4 by 1995, equivalent to a reduction of some 21 Mt CO_2 chiefly as a consequence of reductions in coal production by some 33 per cent from 395 Mt in 1990 to 263 Mt in 1995 (Mezhvedomstvennaya . . ., 1998, pp. 19, 37).

Future CH_4 emissions   Anticipated methane emissions for 2010 are given as some 24 Mt CH_4 (505 Mt CO_2-equivalent) in the Russian *Second National Communication* to the UNFCCC (Mezhvedomstvennaya . . ., 1998, pp. 11, 90). This estimate is 9.3 per cent lower than emissions in 1990, but 22.6 per cent higher than in 1994. The corresponding figures for various branches of the national economy are 15 Mt CH_4 from natural gas, oil and coal extraction and transportation (compared with 19.1 Mt CH_4 in 1990), 6.0 Mt CH_4 from agriculture (compared with 5.4 Mt CH_4 in 1990), and 3.0 Mt CH_4 from solid and liquid waste disposal (compared with 2.0 Mt CH_4 in 1990).

It is expected however, that the production of natural and by-product gas will increase from 584 bcm in the year 2000, to 635–665 bcm in 2010 and 680–730 bcm in 2020 depending on the level of energy saved (Proekt . . ., 2003, p. 50), which could consequently increase Gazprom's losses and leakages by some 0.5 Mt CH_4 (or some 10.5 Mt CO_2-equivalent) in 2010 compared with 1990 unless mitigation measures are taken to reduce these losses to achieve the target figure of 24 Mt CH_4 quoted in Mezhvedomstvennaya . . . (1998, pp. 11, 90). Forecasts from Gazprom however, indicate expected emissions of 7.8 bcm of methane (or 5.46 Mt CH_4) in 2008 and 5.1 bcm (or 3.57 Mt CH_4) in 2012, equivalent to 114.6 Mt CO_2 and 75 Mt CO_2 (Gazprom, 2003, p. 8).

Coal output is planned to increase from 258 Mt in 2000 to some 310–335 Mt by 2010 (Proekt . . ., 2003, p. 50), which could cause methane emissions from coalfields to also increase by some 0.5 Mt CH_4 (or 10.5 Mt CO_2-equivalent) to reach 2.25 Mt CH_4 (or 47 Mt CO_2-equivalent) if the higher target is met, unless as in the case of gas mitigation procedures are carried out. The methane emissions from the coal industry were some 1.38 Mt CH_4 in 2000 and it is planned to reduce further to 1.31 Mt CH_4 in 2010 presumably as a consequence of increased open cast extraction compared with underground mining as well as increased capture of coal mine methane (see below).

Techniques for carbon dioxide and methane mitigation
Carbon dioxide mitigation
Post-combustion processes   There is scope to investigate the implementation of monoethanolomine (MEA)-based post-combustion CO_2 capture systems within the Russian Federation in view of the quantity of possible locations available in coal beds for CO_2 disposal and underground storage and its potential for use in oil extraction or the production of methanol through hydrogenation (Eliasson, 1994). In addition, Russian technologists have some experience of developments in these areas (Bagautdinov et al., 1993 quoted by Eliasson, 1994). At present however, the use of post-combustion processes may not be high on the list of Russian priorities because of their expense and the existence of other options available for the reduction of CO_2 emissions.

Fuel selection   The most significant reductions in Russian CO_2 emissions (apart from those attributed to lower levels of electricity generation and falling industrial production) have come about through the switching of fuels from coal and oil to natural gas, as natural gas produces only some 54 per cent of the CO_2 emissions of hard coal per unit of heat generated, and some 62 per cent of those from oil (Schemenau and Van den Berg, 1990).

The use of natural gas as a fuel in electricity generation is likely to be maintained in the short term for price reasons, particularly in European Russia, and reduction in CO_2 emissions can therefore be achieved by increases in the efficiency of gas combustion processes and operational procedures as discussed below. It is apparent, however, that solid fuel will remain as a major fuel for power generation in many of the industrial areas adjacent to Russian coal deposits, particularly those near to the Kuzbass and Kansk-Achinsk coalfields. Coal's proportion in the fuel mix may increase if gas prices accelerate at a higher rate than those for solid fuels (Hill, 2002b). The continued use of coal will consequently create a demand for 'cleaner coal' combustion technologies (see below) as well as increases in power generation efficiencies particularly in Siberia and the Far East (Saneev et al., 2003).

Operational procedures and supercritical steam boiler technologies   Increased operational efficiency is important for all types of fossil-fuelled power stations to obtain both fuel savings and reduction of CO_2 emissions but it is particularly important in the case of large coal-fired power stations as there is currently no other widely utilized alternative for the reduction of CO_2 per unit of fuel burnt, apart from recent designs of supercritical steam boilers. There is scope therefore, for the transfer of proven energy-efficient operating procedures and supercritical steam boiler technologies from western companies to Russian power stations and power-generation technical establishments (Hill, 1997, pp. 150–88), particularly if old installations currently working at low efficiencies are to be replaced with those of a modern design. In the cases of small and medium-sized gas-fired and coal-fired installations however, various technological alternatives are also available as explained below.

'Cleaner coal' technologies   Gasification combined cycle (GCC) systems, circulating fluidized beds (CFB), pressurized fluidized bed combustion (PFBC) and integrated gasification combined cycle (IGCC) systems are more efficient than pulverized coal combustion in small to medium-sized power stations and can also be retrofitted to existing power plants. CFB systems have potential efficiencies approaching 37 per cent and the capability to reduce CO_2 emissions by some 30 per cent compared with conventional pulverized coal and 'imminent' PFBC systems should be capable of reaching efficiencies of 40 per cent and even higher with the application of ultra-supercritical steam conditions. Such systems should consequently be capable of reducing CO_2 emissions by some 16–36 per cent (Graham-Bryce et al., 1994). There is also evidence of Russian competences in this area as indigenous Russian developments in integrated gasification combined cycle (IGCC) and fluidized bed systems have been reported from the Central Power Engineering Institute (TsKTI) in St Petersburg (Hill, 1997, pp. 179, 180; Prutkovskii and Chavchanidze, 1995).

Gas turbine and combined cycle gas turbine (CCGT) technologies   In the case of gas-fired power stations there are choices in combustion technologies available, namely gas-fired boilers or gas-turbines. Gas-fired boilers are marginally more efficient (39 per cent) than gas turbines (35 per cent), particularly in large installations, and the choice between the two can depend upon the required capacity of the installation and the comparative efficiencies of available boilers and turbines within the particular location. Gas turbines have also been traditionally used as standby units to meet peak loads in power stations and they are also emerging as an attractive option for base load generation in small to medium capacity applications, especially in remote areas provided that there are sufficient servicing expertise and spares availability (Hill, 1997, pp. 77–83).

The most efficient gas turbine configuration for medium-capacity power stations however, is the combined cycle gas turbine (CCGT) system where gas turbines are used to drive one generator and a steam turbine is used to drive another, the latter using steam produced from water heated in a heat recovery steam generator (HRSG) by exhaust gases from the gas turbines. The use of such high-efficiency technologies (up to 58 per cent through the use of high-efficiency gas turbines within a combined cycle configuration) can lead to a significant reduction of carbon dioxide emissions per unit of electricity generated compared with gas boilers and they can be as low as some 40 to 45 per cent of emissions from coal-fired installations as a consequence of the use of a cleaner fuel and operation at high efficiency (Hill, 2000). In addition, combined cycle systems can also be connected to combined heat and power (CHP) or co-generation systems to provide even higher combined cycle efficiencies and used for cost-effective supply of heat to district heating systems (International Chamber of Commerce and World Business Council for Sustainable Development, 1997, pp. 5–10, 16, 17). The Russian Federation already has extensive CHP networks in place installed during the Soviet period: there are therefore consequent advantages to be gained from the use of high-efficiency CCGT systems connected to these networks.

Indigenous 'aero-derivative' gas turbines are produced at aero-engine factories in Samara, Tula, Ufa, Rybinsk, Perm and Ekaterinburg and these machines can be used in a range of installations, including combined cycle installations and compression units in gas pipelines. The Rybinsk factory is currently producing power turbines to Mashproekt (Nikolaev, Ukraine) designs and LMZ is utilizing aero-engine technology developed by the factory in Perm. The capacities of some of these turbines (usually 16–40 MW) are smaller than those of the 100–150 MW 'frame-type' machines produced by LMZ (Hill, 1997, pp. 76–8) but their efficiencies can be higher as a consequence of use of aero-engine technologies and materials (Hill, 2000).

Assimilation of western CCGT technologies is well advanced in St Petersburg, particularly the assimilation of Siemens' gas turbine technologies (mainly focused around combustion chamber design to obtain higher combustion temperatures and consequent enhanced efficiencies) for 'frame-type' machines built by LMZ through a joint venture between the two companies (Interturbo). In addition, gas turbines from Interturbo and steam turbines from LMZ have been installed at the 'North West' combined cycle power station in St Petersburg and this power station is now operational. Cooperation agreements in turbine technologies have also been signed between ABB and the Nevskii Factory in St Petersburg, and GE and the Kirovskii Factory (also in St Petersburg). Domestically-produced turbines could therefore be used for power station and gas pipeline refurbishment although western machines may be preferred in some installations depending upon the comparative efficiency levels, commercial conditions and possible opportunities for technology transfer.

Technology transfer arrangements also exist between Cockerill Mechanical Industries (CMI) (the supplier of HRSG equipment to the 'North West' power station) and the

Ordzhonikidze factory at Podol'sk (near Moscow) for the development of HRSGs (*Energetik*, 1994, pp. 2–3). Furthermore, L&C Steinmuller (manufacturers of boilers and heat recovery steam generators) have a joint venture ('EEE'), which includes the Central Boiler and Turbine Institute (TsKTI) in St Petersburg and a large boiler factory at Taganrog. Each of the Russian partners has decades of experience in boiler construction and previous knowledge of the design and construction of HRSGs: contemporary CCGT systems can therefore be provided by Russian producers.

Gas turbines are also extensively used throughout Russia's gas pipeline system to drive compressor units to maintain gas pressure. The proposed increase in gas production referred to above and consequent pipeline transportation will require the refurbishment of some 25 000 MW of gas compression capacity (probably some 1500 machines if we assume an average turbine capacity of 16 MW) within the present total capacity of 45 300 MW across 251 compressor stations (Minenergo, 2001, p. 37; 'Energy and Climate . . .', 2001). This replacement of gas compression equipment provides the opportunity to further reduce the emissions of greenhouse gases from the fuels industry by the installation of higher-efficiency turbines to obtain lower emissions of CO_2 per unit quantity of gas transported. As 1.0 bcm of gas when combusted emits approximately some 2.0 Mt CO_2 and the gas industry uses some 60 bcm (Mezhvedomstvennaya . . ., 1998, p. 72) to 65 bcm ('Energy and Climate . . .', 2001) of gas per year in gas transportation, it is likely that the quantity of emissions are of the order of 120–130 Mt CO_2 (although Moe and Tangen, 2000, p. 84 provide a figure of some 42 bcm fuel use at compressor stations on high pressure trunk pipelines emitting some 87 Mt CO_2). Increasing the efficiency of gas pumping sets from the present 23 per cent to some 33–36 per cent could save some 8.0–9.0 bcm of gas annually (and consequently some 16–18 Mt CO_2) and even further increases in efficiency using CCGT systems operating at 47–52 per cent efficiency could provide a potential gas saving of some 18–22 bcm of gas per year or some 36–44 Mt CO_2 (Mezhvedomstvennaya . . ., 1998, p. 72).

Data provided by Gazprom (2003, p. 8) for its own gas distribution system reflect substantial reductions in CO_2 emissions from pumping installations, as a figure of emissions of 105 Mt CO_2 from combustion is quoted for 1990 (when production was 637 bcm according to Stern, 1993, p. 14), reducing to 84.0 Mt CO_2 in 2000 (when production was some 584 bcm according to Minenergo, 2001, p. 47). This rate of reduction of CO_2 emissions over the ten-year interval is higher than the reduction of gas production, suggesting an increase in the efficiency of operation, and it is subsequently intended to reduce emissions even further to 70.0–74.0 Mt CO_2 during 2008–12 whilst production is expected to increase to 615–655 bcm (Minenergo, 2001, p. 47).

Techniques for methane mitigation
Gas pipelines   The proposed increase in gas production (see above) will require modernization of distribution facilities, namely the replacement of some 23 000 km of existing pipelines within the present network of some 130 000 km and the construction of some 22 000 km of new pipelines. This programme therefore provides a major opportunity for improved construction, servicing and maintenance of pipelines to reduce methane leakages and thereby save fuel and reduce methane emissions (Hill, 2002a). In addition, improvements can be made to control equipment as well as the use of high-efficiency turbines referred to above, to reduce the quantity of gas combusted in compressor stations (Bruce, 2000; Gale and Freund, 2000).

Coal mines   The Kuzbass coalfields in West Siberia, which produce 70 per cent of total Russian coal output and 55 per cent of mined coal are considered as priority locations for investment in methane extraction technology as some 188–344 bcm of 'virgin coalbed methane' (VCBM) are contained in balance coal reserves in that region (US Environmental Protection Agency, 1996, p. 18). It is likely however, that more attention will be paid to the combustion of 'coal mine methane' (CMM) released from coal seams during the mining operation rather than VCBM, as the capture and subsequent ventilation of CMM is necessary during the coal mining process for safety reasons. The volume of methane liberated from Kuzbass coal production associations is estimated to be greater than 1.0 bcm annually made up of 0.20 bcm of drained gas and more than 0.86 bcm vented (USEPA, 1996, pp. 18, 20), which could possibly be captured and combusted to produce some 3 TWh of electricity. Greenhouse gas emissions could also be reduced by some 12 Mt CO_2-equivalent as the CO_2 emissions from the combusted methane would have a lower greenhouse effect than if it continued to be extracted to the atmosphere. Clearly therefore there is scope for the capture of significant quantities of CMM during mining operations in Russia although many mines in the Kuzbass also contain large faults and severe folding, which can hinder the requisite connections between various parts of the colliery. In addition, significant quantities of 'abandoned mine methane' (AMM) should be available for capture from disused mines as a consequence of the post-1990s major reduction in Russian coal output. Development work on the capture of AMM at a mine at Belovo has been reported by Serov et al. (2000) although capture of significant quantities of methane from exhausted workings may be difficult in the Kuzbass because of flooding.

Commercial factors

The majority of the post-combustion and in-combustion technologies referred to in the previous sections of this chapter have been proven in international practice with evidence provided of some Russian expertise in their development and installation and the assimilation of technology from western organizations. The degree of implementation of Russian technological capabilities and western know-how in the Russian power generation and fuels sectors has been small however, as a result of economic, political and commercial factors influencing recent investment strategies in those industries. These include reduced demand, excess capacity, previous late payments or use of barter instead of cash payments and low levels of energy prices. Furthermore, as Russian CO_2 emissions are currently some 30 per cent below the 1990 baseline figure (Grubb et al., 1999, p. 82) as a consequence of deindustrialization, there is little incentive for the fuels and power industries to invest in clean and efficient technologies as the Russian permitted emissions for 2008–12 were set at the 1990 level in the 1997 Kyoto Protocol. At the time of writing therefore, it is necessary to be circumspect about future environmental improvements in the Russian power generation industry even though there have been some achievements to date such as the production of Siemens' turbines in LMZ facilities.

There is a high probability, however, that fuels and energy prices on the Russian market will be increased to levels similar to their international counterparts, thus providing more funding for investment in the fuels and electricity generation industries. In 1999 the price of Russian natural gas on the Russian internal market was some R388 per 1000m³ (or some US$13.9 per 1000 m³ assuming a rate of exchange of some R28 to the US dollar), which was estimated by Russian energy economists to be some six to seven times lower than the natural gas price on the international market. The sales price for electricity in 1999 was some 26.5 kopeks per kWh (*Elektricheskie stantsii*, 2000, pp. 59–61) (or some 0.95 cents per kWh) and the generators' pool price was some 14.7 kopeks per kWh (Kozhukhovskii, 2000) (or some 0.52 cents per kWh) indicating that the Russian domestic electricity prices were more than five times lower than those on the market of a typical West European country.

By 2010 however, it is expected that the Russian domestic gas price will increase by more than 500 per cent to some R2405 per 1000m^3 or some US$86 per 1000m^3, whilst electricity prices increase by some 324 per cent to 112.3 kopeks per kWh (*Elektricheskii stantsii*, 2000, pp. 59–61) (or some 4.01 cents per kWh) assuming that the Russian cost data is given in 1999 prices and an exchange rate of R28 to the US dollar. These changes will align the domestic price level of natural gas and electricity more closely to those prevailing on the international market, increase income for the gas and electricity industries and thereby increase their attractiveness as locations for investment. Investment should also be fostered in fuel-efficient technologies and the capture of fugitive pipeline emissions should become more economically attractive.

In addition to increases in the price levels for natural gas it is also the aim of Russian energy economists to increase prices for other fuels. Using the thermal capacity of the fuel as a basis the 1999 domestic fuel prices were intended to change from some R530 per standard tonne of fuel for coal, some R340 per standard tonne of fuel for gas and some R925 per standard tonne of fuel for oil (or ratios of 1.0: 0.64: 1.75 based on the thermal capacity of the fuels) to some R1300, R2110 and R2155 respectively (or ratios of 1.0: 1.62: 1.66 based on the thermal capacity of the fuels) in 2010 (*Elektricheskii stantsii*, 2000, pp. 59–61). Coal should therefore become more cost-effective in those fossil-fuelled power stations where coal transport costs are not excessive, thereby providing an incentive for future investment in 'cleaner coal' technologies in both the electricity and coal industries. It should be noted however, that such proposed increases in fuel prices may be opposed by large energy users such as metal smelters, which currently have an advantage on international markets, partly as a consequence of subsidized energy prices, and subsidy may still continue for household prices in view of the low incomes received by a significant proportion of the population.

Fuel savings could also be reinforced by income from consequent reductions in greenhouse gas emissions as a result of lower levels of combustion of fossil fuels, thereby facilitating the sale of emissions allowances through the framework of the Kyoto Protocol. Estimates of income however, are complicated by the lack of any definitive indication of limits that may be placed on carbon emissions trading between countries as a consequence of EU countries wishing to see domestic improvements take priority; consequent caps placed on sales of allowances where reductions have occurred primarily through reductions in industrial output (sometimes referred to as 'hot air') rather than investment in mitigation (Pearce et al., 1998, p. 19); the volume of unused emissions allowances that the Russian government is prepared to sell or to retain as a hedge against increases in emissions if GNP continues to grow, and the probable prices of emissions allowances. A publication by Grubb et al. (2001, p. 23) concluded that optimum revenues for the sale of allowances by the economies in transition will be obtained from the sale of some 20 per cent of available allowances at a price of some $25 per t C (or some $6.81 per t CO_2

equivalent) to meet the potential demand from Europe, Japan, Canada, Australia and New Zealand of some 50–100 Mt C (or 183–367 Mt CO_2 equivalent) (Grubb et al., 2001, pp. 33, 54, 55). Some 50 per cent of these allowances could be provided by Russian suppliers yielding a potential income to the Russian economy of some $0.6–1.3 billion. Since Grubb's publication however, the EU's Emissions Trading Scheme (EUETS) has been established and prices of traded allowances within this scheme were more than €20 per t CO_2 during 2005. These high allowance prices were caused by continuing demand for electricity combined with relatively higher price increases for gas than for coal. High levels of fuel combustion have therefore been maintained, together with proportionate increases of coal in the fuel mix, leading to associated increases in carbon dioxide emissions. If allowance prices through the Kyoto trading mechanisms reflect those within the EUETS, then the potential income to the Russian economy should be higher than the $0.6–1.3 billion cited above.

A compromise may well be reached in which some 'hot air' trading is permitted provided that the income streams generated for the Russian partner are subsequently invested to obtain sustainable reductions in greenhouse gas emissions through a 'green investment scheme' (Tangen et al., 2002). The development of a green investment scheme to channel earnings from emissions allowances into energy-efficient projects should provide an incentive for the implementation of technologies capable of sustaining reductions in greenhouse gas emissions relative to industrial output. The development of such a programme however, depends upon the receipt of income from the sale of emissions allowances.

References

Bagautdinov, A.Z., Zhivotov, V.K., Eremenko, Yu, I., Kalachev, I.A., Misinov, S.A., Potapkin, V.V., Pampushka, A.M., Rusanov, V.D., Strelkova, M.I., Fridman, A.A. and Zoller, V.A. (1993), *High Temperature Chemical Processes*, **2**, p. 47.

Bruce, W.A. (2000), 'Methane Mitigation by In-service Repair of Natural Gas Pipelines', Paper presented at the Second International Methane Mitigation Conference, Novosibirsk, pp. 281–8.

Economic Commission for Europe (ECE) (1995), *Strategies and Policies for Air Pollution Abatement: 1994 Major Review Prepared under the Convention on Long-range Transboundary Air Pollution*, New York and Geneva: United Nations.

Elektricheskie stantsii (2000), 'Nauchno-tekhnicheskii sovet RAO "EES Rossii" i nauchnyi sovet RAN po problemam nadezhnosti i bezopasnosti bol'shihk system energetiki reshili: . . .', *Elektricheskie stantsii*, **8**, 59–61.

Eliasson, B. (1994), 'CO_2 Chemistry: An Option for CO_2 Emission Control?', in Pradier J.-P. and Pradier, C.-M. (eds), *Carbon Dioxide Chemistry: Environmental Issues*, Cambridge: The Royal Society of Chemistry.

Energetik (1994), **10**, 2–3.

'Energy and Climate: Russian-European Partnership' (2001), Workshop organized by The Royal Institute of International Affairs, London, Moscow.

Eremin, L.V. (1999), 'Razvitie elektroenergetiki Rossii i povyshenie ee ekologicheskoi effektivnosti', Paper presented to the Workshop on Energy Prospects and the Implications for Emissions and Climate Policy, Moscow: Royal Institute for International Affairs and Centre for Energy Policy.

Gale, J. and Freund, P. (2000), 'Reducing Methane Emissions to Combat Global Climate Change: The Role Russia Can Play', Paper presented at the Second International Methane Mitigation Conference, Novosibirsk, pp. 73–80.

Gazprom (2003), 'Rol' OAO "Gazprom" v reshenii global'nykh ekologicheskikh problem', Paper presented to the World Climate Change Conference, Moscow.

Gosudarstvennyi komitet SSSR po statistike (Goskomstat SSSR) (1990), *Narodnoe khozyaistvo SSSR v 1989g*, Moscow: Finansy i statistiki.

Graham-Bryce, I.J., Karis, W.G., Kinoshita, A., Sullivan, K.M. and Summers, G.G. (1994), 'Improved Coal Utilization Technology: Potential to Reduce Carbon Dioxide Emissions', in *International Energy Agency (IEA) and Organisation for Economic Co-operation and Development (OECD), The Clean and Efficient Use of Coal and Lignite: Its Role in Energy, Environment and Life*, Paris: OECD, pp. 513–22.

Grubb, M., Hourcade, J.-C. and Oberthur, S. (2001), *Keeping Kyoto: A Study of Approaches to Maintaining the Kyoto Protocol on Climate Change*, London: Climate Strategies.

Grubb, M., Vrolijk, C. and Brack, D. (1999), *The Kyoto Protocol: A Guide and Assessment*, London: Royal Institute for International Affairs.

Hill, M.R. (1997), *Environment and Technology in the former USSR: The Case of Acid Rain and Power Generation*, Cheltenham UK, Lyme, US: Edward Elgar.

Hill, M.R. (2000), 'Technological Options for the Reduction of Russian Acid Rain and Greenhouse Gas Emissions', *Environment and Planning B: Planning and Design*, 27, 393–410.

Hill, M.R. (2002a), 'Russian Methane Emissions', *Energy and Environment*, **13**(1), 57–80.

Hill, M.R. (2002b), 'Russian Power Generation: Between a Rock and a Hard Place', *Environment and Planning B: Planning and Design*, **29**, 819–39.

International Chamber of Commerce (ICC) and World Business Council for Sustainable Development (WBCSD) (1997), *Business and Climate Change: Case Studies in Greenhouse Gas Reduction*, Kyoto.

Kozhukhovskii, I.S. (2000), 'Itogi deyatel'nosti RAO "EES Rossii"', *Elektricheskie stantsii*, **6**, 3–6.

Makarov, A.A. and Bashmakov, I.A. (1990), 'USSR: A Strategy of Energy Development with Minimum Emission of Greenhouse Gases' in *Carbon Emissions Control Strategies: Case Studies in International Co-operation*, Washington DC: Battelle Memorial Institute.

Mezhvedomstvennaya komissiya Rossiiskoi Federatsii po problemam izmeneniya klimata (1998), *Vtoroe Natsional'noe Soobsheniye Rossiiskoi Federatsii*, Moscow.

Mezhvedomstvennaya komissiya Rossiiskoi Federatsii po problemam izmeneniya klimata (2002), *Tret'ye Natsional'noe Soobsheniye Rossiiskoi Federatsii*, Moscow.

Ministerstvo topliva i energetiki Rossiskoi federatsii (Mintopenergo), Institut energeticheskoi strategii (2000), 'Osnovnye polozheniya energeticheskoi strategii Rossii na period do 2020 goda (Proekt)', *Energeticheskaya Politika (Prilozhenie)*, Moscow.

Ministerstvo energii Rossisskoi Federatsii (Minenergo) (2001), *Osnovnye polozheniya energeticheskoi strategii Rossii na period do 2020 goda*, Moscow.

Moe, A. and Tangen, K. (2000), *The Kyoto Mechanisms and Russian Climate Politics*, London: The Royal Institute of International Affairs Energy and Environmental Programme.

Pearce, D., Day, B., Newcombe, J., Brunello, T. and Bello, T. (1998), *The Clean Development Mechanism: Benefits of the CDM for Developing Countries*, London: Centre for Social and Economic Research on the Global Environment, University College.

Proekt k zasedaniyu Pravitel'stva Rossiiskoi Federatsii 22 maya 2003 (2003), *Energeticheskaya strategiya Rossii na period do 2020 goda*, Moscow.

Prutkovskii, E.N. and Chavchanidze, E.K. (1995), 'Combined Cycle Steam and Gas Units with the Clean-up of the Flue Gases from Carbon Dioxide', *Heat Recovery Systems and CHP*, **15**(2), 215–30.

Russian Federation (1995), *First National Communication*, Moscow: Interagency Commission of the Russian Federation on Climate Change Problems.

Saneev, B.G., Lagerev, A.V., Khanaeva, V.N. and Tchemezov, A.V. (2003), 'Prospects for Electric Power Sector Development in Russia in the 21st Century and Greenhouse Gas Emissions', Paper presented to the World Climate Change Conference, Moscow.

Schemenau, W. and Van den Berg, C. (1990), 'The Future of Coal-fired Power Plants', in *Power Generation and the Environment*, London: I.Mech. E./E.M.E.P.

Sedikh, A. (1995), 'Environmental Problems: The Gas Industry and the Environment', in *Natural Gas: Trade and Investment Opportunities in Russia and the CIS* (Fourth International Conference convened by the Royal Institute of International Affairs, London, 26–27 October 1995), London: Royal Institute of International Affairs, pp. 28–30.

Serov, V.I., Vinogradov, V.P. and Gosteva, S.N. (2000), 'The Reduction of Opportunities of Coalbed Methane Emission to the Atmosphere at Russian Coal Mines', Paper presented to the Second International Conference on Methane Mitigation, Novosibirsk, pp. 553–6.

Stern, J.P. (1993), *Oil and Gas in the Former Soviet Union: The Changing Foreign Investment Agenda*, London: Royal Institute of International Affairs.

Tangen, K., Korppoo, A., Berdin, V., Sugiyama, T., Egenhofer, C., Drexhage, J., Pluzhnikov, O., Grubb, M., Legge, T., Moe, A., Stern, J. and Yamaguchi, K. (2002), *A Russian Green Investment Scheme: Securing Environmental Benefits from International Emissions Trading*, London: Climate Strategies.

United Nations Framework Convention on Climate Change (UNFCCC) (1997), 'Activities Implemented Jointly (AIJ) List of Projects: USIJI Uniform Reporting Document', in *Activities Implemented Jointly under The Pilot Phase*, RUSAGAS Fugitive Gas Capture Project.

US Environmental Protection Agency (USEPA) (1996), *Reducing Methane Emissions from Coal Mines in Russia: A Handbook for Expanding Coalbed Methane Recovery in the Kuznetsk Coal Basin*, Report No. EPA 430 D95–001, September.

30 Environmentally friendly energy development and use: key capacities in the building and construction industry in Australia

Jane Marceau and Nicole Cook*

Introduction

The building and construction (B&C) sector worldwide is increasingly recognized as a major industry that must improve the ways in which it operates because buildings and the products used and their construction have major impacts on the environment. A decade or so ago analysts calculated the magnitude of that impact, saying for example:

> Buildings account for one-sixth of the world's fresh water withdrawals, one-quarter of its wood harvest, and two-fifths of its material and energy flows. This massive resource use has massive side-effects: deforestation, air and water pollution, stratospheric ozone depletion, the risk of global warming. (Malin and Lesson, 1995, p. 5)

In 1995 Ballinger et al. (1995, p. 101) suggested that 'the operational energy and initial embodied energy [of buildings] represents approximately 28% of national energy consumption, comprising 10% embodied energy and 18% operational energy', while Tucker and colleagues estimated that the embodied energy component can contribute up to 20 per cent of national energy consumption (Tucker et al., 1996, p. 56; see also Langston, 1997). These high figures arise both because there is much embodied energy in built products and because building designs, construction methods and use are often wasteful in environmental terms.

Energy efficiency in building design and construction has now emerged as an international issue facing public policy and research. In Australia several policy initiatives indicate the growing importance of energy efficiency concerns in the construction industry. Initiatives taken include the Australian Greenhouse Office's (AGO) Voluntary Building Industry Initiatives, notably the Australian Building Energy Council's (ABEC) voluntary code of best practice; the Housing Industry Association's *Nationwide House Energy Rating Scheme* (NatHERS); the Building Designers' Association of Australia's Energy-efficient Design Award; and the Window Energy Rating Scheme developed by the Australasian Windows Council. The AGO has also been instrumental in mobilizing local councils to develop their own activities and has partnered ABEC and the Australian Building Codes Board in developing mandatory energy efficiency standards to eliminate construction industry worst practice (see for example, AGO, 1999a, 1999b; see also the websites given in the references to this chapter for further information on different initiatives, both state and federal).

At the state level, energy policies have been developed and promoted through a range of organizations, such as the Sustainable Energy Development Authority (SEDA) in New South Wales and Energy Efficiency Victoria, which have had a key role in driving more energy-efficient building and in shifting government procurement practice, although the role of SEDA especially has been modified in recent years. In practical terms government procurement has also played a role; New South Wales (NSW) for example has built schools, hospitals, airports, army barracks and other developments using energy-saving design features to reduce energy requirements (NSW Department of Public Works, 1998). Then, of course, there was the commitment to the 'Green Games' as the Sydney Olympics were designated, together with the associated rules (see Department of Urban Affairs and Planning, 1995).

The effectiveness of policies aiming to improve environmental management and sustainability through better industrial practices depends greatly on the capacity of organizations operating in affected industries to work in the required new ways. For policy initiatives concerning it to carry through successfully into practice, the building and construction industry, like any other, must have the capacity to translate rules into complying construction products and processes. This chapter uses the results of a small study we carried out through the Australian Expert Group in Industry Studies (AEGIS) in 1999 and 2000 to examine the industry's capacity to address energy efficiency issues in the B&C industry in Australia. Environmental issues are multiple in the industry: our study focused on energy use as a focus from which to indicate where policy-makers should look to assess whether the capacities needed for new methods and products are available when regulating for environmental concerns.

The Olympic 'Green Games' generated a bevy of energy-conscious developments, including the solar-powered Athletes' Village, the international award-winning Superdome and other flagship projects showcasing energy-efficient design and development. Incorporation of environmentally sustainable development (ESD) principles into Olympic projects also generated new laws, guidance systems, tools and definitions in environmental management as well as innovation in development (Olympic Coordinating Authority, 1997 and 1998). Our study focused largely on the building of the Athletes' Village, hereafter referred to as Newington, the present name for the housing area, to show what can happen when the efforts of players in the building and construction industry are focused and coordinated within a given regulatory framework. The data for the study reported here were gathered through interviews with 21 B&C companies, notably architects, engineers, construction firms and developers, and a further 11 interviews with utilities, research and development centres, industry associations and regulators. The interviews explored the capacities of the industry to shift to more energy-efficient and sustainable designs for constructed items and construction practices, while the focus on Newington provided insights into how the key capacities worked in practice. While this sample is too small to assume any general significance in the responses, the interviewees were 'informed industry observers and participants' and the key players in a major project, making their views worthy of wider consideration.

Innovation in a complex product chain requires ten key capacities

Our analyses of the drivers of and skills needed for successful innovation in industry suggested that ten capacities are key when encouraging innovation through a whole product chain rather than an individual firm. The B&C industry is one that centrally involves coordination of many products and services into semi-permanent, and often large-scale finished products so considerable coordination capacity is vital.

We suggest that the ten key capacities needed to ensure good outcomes from environmental regulation of the kinds that affect the B&C industry are:

1. capacity to generate new technologies;
2. capacity to build up knowledge of energy-efficient design and products;
3. capacity to fund new initiatives in the field;
4. capacity to produce relevant technologies (in Australia in this case);
5. capacity to diffuse the technology;
6. capacity to coordinate energy products and their use;
7. capacity to incorporate new technologies into existing businesses;
8. capacity to develop skills;
9. capacity to export technologies so as to recoup innovation costs; and
10. capacity to regulate.

The first nine of these relate to the capacities of companies in the different segments of the B&C field while the last relates to the capacity of regulators to understand the dynamics of the industry and hence make regulation effective in current and desired future circumstances.

1. Capacity to generate new technologies

The capacity of the building and construction product system to generate new technologies for improving energy efficiency through firm and public sector R&D seems to be good in Australia. The B&C product system benefits from the in-house R&D capability of suppliers and lead firms, assisted by the utilities especially, and their linkages to dedicated external R&D centres (public and private) working in energy-related and building-related technologies. The whole field is quite R&D-intensive by Australian industry standards.

The growth of in-house knowledge is partly fuelled by purchasing patents. The two major players in our study had bought patents, effectively guaranteeing a quasi- or complete monopoly in the commercialization of those technologies. Patents are expensive and protect only for a limited time so their owners place a high premium on their skilled researchers and product development teams quickly interpreting and converting the knowledge embodied in the patent to a marketable, even mass-produceable, product.

A key feature of technology generation by the supplier firms in our sample was the high degree of collaboration and partnering between the in-house R&D divisions of suppliers and external public and private research capabilities. The public sector provides both technological breakthroughs and a knowledge safety net in monitoring and evaluating the various products available and providing preliminary advice and expertise in setting up renewable energy projects. The two large players both had strong links with universities, buying patents from the university sector as well as the know-how embodied in people recruited through their secondment to the company. Some firms also linked with their own suppliers in new technology/product development: one for example, developed a patent for attaching photovoltaic cells (PVC) to roofs 'in conjunction with a plastics manufacturer' and two reported that 'as we got more involved with product suppliers, we built up our knowledge that way'. In this, B&C experience matches that of many industries. One major producer gave a typical account of the multiple collaborations resulting from coordination by a supplier at the centre of the product development 'team':

> With the Skytube, we worked closely with UTS [University of Technology Sydney] in Sydney – they did a lot of the testing for the project. We also worked with 3M and developed what we regard as the highest performing product of its type in the world. Subsequently, we developed a three-in-one Skytube that combines the benefits of natural light through the day, artificial light at night and a wind-driven air extractor. It's almost 100 per cent renewable energy, bar the night light.

For some suppliers, the generation of new technologies occurred in the R&D centres of their parent (often overseas) companies; local branches would import new material to build domestic market and then begin manufacture locally, using the technical data developed by the parent. The R&D and developmental relationships with parent enterprises can also work both ways, with the local firm in some cases using in-house R&D capabilities in Australia as well as external connections and the parent doing the same. In this way they were thus helping to generate technology rather than just receive it from elsewhere. More generally, most of the technologies generated through R&D arise from 'one-off' projects or collaborations between suppliers and research centres and result rather more often in product or material refinements than in high-tech innovations that reverberate through the supply chain.

Much new technology comes into the B&C product system 'sideways'. Technology generation in construction-related firms seemed to be most concentrated and highly resourced in the commercialization of a key 'component': low-cost silicon photovoltaic technology. This technology was not funded by the B&C product system itself however, but grew from the Photovoltaics Special Research Centre (PSRC) at the University of New South Wales, funded by the Australian Research Council and Pacific Power, although the technology has construction applications. In addition Pacific Solar was set up with the specific aim of commercializing the thin film PV (photovoltaics) technology invented at the PSRC.

Industry-led research in energy efficiency carried out by university departments and centres and the Commonwealth Scientific and Industrial Research Organization (CSIRO), at least at the time of the study, generally seemed to focus on the application of products rather than the generation of new ones, on the energy evaluation or rating of products, and on the development of energy rating software such as the National Housing Energy Rating Software. These are elements of the creation of the packages of products and services needed to convince suppliers and clients of the industry to take risks, as some of our interviews showed. Some of the technologies nominated included tools to evaluate building product performance, direct digital control, better lighting systems, luminaries, ballasts and control systems for lighting, daylighting systems to reduce energy in commercial buildings, variable speed drives for fans in air-conditioning systems and improved glazing and computer design tools.

Generally it was felt by participants in our study that new technologies seem to have been 'embraced [only] sporadically' by the industry, which concentrated on 'technology that [had] short payback periods'. This seemed to be true whether the technologies were local or imported. Thus, while two of the larger suppliers interviewed were developing more revolutionary products, both were finding it hard to interest potential partners; in one case, after two attempts to gain industry support and two mergers, the firm had still not secured an industry partner.

Despite the general dearth of developmental-type research carried out by the industry itself, the present study found a number of 'one-off' collaborations between suppliers, industry associations and public sector research centres that generated patented application-based and low-technology product innovations. Thirty-two linkages between 16 interviewees (excluding those suppliers commercializing R&D) and external R&D centres were found. Links were mostly with specialist university research centres and university departments.

2.   *Capacity to build up knowledge of energy-efficient design and products*
This capacity includes the extent to which the industry builds up knowledge of technologies available, in this case in energy-efficient building design and construction. Responses from the 21 companies interviewed suggest that, as well as in-house and collaborative R&D, there are three main ways of building knowledge in this field:

- indirect R&D;
- building expert teams and gaining experience on projects and in industry;
- general 'scanning' of the environment through taking research trips, reusing journal articles and books, and attending conferences and fairs.

Indirect R&D   Indirect R&D (the results of R&D conducted by others) was used by all types of firm and by all utilities in the sample. Indirect R&D was considered to be very important, despite meeting problems related to poorly spread expertise and distorted or incomplete information in the industry as well as competition for knowledge as a competitive edge. Some firms felt that suppliers were not up to date, saying for example, that: 'We call some of the glass manufacturers and they have never done double-glazed, integrated PV. They really haven't got the sophistication of projects overseas' or the 'lateral thinking' knowledge that could reduce reliance on overseas conditions. Sometimes difficulties arise from the structure of the industry, concentrating knowledge in 'too few hands':

> [We] always have an expectation of close involvement with . . . suppliers–a lot of expertise lies with them. There isn't a well-established consultant base. One of the problems is, you're playing into their hands but if there's only a few suppliers you really have to specify their products and rely on their expertise. There is some risk that monopolies could form that restrict that level of competition.

Expert teams   The creation of expert teams of key suppliers, architects and other professionals and researchers together with 'experience on projects' are key strategies for building knowledge within the industry. Experience on projects seemed particularly relevant to the photovoltaic suppliers. They enabled the main player to access the wide range of knowledge sources needed for successful innovation in this new field. For the construction of a major scientific research centre, for example, the composition of the expert team enabled commercial and ESD principles to be married together while suppliers were critical at the building stage and the energy performance of the completed building was monitored by a university research centre.

The role of specialist consultants is also central in the teams. One firm used 'one [consultant] for PV, one for lifts, one for mechanical, one electrician, one for communication and one for computer systems' so as to include knowledge of alternative energy sources and management systems in concert with the more traditional areas of construction expertise.

The process of building expert teams for projects enables companies both to develop and to share information about product energy performance and feasibility, which is especially important in the application of existing technologies in new settings. The type of knowledge generated relates to new product characteristics as well as to information about how to install and incorporate the product into existing or new designs. Modification and installation may also generate new knowledge, which, through the consultant teams, may build up in and travel from suppliers to the lead firms. The solutions-based approach of engineers on commercial projects was critical while project experience was particularly important for architects and engineers as 'a lot of what we do, we try to refine. If we do something good on one project we try to refine it for the next'.

Solar technologies can be quite difficult to install so the importance to the PV suppliers of building up knowledge while engaged on projects was also clear. Both crystalline and amorphous PV technologies for example, have been problematic in terms of their application, often requiring unique building solutions, special casings and waterproofing.

Scanning the environment   Scanning the environment in both formal and informal ways is as important in the B&C sector as it is in most industries, and often represents a considerable investment. Nine of the 11 suppliers interviewed used international experience and specialized training to ensure the right knowledge because 'the only place you learn about energy efficiency is outside of Australia'. At the time of the study '[o]ne major architectural firm, for instance, provide[d] an internal travel scholarship for its employees to locate new technologies and applications overseas' while another had recently undertaken a world tour to locate technologies not yet available in Australia.

3.   Capacity to fund new initiatives in the field

The capacity to find funding for innovation is critical. Some funds come through general financial institutions or through the firms in the field themselves, especially those in the energy-conservation field and among the utilities. Some firms have a greater proclivity to take financial risks and fund new technologies than others. Finding funding partners upfront to share the risk of investing in unfamiliar energy-related technology was an important strategy for lead groups and utilities. One firm linked such funding to the sale of one of its related products:

> the industry does not have a quick uptake of new technology because our traditional role is to manage risk for our client and ourselves . . . Historically [however] we are not hesitant in trying certain products or materials if it's done as a 'partnership'. If a product or material does not perform or deliver, then there might be a deal made with the supplier to replace [the product with one that] does perform. The industry will try innovative solutions where the risk is seen to be manageable or the risk is shared in partnership or if a prototype scenario can be created and we have enough time to work with the new product.

Acquiring venture capital per se was more difficult and had strings attached. The development of mandatory minimum performance standards was seen by some, though not all, to help in this respect as standards could help link the recurrent cost of buildings more closely to their initial capital cost:

> Our whole system of bank financing and government system of financing is all based on initial capital cost. Recurrent cost is left for someone else's budgets. How much [a building] costs to run is an energy and environmental issue [but] we don't bring those two together. If we start to get [standards put] into the Building Code of Australia then people have to start doing it.

In the residential market suppliers joined the lead players in a collective marketing strategy to help fund energy-related initiatives once products had become available. The strategy involved building a web of experts and suppliers to market and sell products to consumers from their combined energy-efficient range. Thus, for example, Supplier 6 built a consortium of building products producers and solar hot water suppliers, which it felt was 'the classic example of forming a consortium to gain momentum in the industry'.

In risk-reduction 'big investors' are particularly important, emphasizing the significance of demonstrable market demand in attracting their funds:

> Who are the big investors in commercial [property]? It's the superannuation funds, pension funds. They buy a building, they operate the building for perhaps five or ten years maximum and they sell the building. Now, unless we can get statistics to say that this building . . . will be worth [more] in five years because it's going to get recognition because the marketplace is going to be hunting out buildings like that. Unless you've got that information on the table, then that client is not necessarily interested. They're not the operators of the building.

Client commitment to energy efficiency varies. 'Major corporations, major investment houses [and] lawyers' felt it was safer to 'stick with proven technology' and especially American clients were dismissive of low energy alternatives. This creates 'a real problem with only one or two doing it in Australia – a real worry. There have been instances where US companies have dismissed buildings by virtue of energy-efficient measures whereas Germans and Scandinavians tend to be more interested'. Despite this, some players saw energy efficiency as a 'marketable point of difference' and one respondent identified a 'growing band of architects in Sydney' interested in energy-efficient design and products, indicating that the industry may be 'maturing fast and see[ing] the environment as an opportunity and not just a cost'. Indeed some clients are deliberately seeking low-energy options, especially in the public sector. In the private sector energy efficiency was marketable to clients with 'tourist projects in sensitive areas . . . and commercial [clients seeking] cost-efficiency . . .'.

In this area especially, consultant and supplier teams put together by a lead firm can evaluate the energy and cost performance of low-energy alternatives before, during and after construction, enabling the lead firm to expand and consolidate its baseline knowledge for future developments – but this is not always the case. The problem may diminish when the new range of low-energy design options and different packages comes on-line to meet the short- and long-term funding goals of individual clients. Energy and cost performance data are critical here, indicating again the importance of key suppliers and consultants who are able to evaluate as well as supply low-energy products.

4.   *Capacity to produce relevant technologies in Australia*

This capacity concerns the degree to which the innovative products specified in buildings through the use of new designs are manufactured in Australia and hence open to innovations suggested by local clients. If most products can easily be imported, further local innovation may be stifled. The capacity to produce relevant technologies and products to enhance the energy performance of buildings may be an important part of the industry's capacity to drive energy-efficient development.

The literature on innovation has long emphasized the importance of close interaction between clients and suppliers in developing marketable products (see for example, Marceau, 2001; Shaw, 1994; von Hippel, 2001), implying that local manufacture is often helpful in product development. Respondents to our study were asked to nominate the most innovative products they had used in buildings and indicate whether these products were manufactured in Australia. Some were manufactured locally (including PV, wind turbines, chilled beams and geothermal heat pumps) but a 'large chunk of renewables came from overseas', as one interviewee pointed out. Another underscored the resulting three- to four-week wait for supply. Some of these responses may reflect individual perceptions, however, as interviews with local suppliers of some nominated products suggested that these products are in fact manufactured locally.

The extent to which the lead firms in our sample encouraged suppliers to produce new technologies or products differed among the firms. For one lead firm collaboration with suppliers on product or process development was critical because it eliminated some of the risk associated with using unknown technology as well as providing other returns. The returns achieved encouraged the firm to develop a national procurement strategy in which the supplier of environmental policies had a substantial say. In return, '[i]f the supplier has problems with this requirement, we are interested in seeing how we can help. If it's a matter of dollars that's okay because it costs us money to dispose of this plasterboard in landfill and we would prefer to work in partnership for a better triple bottom line'.

Many aspects of this lead firm's partnering strategy were mirrored elsewhere in the sample as firms actively encouraged suppliers to develop new products and practice and share the financial risk of innovation. In managing this risk however, the capacity of this lead firm to encourage product development is also constrained because it needs to be sure that innovations that do not work can be easily replaced by conventional products and are hence unlikely to have a dramatic impact on the supply chain or design process.

A second lead firm had a different approach but also recognized the importance of its principal engineering consultant in ensuring local supply of the highly innovative chilled beam product used in its commercial office development. Dependence on this technology means that 'we may insist on it being made locally because long supply chains from overseas can be problematic. If we use chilled beam technology, we want to see it made here'. This factor will limit the capacity of local R&D producers to commercialize their results if there is no local manufacturer, especially if the products concerned are radically new.

There are important skill gaps in manufacturing capability in Australia. As one respondent said: 'We call some of the glass manufacturers and they have never done double-glazed integrated PV. They really haven't got the sophistication of some projects overseas. Overseas glazing and curtain walling manufacturers have teamed up with PV manufacturers and are already doing sophisticated work in the area.'

There is thus an unrealized opportunity for some of the glazing and curtain-walling suppliers to form alliances with the PV suppliers and develop a value-added product, which might serve as a strong basis for architects and engineers to specify new energy-efficient technologies in their designs. All suppliers interviewed could develop new or custom-made products for clients. One supplier for instance would frequently 'produce a special density that wouldn't end up in [our normal] product range'. Two suppliers are also intending to build plants near clients to consolidate their local supply.

5. Capacity to diffuse the technology

This capacity includes firms' ability to commercialize R&D developed locally, together with the principal mechanisms for 'spreading the news' about new technologies in the energy-related field in Australia and evaluating their adequacy. The capacity of any industry to diffuse a new technology may depend greatly on its ability to commercialize its own or others' R&D to develop new markets. The path of development from research results to product however, can be long and circuitous since several additional steps are needed, including perfecting ways to manufacture and use the new product.

The products commercialized by the suppliers interviewed are usually 'add-ons' in terms of the construction process and relate either to the performance of existing products or to improvements in their application. More radical innovations can be harder to 'sell', as one respondent pointed out when talking of thin-film or building integrated PV. Almost everywhere the B&C industry's ability to diffuse technology seemed to be restricted to 'low-tech' innovation for which there is a readier market. In some cases the research centres generating new technology may be better at commercializing it than local firms had proved to be.

One of the other key steps in successful diffusion of new technology is 'spreading the word' within and across industries and different parts of the supply chain, from manufacturer to consumer. Firms use the familiar newsletters/journals/Internet, work on projects, advertising and marketing, government or industry bodies, conferences, lectures and award programmes. Suppliers seem to have been most active in spreading the news about new technologies.

The effectiveness of written, government and supplier-based mechanisms in knowledge diffusion was seen as greater in industry circles than among the broader audience of final consumers who may be equally important in the adoption of new technologies. One among several respondents remarked that he was 'constantly surprised at how many people, because they're not involved in the industry, don't know about it. It's not mainstream yet'.

One of the problems was seen to be with the diffusion activities of other critical players in the supply chain:

> There is an awareness problem . . . We don't do any consumer advertising. There are other areas of the supply chain that are sometimes road blocks to change. Many suppliers are interested [only] in price because of the large builders. We offer incentives to some clients . . . with energy-efficient products. Initially we say 'you grow your range in energy-efficient products and we'll reward you'. This is very inadequate on the consumer end–it's a struggle to get to the market. We go through the chain and address each organization, using the customer base and distribution systems.

One of the architects saw the impact of the 'built example' as extremely important in conveying the 'news' about energy-efficient building development and creating new awareness among consumers, suggesting the need for demonstration projects in the spread of acceptance of new technologies available. Others suggested the importance of mandatory standards.

One of the key barriers to successful technology diffusion in the B&C product system is the highly fragmented structure of both industry and clientele: 'There are many different stakeholders in the B&C industry – financiers, clients, different consultants . . . then the builder, then the builder being subcontracted, then the building being commissioned, then handed over to a client and the client appointing a building manager and the building manager is supposed to operate the building.'

6.   *Capacity to coordinate a range of new energy products*

This capacity includes firms' ability to coordinate the supply of innovative energy-related products and overcome the potential barriers to coordination and client and supplier involvement in specifying new products. Since building and construction is an 'assembly' industry the capacity to find and coordinate the use of new products is especially important. Lead firms are particularly critical here but opinions varied greatly among respondents as to which these firms might be.

Suppliers are also involved in these projects, sometimes in critical roles. A geothermal pump supplier 'assisted with modelling and they were very involved with on-site installations'. An architect talked of innovation at the design stage when the firm looks at the equipment available, incorporates the equipment into the design and specifies in generic terms the equipment required, thus providing a basis for mechanical contractors to obtain a quotation for supply. The mechanical contractors then match the specifications with available equipment, subject to the designer's approval. Many suppliers also coordinate through their distributors and installers.

It is thus possible for firms in different places in the supply chain to coordinate energy product supply effectively. Some of the potential barriers to coordination identified by the lead firms and consultants related to the availability of customized products. Poor local availability sometimes meant protracted delivery times and cost issues. One designer nominated as a major problem finding three suitable suppliers to provide quotations and proper supply: 'You have to write a tight performance measure. A lot of suppliers say they comply and they don't.'

7.   *Capacity to incorporate new technologies into existing businesses*

Firms introducing new technologies need to incorporate these into their existing business practice. This is not always simple and companies interviewed adopted diverse strategies. In our sample strategies were focused around creating expert teams for projects, topping up in-house R&D with external research, buying patents and building in-house teams around these, or working on projects – one where considerable development was required. Where the technological change was important or disruptive, several players created special units or subsidiaries while one firm created a new division.

8.   *Capacity to develop skills*

The development and availability of skills needed for dealing with the new energy-efficient products available in Australia are critical to success. Most respondents had developed strategies for ensuring that the skills needed were developed, ensuring for instance that workers had the skills to install new technologies for the first time. The strategies ranged from on-site technical support to simple 'stickers' to show which way was 'in' or 'out'. Specialist suppliers were likely to train on-site or through their own installers or distributors who were frequently the target audience for technical manufacturing and installation instructions, information nights and on-site technical help.

In some cases the supplier had several products that each generated different training strategies. One supplier for example, did most of the large glazing projects for individual clients but 'designed [its smaller domestic products] specifically to make them user-friendly and [easily] installed by a handyman'. For market development this supplier saw itself as an 'industry trainer':

> We train [sales] representatives. We're about to start on a one-month training programme [with a sales company] – we're having multiple sessions in every capital city. We're value-adding, showing them how to sell energy efficiency. With [the consortium of suppliers] we have a full-time project manager who goes around training Housing Institute of Australia (HIA) members and industry about how to sell energy efficiency.

Installation was identified by the second lead firm as an important area for training the workforce in the use of new technology in commercial office sites. A representative said that 'installation is critical – if efficiencies fall away you lose the advantage of putting them in'. This comment also applied to Newington and the Superdome in particular where the process of installing (and waterproofing) PV laminates absorbed considerable resources. It also echoes the comments and work of a number of companies and research centres in trying to simplify the installation process, particularly in the area of PV. As pointed out by the Photovoltaics Special Research Centre, '[m]anufacturers are taking the initiative in relation to training. In this way, they are developing systems that can be largely installed by untrained staff or, more correctly, by staff that are not electrical contractors'.

Energy-related authorities such as SEDA and Energy Efficiency Victoria also have a key role in skills development for different parts of the B&C product system. To support its 'Energy Smart Homes Industry Partnerships', which encourages builders, industry associations and architects to 'commit to designing energy-efficient houses as measured by NatHERS', SEDA has developed training manuals incorporating energy-efficient design principles for land and property developers' sales staff, architects and building designers, residential building companies' sales staff, land and property developers' design staff and product suppliers. The partnerships target the large project homes builders:

> We are trying to sign-up large volume builders in project housing and medium–high density developers. When they sign up, we offer a range of support services such as training. Some have architects on-staff, others contract architects for specific projects. Industry support for the Energy Smart Housing programmes has been unanimous. All of the peak industry bodies have become partners under the SEDA Industry Partnerships Programme.

In terms of national training, there has also been some incorporation of energy-efficient competences in National Training Packages, although these are in a diverse range of trades, not all of them related to building and construction. In this area SEDA has reported that sustainable energy competences have been incorporated into electro-technology, transmission and distribution trades. At the tertiary level, the University of New South Wales is now offering an undergraduate degree in Photovoltaics and Solar Energy, a 'world first', to 'respond to [the] rapid growth in both manufacturing capacity and job creation in the photovoltaics industry'.

It seems therefore that the availability of skills both on- and off-sites is not a major barrier to the diffusion of energy-efficient technologies in the B&C sector, although it seems that some managers of firms are either ignorant of available technologies, or reluctant to make the necessary changes.

9. Capacity to export technologies

The capacity to export was also seen as a key factor in the industry's ability to develop and drive change because exporting provides larger markets and greater viability for new and innovative products. The greatest export strengths were in renewable energy products, with the two manufacturers in the study reporting export ratios of 70 per cent and 50 per cent, a remarkable success; the average in the sector for PV was 64 per cent, which is itself high by the standards of other industries.

The strong performance of these firms may reflect the overall export strength of PV within the renewable energy industry in Australia. Redding (1999) also identified PV as both the major export item for Australia in the renewable energy sector and 'a significant component of Australia's total exports in high technology manufacturing' (Redding, 1999, p. 55). Worldwide grid-connected applications grew from 14 per cent in 1996 to over 35 per cent in 1997, indicating good market potential and a shift to 'building integrated solar panels for grid connect' (AGO, 1999c, p. 1) as well as for solar homes (roofs) in Europe, Japan and the US. The makers of parts for the systems (wool for insulation, roof tiles and so on) were also exporting.

Until recently renewable energy products were used predominantly in remote or off-grid projects but in the case of PV there has been a recent shift towards grid-connected systems, applicable in residential and commercial urban settings, largely through projects initiated by governments around the world to address greenhouse gas abatement commitments. The B&C industry has played only a secondary role in the growing PV grid-connect market because of the high capital cost of PV manufacture and installation when compared to the connection of houses and offices to mains electricity systems and the industry is still cautious about consumer acceptance of the up-front cost of PV.

10. Capacity to regulate: expertise in the public sector (regulators)

Expertise in the public sector enables sensible and progressive regulation to be developed and implemented and can take a lead role in innovation within the industry. The regulatory framework surrounding the B&C product system in energy-efficient design and construction is in a state of transition and development in Australia. Numerous industry associations have developed initiatives, seed-funded through specialist regulatory authorities – including the AGO, SEDA and Energy Efficiency Victoria – as part of their partnership initiatives, have helped develop mandatory standards through the energy efficiency in building body, ABEC, and have contributed to the elimination of 'worst practice'. Similarly, Victoria amended the Building Code of Australia to add general national mandatory minimum energy performance modifications and some local councils have cooperated to eliminate worst practice through their Development Control Plans.

These mandatory initiatives are complemented by voluntary government and industry initiatives. One of the key features of the regulatory growth in the B&C industry with regard to energy efficiency has been ABEC's development of minimum energy standards for buildings. Industry associations have committed to agreed standards in this area, especially where these link design and management elements. The development of standards is also important in gaining consumer change as standards provide a mechanism for ensuring that new information eventually filters through to the consumer.

Regulation and the encouragement of change can be enormously facilitated by the use of new IT tools. One such tool has been energy rating software used by Councils (for example, the National Housing Energy Rating Scheme) or industry such as the Window Energy Rating Scheme to estimate energy consumption of houses and their component parts. These tools first establish a 'standard' and then provide a way to measure it.

Some project budgets are quite large. In one case in this present study, the project budget was four times the annual budget of the largest R&D centre, the Photovoltaics Special Research Centre. While funding for PV is expected to increase through the AGO in the near future, this difference in budget highlights the emphasis in the recent wave of B&C energy-related reform on 'applying the simple available products' (Energy Efficiency Victoria) rather than encouraging developmental research or commercialization by the B&C industry of low-energy products and equipment. Indeed the industry's capacity and willingness to drive change at a basic level has perhaps been the most prominent feature of regulatory reform in this area. As summed up by the Property Council of Australia (1999): 'We need an integrated approach led by an alliance of respected industry leaders that is marketed in a language that is understood by the various target organizations. We're creating a cookbook with simple recipes with pictures that people can look at and say, "I want that". At the moment, there is no cookbook.'

Conclusions
This study tells us something about the capacity of the building and construction industry in Australia to drive energy efficiency through the analysis of performance in ten key capacities. It shows that some of the critical technologies are ready to go, although their cost may still constitute a barrier to general use, while the next generation technology – thin film technology – is now reaching the market. The study suggests that all key aspects of the development and use of the new solar technologies and other energy-saving devices are in place, that the industry has the capacity to develop new ones as needed and that the skills needed to incorporate their design and use are already available. The Newington case study also indicates that both the high calibre lead firms and suppliers required to deliver design, funding, skills and components to a large mainstream housing project or similar buildings are also established in Australia.

In concluding this chapter we summarize the findings of our project in terms of some opportunities linked to the capacities discussed above. Current 'gaps' are presented as suggestions about ways that policy-makers and industry players can encourage further development and use of energy-efficient technologies. They are presented simply as bullet points to direct the attention of readers:

1. Capacity to generate new technologies
   - Diffuse the results of R&D to suppliers and contractors more evenly.
   - Develop industry-wide definitions of energy efficiency so energy efficient components are easily specified by developers and building owners, and develop a mechanism to enable knowledge-building to spread much further through the industry.

- Encourage all players to work more closely with the renewable energy industry to 'prove' to the B&C industry as a whole that energy-efficient products and design do perform.
- Monitor and promote energy performance by energy-efficient buildings through the further development of sophisticated measuring tools.
- Encourage large-scale homebuilders to develop energy-efficient designs.
- Ensure owner-developers have the latest information about the costs and energy performance of energy-efficient design features, which investors require for performance modelling by consultants and key suppliers, to link low-energy features with client budgets and reduce ongoing operational costs.

2. Capacity to build up knowledge of energy-efficient design and products
   - Facilitate more collective marketing and compile market research about the acceptance of new energy-efficient design features and technology in the market.
   - Explore options for the industry to get behind developmental research in energy-efficient technologies.

3. Capacity to fund new initiatives
   - Encourage better awareness of energy efficiency as a 'marketable point of difference' to superannuation and other major funding agencies, including the more traditional mortgage suppliers in the residential market.

4. Capacity to produce relevant technologies in Australia
   - Encourage industry to support 'building integrated solar panels for grid connect' research. Most of the low-energy products sought by architects in Australia are available locally within four weeks but building-integrated PV is not, which represents a 'gap' in supply.
   - Encourage partnerships and innovation-related procurement practice. For specialist engineers encouraging local supply of high-tech products could be critical to guaranteeing delivery, performance and wider acceptance. The risk is that otherwise the technologies developed remain low-tech and low risk, and the use of new technologies is slow as a result. The technologies used at Newington were not really new.

5. Capacity to diffuse technology
   - Develop strategies to diffuse technologies developed or used in demonstration projects. Australia lacks 'middle-ground' in the commercialization of energy-efficient construction-related technologies: at present, highly customized, design-based products are integrated by architects, engineers and key suppliers into design-led construction for clients with specific energy-related goals. There results a 'gap' between the niche commercial sector and the rest of the industry in technology diffusion in building-related energy-efficient design and construction.
   - Encourage further alliances among suppliers.

6. Capacity to coordinate energy products
   - Encourage suppliers to 'step-forward' with low-energy alternatives early on in a project and become the first point of call for low-energy design solutions for their clients.

7. Capacity to incorporate new technologies into existing businesses
   - Encourage companies to expand their coordination capacity to access new technologies and become 'one-stop-shops' for energy-efficient equipment and design.

8. Capacity to develop skills
   - Encourage the industry to incorporate hallmark projects into its training pro-grammes as competencies in energy-efficient design and construction still need incorporation into National Training Packages to target smaller players.
9. Capacity to export technologies
   - Encourage expertise that is potentially transportable elsewhere when large-scale building and construction firms can work with Australian-based suppliers on overseas projects. In our study the strongest exporters were the PV and solar hot water heater suppliers. Australia may be poised to benefit from the falling cost of PV systems and their installation, opening export markets further. The Newington project demonstrates the coordination capacity of the Australian industry in delivering a large energy-efficient residential development, which is an exportable expertise.
10. Capacity to regulate
    - Develop further and more consistent regulation. To achieve the desired shift in practices relating to energy efficiency in building, a mix of mandatory regulations (including those set down in the Building Code of Australia as overseen by the AGO) and voluntary government and industry initiatives have been set in place. Both mandatory and voluntary types of regulation however, seek primarily to encourage best practice and eliminate worst practice rather than to enforce indus-try compliance with set, industry-wide measures and standards and require B&C firms to improve their energy performance. The current regulatory framework may serve to foster inconsistent and isolated improvements in energy efficiency within the industry rather than a major and coherent shift.
    - Develop common standards using newly developed IT tools to measure com-pliance.

The study reported here suggests that the B&C industry in Australia has much potential as well as considerable achievements in all ten of the capacities key to successful industry inno-vation in energy efficiency in built products and processes used in the industry. That it has not yet achieved its potential and that risk aversion is commonplace, is also clear. It seems that all of the elements of success are present but that much of the industry currently lacks the internal dynamism and external incentives to use the capacities it has without further regulatory encouragement. Energy efficiency as the 'marketable point of difference' has yet to gain widespread recognition. The study reported here suggests that regulation can kick-start that recognition and, where it does so, the industry as a whole has the capacity to respond, especially under the guidance of some lead firms, whether they are the coordinat-ing developers/constructors or the architects and suppliers of the products needed.

Note

*   The study was also assisted by several AEGIS team members, notably Bronwen Dalton and Brian Wixted. Jane Marceau and Nicole Cook were the principal designers and authors of the project and its reporting.

References

Australian Expert Group in Industry Studies (AEGIS) website: http://www.uws.edu.au/about/acadorg/clb/ research/aegis

Australian Greenhouse Office (1999a), *Australian Residential Building Sector Greenhouse Gas Emissions 1990–2010 Executive Summary Report*, Canberra: Commonwealth of Australia.

Australian Greenhouse Office (1999b), *Baseline Study of Greenhouse Gas Emissions from the Commercial Buildings Sector*, prepared by EMET Consultants and Solarch Group, Canberra: AGO.

Australian Greenhouse Office (1999c), 'Renewable Energy Technologies', website: http://www.renewable. greenhouse.gov.au, accessed 30 November 2005.

Ballinger, J., Prasad, D., Lawson, B., Samuels, R. and Lyons, P. (1995), 'R&D of Current Environmental Technologies in Australia', in *Proceedings: Pan Pacific Symposium on Building and Urban Environmental Conditioning in Asia*, Nagoya, Japan.

Department of Urban Affairs and Planning (1995), *Environmental Planning for ESD: Guidelines for Compliance with the Environmental Guidelines for the Summer Olympic Games*, prepared by Manidis Roberts Consultants, Sydney: DUAP.

Langston, C. (1997), *Sustainable Practices: ESD and the Construction Industry*, Sydney: Envirobook.

Malin, R. and Lesson, N. (1995), *A Building Revolution: How Ecology and Health Concerns are Transforming Construction*, Washington DC: World Watch Institute.

Marceau, J. (2001), 'Managing Medical Technology: Industry-Hospital Relations in the Biomedical Industry in Australia', *International Journal of Healthcare and Technology Management*, **2**(1/2/3), 281–95.

NSW Department of Public Works and Services (1998), 'Economic and Environmental Life Cycle Costs of Buildings', Report prepared for the Department of Industry Science and Resources – unpublished.

Olympic Coordination Authority (1997), *Compendium of ESD Initiatives and Outcomes for OCA Facilities and Venues*, Sydney: OCA.

Olympic Coordination Authority (1998), *Environmental Practices Code*, Sydney: OCA.

Redding Energy Management Energy and Environmental Management Group (1999), *Redding Energy Management Report on the Renewable Energy Industry*, Canberra: Australian Greenhouse Office.

Shaw, B. (1994), 'User-supplier Links in Innovation', in M. Dodgson and Rothwell, R. (eds), *The Handbook of Industrial Innovation*, Aldershot, UK and Brookfield, US: Edward Elgar, pp. 275–84.

Tucker, S., Salomonsson, G., Ambrose, M., Treloar, G., Hunter, B., Edwards, P., Stewart, P., Anton, S. and Crutchley, G. (1996), *Development of Analytical Models for Evaluating Energy Embodied in Construction*, Report prepared for the Energy Research and Development Corporation (DBCE DOC 96/84M), Canberra: CSIRO.

Von Hippel, E. (2001), 'User Toolkits for Innovation', *Journal of Product Innovation*, **18**, 247–57.

31 Tools for environmental impact assessment
Sharon Jones

Introduction

Environmental impact assessment (EIA) embodies that wise old saying that 'prevention is better than cure'. The process is commonly defined as 'the systematic identification and evaluation of the potential effects of proposed projects, plans, programs, or legislative actions relevant to the physical, chemical, biological, cultural, and socio-economic components of the total environment' (Canter, 1995, p. 2). In other words, EIA forces the responsible parties to evaluate the potential impacts that their project may have for the environment *before* the project is implemented. Evaluation includes identifying both positive and negative consequences to ensure that the project objectives can be met and that unforeseen problems are avoided. Such up-front consideration of environmental impacts as part of an EIA should result in projects that include provisions to prevent or minimize potential negative impacts. Also important is that EIA is based on a very broad definition of environment that encompasses all of earth's spheres including the lithosphere, atmosphere, hydrosphere, biosphere and anthroposphere. The 'spheres' may be defined as follows:

- lithosphere: all land masses including soil strata and minerals on earth;
- atmosphere: mix of gases held in several large layers surrounding earth by earth's gravitational forces;
- hydrosphere: all natural water systems on earth as balanced via the hydrologic cycle;
- biosphere: all life on earth within a sphere that ranges from the very depths of the oceans to several thousand metres into the atmosphere;
- anthroposphere: the environment built by human society.

Environmental impact assessment is a natural part of project planning and implementation. For example, whenever someone includes a mitigation measure as part of implementing a project, that person has performed a partial EIA (even if it was not called an EIA) that identified a potential negative impact for which the mitigation measure addresses. However, formal EIA is attributed to the 1969 National Environmental Policy Act (NEPA) enacted in the United States. The Act mandated that a federal project in the US must be evaluated for its environmental impacts before it can be implemented.

Following NEPA, many states within the US implemented their own NEPA processes, as did many developed countries. In the 1990s, development aid and lending agencies incorporated similar EIA requirements for sponsored projects. As such, many developing countries are also using EIA. Although EIA has traditionally been used during the planning stages of a new project, the EIA methodology can be used in conjunction with all stages of a project from planning to implementation to long-term operation.

Given the widespread use of EIA across the world, those associated with managing the environmental impacts of technology should be well versed with the fundamentals of the

EIA method and the corresponding tools that can improve the effectiveness of the EIA. This chapter begins by describing the scope of a traditional EIA. There are several tools that are used routinely to complete an EIA and the three most common discussed include: cost–benefit analysis, environmental systems modelling, and risk assessment and management. A number of the new tools being used in conjunction with EIA are also discussed. These additional tools, which address some of the non-traditional ways in which EIA is being used, include lifecycle analysis and geographical information systems.

Throughout this chapter the examples used to illustrate definitions, methods and so on are primarily taken from a completed EIA for a health centre to be located on a Native American Reservation in south-western US, although examples from other projects are also drawn upon where needed.

The methodology and scope of EIA
As stated, EIA has traditionally been used during the planning stage of a new project to inform the stakeholders of the potential positive and negative environmental impacts of a project and its alternatives so that negative impacts may be reduced or eliminated. Negative impacts may be associated with the preferred alternative because of the site location, the chosen technology, the method of implementation and so on. The traditional process used to conduct an EIA includes the following steps:

- Understand the proposed project across its lifecycle.
- Identify alternatives to the proposed project and understand those across their lifecycle. One of them should be the 'no action' alternative so as to illustrate clearly both the positive and negative impacts of the proposed project.
- Identify the baseline environmental conditions that existed before the project is implemented.
- Identify potential environmental concerns associated with the overall project (and all alternatives) across its lifecycle including development, implementation, operation and closure. Potential environmental concerns are not limited to one aspect of the environment but include consideration of air, water (surface and ground), land, animals, plants, built environment, archaeology and other cultural resources, and impacts on human communities.
- Determine (for those potential environmental concerns) whether the project and its alternatives will result in positive and/or negative environmental impacts. This involves using the best available science and modelling tools to quantify if possible, the negative (or positive) change to the environment that was predicted to result from the project. The change to the environment (or environmental impact) is determined for both the short and long term.
- Identify mitigation measures for each negative environmental impact for the proposed project and determine to what degree the measure minimizes the impact. Some mitigation measures may result from including components of the various alternatives that were considered as part of the EIA.
- Consider the cumulative environmental impacts predicted to result from the proposed project, in addition to impacts that may arise as the result of pre-existing conditions.
- Consider the impact of the proposed project on irretrievable and irreplaceable resources.

- Involve the affected stakeholders throughout the process to ensure that all relevant issues are considered and that the EIA is as thorough as possible.

Table 31.1 gives an example to show the scope of an EIA.

Traditional tools for EIA

Environmental systems modelling

Since traditional EIA is done in the planning stages of a project, actual environmental impacts cannot be measured as part of the EIA process. Instead, potential environmental impacts are determined from either similar experiences that have occurred or by using environmental systems modelling (ESM). This tool involves the expert use of mathematical models to simulate what may occur if a project is implemented. It includes the use of all medium-specific models, which are used to determine the fate and transport of pollutants within the environment. Because these models are based on theory and/or empirical evidence, their accuracy depends on how well the particular environmental system is understood. The outcome of the modelling effort is an estimate of the quantity of a pollutant predicted to be present across the pathway it may travel during its lifetime.

The steps in the EIA methodology can be completed qualitatively. However, quantitative impacts are often easier to compare across alternatives and across mitigation measures. An EIA relies on ESM to quantify the potential environmental impacts, since the impacts (in terms of human health or ecosystem health) depend on the quantity of the pollutants present. An ESM exercise may be used, for example, to evaluate the downwind atmospheric concentration of chemicals released from a factory stack, the downstream surface water concentration of a chemical released from a factory's wastewater system, the flood levels due to increased development in a former wooded area, the noise levels in neighbourhoods adjacent to a planned airport and so on. For an EIA, a comparison of alternatives may use ESM to demonstrate quantitatively that implementing a mitigation measure may reduce a negative environmental impact by a certain percentage.

There are limitations to ESM (as there are with most tools) since all models are simplifications of reality and thus depend on assumptions about how a system works. No model can predict with complete accuracy at all times and no model can account for all of the possible scenarios that may actually occur. These limitations are particularly important since quantitative results are often preferred by decision-makers and may limit how thoroughly a particular project is evaluated. In addition, different alternatives often have different impacts that remain difficult to compare even if quantified. An example is the difference (or lack thereof) between project X that may result in 20 per cent more flooding versus project Y that may result in 20 per cent higher noise levels. Example B in Table 31.2 illustrates how ESM was used for the Westside Health Center discussed in Example A, Table 31.1.

Cost–benefit analysis

Cost–benefit analysis (CBA) is a method to consider quantitatively alternative development proposals in terms of their negative and positive impacts (known as costs and benefits). Costs and benefits are typically quantified in monetary terms. Although many CBAs only present the direct tangible costs of a project (and many only show the capital costs), the analysis should also include lifecycle costs and impacts not currently assessed

Table 31.1   Example A. Scope of an environmental impact assessment: to build or not to build?

	Alternatives Considered				
	A (No action)	B (Proposed action)	C	D	E
Development footprint	N/A	30 acres	30 acres	30 acres	30 acres
Staff quarters	N/A	27 people	27 people	27 people	27 people
Water storage tank	N/A	400 000 gallons	400 000 gallons	400 000 gallons	137 000 gallons
Wastewater system (WW)	N/A	Expanded and upgraded High School WW System	Expanded and upgraded High School WW System	Three acres on an off-site parcel of at least ten acres	Expanded and upgraded High School WW System
Floodplain mitigation	N/A	Elevated buildings; bermed wastewater system	Elevated buildings; bermed wastewater system; upgraded levee	Elevated buildings; bermed wastewater system	Elevated buildings; bermed wastewater system
	Possible Impacts				
Environmental resources	A (No action)	B (Proposed action)	C (B with upgraded levee)	D (B with regional WW site)	E (B with site-specific water tank)
	Physical resources (ambient)				
Air quality	No impact	Possible but unlikely odours or increases in vehicle pollutants	Possible but unlikely odours or increases in vehicle pollutants	Possible but unlikely odours or increases in vehicle pollutants	Possible but unlikely odours or increases in vehicle pollutants
Groundwater quality	No impact	Eliminates potential for groundwater contamination	Eliminates potential for groundwater contamination	Potential contamination source remains at High School	Eliminates potential for groundwater contamination
Sediment control and stormwater management	No impact	Short-term soil disturbance during construction	Increased short-term soil disturbance during construction	Short-term soil disturbance during construction	Short-term soil disturbance during construction

Table 31.1   (continued)

Environmental resources	A (No action)	B (Proposed action)	C (B with upgraded levee)	D (B with regional WW site)	E (B with site-specific water tank)
Possible Impacts					
Visual/open space	No impact	Decrease in open space, visual impact of an elevated tank	Decrease in open space, visual impact of an elevated tank	Decrease in open space, visual impact of an elevated tank	Decrease in open space, visual impact of several elevated tanks
Noise	No impact	Increased for construction and traffic	Increased for construction and traffic	Increased for construction and traffic	Increased for construction and traffic
Floodplain	No impact	Low safety risks during flood events	Increased protection for area structures	Low safety risks during flood events	Low safety risks during flood events
Physical resources (built)					
Water supply	No impact	Regional supply for the area	Regional supply for the area	Regional supply for the area	No impact
Wastewater disposal	No impact	Improves existing system	Improves existing system	No change to existing system	Improves existing system
Energy use	No impact	Increased use of fossil fuels	Increased use of fossil fuels	Increased use of fossil fuels	Increased use of fossil fuels
Transportation/ traffic	No impact	Increase in area traffic flows	Increase in area traffic flows	Increase in area traffic flows	Increase in area traffic flows
Socioeconomic resources					
Health care	No impact	Major improvements in availability and services	Major improvements in availability and services	Major improvements in availability and services	Major improvements in availability and services
Public safety	No impact	Medical services closer to people	Medical services closer to people	Medical services closer to people	Medical services closer to people
Demographics	No impact	Employment will bring people to the area to live and work	Employment will bring people to the area to live and work	Employment will bring people to the area to live and work	Employment will bring people to the area to live and work

Note: Proposed Action: construction and operation of the Westside Health Center (WHC) to be located on the XYZ Nation adjacent to the Regional High School.

Source: Indian Health Services (2002).

Table 31.2   Example B. Environmental systems modelling: flooding the desert

Modelling Tool	Floodplain hydrologic modelling using the US Army Corps of Engineers HEC software
Potential Impact	The proposed Westside Health Center (Alternative B from Table 31.1) would result in a built facility that is completely within the 100-year floodplain and subject to erosion, loss of use, property damage and so on
Mitigation Measure	The following mitigation measures will allow stormwater conveyance through the site while protecting the buildings from flooding: • elevating the building pads and one traffic lane one foot above the 100-year flood water surface elevation to allow the use of the facilities during flood events, while minimizing changes to regional drainage patterns and off-site impacts • provision of retention/detention basins with water harvesting techniques and native plant communities to prevent off-site drainage impacts to surrounding land uses • slope building pads away from the structure and secure with slope protection to prevent erosion damage to the elevated building pads

Source: Indian Health Services (2002).

by market mechanisms (otherwise known as externalities or social costs). This means that costs such as future environmental liability, and benefits such as enhanced public relations, may be included in the analysis. A thorough CBA involves determining the types of benefits and costs (direct versus indirect, tangible versus intangible), quantifying the costs and benefits of each alternative, and applying a decision criterion – such as a cost–benefit ratio – to relate the costs to the benefits. A traditional EIA often does not make a conclusion about which alternative is the preferred one. Instead, the EIA becomes a summary of costs and benefits of the various alternatives for decision-makers and stakeholders to consider.

Cost–benefit analysis also has its limitations. One of the primary limitations is that many indirect costs and benefits are difficult to quantify monetarily. While some costs and benefits may be tangible, others are intangible. With regard to air pollution, for example, the health impacts may be quantifiable but the aesthetic impact may not. Another important limitation is that, because CBA is considered to be an 'objective' tool, it does not deal with issues considered to be 'subjective'. Thus equity concerns associated with the impact of a proposal on a particular group of people, compared with the impact across a broader population, are not considered. For example, a highway may result in the closing of a restaurant in one location and the opening of a new fast food chain restaurant at another location.

As with ESM, CBA should be used as a tool to help illuminate the issues for decision-makers and stakeholders. It should not constitute the only consideration however. Example C in Table 31.3 illustrates the direct costs, as well as the direct and indirect benefits, of the Westside Health Center discussed in Example A in Table 31.1. The only costs quantified in the example are the capital costs. However, it is possible to quantify the operation and maintenance costs. Quantifying the benefits is more difficult as it depends

Table 31.3 Example C. Valuing the costs and benefits

	A (No Action)	B (Proposed Action)	C (B with Upgraded Levee)	D (B with Regional WW Site)	E (B with Site-specific Water Tank)
Costs – direct	N/A	$11.5 million + operation/ maintenance	$12.5 million + operation/ maintenance	$12 million + operation/ maintenance	$11.2 million + operation/ maintenance
Benefits – direct	N/A	Health care and ambulatory services	Health care and ambulatory services	Health care and ambulatory services	Health care and ambulatory services
Benefits – indirect	N/A	Removal of current environmental concern (at existing lagoon)	Removal of current environmental concern (at existing lagoon)	Potential for regional wastewater system for future development	Removal of current environmental concern (at existing lagoon)
		Regional water supply system	Regional water supply system	Regional water supply system	
			Reduced flooding for San Simon Village		

Source: Indian Health Services (2002).

on valuation of people's travel time, improved health access and other indirect benefits related to improvements to water and wastewater infrastructure.

Risk assessment and management

Both CBA and ESM can benefit from the use of risk assessment as a complementary tool. Used in conjunction with ESM, risk assessment can take the modelled pollutant quantity generated by ESM and estimate the probabilistic ecological risk and the human health risk. More generally, risk assessment can help to determine the degree of uncertainty about whether an environmental impact will occur as a result of a particular project. Thus the benefits of risk assessment are similar to those that accrue from financial risk assessment, which firms use to address the uncertainties associated with investment decisions.

The risk assessment method involves using existing information to determine if there is a risk potential. If the affected entity can be exposed to the environmental risk – for example, via dermal, ingestion or inhalation pathways – risk assessment quantifies the actual dose the affected entity may receive and calculates the risk with appropriate formulae. The final step is to determine what the calculated risk means, both quantitatively and qualitatively. Risk assessment can also be used in conjunction with EIA to enhance the quantitative basis for making conclusions and recommendations.

Once the risk is quantified, there are several categories of risk management (or mitigation) alternatives, including modifying the environment, modifying the exposure,

Table 31.4   Example D. A risky process: hypothetical farm use of fertilizer

Receptor	Pathway	Actual Risk	Risk Perception	Risk Management
Field workers	Inhalation; dermal	High probability	Moderate probability	a. change exposure: use safety masks, gloves b. change environment: use less toxic fertilizer c. compensate victims: hazard pay for workers d. modify effects: treat illness
Surrounding human community	Ingestion via groundwater/ surface water	Low probability	Low probability	a. change exposure: alternate drinking water source b. change environment: treat contaminated water c. compensate victims: community projects d. modify effects: treat illness
Surrounding human community	Ingestion via purchased food crop sold	Low probability	High probability	a. change exposure: alternate food source b. change environment: wash food crop c. compensate victims: cheaper food crop d. modify effects: treat illness
Surrounding ecosystem	Inhalation; dermal; Ingestion depending on species	Low probability	Low probability	a. change exposure: move affected species b. change environment: prevent fertilizer runoff with detention ponds c. compensate victims: not possible d. modify effects: treat ill species

modifying the effects or compensating for the effects. The actual option chosen depends on a variety of criteria including costs, benefits, values, legislation and standards among others. The choice may also depend on stakeholders' perception of the risk, which can be quite different from the calculated risk.

Understanding and accounting for risk perception is thus a key element of risk assessment. The perception of risk may be influenced by a variety of factors, including the immediacy of discussing the issues, the fact that human beliefs change slowly, the memory of similar events, population versus individual risk, the way the information is presented, voluntary versus involuntary risk and natural versus artificial risk among others. Moreover, the success of the risk management plan depends on the firm's expertise in risk communication and in recognizing public perception. Therefore the risk management component of a risk assessment is useful for deciding on the mitigation measures that are an essential part of all EIAs.

A hypothetical example of fertilizer application to agricultural lands (Example D in Table 31.4) serves to illustrate risk assessment and risk management processes. In reality, the quantitative risk varies depending on the actual scenario. For example, chemical type and concentration, distance to water body and so on. Nonetheless, the example presents a framework for including risk considerations as part of EIA.

EIA and industry
The EIA methodology described at the beginning of this chapter is based on the traditional scheme set up under the regulatory model used in the US. This approach is similar to that used in other countries; however, it applies primarily to projects in their planning stage and often applies to large-scale capital projects such as dams and highways. A major problem with EIA is that once the project is approved, it is often implemented without adequate attention to the mitigation measures recommended in the EIA (Sanchez and Hacking, 2002). However, EIA can continue to be used throughout the implementation of a project and ultimately across its lifecycle. It can and is being used at manufacturing plants and within operating departments, for consumer product design and so on. In particular, ISO 14001 and the growth of Environmental Management Systems (EMSs) are encouraging the innovative application of EIA.

ISO 14000
The 1996 international standard, ISO 14000, is the clearest source of international guidance on environmental management for those companies that choose to follow this voluntary standard. ISO 14000 is modelled after the more widely known ISO 9000 set of standards for consumer product quality. However, the focus of ISO 14000 on environmental issues means that it is wider in scope because environmental issues affect more than just consumers. ISO 14000 includes voluntary standards for the following (Crognale, 1999):

- environmental management systems;
- environmental auditing;
- environmental performance evaluation;
- lifecycle assessment;
- environmental labelling; and
- environmental aspects in product standards.

An evaluation of the specific role of EIA in terms of ISO 14001, one of the ISO 14000 standards, is given by Sanchez and Hacking (2002). ISO 14001 describes how companies can become certified as having an adequate EMS in place. This standard does not prescribe the company's goals but allows companies to set their own goals, and consequently to select the environmental issues to be addressed. In practice, the objective of the standard is for companies routinely to inventory, prioritize and assess all environmental issues (called 'aspects' in the standard) associated with the operations, products and services provided. In other words, companies following ISO 14001 are also performing EIAs.

Environmental Management Systems
ISO 14001 is just one example of the change that has taken place over the past 25 years regarding private sector environmental management. Instead of reacting to regulatory mandates, many managers now see environmental issues as part of the overall business enterprise (Crognale, 1999). This evolution of environmental management is not limited to developed countries (Wheeler, 1999) but has become an important consideration around the world, due in part to international environmental policy requirements prompted by the globalization of companies.

Environmental management is also becoming more relevant for developing countries since international development aid agencies incorporated environmental requirements into their criteria for funding decisions. For the private sector, the World Bank has taken a leadership role in the drive for sustainable and equitable private sector investment in the developing world. According to the United Nations Environment Programme's (UNEP) Executive Director, it is recognized that the private sector has a significant impact on environmental trends through investment and technology decisions (World Bank, 2000). As an example, the following recent projects in Latin America and the Caribbean recognize the role of environmental management within the private and public sectors:

- The Brazil NEP II project provides grants to stakeholder coalitions that manage environmental assets (natural or built) such as airsheds, water bodies, nature preserves and so on, which provide services to human societies. Potential recipients need to demonstrate improvement in environmental management capacity (World Bank, 2000).
- The Clean Air Initiative promotes development and enhancement of clean air action plans, the exchange of information, public participation and the involvement of the private sector (World Bank, 2000).
- The Furima Iron Foundry, a small foundry in Medellin, Columbia, faced increased pressure from environmental regulations passed in 1995 and participated in a sustainable development programme, which included an EIA of the metallurgy sector. The foundry implemented the recommendations and evaluated how much the environmental impact was reduced. The results showed that the foundry was not only successful at reducing air emissions and coke consumption but also increased productivity (International Network for Environmental Management, 2000).
- The Favorita Banana Company, located in Ecuador, requested funding of US$15 million from the International Finance Corporation in 1998 to modernize its seaport facility. To fulfil the conditions of the investment, the company was required to evaluate the environmental, social and technical impacts of the project and develop mitigation measures. As a result, the company became the first banana grower to receive certification under the Rainforest Alliance using an independent audit of 200 sustainability criteria. This certification was an added marketing benefit for the company (World Bank, 2000).

Strategically managing environmental issues during the ongoing operation of a firm (instead of simply reacting to them) requires that managers and their staff are aware of

EIA and associated tools. In particular, the role of EIA in conjunction with EMS has been appraised (Jones and Mason (2002)). The methodology of EIA was found to complement EMS by considering the short-term and long-term impacts on the overall environment, and the health of employees and the community. However, if it is to complement EMS constructively, EIA has to be conducted in an ongoing fashion to identify environmental deficiencies and opportunities for improvement. Ongoing assessment activities can also help in the design of more effective products and processes as well as environmental controls. A benefit of EMS is that it highlights the strategic advantages of 'green' approaches so that firms become aware that efficient resources save money, healthy environments contribute to more productive workers, global customers respond to news about environmental impacts and sound approaches imply cost-effective regulatory compliance (Wever, 1996).

Lifecycle assessment
For green design and EMS in general, lifecycle assessment (LCA) is a tool that can identify and compare alternatives to reduce environmental impacts across the lifecycle of a consumer product. The approach is very similar to lifecycle costing, where the capital and operating costs of different alternatives are evaluated using economic techniques such as present worth analysis to convert costs to compatible units. Lifecycle assessment can be used to design green products for the marketplace (products with minimal environmental impacts) as well as to evaluate existing industrial processes to identify the most cost-effective methods to mitigate environmental impacts.

The LCA method involves establishing the system boundaries, identifying the environmental impacts at each step of the lifecycle from raw material extraction to final disposal, characterizing and prioritizing those impacts, and redesigning the product/process to reduce environmental impacts in a manner that is both cost-effective and compatible with facility goals. By definition, the LCA method is based on EIA, where the EIA is focused on the impacts that occur during the lifecycle of a consumer product. In addition to computer-based tools and formal methodologies, LCA is also a way of thinking that encourages those involved with a project to take a big picture approach and to more thoroughly consider the environmental impacts of that project across its lifecycle.

As with all impact assessment methods, LCA has limitations. In its pure form, LCA is very data-intensive as it theoretically asks the analyst to inventory environmental issues across all media and all stages of the lifecycle. It is even more difficult to then convert all the environmental issues into potential impacts. The most complicated step may be the final comparison of the various diverse impacts while considering the many related management issues.

Example E in Table 31.5 presents a qualitative summary of a LCA completed by Harvard Medical School (Epstein and Selber, 2002). It shows how the overall framework can be used to ensure the EIA is thorough, as well as illustrating the diversity of impacts at various LCA stages.

The new player: geographical information systems
As discussed by Jones (2003), a geographical information system (GIS) is a spatial data management tool comprised of 'intelligent' computer maps. Databases describing the attributes of physical assets are linked to map features, which include the graphical lines, polygons and

Table 31.5   Example E. Recovering oil: a lifecycle assessment

Stage	Effect	Subcategory
Exploration	Deforestation	Emerging infectious diseases
Drilling and extraction	Chronic environmental degradation	• Discharges of hydrocarbons, water and mud: increased concentrations of naturally occurring radioactive materials
	Physical fouling	• Reduction of fisheries • Reduced air quality resulting from flaring and evaporation • Soils contamination • Morbidity and mortality of seabirds, marine mammals and sea turtles
	Habitat disruption	• Noise effects on animals • Pipeline channelling through estuaries • Artificial islands
	Occupational hazards	• Injury, dermatitis, lung disease, mental health impacts, cancer
	Livestock destruction	
Transport	Spills	• Destruction of farmland terrestrial and coastal marine communities • Contamination of groundwater • Death of vegetation • Disruption of food chain
Refining	Environmental damage	• Hydrocarbons • Thermal pollution • Noise pollution • Ecosystem disruption
	Hazardous material Exposure	• Chronic lung disease • Mental disturbance • Neoplasms
	Accidents	• Direct damages from fires, explosions, chemical leaks and spills
Combustion	Air pollution	• Particulates • Ground level ozone
	Acid rain	• NO_x, SO_x • Acidification of soil • Eutrophication; aquatic and coastal marine
	Climate change	• Global warming and extreme weather events with associated impacts on agriculture, infrastructure and human health

Source: Epstein and Selber (2002).

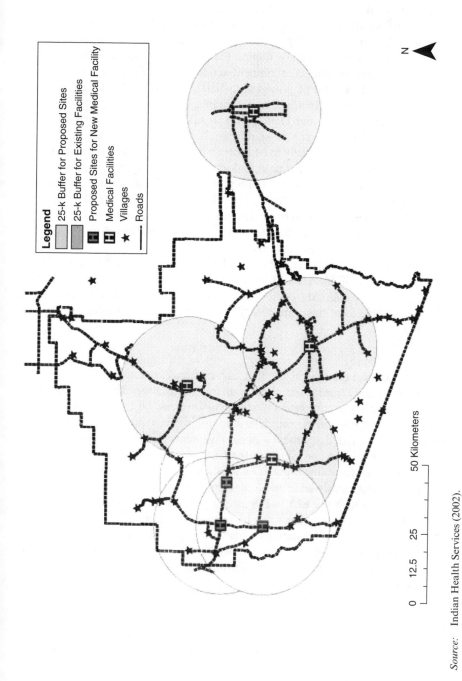

Source: Indian Health Services (2002).

Figure 31.1 Example F: Using GIS – Tohono O'odham Nation existing and proposed clinic sites

points that represent the important geographic locations of the physical assets. Ideally the graphic data are arranged as self-contained layers, with each layer containing similar features. Many GISs also include raster-based imagery to enhance visualization.

Inherent to EIA is access to information, the management of information and the analysis of information. A GIS is a tool that can allow all of these tasks to occur. Even more appealing is that GIS locates these tasks in a spatial environment, which provides a useful communication tool, allows for multiple scenarios to be evaluated and affords another consideration to aid decision-making (Warner and Diab, 2002). In addition, a GIS can be updated over time to allow for impact assessment across the life of a project.

Although GIS has a great deal of potential for EIA, it is still in its infancy in many countries. A GIS is only as good as the data it contains, and many databases are incomplete for EIA purposes (Warner and Diab, 2002). A GIS also requires fairly advanced technical skills to perform the types of spatial analyses that may be useful for EIA. However, all of these concerns are likely to be short term as more organizations and countries build GIS capacity given its great potential in so many areas. In fact the benefit of GIS for tasks other than EIA may result in the centralization of data so that EIA analysts can more efficiently access relevant information. However, even when GIS capacity is more widespread, it remains a tool to evaluate existing conditions and potential scenarios. As with all of the tools discussed in this chapter, decision-makers ultimately need to consider multiple issues to formulate final recommendations.

Example F in Figure 31.1 illustrates the use of GIS for the Westside Health Center described in Example A in Table 31.1. As shown, GIS was used to determine how many people would be positively affected by the Westside Health Center depending on where it was built. Three possible locations were compared and two existing health centres were used as a baseline.

Conclusion

This chapter has discussed the central role that EIA can, and arguably should, play in the management of technology from planning through implementation. There are many tools available to an EIA analyst to complete a thorough assessment. Of those discussed in this chapter, three tools are integral to the determination, estimation and comparison of impacts: ESM, CBA and risk assessment and management. Newer tools, including LCA and GIS, have the potential to contribute to the EIA process in a different manner. For example, even if it is not formally applied, LCA is a way of framing the EIA so that it thoroughly addresses both short-term and long-term impacts. On the other hand, GIS is a practical tool that can help the EIA to manage large amounts of data and may therefore serve to ensure a more comprehensive EIA process over time.

Throughout this chapter the limitations of each tool have been highlighted so as to underline the message that although these tools aid the completion of an EIA, the decision-maker is ultimately responsible for choosing the best project alternative with the minimum environmental impacts.

References

Canter, L. (1995), *Environmental Impact Assessment*, New York: McGraw Hill.
Crognale, G. (ed.) (1999), *Environmental Management Strategies: The 21st Century Perspective*, New Jersey: Prentice Hall.

Epstein, P.R. and Selber, J. (ed.) (2002), *Oil: A Life Cycle Analysis of its Health and Environmental Impacts*, Massachussetts: Center for Health and the Global Environment, Harvard Medical School.

Indian Health Services (2002), *Environmental Impact Assessment for the Westside Health Center*, USA Department of Health and Human Services.

International Network for Environmental Management (2000), *Case Studies*, www.inem.org/htdocs/inem_casestudies.html, accessed 29 November 2005.

Jones, S.A. (2003), 'Challenges of Developing a Water and Wastewater Infrastructure GIS for the Tohono O'odham Nation', *Public Works Management and Policy*, **8**(2), 121–31.

Jones, S.A. and T.W. Mason (2002), 'Role of Impact Assessment for Strategic Environmental Management at the Firm Level', *Impact Assessment and Project Appraisal*, **20**(4).

Sanchez, L. and Hacking, T. (2002), 'An Approach to Linking Environmental Impact Assessment and Environmental Management Systems', *Impact Assessment and Project Appraisal*, **20**(1), 25–38.

Warner, L.L. and Diab, R.D. (2002), 'Use of Geographical Information Systems in an Environmental Impact Assessment of an Overhead Power Line', *Impact Assessment and Project Appraisal*, **20**(1), 39–47.

Wever, G. (1996), *Strategic Environmental Management*, New York: John Wiley and Sons.

Wheeler, D. (1999), *Greening Industry: New Roles for Communities, Markets, and Governments*, Washington, DC: World Bank.

World Bank (2000), *Environment Matters*, Washington, DC: World Bank.

32 Technology transfer and adoption by small-scale women farmers: a case study in Qwaqwa district in South Africa

Stanley Yokwe, Wilhelm Nell and Dora Marinova

Introduction

Agriculture has long been the dominant sector in terms of output, employment and export earnings in most developing countries, particularly in many regions of Sub-Saharan Africa. About 30 per cent of Gross Domestic Product (GDP) and 40 per cent of export earning in Sub-Saharan Africa derive from agriculture (West Africa Research and Development, 2001). Yet at a time when agricultural productivity has increased in other regions of the world, agricultural performance in Sub-Saharan Africa has been far from satisfactory, with the rate of agricultural output growth lagging well behind population growth (Saito, 1994; Sanders et al., 1996). This creates serious challenges for the sustainable development of these countries where environmental considerations are only a non-avoidable background for major economic and social disparities.

The continuing productivity decline in Sub-Saharan agriculture may be linked in part to the neglected role of women in agriculture. Women farmers are the mainstay of the food supply for millions of people in Sub-Saharan Africa (Delgado, 1997) despite the success achieved by extensive commercial farms in this region. Much of the land under relatively intensive food production consists of small-scale farms managed by women. Their contribution and importance for sustaining the life of their families and the local economies would suggest that women farmers should be the main focus for the implementation of new technologies, however, this is not the case. Significant barriers remain that stem from comparatively low investment in technologies, low levels of education and health and restricted access to services and assets (Kumar, 1987; Quisumbing, 1993; Saito, 1994; Schultz, 1993).

These barriers are perhaps most noticeable in South Africa, especially in the traditionally black rural areas (formerly homeland areas) where the majority of farmers are women. About 87 per cent of all agricultural land in South Africa is owned by a handful of large-scale white farmers who gained access to the land during the former apartheid regime. Small-scale African farmers are found mostly in the former homeland areas, which make up only 13 per cent of agricultural land in South Africa. These areas were established under the Native Land Acts of 1913 and 1936 and are characterized by traditional forms of land tenure that were regulated by a series of laws and regulations proclaimed under the Black Administration Act of 1927. With escalating population growth in the former homeland areas, this land is increasingly used for settlements, resulting in the decline of agricultural land available for the majority of African small-scale farmers.

Despite such constraints however, most rural women still depend on traditional farming to provide a livelihood for more than a million people from relatively small plots of land (National Department of Agriculture, 2001). In response to this situation the current South African government has proclaimed rural agricultural development a priority (African National Congress, 1994). It is believed that the potential for increased food production would be tremendous if yield increments could be achieved on each hectare through the use of improved technology. While such objectives are imperative, the identification of factors that contribute to and hinder the adoption of new technological inputs will be vital to the success of interventions designed to achieve them. It is also important that women have the ability and power to influence the type of technologies to be introduced and to be equal participants in the decision-making processes.

The case study reported on in this chapter characterizes the current situation of small-scale women farmers in the Qwaqwa district in South Africa and identifies the factors that limit the adoption of appropriate agricultural technologies. Strategies aimed at improved technology adoption and farm sustainabiliity are also proposed. The results of this study hold the potential to make a timely contribution to contemporary agricultural production in South Africa in the current context of decentralization, privatization and liberalization.

Motivation and context

South Africa is characterized by high levels of poverty, especially in rural areas where the majority of the population reside. Approximately 70 per cent of South Africa's poor people live in rural areas, of which about 70 per cent are poor (Kirsten et al., 2000). Subsistence agriculture and other natural resource-based activities provide the basis for many livelihoods (Kirsten et al., 2000). According to Parker (1998), many rural poor view agriculture as a panacea to escape the problem of poverty and unemployment, despite the fact that productivity remains very low. The South African government's conviction is to improve women's agricultural productivity through improved technology (African National Congress, 1994). To date small-scale women farmers in Qwaqwa (one of the major former homelands) have been neglected by existing developmental programmes due to inadequate consideration of their needs and the important role they play in the agricultural sector. Women have also been excluded from active participation in the processes that potentially determine the future of the current and future generations on the land.

Rural black South African women farmers

In Southern Africa, and probably in South Africa, small-scale black women farmers are the main food suppliers for the majority of people in rural areas in spite of the fact that commercial farms are doing extensively well in many parts of the country (Kirsten et al., 2000). This is evidenced by the fact that when men leave the villages and transfer to employment in plantations or non-agricultural enterprises, women become the head of the household, taking on the role of working the fields – including sowing, planting, weeding, watering if necessary, harvesting and carrying the crops back to the village (Development Bank of Southern Africa, 1985). It is apparent that such trends have important implications for technology adoption and household agricultural productivity, for the division of labour and for household income.

Parker (1998) has indicated that small-scale farmers in South Africa are spatially dispersed with a low population density. For example, in traditional black areas the populations range from 500 to 10 000, with an average household size of seven members

(Kirsten et al., 2000). This makes provision of agricultural support services extremely difficult. Matata et al. (2001) stressed that the deepest poverty in South Africa is found in rural areas and among women farmers, with female-headed households being the most disadvantaged, experiencing high rates of poverty and low levels of development. Kirsten et al. (2000) showed that small-scale farmers have low levels of resource mobilization and rely on support from central government. However, this support is characterized by an urban bias. Small-scale farmers are also marginalized politically and their manufacturing base is weak due to poorly-developed infrastructure, all of which makes it difficult for this group of farmers to access the essential resources.

Thus small-scale farmers in South Africa have meagre resources to help them fulfil their critical role as food producers in rural areas (Parker, 1998). They have a pitifully low per capita contribution to Gross National Product compared with white commercial farmers. Moreover, they spend their livelihood on cultivating small land holdings. These meagre financial and land resources of small-scale women farmers limit their capacity for technology adoption. According to Kirsten et al. (2000), the small size of most farms in South Africa provides a resource base that is far too limited to enable the use of most new technologies, including those that can provide a more sustainable way of living for the farmers and their families. In sum, Snyder and Tadesse (1995) characterize small-scale women farmers as producers who, with the help of simple equipment and the labour of their families, produce mostly for their own consumption (direct or indirect) and to fulfil their obligations to holders of political and economic power.

In South Africa black women farmers outnumber men in the agricultural labour force. They work more hours in agriculture than men. They are also responsible for other household chores; for example, in the changing environmental conditions women take more time collecting water in the dry season because nearby water supplies have dried up (Snyder and Tadesse, 1995). These numerous activities mean that women are unable to gain sufficient income from a single non-farm activity and the severe constraints on their time can make it extremely difficult, if not impossible, for them to explore new technological innovations.

Despite the fact that black women farmers in South Africa have been constrained by many factors, they are the main generators of primary and secondary employment in rural areas (Kirsten et al., 2000). Although some rural income comes from outside industries, they represent the main source of income for households. Black women are the largest component of the rural economy, and thus hold the potential to alleviate the flow of resources to the industrial sectors and to bring about equity and stability in rural households, should their activities be supported.

Fényes et al. (1988) stress that women small-scale farmers are unable to obtain and use new technologies, including environmental technologies, due to a lack of knowledge, financial capacity and necessary infrastructure. If the growth and equity goals espoused by government are to be achieved, it is important that these constraining factors be reversed. Women farmers and their specific role in agricultural development should be recognized and supported in the planning and development stages of new technologies, including through direct women's involvement. Improved access to formal and informal financial services could drastically increase women's participation in the adoption of the right agricultural inputs.

Technology adoption by South African small-scale women farmers

Numerous national and international surveys on women's agricultural productivity in South Africa have indicated that 80 per cent of South African small-scale women farmers favoured modern agricultural technologies (Delgado, 1997; Development Bank of Southern Africa, 1985; Fényes et al., 1988). Environmental considerations are not a top priority for their sustainability. Women cited increasing food productivity in both quantity and quality, and reduction of food losses, as major advantages derived from the use of modern technologies (Saito, 1994). The spread of modern agricultural technology has also increased the demand for labour and credit that has mainly affected African women farmers (Celis et al., 1991). Sanders et al. (1996) suggest that the more intensive or yield-increasing technologies are employed, the more labour demands increase per hectare – labour normally provided by women.

According to the Development Bank of Southern Africa (1985), limited access to critical inputs such as information, credit and purchased goods (such as fertilizers and other chemicals) are some of the factors making the adoption of new technologies undesirable for many South African women farmers. The Development Bank stressed that the lack of these resources has led to low levels of agricultural productivity being achieved by women and has reduced the returns when new technologies are introduced. To remedy these problems, development economists have suggested that future technology transfer strategies should be based on a targeted approach in order to reach progressive, low-access and resource-poor women farmers (Usherwood et al., 1985; Saito, 1994; Hemmati, 2000). Small-scale women farmers should be provided with new technologies as well as environmental awareness and management skills to make the best use of existing technologies. It is also important to take into consideration the local and traditional knowledge that women have when technology adoption decisions are made in order to achieve long-term sustainable changes.

In South Africa, as in many parts of Africa, research reports have demonstrated the potential effectiveness of new technology for increasing crop yields and agricultural productivity. However, while technology has contributed greatly to increased production more generally, it has not had a broad impact on small-scale traditional black women farmers in particular. In the rural areas of South Africa the average crop yield for black women farmers remains low and consequently output per worker and per unit of land area are also low (Kirsten et al., 2000). This implies that appropriate and acceptable technologies have not been available to these farmers, or that the technology delivery system has not favoured this group of farmers. Of course, other factors – such as the lack of a market, high input costs and inadequate infrastructure – can also depress the adoption or development of the right technologies.

Qwaqwa district case study

A survey on the adoption of new technology in agriculture on small-scale women farmers was conducted in the Qwaqwa district of the Republic of South Africa between February and May 2001 (Yokwe, 2003). This district lies in the south-eastern part of the Free State province. It borders the Kingdom of Lesotho in the south and Kwa-Zulu Natal province in the east. It is situated between latitude 28°S and 30°S and longitude 28°E and 30°E (Nell, 1998). The Qwaqwa district is an interesting case study for two main reasons: (1) the explosion of its population – hence there is a need for promotion of agricultural technologies to assist crop production for improved farm income and poverty alleviation; and (2) the external environment (including market conditions, entry of potent new competitors, new regulations and the growing bargaining power of customers or suppliers), which has a direct influence on the performance of the farming community as well as on the diffusion and adoption of agricultural technologies by small-scale farmers in the region.

The results discussed below were obtained through a survey of 52 male (27) and female (25) small-scale farmers who were interviewed (Yokwe, 2003). The discussion that follows sets out an analysis of the characteristics of the respondents surveyed in order to determine the extent to which the farmers were exposed to new methods and technologies and the extent to which they will be able to access new ideas and technologies to improve their productivity in a sustainable way. The analysis will proceed in four sections: human capital endowment, management skills, financial resources and technology inputs.

Human capital endowment
Human capital refers to the bundle of skills and abilities that a person carries into the labour market (Mijindadi, 1995). According to Nell (1998), human capital is perhaps one of the most important predictors of new technology development and adoption. Formal education and experience are strongly related to knowledge and adoption of production and environmental technologies. In general, the less human capital a farmer possesses, the lower is her/his potential farm productivity. Human capital is built through education as well as through experience.

Previous studies in South Africa (Nicholson and Hirschowitz, 1988) have indicated that younger farmers tend to be more successful than older ones since younger people may be more adaptable, and therefore more willing to try out new methods or technologies than older people. Younger people may also have had more exposure to innovations, contributing to their willingness to adapt. The female heads of households in the Qwaqwa district were younger on average (48 years of age) than male heads of households (66 years of age). One reason for the higher average age of male heads of households is their preference for remunerative off-farm jobs before they reach 64 years or pension age, when they become interested in the farming business. The younger age of women farmers also indicates that women are likely to be more committed to the farming business compared with their male counterparts.

Female-headed households however, had fewer members, approximately five compared with seven for the male-headed households. A household with more members can provide more labour and other support services, such as funds for farm inputs, compared with one with fewer members. Previous reports have also indicated that a greater number of household members can have positive influence on risk management, as they can render assistance during harvesting, storage and marketing (Kirsten et al., 2000). Hence, women with smaller households were exposed to more uncertainties and difficulties in farming the land. As polygamy is practised in Qwaqwa, men also have the advantage of mobilizing more children from polygamous marriages.

A farmer's level of education could positively and significantly affect the farmer's probability of adopting new technologies (Nell, 1998; Saito, 1994). The level of education achieved by a farmer's children is also taken into account, because studies have shown that these factors combine with farmer education to constitute the most influential factor in

Table 32.1 Educational level of Qwaqwa respondents, 1999–2000 (percentage)

Educational Level	Male Farmer	Female Farmer	Male Farmers' Children	Female Farmers' Children
None	26	52	3	16
Grade 5 and lower	33	24	15	11
Grade 6 to 7	26	20	34	37
Grade 8 to 9	11	4	13	16
Grade 10 to 12	0	0	20	12
Post-matriculation	4	0	15	8

Source: Yokwe, 2003.

the adoption of new technologies (Bonnen, 1990). In the Qwaqwa sample, the most noticeable difference is that male-headed households are more educated than female-headed households (Table 32.1). However, the level of education of farmers is much lower than that of the children. This difference, according to Nell (1998), is due to the major improvement in educational services and facilities that took place in the period between 1980 to 1991.

If Grade 6 and above – that is, the grade level in South Africa in which a person can be able to read, write and do simple arithmetic – is considered to qualify a person as being literate (ANC, 1994), then only 41 per cent of male and 24 per cent of the female farmers in the sample qualify as being literate. This indicates a big disadvantage for female farmers; moreover, none of them had education beyond Grade 10. These results are consistent with earlier findings on education and gender (see for example, Saito, 1994; World Bank, 1994). Also, more male farmers' children (97) than the female farmers' children (84) were attending school (Table 32.1).

According to Rogers (1983), an important characteristic of technology adopters is their ability to understand new techniques and try them out, which is strongly facilitated when farmers and their children are educated. For example, they will be better able to make use of resources, such as banking facilities and radio reports on the market prices of farm produce, and are more likely to have basic bookkeeping skills to run their enterprises and the skills to manage risks than those with less (or without) education. Better education of farmers and their offspring may also ensure that they are able to choose and apply sound farming innovations, particularly if their education had provided them with a better understanding of modern farming methods within a sustainability context.

Management skills

While pertinent experience is definitely an asset, modern farm technology also requires management knowledge and skills to avoid continuous risk problems. According to Heinze (1996), management skills can be used to identify, analyse, evaluate or assess risks and to respond to uncertainties. In agriculture, extension services also play an important role for the adoption of new technologies.

In the Qwaqwa case all small-scale farmers exhibited risk-aversive behaviour in making use of agrochemicals (including new varieties of seeds, pesticides, fungicides, and herbicides) in order to kill insects on the crops, control pest, cutworms and diseases, and to

Table 32.2   Farmers' response to the services rendered by extension agents in Qwaqwa, 1999–2000

	Yes		No	
	Male No. (%)	Female No. (%)	Male No. (%)	Female No. (%)
Availability of extension agents on farm	26 (96)	24 (96)	1 (4)	1 (4)
Accessibility of extension agents				
Government extension agents	26 (96)	22 (88)	1 (4)	3 (12)
Input supplier extension agents	–	–	27 (100)	25 (100)
Output markets	–	–	27 (100)	25 (100)
Banking service	6 (22)	3 (12)	21 (78)	22 (88)
Training of farmers on new technologies by extension agent				
Farmer received training on technological inputs	18 (67)	7 (28)	9 (33)	18 (72)

Source: Yokwe, 2003.

Table 32.3   Type of training received by farmers from extension agents in Qwaqwa, 1999–2000 (average numbers)

	Male	Female
Contact with extension agents received per cropping season	12	7.2
Vocational training received	6.43	3.95
Fertilizers advice received	15.12	9.0
Pesticides advice received	6.05	2.19
Improved seed advice received	1.5	1.42
Land use advice received	–	–

Source: Yokwe, 2003.

reduce losses in the crops to generate a higher income. The fact that most of their agricultural activities are of subsistent nature makes risk-seeking behaviour not an option. However, women's use of agrochemicals was lower (40 per cent compared with 52 per cent for male farmers) exposing them to greater risks. A possible explanation is their poorer reading and writing abilities.

According to Saito (1994), a farmer's contacts with extension agents can have a positive influence on the farmer's decision to use improved seeds, fertilizers and pesticides. Saito (1994) reported that extension contact contributes positively and significantly to the adoption of modern farm inputs by both male and female small-scale farmers. However, the mere fact of having extension contacts will not automatically result in increased output; it is only when a farmer applies the information provided by extension agent, and develops the relevant skills that the increased farm performance can be realized.

Table 32.4 *Gross output value gained from different crop practices in Qwaqwa,*
1999–2000 (in rand per 1.65 hectares of land)

Crop Name	Average Yield per Household	Household Consumption	Quantity Sold Rand/year	Crop Market Price Rand/year	Total Rand Value *
Males 1999–2000					
Maize	200 kg	80 kg	120 kg	1.55	310.00
Cabbage	103 heads	30 heads	75 heads	1.00	103.00
Spinach	85 bundles	25 bundles	70 bundles	2.00	170.00
Onions	50 kg	10 kg	40 kg	3.25	52.50
Carrots	25 kg	5 kg	20 kg	3.00	75.00
Beetroots	9 kg	9 kg	–	3.25	29.25
Beans	55 kg	20 kg	35 kg	3.50	192.50
Tomatoes	42.5 kg	2.5 kg	40 kg	4.00	170.00
Pumpkins	11 pumpkins	8 pumpkins	3 pumpkins	4.75	52.25
Total sales value					1254.50
Females 1999–2000					
Maize	240 kg	80.15 kg	159.85 kg	1.25	300.00
Cabbage	80 heads	50 heads	30 heads	1.00	80.00
Spinach	15 bundles	15 kg	–	1.50	22.50
Onions	35 kg	10.20 kg	24.80 kg	2.50	87.50
Carrots	15 kg	4.5 kg	10.50 kg	3.00	45.00
Beetroots	9 kg	9 kg	–	3.25	29.25
Beans	65 kg	24 kg	41 kg	3.50	227.50
Tomatoes	36 kg	15.43 kg	20.57 kg	3.50	126.00
Pumpkins	16 pumpkins	6 pumpkins	10 pumpkins	4.75	76.00
Total sales value					993.75

Note: * Calculated as average yield per household multiplied by market price.

Source: Yokwe, 2003.

In the case of Qwaqwa, male farmers again had an advantage compared with their female counterparts with 67 per cent of men having attended a training by extensions services on technological inputs (such as agrochemicals, seeds, equipment and tools) compared with only 28 per cent for women (see Table 32.2). Table 32.3 specifically addresses the types of training delivered to farmers. It is interesting to note that the focus was on the technical aspects, such as the use of fertilizers, with no provision made for the social, economic or environmental implications from the modern technologies. More specifically, there was no advice given on the use of land (see Table 32.3).

The agricultural output by male and female farmers in Qwaqwa was also in line with the comparative advantage that men have (see Table 32.4). The average gross farm income (in rand) from 1.65 hectares of land earned by female farmers (R993.75) was also less than that earned by male farmers (R1254.50) for the 1999–2000 period (Table 32.4).

According to Merle et al. (2000), farmers with interest in technology are those with farming as the main source of income. Thus a household head whose main source of income is farming should be more motivated in adopting new cash crop technologies or improving current technological practices than those with off-farm activities as the main source of income. In Qwaqwa, 59 per cent of male-headed households earned their income from the farm, 33 per cent from off-farm employment and 8 per cent had no income. In comparison, 8 per cent of female-headed households' earnings came from the farm, 27 per cent from off-farm business and a huge 65 per cent had no income at all. Once again, the difference between men and women is striking. This disparity is also reinforced by the fact that female farmers have been involved with farming for a longer period of time (22 years on average compared with 12 years for male farmers).

Financial Resources
Against this meagre economic background, the lack of access to credit financial facilities is an important factor hampering the farming operations of black farmers in South Africa (Lyne, 1996; Nicholson and Hirschowitz, 1988). On the other hand, Fényes (1982) has argued that black farmers are resistant to accepting credit, partly because in the past credit was given to them without sufficient checks and controls, and without any training regarding the use and implications of having credit facilities.

This is apparent in the Qwaqwa case study where farmers indicated that they are not well served by formal financial institutions. Of the total 52 farmers (male and female), only 11 per cent of male farmers (three in total) and no female farmers had obtained credit from a land bank. Moreover, both male and female respondents stated that none of the commercial banks had offered them any sort of credit. The main source of credit for both men and women farmers was informal, mostly from friends and family members.

The lack of funding (or any other economic incentives) makes it extremely difficult for these small-scale farmers to undertake technological adoption or improvement. In general, the problems associated with formal financial institutions in providing services to small-scale farmers in Qwaqwa stem from the general approach of credit providers: 'the financial institutions deal mostly with medium and large-scale farmers while neglecting small-scale holders' (Majake, 2001, p. 4). According to Majake (2001), there is strong competition from emerging white and black commercial farmers in the area who are operating more profitably on larger farms. The neglect of small-scale farmers in Qwaqwa by financial institutions leads to weak adoption of new cash crops or technology.

The general perception held by financial institutions appears to be that small-scale farmers are too poor to engage in meaningful production. In addition, they lack collateral. According to Snyder and Tadesse (1995), formal financial institutions often deem the type of valuables held by women (such as jewellery) to be unacceptable. In South Africa, the commercial banks are not active in rural areas because they have a perception that lending in such areas demands extensive outreach, tends to accrue higher transaction costs, and is more risky than medium and large-scale farm businesses (Kirsten et al., 2000).

Low farm income together with the low level of education, particularly of women, are likely to limit significantly the ability of farmers to obtain credit. Moreover, transportation costs and the time spent waiting to obtain credit may be higher for women than men, owing to higher opportunity costs from forgone activities.

Table 32.5   Technology input used by small-scale farmers in Qwaqwa, 1999–2000

Inputs	Yes		No	
	Male No. (%)	Female No. (%)	Male No. (%)	Female No. (%)
Seeds	27 (100)	27 (100)	–	–
Fertilizer	23 (85)	18 (72)	4 (15)	7 (28)
Insecticides	21 (78)	17 (68)	6 (22)	8 (32)
Pesticides	20 (74)	17 (68)	7 (26)	7 (32)
Machinery	2 (7.4)	–	–	–
Tools (hoe, spade, rake, fork)	16 (59.3)	24 (96)	11 (40.74)	1 (4)
Animal traction	9 (33.3)	1 (4)	23 (85.19)	24 (96)

Source: Yokwe, 2003.

Technology inputs used

New or improved technologies can increase the productivity of women small-scale farmers. Econometric analysis of data from women's agricultural productivity surveys in Kenya and Nigeria, and evidence from other analyses in Kenya and Zambia, has confirmed the positive effect that adoption of fertilizer and agrochemical technologies can have on the total value of output (Bindlish and Evenson, 1993). The majority of farmers in Qwaqwa used fertlizers, pesticides and insecticides (see Table 32.5). Thus agrochemical inputs are used in most of the farms in Qwaqwa, except a few where kraal manure was used in the place of chemical fertilizers. In contrast, chemicals such as herbicides and fungicides are not used at all indicating some environmental awareness. Despite the resource constraints experienced by all small-scale farmers, they are able to obtain seed, which is essential for their survival.

Very few farmers use agricultural machinery such as tractors; however, a larger share of male farmers rely on animal traction while female farmers use predominantly tools such as hoes, spades, rakes and forks (Table 32.5).

The economic sustainability of the small-scale farmers is an issue of major concern. If the amounts paid for technological inputs (Table 32.6) are compared with the total farm gross income gained (in Table 32.4), it becomes apparent that these farmers are operating with very low returns. The same is true for vegetable farmers in many Sub-Saharan Africa countries. Thus expenditure on farm inputs renders it extremely difficult for farmers to adopt new technology, as farm returns do not provide them with incentives to purchase new technological inputs. A number of factors may help to explain the low production per hectare achieved by these farmers, including ineffective extension services, the discouraging ratios of input to yield, poor accessibility to credit and low levels of education.

To assess the farmers' attitudes toward the adoption of new technological inputs, farmers were asked why they do not use chemical inputs (Yokwe, 2003). The following results were recorded from the respondents: 21 per cent indicated that the chemicals are too expensive, hence they can't afford to buy them; 88 per cent indicated that they can't get credit to buy the chemicals; 58 per cent indicated that they have no transport to get the chemicals to their farm; and 21 per cent indicated that they are involved in organic

Table 32.6   Average amount paid for technological inputs used per hectare in Qwaqwa, 1999–2000 (in rand)

Inputs	1999	2000	Average
Fertilizer	62.50	61.00	61.75
Insecticides	30.00	55.25	42.63
Seed	42.00	46.50	44.25
Pesticides	22.10	40.00	31.05
Machinery	100.00	50.00	75.00
Animal traction	50.00	50.00	50.00
Tools (spades, rakes, forks, etc.)	105.20	96.00	100.60
Total	369.80	352.25	361.03

Source: Yokwe, 2003.

farming (Yokwe, 2003). It is interesting to note that one in five farmers were engaged in organic farming. However, they (as well as the rest) should be in a position to afford to use machinery and tools to facilitate their work.

In addition, only one-third of total farmers in Qwaqwa indicated that they have access to roads, transport, local markets and electricity, while the remaining two-thirds indicated that they have no access to any of the mentioned infrastructure variables. Therefore it can be assumed that these farmers are walking long distances to market their products (some indicated that they use horses to carry their product to the local market) and local markets provide the only opportunity for them to sell their produce. Women farmers are generally more affected than men as this puts extra pressure on their family obligations and they have a lower ability to walk longer distances (World Bank, 1994). According to Saito (1994), the availability of road, transport and telecommunications infrastructure has a positive influence on farmers' adoption of technology but this was not the prevailing situation in Qwaqwa.

Conclusion

As the Qwaqwa case study shows, women farmers in Sub-Saharan Africa are in general the most disadvantaged in their access to resources and technology, compared with their male counterparts, despite the fact that they represent the majority of small-scale farmers in rural areas. Moreover, the education levels of female farmers are lower than those of male farmers, which results in poor technological adoption and poor farm income. According to the World Bank (1994), if women farmers have had the same education or reading ability, and the same amount of resources as their male counterparts, the value of their output would actually be more than that of their male counterparts. Increasing the literacy rate of female farmers is essential for achieving a more sustainable way of living.

As women farming businesses become increasingly important for the survival of impoverished families in the Qwaqwa district, initiatives to raise the sector's productivity are urgently required. Access to appropriate agricultural technology can play a major role in improving the quality of life in this area of the world. Efforts to increase the mechanical efficiency of the ploughing implements, which would also increase the labour productivity of women, should be explored. With improved agricultural productivity and crop diversification, possibilities for employment and capital formation in Qwaqwa exist. The choice of technology should be rationalized, however, if the potential gains from forward and backward linkages in the production-marketing process are to be maximized. The removal of barriers, such as low level of education, poor infrastructure, limited access to extension services and credit, could contribute to better technological choices, including diversification to other farming activities, such as poultry and dairy, in order to increase the farmers' income. This should:

- reduce poverty through substantial economic and social pay-offs;
- contribute to environmentally sustainable development; and
- produce significant social gains, including lower fertility, better household nutrition and reduced infant and maternal mortality.

References

African National Congress (ANC) (1994), *The Reconstruction and Development Programme*, Umanyano, South Africa: Bloemfontein.

Bindlish, V. and Evenson, R. (1993), 'Evaluation of the Performance of T and V Extension in Kenya', *Agricultural and Rural Development Series, Number 7*, Technical Department, Africa Region.

Bonnen, J.T. (1990), 'Agricultural Development: Transforming Human Capital, Technology And Institution', in Eicher, C.K. and Staaz, J.M. (eds) *Agricultural Development in the Third World*, 2nd Edition, Baltimore, USA: John Hopkins University Press, pp. 217–300.

Celis, R., Milimo, J.T. and Wanmali, S. (eds) (1991), *Adopting Improved Farm Technology: A Study of Smallholder Farmers in Eastern Province*, Zambia, Lusaka: University of Zambia, GOZ and IFPRI.

Delgado, C.L. (1997), 'Africa's Changing Agricultural Development Strategies', *2020 Brief 42*, Washington, DC: International Food Policy Research Institute.

Development Bank of Southern Africa (DBSA) (1985), *Qwaqwa Development Information, Section 1–10*, Midrand: DBSA and Qwaqwa Government.

Fényes, T.I. (1982), 'A Socio-economic Analysis of Smallholder Agriculture in Lebowa', DSc (Agriculture) thesis, Pretoria, South Africa: University of Pretoria.

Fényes, T.I., Van Zyl, J. and Vink, N. (1988), 'Structural Imbalances in South African Agriculture', *South African Journal of Economics*, **56**(2/3), 181–95.

Heinze, K. (1996), *Cost Management of Capital Projects. Project Initiation*, New York: Marcel Dekker, pp. 108–15.

Hemmati, M. (2000), 'Women & Sustainable Development: From 2000 to 2002', in Dodds, F. (ed.) *Earth Summit 2002 – A New Deal*, London: Earthscan, pp. 65–83.

Kirsten, J., Perret, S. and Rwelamira, J.K. (2000), *Educational Manual for Postgraduate Students in Rural Development: Focus on Small-scale Agriculture in Southern Africa*, Programme for Institutional University Cooperation, Flemish Inter-University Council and Partners, Pretoria, South Africa.

Kumar, S.K. (1987), 'Women's Role and Agricultural Technology', in Mellor, J.W., Delgado, C.L. and Blackie, M.J. (eds) *Accelerating Food Production in Sub-Saharan Africa*, Baltimore, USA: Johns Hopkins University Press, pp. 135–47.

Lyne, M.C. (1996), 'Transformation Developing Agriculture, Establishing a Basis for Growth', *Agrekon*, **3**(4), 188–92.

Majake, C. (2001), 'Women and the Agriculture Sector', Paper presented at the National Workshop on Women and Agriculture, Commission of Gender Equality, Johannesburg, South Africa, pp. 1–10.

Matata, J.B., Anadajayakeram, P., Kiriro, T.N., Wandera, T. and Dixon, J. (2001), *Farming Systems, Approach to Technology Development and Transfer*, Harare, Zimbabwe: FARMESA.

Merle, S., Oudot, S. and Perret, S. (2000), *The Technical and Socio-economic Circumstances of Family Farming Systems in a Small-scale Farming Scheme of South Africa (Northern Province)*, Synthetic Report, CIRAD Tera, No. 79/00, Montpelier, France.

Mijindadi, N.B. (1995), 'Agricultural Technology Transfer in Nigeria, Emerging Issues and Lessons For Other African Countries', in Doward, P.T. (ed.) *The World of Farm Management, An International Exchange, Proceedings of the 10th International Farm Management Congress at University of Reading*, Reading, UK: University of Reading.

National Department of Agriculture (2001), *The Strategic Plan for South African Agriculture*, Pretoria, South Africa.

Nell, W.T. (1998), 'Transfer and Adoption of Technology, the Case of Sheep and Goat Farmers in Qwaqwa', PhD thesis, Bloemfontein, South Africa: Centre for Agricultural Management, University of Orange Free State.

Nicholson, C.A. and Hirschowitz, R. (1988), 'Characteristics of a Group of Small-scale Farmers in Gazankulu, An Exploratory Study', *Research Finding MN 128*, HSRC, Pretoria, South Africa.

Parker, A.N. (1998), 'Decentralisation, Rural Development and Local Government Performance: A Case Study of Rural Municipalities in Northeast Brazil', PhD thesis, Pretoria, South Africa: Department of Agricultural Economics, Extension and Rural Development, University of Pretoria.

Quisimbing, A. (1993), *Women in Agriculture*, Washington, DC: World Bank, Education and Social Policy Department.

Rogers, E.M. (1983), *Diffusion of Innovations*, New York: The Free Press.

Saito, K.A. (1994), 'The Raising of the Productivity of Women Farmers in Sub-Saharan Africa: The Role of Women in Agriculture', Washington, DC: World Bank Discussion Papers.

Sanders, J.H., Shapiro, B.I. and Ramaswamy, S. (1996), *The Economics of Agricultural Technology in Sub-Saharan Africa*, Baltimore, USA: John Hopkins University Press.

Schultz, P. (1993), 'Returns to Women's Education', in King, E. and Hill, M.A. (eds) *Women's Education in Developing Countries: Barriers, Benefits, and Policies*, Baltimore, USA: Johns Hopkins University Press.

Snyder, M.C. and Tadesse, M. (1995), *African Women and Development: A History*, Johannesburg: Witwatersrand University Press.

Usherwood, N.R., Kral, D.M. and Cousin, M.K. (1985), 'Transferring Technology for Small-scale Farming', *American Society of Agronomy (ASA) Special Publication No. 41*, ASA Wisconsin, USA.

West Africa Research and Development Association (2001), Newspaper of the West Africa Research and Development Association, July 2001.

World Bank (1994), *A World Bank Policy Paper: Enhancing Women's Participation in Economic Development*, Washington, DC: The World Bank.

Yokwe, S. (2003), 'The Adoption of Cash Crop Technologies: A Case Study of Small-scale Women Farmers in Qwaqwa District in South Africa', Masters thesis, Bloemfontein, South Africa: University of Free State.

33 Waste streams to value streams

Compiled by Karlson 'Charlie' Hargroves and Michael H. Smith of The Natural Edge Project, with Chris Page, Caroline Plunkett, Gerry Gillespie and Alexis Nelson

Introduction

Several obstacles must be overcome for any industry to make a transition towards sustainability, such as achieving zero resources wasted. Sound impossible? Consider the following: some of business's most significant costs are capital and inputs, such as costs of building infrastructure, sourcing raw materials, energy, water and the costs of transportation. It is therefore in business's best interests to minimize these costs, and hence the amount of raw materials and other inputs needed to create the product or provide the service. Business produces two main outputs: useful products and services, and waste. How does it assist a business to have capital, plant, equipment and labour tied up in generating waste? It is in business's interests to find markets for this 'waste' and/or design processes so that waste is minimized and that which is produced can be used or sold elsewhere. According to the Rocky Mountain Institute (RMI), the majority of raw materials used in production are transformed into waste products within six months of being sold. Hence, as Amory Lovins (2001) states:

> It is extremely profitable to wring out waste, even today when nature is valued at approximately zero, because there is so much waste – quite an astonishing amount after several centuries of market capitalism. In the American economy, the material that we extract from the planet, that we mobilise for economic purposes, and process and move around and ultimately dispose of, totals about 20 times your body weight per person per day. So worldwide this resource flow is in the order of a half-trillion tons per year. And what happens to it? Well, only about 1% of it ends up in durable goods; the system is about 99% waste. That's a business opportunity.

So why then do we have a waste problem? The trouble is that in the past the costs to business of dumping waste into landfill have been negligible along with access to abundant cheap raw materials. Hence, there has been little economic incentive for industry to minimize or recycle waste. Things have changed however, and the following case studies are examples of what can be done.

Sustainability now pays in ways no one could have predicted
Chris Page[1]
The Mayor of New York City recently suspended recycling of glass and plastic, declaring that it was far cheaper to incinerate or landfill these products. Electronic waste exported to China and Thailand has been mishandled, resulting in health and environment problems and an outcry for stricter regulations on the international management of hazardous waste. However, a funny thing is happening to municipal recycling on the road to obsolescence. A combination of government pressure, private enterprise and global demand for scrap materials may make recycled materials financially attractive after all. As the planet's population grows, space becomes an increasingly precious commodity. New York

used to be able to send much of its waste to the adjoining state of Pennsylvania at a minimal cost. After New York City suspended glass and plastic recycling, Pennsylvania began charging more for waste heading to its landfills. This was an unanticipated cost that New York City's overburdened budget could not easily afford. Suddenly, the 'away' that the City assumed it could throw its waste to became much more expensive.

Meanwhile a company called Hugo Neu Schnitzer East saw an economic opportunity in all that waste. The metal salvage company has offered to take on recycling the City's metal and plastic for a significantly lower fee than other waste management companies were offering and will *pay* the city US$5.10 per ton for metal and plastic. Glass is the hardest product to recycle in a financially successful manner. Hugo Neu has found a partner company that will turn the glass feedstock into manufacturing construction materials and other items. For metal waste Hugo Neu has an ever-increasing level of demand overseas. Demand for US waste paper and scrap metal is a rapidly growing phenomenon. The newsletter *Resource Recycling* comments that exported plastic from the United States to China has increased more than sevenfold since 1997. A *Wall Street Journal* article quoted an industry expert as saying that the United States is now 'the Saudi Arabia of scrap'. Massive deforestation and desertification in China has made wood a precious commodity. In order to manufacture paper for its 1.3 billion population, China is paying increasing amounts for US paper waste.

It is here that the fate of recycling in New York City becomes particularly interesting. When glass and plastic recycling was suspended, citizens stopped sorting paper out of their trash as well. Paper has consistently been the most profitable material to recycle and, with increasing demand abroad, prices for recycled paper feedstock should continue to rise. The City was thus hit with a double blow: it suddenly had less of the profitable waste paper available, and more of the paper was mixed into waste for disposal, leading to *more* trash to be landfilled and incinerated . . . again, at higher prices.

Obviously there are still problems to be addressed. Scrap demand may reduce pressure on companies to dematerialize their products in the first place: if the cost of disposal is reduced, designing things so that they do not produce waste at the end of their lives loses some of its urgency. Proper treatment of exported scrap in the countries importing it is a concern. And it is easy to overlook the imbedded energy issue: how much more energy does it take to ship scrap halfway around the world – especially if it is converted to a consumer product and turns right back around again?

At the same time, there are two valuable lessons. First, in a city of eight million, recycling is a cheaper alternative to landfilling, possibly for the first time. Sustainability now pays in ways no one could have predicted. Secondly, a city government and numerous companies are beginning to see waste not as waste, but as a stream of re-saleable goods and a genuine, internationally tradable economic opportunity. In a world of shrinking space and diminishing raw resources, it is increasingly being understood that our waste is, in fact, an economic treasure if we can learn to manage it correctly.

The governments commit to zero waste
The Natural Edge Project (TNEP)[2]

Internationally, changes are being initiated by governments and business who understand this. Gerry Gillespie (2002), from the NSW Government Department in Australia, NSW Resources writes:

The past five years have seen remarkable changes in the Waste Management industry to the point where it is becoming common practice to refer to the development of new waste management programs as 'resource management' strategies, and unless programs contain a strong element of resource recovery they are considered immature and ill-conceived.

This was highlighted in the Australian Capital Territory (ACT) when the ACT Government became the first local authority in the world to release a strategy with the title 'No Waste by 2010': at the time the target was considered foolish and politically naive in waste industry circles. Since that time the ACT has reached a point where it now recycles 70 per cent of its waste stream and in the process has created over 300 new jobs. The Canberra communities' brave move is in line with many other local government authorities in various parts of the world. In New Zealand the Tindall Foundation established the Zero Waste New Zealand Trust the year after Canberra's document was released. Stimulating interest in the use of wastes as resources, the Trust now has 36 councils registered as local authorities with a zero waste focus. At the state government level South Australia has set up a Zero Waste Office and Western Australia has implemented a Waste 2020 Strategy across the state with regional initiatives to achieve zero waste.

There are three Zero Waste organizations in the United States that are pushing to establish a National Zero Waste programme. In the UK, the Mayor of London, Ken Livingstone, is seeking advice on the establishment of a Zero Waste agency and a local recycling forum in Medway in Kent has just set its policy in a zero waste direction.

This shift in government thinking reflects a community awareness of the frightening impact that we, as a species, are having on our environment. However, such is the scale of the 'waste problem' that many feel it is inevitable, but in fact, as we will show, it is possible to achieve significant progress towards zero waste: we just need a holistic multi-levelled approach this century for our countries to become waste-free societies.

There are numerous success stories from which we can learn. Twenty-five years ago South Australia implemented a levy on soft drink, beverages (such as beer bottles) and milk cartons that has resulted in 85 per cent being recycled. In most European countries landfill levies are significantly higher than for most USA and Australian cities. Northern Ireland introduced a levy on plastic bags payable at the checkout at the shop, which reduced their usage by 90 per cent within six months. Numerous European countries are introducing product stewardship legislation that places some responsibility on the manufacturer to take back their product and recycle it. European law already requires 90 per cent of certain appliances and cars to be designed to be recyclable. Singapore, in its 10-year Green Plan for 2002–12, has committed to a 60 per cent recycling target by 2012. Japan's 'Home Appliances Recycling Law' requires retailers and manufacturers to take back used home appliances and customers are responsible for paying take-back and recycling fees. In 2006, computers will be added to the list of regulated equipment.

Legislation and regulation are an integral aspect of successfully transitioning to waste-free societies. In 2003 the European Union Commission passed into law the waste from electrical and electronic equipment (WEEE) and restrictions on hazardous substances (RoHS) directives that go into effect in 2006. Electrical waste is one of the fastest growing sources of municipal solid waste in landfills.[3] For example, e-waste is currently growing more than three times faster than general municipal waste in Australia (Kyocera Mita,

2005, p. 43). The WEEE directive enforces the take-back and responsible reuse, recycling and recovery of computers and various electrical equipment. In 2006 the WEEE will require retailers and manufactures in the EU to provide customers with product take-back and recycling options similar to Japan's Home Appliance Law. The RoHS directive targets the urgent need to re-design products without the use of toxic heavy metals that leach from landfills and contaminate ground water supplies.[4] Both initiatives encourage a whole-systems approach to promoting a shift in design, development and effect change in the products' lifecycle. Leaders in the field, Fuji Xerox, report that 0.5 per cent of copier and printer products will go to landfill (Kyocera Mita, 2005, p. 43).

Five New England states participated in the e-Cycling Pilot programme during summer 2004. The US EPA, Staples Inc. and Product Stewardship International (PSI) collaborated to execute and analyse the cost-effectiveness of widespread consumer recycling of electronic equipment. Analysis concluded that computers can be recycled at low cost to customers and retailers.[5] Manufacturers like Dell, HP, Apple Corporation and Sony participated in the programme and enhanced their public image by proactively reducing electronic waste reaching landfills. A business's environmental image is important in a growing world of environmentally concerned consumers.

California has led the United States in minimizing waste streams since the 1980s with the passing of the Integrated Waste Management Act and more recently with the implementation of the State Agency Buy Recycled Campaign (SABRC). The SABRC was developed to implement state law requiring state agencies and the legislature to purchase products with recycled content. Benefits of buying recycled-content products include creating jobs, improving markets for recycled contents, diverting waste from landfills, reducing manufacturing waste and pollution, and reducing energy consumption.[6] 'Buy recycled' business, government, school and university networks are becoming more common globally. For instance, 30 of Australia's top 100 corporations have formed the 'Buy Recycled Business Alliance'. Sporting events or major national festivals can also be high-profile places for governments to demonstrate how to achieve zero waste.

In the shorter to medium term, we know how to cost-effectively address the major waste streams. There are cost-effective recycling opportunities for paper, organics, plastics, oils, packaging, many chemicals and metals. There are also ways of designing plastics to biodegrade efficiently and even recycle construction waste, including concrete. Further, increasing proportions of components in electrical goods can be recycled.

Another major issue is how industry can design future industrial plants and products to produce as little waste as possible in the first place. This is linked to the sustainable consumption and production framework (Hargroves and Smith, 2005, p. 407) that integrates everything from product policy, creating markets for recycled products, government frameworks, education, sticks and carrots to assist this process. Some may be sceptical that any government would risk introducing such a comprehensive sustainable consumption and production framework. However, several European countries are already using this UN framework and the Japanese Government has over the last five years introduced many aspects of it into law.

Japan commits to create a recycling-based society
Caroline Plunkett[7]

Japan is one of the first countries to be a signatory to a raft of legislation committing itself to becoming a recycling society and truly moving to a closed-loop economy. The Japanese first step towards creating a 'recycling-based society' began in the 1990s with the consolidation of a package of laws. The package stems from the Basic Environment Law, which became effective on 13 November, 1993. Underlying this law is the Basic Law for Establishing the Recycling-based Society, which came into force 6 January, 2001 and is also known as the 'Basic Framework Law'. The reform process in Japan has the potential to be very significant as waste in Japan is a major environmental and social problem. Even compared with the more congested industrial nations of Western Europe, Japan faces a much higher risk of pollution per unit area and pollution per capita (McKean, 1981, p. 17):

> Between 1966 and 1975, the volume of domestic wastes rose from less than 18 million to almost 32 million tons/yr; since then the volume has continued to increase, although less rapidly, to 38 million tons in 1985. Over time these wastes included an increasing proportion of plastics, which increases the difficulty of disposal. The cost per unit weight of disposing of domestic waste more than doubled every ten years, and the total cost of domestic waste disposal rose 100-fold between 1960 and 1980.

The 127 million Japanese consumers place an overwhelming strain not just on Japan's shrinking landfill reserves, but also on other nations' ecosystems from which they import a significant percentage of raw materials (Nakano, 1986). By approximately 2010 Japan will no longer have landfill space in which to store waste materials of any description, according to Japan's Ministry of the Environment (2003). Previously much of Japan's waste was incinerated. However, massive popular outcries against toxic emissions mean that governments are no longer viewing large-scale incineration as a viable waste management strategy (American Chamber of Commerce in Japan, 1998). In addition, increasing scientific and popular recognition of the serious consequences of waste generation has led to Japan's setting of a new benchmark, which is attracting worldwide attention (Bleischwitz, 2002).

Germany has put forward progressive environmental policy options that could also be adapted to Japan's policy environment. In particular, the German-devised concept of Extended Producer Responsibility (EPR) (OECD, 2001), which encourages manufacturers to take responsibility for the disposal of a product beyond its consumer lifespan, attracted the attention of Japanese legislators (Fowler, 2003, p. 451). Systemic transformation is the overarching mechanism used to achieve this goal. That is, Japan has pledged to restructure economic and social systems based on an integrated revamp of the production to consumption chain, with an emphasis on closing the loop in order to ameliorate the 'waste problem' (Environment Agency, Japan, 2000). The creation and amendment of progressive recycling related laws and regulations, commencing in the early 1990s, are an integral component of this change.

Several Japanese recycling laws that fall under the heading The Basic Law for Establishing the Recycling-based Society include the: Waste Management and Public Cleansing Law; Law for Promotion of Effective Utilization of Resources; Container and Packaging Recycling Law; Electric Household Appliance Recycling Law; Construction Material Recycling Act; Food Recycling Law; and Law Concerning the Promotion of

Procurement of Eco-friendly Goods and Services by the State and Other Entities (Law on Promoting Green Purchasing). Each of these sub-laws has a different history, with the first four either already enacted or revised. The latter three laws were newly enacted to fulfil the aims of encouraging state, local government, corporations and consumers to become a 'recycling-based society'.

The Law Concerning the Promotion of Procurement of Eco-friendly Goods and Services by the State and Other Entities (the 'Green Purchasing Law') was enacted in 2000 with the purpose of encouraging the national government, independent administrative institutions and local governments to procure eco-friendly goods, while giving consideration to the appropriate use of the budget.[8] The overarching goal dovetails with that of the Law for Promotion of Effective Utilization of Resources, namely 'to establish a society which can enjoy sustainable development with a lower environmental impact'.[9] The Law for Recycling of Specified Kinds of Home Appliances came into effect in April 2001, with the purpose of promoting the reuse and recycling of used household electrical appliances, specifically air-conditioners, televisions, refrigerators and washing machines. The broader goal is to create a 'closed-loop economy' in which used materials can be made into new products, and is based on new obligations on manufacturers and retailers (Ministry of International Trade and Industry, 2001, s 1(2)).

The law comes under the jurisdiction of the Ministries of Welfare, International Trade and Industry, and the Environment. Their duties are to decide the basic plan and then publicize it (Ministry of International Trade and Industry, 2001, s(2)). One of the seven sub-laws, the Law for Promotion of Effective Utilization of Resources, was originally promulgated in 1991, and was significantly revised for the third time in June 2000.[10] Article 1 outlines the aims of the law, which are manifold and ambitious. A reduction in the usage of resources by industries is described as a way to both protect the environment and promote economic development. Waste minimization is a third and vital aim, as the law acknowledges that most of Japan's resources are imported and discarded without either recycling or reuse.[11] Thus, the law establishes guidelines for certain 'specified resource-saved products' to be produced in a way that uses fewer materials and resources.[12] Concurrently certain 'designated resource-reutilizing industries' for each type of recyclable resource or reusable part are to institute the use of recyclable resources or reusable parts in the manufacture of that product.[13]

There are still many who are cynical about zero waste targets but, as seen here, there are numerous win–win ways to make significant progress toward achieving these goals that have not been taken advantage of in the past. Governments are increasingly providing comprehensive online resources to demonstrate how we can now recycle everything from mobile phone batteries to cars.

Paper recycling
The Natural Edge Project[14]
In recent years several publications have drawn attention to how poorly we are managing what remains of the world's native forests (Gibbons and Lindenmayer, 2002; Mackey et al., 2002). The continuing decline in native old growth forests globally[15] has created great interest in ways to recycle paper and cardboard to help take the pressure off what remains of these forests.[16] Furthermore, the continuing increase in use of paper and timber products is often creating balance of payments deficits in OECD countries from

imports.[17] This is generating great interest from governments in promoting recycling of these products to reduce the level of imports.

Forests provide numerous ecosystem services of tremendous value in their own right. In 1999 the Ecological Economics Unit of the Australian Government's Department of Environment and Heritage[18] completed a comprehensive overview and study on the 'Real Value of Australian Forests'. They quoted from the Read-Sturgess report (1992) that made it clear that it was more financially valuable for one of the major water catchments in Australia, the Upper Thomson catchment, to supply water than to supply timber, and consequently more profitable for the economy. They wrote that the overestimation of the worth of forestry would tend to hide this potential economic 'free lunch' (Read and Sturgess, 1992). Other competing uses for the forest may be similarly affected. More generally, the over-valuation of forestry financial values disguises what may be net losses in this sector, which are actually diminishing the wealth of nations, unlike other commercially viable industries. The poor measurement processes for forestry blind the commissions to other profit centres such as leasing forests to supply water to regional communities with expanding industries. The impact of miscalculating the financial value of timber production is significant for Australia's macro-economic performance. It overestimates the financial worth of timber production with regard to some other form of commercial activity.

In 1985 the Australian Government stopped subsidizing land clearing, yet Australia still subsidizes the logging of old growth forests. The Productivity Commission has published a major report to address the state forests' logs being under-priced, recommending competitive neutrality plus a raft of other measures to remove these government subsidies.[19] In addition, Forest Policy in Australia contravenes the 1995 competition policy laws due to the maintained subsidies for old growth logging. Equivalent or increased employment could easily be created if similar subsidies were put into plantations and/or paper recycling. In *Natural Capitalism* (Hawken et al., 1999), the authors reported studies that showed that if all of the agricultural 'residues' in the USA were used to make paper, it would supply the world's paper needs. Hence agricultural residues that are often simply burnt, offer another potential solution to supplying the world's paper needs. So far cotton waste and sugar cane bagasse (sugar cane waste fibre) have been used in recycled papers. However, both crops have a significant impact on the environment, via water usage and fertilizer run-off respectively. Nevertheless, PlanetArk is promoting 'Cane Fields' paper that is produced from a minimum 75 per cent bagasse and 25 per cent certified plantation-sourced eucalyptus wood fibre.[20]

Reducing waste in the chemical industry
The Natural Edge Project[21]
As TNEP wrote in *The Natural Advantage of Nations*:

> The costs of cleaning up the existing hazardous waste sites are also estimated to be in the hundreds of billions of dollars range in the USA economy . . . Many individual chemical companies have budgets for environmental compliance programmes that are as large as their budgets for R&D, and for some of the largest chemical companies, environmental compliance budgets can approach AU$1 billion per year . . . A realization that is currently occurring throughout the chemical and related industries is that re-designing waste out of the initial process will not only save significant costs but can also result in greater profits. With the challenges of increased global

competition facing industry, both domestically and internationally, and increased regulatory requirements, it is clear that every company is redoubling its efforts to find new ways of turning cost centres into profit centres. Solutions for bypassing the environmental and economic hurdles associated with waste treatment and disposal in the chemicals industry are now an increasingly high priority. The chemicals industry has turned to research institutions for guidance. (Hargroves and Smith, 2005, p. 97)

A great example of this has occurred in Massachusetts, USA. The chemical industry there along with the other industries said to government that they were keen to work with government and the universities and research institutions to find cost-effective alternatives to currently hazardous chemical approaches. Out of a partnership with this group of responsible industries, citizen groups and the University of Massachusetts, came in 1989, the Commonwealth of Massachusetts Toxic Use Reduction Act.[22] The stretch goal was a 50 per cent reduction in toxics by 1997 without any loss of costs, profits or competitiveness of businesses. They have worked together to show that it is possible to make products safely and to phase out the use of carcinogenic chemicals and replace them with safe chemicals.[23] Dr Samuel Epstein, respected world authority on environmental and health and safety issues related to the chemical industry, states that: 'These industries said, "Look, if you can show us how to go on about our business without losing money, then we'll work with you." The results of the law have been phenomenal. Over the past decade, for example, hazardous organic solvents have been substantially phased out and replaced by safer alternatives' (Epstein, 2000). The success of this programme provides hard indisputable evidence that it is economically possible to phase out a variety of hazardous products and processes already on the market.

What is Europe doing? In Europe significant changes are also occurring. One of the most important developments is the new registration, evaluation and authorization of chemicals (REACH) legislation, proposed by the European Union. This initiative attempts to remedy current practice that sees tens of thousands of chemicals to be used without sufficient knowledge about their environmental and health effects. Not only does REACH explicitly restrict the use of carcinogens and mutagens, but it has an additional key element that should lead to significant change. It forces consideration of alternatives as part of the chemical licensing process. REACH therefore will catalyze cleaner technologies and safer products globally. In preparation for its enactment in 2006, many companies in Europe and North America have already started phase-outs of chemicals of high concern.

Individual members of the European Union have gone farther still, adopting legislation that simply bans the commercial production and use of carcinogens. One of the leading examples of this occurred in Sweden in 2001. There they passed a new sustainable chemical policy that requires that all new chemicals proposed for use must now be accompanied by evidence that they do not pose carcinogenic risk. The collective work of these leading programmes, and advances in chemistry such as the field of green chemistry and green engineering suggest that all governments could commit to significant targets to reduce chemical waste over the coming decades.

Unappreciated benefits of organic recycling
Gerry Gillespie[24]

One of the next significant areas to develop in recycling will be organic recycling: an area that presents significant long-term business opportunities. Over 15 per cent, and in some places as much as 50 per cent of all the materials regions put into landfill are organic (food wastes, clippings from lawns and ovals, agriculture wastes)[25] and a significant part of this has come from our own national farming processes. Farming is a mineral extractive industry, which progressively removes not only the organic fraction, but also minerals and trace elements from the soil. All of this material is either exported or carried into the cities where it is processed through people, passes through a waste management system to end up either in landfill or in sewage treatment works.

Plants cannot make minerals and trace elements, and these important structures in healthy plant growth are not put into our soils through the application of fertilizer. The process of degradation of our soils costs Australia millions of dollars per year. At the same time one of the factors in this degradation, chemical fertilizer, is constantly rising in cost both to the farmer and to the broader community. Australia's national fertilizer bill at the farm gate every year is in excess of AU$4 billion. Over the past ten years the government has attempted to address some of the soil- and water-related issues through a variety of means such as the National Landcare Program. While the work of Landcare has been enormous in its scope and brilliant in its execution, at every turn it is soundly defeated by the size of the problem and the vastness of the country. Much of this work currently relies on direct funding from government budgets or on the sale of utilities such as the national telecommunications utility, Telstra. As a funding effort it provides large quantities of resources, but on a national basis it barely touches the sides. Landcare, as an entity, needs an enduring funding source that provides it with an ongoing income: one that will be there for the duration of the problem. It needs an income that derives from the provision of a service: one that is business based. A business that can last forever. That business is waste management. Or, to be more precise, waste reduction. The focus of this new business would be the diversion of waste from landfill to farming.

From landfill to Landcare
Landfill in all its forms has become one of the largest long-term problems facing urban society today. It steals our space, devalues our property, threatens our waterways and contaminates the future. It is the graveyard of sustainability and it compromises the very survival of future generations. At the same time, the other end of the process, farming, depletes our soil, pollutes our waterways and increases our foreign debt. Depletion of soil quality is a problem that hits the headlines in newspapers globally almost every day. A report from Britain stated that in excess of 30 per cent of farm soils in the UK were deficient in organic material.

Yet the greatest contaminant in landfill is organic material. It is organic material that leaches through the landfill to create further problems of contamination and pollution. There is a constant cry from composters that there is no market for products. At the same time our soils cry out for the application of the organic materials, micronutrients and the microbial activity that is compacted into our landfills every day. If this organic material could be returned to the food chain we could eliminate many of the problems associated with landfill, create local employment programmes and go some way to relieving the destruction of our soils through the overuse of chemical fertilizer and unsustainable farming practices. Every nation needs a programme that is focused on the removal of organic materials from the waste stream and the processing of this material into a viable, balanced organic product for use on farms.

In favour of the farmer
There is not a farmer in this world who wishes to leave their children acres of desolation and destruction, but the farmer has no choice. They are the keepers of the nation's soil, and it is their activities that will determine the long-term sustainability of our agricultural base. It should not be expected that a farmer susceptible to flood, famine, fire and drought should be driven to using methods that maintain production, while destroying the soil. Yet they seem to have no choice: who will pay the farmer's mortgage while s/he changes to more 'organic' practices? Who will provide the financial breathing space to allow for the luxury of change? Who can provide the farmer with a viable, productive alternative to the constant use of chemical fertilizer?

True landfill costs
In most urban societies around the world, the cost of landfill is skyrocketing. Yet landfill fees only cover a small part of the cost of landfill. The true costs of landfill when all burial, amenity, administration, security, replacement and on-costs are included is, in most cases, at least three times the cost charged at the gate. Even in small unattended country landfills, when all costs are included the price per ton is often around AU\$50 to AU\$70. If these funds were redirected, this money could be used for the processing of our organic materials into a compost suitable or *even specifically designed for* farm use. In most instances the cost of this process would be far less than the current cost of landfill. It would have the additional benefits of reducing the fertilizer bill for local farmers, increasing the organic levels in the soil, raising the microbial density of the soil and at the same time, the ability to produce quality products. Once organic material is removed and used in this way, all other products in our waste streams become available for reuse.

The 1 per cent of hazardous waste in any stream of material could be removed before it becomes a problem; the remaining inert wastes and packaging could be used in local industrial processes or transported to national markets when sufficient material become available. Other changes in the packaging industry could bring even larger benefits to the Landcare/Waste Management nexus. A leading example is that of the new plasma technology developed in Brazil[26] to separate the aluminium and plastic layers of long-life cartons, having attracted international investment of over US6.5 million dollars, and currently processing 32 000 tons of packaging a year.

Packaging for community profit
As the humble brown paper bag taught us many years ago, packaging does not need to be complex in its make-up to be effective, and indeed even when it is complex, it does not need to be antisocial. It can be designed to be recycled to paper or recycled to compost depending on its clean or contaminated state. Combined with the major corporate commitments to the development of safe biodegradable plastics, this will see us growing increasing quantities of our compostable or recyclable packaging within the coming years.

The recent work of VISY Closed Loop, as seen at the Sydney Olympics, demonstrated that it is possible to make safe, functional packaging out of materials that can be cleanly composted back to a soil-enhancing product. A product that is then capable of going back to the agriculture, which, in turn, is growing the materials to make more packaging. This type of product, having carried its packaged contents to a rural centre can then be composted for the benefit of the local community: to create local jobs, to regenerate local soils and to grow local crops.

The last few years have seen significant corporate commitments and investment in bio-materials. That is, materials that can be readily composted or biorecycled following their use in the community. Massive opportunities exist for VISY to move forward from the gains at the Sydney Olympics and begin to design packaging for organic reuse across the board. Eastman, BASF, Mitsubishi, Cargill Dow, Toyota, ADM and Dupont have all recognized the need to change and other companies such as Ford, 3M, Daimler Chrysler, Proctor & Gamble, Fujitsu, NTT and Sony have either made or expanded their commitment to the adoption and use of biodegradable plastics in their product lines. There is a nexus between waste management, soil management and landfill. The nexus is that part of the solution to both problems has a common source: the soil. We can have pollution, desertification, contamination and waste, or we can have employment, good food, clean air and health. The cost will be about the same.

Construction and demolition materials recycling
Alexis Nelson[27]

As we said at the beginning of this chapter, even construction materials and cement can be recycled into useful products. Alex Fraser is one of the leading companies in Asia doing this.[28] When they pioneered recycling concrete rubble in Australia nearly 20 years ago, the company wanted to save what it considered was excellent fill from going to a landfill site. At that time, solid waste materials were one of the main contributors to shrinking landfill sites. The company has grown from the first project involving 50 000 tons of concrete, brick and masonry, to recycling and selling over seven million cubic metres of construction and demolition (CD) material in 2003. Material is processed into a range of high-quality, specification road and civil construction recycled products that match and in many cases, out-perform virgin quarried materials. Even the arguably most technically precise roadway in Australia, the Formula One Grand Prix Track, is substantially built from material that was once considered waste.

While these days we take recycling for granted, in the mid-1980s there was a high degree of scepticism about the quality and performance of such materials. Alex Fraser invested millions of dollars (and hours) testing procedures in its new recycling endeavours and worked closely with scientific authorities such as the CSIRO and the end users to develop economically viable, high-quality, relevant, recycled products that fully satisfy market demands. Its recycling centres, technology and products have set a benchmark for many overseas commercial business and government authority delegations in recent years, and several countries have invited them to share this knowledge by establishing recycling technology in overseas markets such as Singapore.

Construction and demolition materials recycling is one of the fastest-growing sectors of the recycling market according to the US Environmental Protection Agency (EPA). There are many benefits to reducing CD waste, both cost-saving and environmental. Benefits include avoiding disposal costs and generating revenue from material sales. Waste from landfills is diverted and valuable materials can be reused. Recycling also helps contractors and building owners comply with increasingly strict local and state regulations. In support of the green building movement in the US, the Green Building Council's Leadership in Energy and Environmental Design (LEED)[29] building rating system and the National Institute of Standards and Technology's Building for Environment and Economic Sustainability (BEES)[30] and Seattle's Built Green organization[31] promote the use of recycled-content materials in building construction and provide resources for achieving sustainability in building and construction.[32]

Recycling and reusing CD materials greatly reduces disposal costs. In *Natural Capitalism* (Hawken et al., 1999), case studies highlight the financial opportunities of recycling buildings. In 1992 the National Audubon Society recycled a century-old, 98 000 square foot building at a cost approximately 27 per cent below that of building anew. In 1996 the City of San Diego retrofitted a 73 000 square foot municipal office building using low or no toxic materials, high recycled-content materials, as well as recycling over 40 tons of construction debris. This efficient, commercial structure yielded a 60 per cent saving in energy costs and a four-year payback. Construction companies also stand to realize a profit from recycling and reuse of materials. While building the Rose Garden arena in Portland, Oregon, Turner Construction transformed estimated disposal costs of $190 000 into income by rerouting 45 000 tons of concrete, steel, gypsum, paper and other construction waste to recyclers (Hawken et al., 1999).

The European Council of Civil Engineers (ECCE) was created in 1985 to assist Europe advance its built environment and protect the natural environment, and advises governments and institutions in formulating standards and regulations for the construction industry. The ECCE believes that the Landfill Directive of Europe will place pressures on the construction industry to minimize and reuse construction by advocating designing with whole-life costs in mind to minimize waste; specifying and using reclaimed or inert materials in construction; using techniques that avoid creating waste; reusing materials on-site for other purposes or finding profitable uses off-site; and disposal of inert waste on-site.[33] Beginning in July of 2006, the Landfill Directive will progressively limit the amount of biodegradable municipal waste that can be disposed in landfills in member states throughout Europe over the next two decades. Member states are required to set up national strategies for the implementation of the reduction of biodegradable municipal wastes ending up in landfills.[34]

Markets for CD recycled materials are growing rapidly as business, government and industry realize the benefits of minimizing and ultimately eliminating waste streams from processing products and by buying recycled-content products. State governments in the United States have developed extensive information on purchasing recycled-content products. The Department of Environmental Protection of the Massachusetts State government is committed to attaining an 88 per cent reduction in CD waste by 2010 and provides market resources on its website.[35] The West Virginia State government developed an Internet-based materials exchange programme providing businesses, industry, state and local government and the general public with resources and contacts in recyclable materials markets.[36]

Non-profit organizations like WasteCap Wisconsin provide state-wide resources and support for turning waste into resources. WasteCap Wisconsin is leading the way in developing new CD recycling markets by identifying the most abundant and harmful components of construction materials. Drywall is the second largest component of the commercial construction waste stream. A case study of the Alliant Energy Company exemplifies the possibilities for identifying and developing recycling markets. During the construction of the company's 325 000 square foot headquarters in Madison, WI, Alliant Energy fully committed to waste reduction and recycling of construction materials, specifically drywall. In collaboration with Green Valley Disposal and Wisconsin's Department of Natural Resources (DNR), Alliant Energy conducted a study to investigate the viability of applying crushed drywall as fertilizer on farmland soil. Residential drywall has been used in the past in place of commercial gypsum fertilizer providing essential nutrients to farmland soil. Construction drywall (Type X) contains toxins that may have inhibited its safe application to soil but Alliant Energy conducted research that exposed earthworms to crushed construction drywall. Based on results showing no harmful effects on the earthworms, the Department of Natural Resources approved the first land application of Type X drywall in the state of Wisconsin, showing the potential for drywall diversion in new recycling markets.[37]

Conclusion

All of the above examples demonstrate that what appears to be impossible can be resolved and waste streams can be transformed into value streams. The most important thing is to change the way of thinking about 'waste' and to eliminate it from the production process or fully exploit its properties and benefits.

Notes

1. Chris Page is a member of the Integrated Design Practice team and project leader for RMI's educational initiatives. For more information on Chris Page please see http://www.rmi.org/images/other/StaffBios/BioCPage.pdf.
2. TNEP Secretariat members Karlson 'Charlie' Hargroves, Michael Smith, Cheryl Paten and Nick Palousis. For more information see http://www.naturaledgeproject.net/Secretariat.aspx.
3. Computer Takeback Campaign, http://www.computertakeback.com/the_problem/index.cfm.
4. European Environment Agency, http://europa.eu.int/comm/environment/waste/weee_index.htm.
5. Greenbiz News.com, http://www.greenbiz.com/news/news_third.cfm?NewsID=28397.
6. California State Government, http://www.ciwmb.ca.gov/BuyRecycled/StateAgency/.
7. Caroline Plunkett is a Judges Associate at the ACT Supreme Court (material adapted from Plunkett, 2003).
8. Article 1(1).
9. Article 1(1).
10. The law consists of 11 chapters including General Provisions, Basic Policy and specification of resources-saving/reutilizing industries, products and by-products. Chapter 11 is dedicated to penalty provisions.
11. Article 1.
12. Article 2.9.
13. Article 2.8.
14. TNEP Secretariat members Karlson 'Charlie' Hargroves, Michael Smith, Cheryl Paten and Nick Palousis. For more information see http://www.naturaledgeproject.net/Secretariat.aspx.
15. http://www.worldwatch.org/alerts/pr980402.html; http://www.worldwatch.org/pubs/paper/117.html. Likewise the Amazon, as reported by *Science* is in serious peril, http://www.abc.net.au/science/news/enviro/EnviroRepublish_235510.htm.
16. Numerous reports in highly reputable journals such as *Science* suggest that the destruction of forests is occurring disturbingly rapidly: http://www.abc.net.au/science/news/enviro/EnviroRepublish_289408.htm.
17. Australia for instance has had a $2 billion trade deficit in paper and timber products for many years.
18. Francis Grey Consulting Economist At Large and Associates for the Department of the Environment, Sport and Territories (1996).
19. Productivity Commission (2001).
20. http://www.abc.net.au/science/news/enviro/EnviroRepublish_257520.htm; http://www.tnpl.net/.
21. TNEP Secretariat members Karlson 'Charlie' Hargroves, Michael Smith, Cheryl Paten and Nick Palousis. For more information see http://www.naturaledgeproject.net/Secretariat.aspx.
22. http://www.abc.net.au/science/news/stories/s257s20.htm.
23. http://www.turpa.org/about/the_tura_program/about_the_toxics_use_reduction_act.html.
24. Gerry Gillespie is Chair, Zero Waste Australia. He has published booklets on waste reduction for local councils in NSW in the 1980s, was involved in the development of ACT communities 'No Waste by 2010' strategy and was Founding Manager Zero Waste New Zealand Trust 1997–98. To contact Gerry Gillespie please email to: gerry.gillespie@resource.nsw.gov.au.

25. http://www.act.gov.au/nowaste/nextstep.pdf.
26. Developed by the Institute of Technological Research at the University of Sao Paulo and now supported by companies such as Alcoa, Tetra Pak, Klabin and TSL Ambiental, as reported in Kyocera Mita (2005, p. 43).
27. Alexis Nelson is a member of The Natural Edge Projects Working Group. Alexis is currently pursuing an undergraduate degree in Environmental Studies from the University of Colorado. Alexis is focused on aiding society by assisting the mitigation of conflicts between science and policy that slow our progress towards a sustainable world. She has completed an internship at the Natural Resources Law Center researching the effectiveness of water rights language on protecting wilderness areas in Colorado and Arizona. For more information on Alexis see: http://www.naturaledgeproject.net/WorkingGroup. aspx#AlexisNelson).
28. http://www.alexfraser.com.au/news.htm.
29. http://www.usgbc.org/DisplayPage.aspx?CategoryID=19.
30. BEES Version 3.0, http://www.bfrl.nist.gov/oae/software/bees.html.
31. http://www.builtgreen.net/studies.html.
32. Jobs through recycling construction materials: EPA, http://www.epa.gov/epaoswer/non-hw/recycle/jtr/comm/construc.htm.
33. European Council for Civil Engineers, http://www.eccenet.org/Papers/waste.html (paper regarding a EU construction waste proposal).
34. European Environment Agency report on biodegradable waste management in Europe, http://reports.eea.eu.int/topic_report_2001_15/en/tab_content_RLR.
35. Massachussetts State Government, http://www.mass.gov/dep/recycle/cdhome.htm.
36. West Virginia State Government, http://www.state.wv.us/swmb/RMDPS.htm.
37. WasteCap Wisconsin Leading New C&D Recycling Markets, http://www.wastecapwi.org/drywall.htm; http://www.wastecapwi.org/drywall/LeadingNewCDRecyclingMarkets.doc.

References

American Chamber of Commerce in Japan (ACCJ) (1998), 'Business and the Environment: Can Foreign Companies Profit in Japan?', *American Chamber of Commerce in Japan Journal*.
Bleischwitz, R. (2002), *Governance of Eco-efficiency in Japan: An Institutional Approach*, Wuppertal Institute for Climate, Environment and Energy, Germany.
Environment Agency, Japan (2000), *The Challenge to Establish the Recycling-based Society*, Environment Agency, http://www.eic.or.jp/eanet/, accessed 19 July 2003.
Epstein, S. (2000), 'An Epidemic of Cancer Deception: The Establishment, Why We Can't Trust Them', *The Sun*, March, http://www.healthy-communications.com/prepedemic3-02-02.html, accessed 5 December 2005.
Fowler, L. (2003), 'From Technical Fix to Regulatory Mix: Japan's New Environmental Law', *Pacific Rim Law and Policy Journal*, **12**, 441–66.
Francis Grey Consulting Economist At Large and Associates for the Department of the Environment, Sport and Territories (1996), 'Estimating Values for Australia's Native Forests', Environmental Economics Research Paper No. 4, http://www.deh.gov.au/aboutpublications/economics/estimating/index.html.
Gibbons, P. and Lindenmayer, D. (2002), *Tree Hollows and Wildlife Conservation in Australia*, Melbourne: CSIRO Publishing.
Gillespie, G. (2002), 'Waste Management or Soil Management', 3rd Shed a Tier Congress, Canberra, 22 March, http://www.beyondfederation.org.au/Gerry_Gillespie_paper_delivered_27_Dec_2001.html, accessed 5 December 2005.
Hargroves, K. and Smith, M. (2005), *The Natural Advantage of Nations: Business Opportunities, Innovation and Governance in the 21st Century*, London: Earthscan.
Hawken, P., Lovins, A. and Lovins, L.H. (1999), *Natural Capitalism: Creating the Next Industrial Revolution*, London: Earthscan.
Kyocera Mita (2005), 'Kyocera Mita 2002', *Waste Management and Environment Journal*, **16**(5) and also 'Business e-Waste Survey Results September 2002', www.kyoceramita.com.au/files/1/2002%20e-Waste%20Survey%20Results.pdf, accessed 5 December 2005.
Lovins, A. (2001), *Natural Capitalism*, http://www.abc.net.au/science/slab/natcap/default.htm, accessed 5 December 2005.
Mackey, B., Lindenmayer, D., Gill, M., McCarthy, M. and Lindesay, J. (eds) (2002), *Wildlife, Fire and Future Climate: A Forest Ecosystem Analysis*, Melbourne: CSIRO Publishing.
McKean, M. (1981), *Environmental Protest and Citizen Politics in Japan*, Berkeley: University of California Press.
Ministry of International Trade and Industry (2001), *Policy Information: Law for Recycling of Specified Kinds of Home Appliances*, Ministry of International Trade and Industry, www.meti.go.jp/english/information/data/cReHAppre.html, accessed 28 September 2003.
Ministry of the Environment, Japan (2003), *Japan's Environment at a Glance: Wastes, Ministry of the Environment*, http://www.env.go.jp/en/jeg/waste/wastes.html, accessed 25 September 2003.

Nakano, T. (1986), 'Environmental policies in Japan', in Park, C. (ed.), *Environmental Policies: An International Review*, London: Croom Helm, pp. 259–92.

Organisation for Economic Co-operation and Development (OECD) (2001), *Extended Producer Responsibility: A Guidance Manual for Governments*, Paris: OECD.

Plunkett, C. (2003), 'Eco-enlightenment? How Japanese Laws will Create a "Recycling-based Society"', Research Paper submitted to the Faculty of Law, Canberra: Australian National University.

Productivity Commission (2001), 'Competitive Neutrality in Forestry', http://www.pc.gov.au/agcnco/reports/researchforestry/forestry.pdf.

Read, M.P. and Sturgess, N.S. (1992), *Evaluation of the Economic Values of Wood and Water for the Thomson Catchment*, Prepared for Melbourne Water and the Department of Conservation and Environment, Read Sturgess and Associates (unpublished).

PART VI

ENVIRONMENTAL TECHNOLOGY MANAGEMENT AND THE FUTURE

34 Environmental technology management: insights from holistic science
Stephan Harding

Introduction

There is now less and less doubt that Western industrial culture is precipitating the massive and seemingly overwhelming global ecological and social crises that we face in our times. Climate change, widespread species extinctions, social disruption, economic instability, increasing violence and psychological and physical ill health loom ever larger in our world (Morris, 1995; Wilson, 2002; Worldwatch Institute, 2002) and yet despite the deployment of ever more sophisticated technologies such as those used in environmental management, we have to date been incapable of making much significant progress in stemming the rising tide of degradation and destruction. It seems that the science and technology we have become used to in our culture are inadequate to the task of solving the many problems we now face.

So how are we to contribute to solving these crises as managers of environmental technologies? If Einstein was right in saying that it is not possible to solve a problem from within the worldview that created it, we can only address this question by understanding and re-appraising the worldview, or 'guiding metaphor', which has led our culture to be so destructive of nature. The guiding metaphor in mainstream western culture as a whole, and in western science in particular, simply put, is that the entire cosmos, with all its stars, solar systems, planets and earthly living beings, is no more than a vast, dead machine consisting of nothing more than a collection of essentially isolated and unchanging objects in interaction, much like colliding billiard balls. Because these objects are 'dead', they are deemed to have *instrumental* value, but only when they are put to work in the service of the human enterprise; the view that all entities are imbued with *intrinsic* value because of their innate existence is antithetical to mechanistic thought. This view of world-as-mechanism has given our culture licence to exploit nature with complete ruthlessness and apparent impunity for the last 400 years (Merchant, 1989). Is it any wonder then that we have unleashed ecological havoc on a scale that has not been seen on Earth since the last great extinction 65 million years ago? Our challenge today, both as a culture and as environmental technology managers, is to understand how it was that the mechanistic perspective took hold and then to search for a worldview more in tune with the development of an ecologically sustainable science and technology.

The scientific revolution

So how did the mechanistic worldview become prominent in our culture? Most historians of science suggest that it became predominant in sixteenth- and seventeenth-century Europe in the wake of war, plague and famine in which millions of people had perished. During the sixteenth century, the Church, which had fragmented during the Reformation with the bloodiest of consequences, seemed increasingly incapable of providing reliable, certain knowledge or indeed any means for controlling a wild and wilful nature or the tempestuous violence of the human psyche. The vast epistemological vacuum left by the disappearance of the old religious certainties created an urgent need for a new way of knowing, which was filled by what we now know as modern science. Its earliest proponents, amongst them Galileo, Bacon and Descartes, were convinced that the new approach to knowledge must be based on rational thought and empiricism rather than on blind faith in established religious dogmas (Toulmin, 1990).

Galileo (1564–1626) believed that reliable knowledge could be gained only by narrowing the focus of inquiry down to quantifiable aspects of the world, which were later called 'primary qualities' by Locke, the English philosopher. Subjective experiences, which had formed so large a part of the old pre-mechanistic epistemology, were considered by Locke to be merely 'secondary qualities' because they had proven unreliable, fallible and prone to gross errors of judgement.

Francis Bacon (1561–1626) believed that it was legitimate to bind and constrain nature using mechanical inventions in service of systematic scientific investigation, so that nature could be 'tortured' into revealing her secrets. His aim in using this method was to extend human dominion over the entire physical universe. The image of nature as machine was revealed to Rene Descartes (1596–1650) in November 1619 at Nenberg on the banks of the Danube in a vision that led him to describe the Earth and indeed the whole visible universe as nothing more than vast machines, the '*res extensa*', a realm utterly separate from human consciousness and reason, the '*res cogitans*'. Descartes took his mechanistic views so seriously that he taught his students to consider the screams of vivisected animals to be no more than the creakings of mere machines, a view that has become so deeply entrenched in our culture that scientists to this day are taught to disregard their own feelings of distress when carrying out such experiments. The work of the great English scientist Isaac Newton (1642–1727) seemed to validate the newly emerging mechanistic worldview. His equations stunned his contemporaries with their ability to precisely predict the trajectories of moving bodies, confirming once and for all that the world was indeed a vast machine that could be predicted and controlled through the use of quantification, reductionism and objective, systematic experimentation.

These pioneers of modern science had broken through into intellectual territory that was largely uncharted at the time and which was extraordinarily useful, productive and exciting. Quite apart from its undoubted intellectual benefits, there was also a powerfully liberating social dimension to this new way of knowing, for in science it was no longer necessary to accept truths handed down by authorities such as the Church or the Pope. Science was, and still is, open (at least in theory) to contributions from anyone, and it was possible for any scientist to challenge and overturn an accepted paradigm if they could convince their peers of the validity of their evidence (Kuhn, 1962). Given these considerable intellectual and social benefits, it is not surprising that scientific materialism swept through the western world like intellectual wildfire, gathering momentum as more and more phenomena in nature fell under its sway. Looking back from the vantage point of the present moment there is no doubt that mechanistic science and technology have provided many benefits, without which our lives would be greatly impoverished, both practically and intellectually. However, it is now becoming more and more apparent that the impulse to control nature has become highly dangerous and inappropriate, coupled as it is with the overbearing technological might of our species.

Clearly the mechanistic metaphor has had its day and must now be replaced, but with what? If seeing the cosmos as a machine has led us into a tremendous crisis, perhaps the alternative is to re-consider the ancient understanding of the cosmos as *organism*, as living process. In this perspective, in the words of Thomas Berry (1999), the cosmos is understood to be not a collection of objects, but rather a 'communion of subjects'. The radical insight here is that subjectivity is not unique to humans; it occurs in other organisms and, according to most non-western cultures and even to the pre-scientific West, in so-called inanimate entities such as landscapes, mountains and rivers. Because in the organismic view all subjects are by definition imbued with subjectivity, we recognize their intrinsic value, irrespective of their usefulness to humans (Naess, 1986; Sessions, 1995). According to the ancient wisdom that underlies the organismic view, each subject has the right to fully develop its unique inner potential as far as possible, so if we are to live in ecologically sustainable ways we must re-discover the humble art of participation in the biotic communities in which we are embedded, not as their controllers, masters or stewards, but simply as 'plain members' within them (Leopold, 1949). In the fields of science and technology this new participatory approach is being developed in the emerging discipline of holistic science.

Holistic science

Holistic scientists recognize that a genuinely participatory science and technology require us to engage with the world as whole human beings, not as detached, uninvolved observers (Goodwin, 1994). So how can we be whole as practitioners of science and technology? I have developed an understanding of both conventional and holistic science that is based on the work of the psychologist C.G. Jung, who proposed that wholeness depends upon the conscious integration of four key 'psychological functions' common to all humanity, namely: thinking, feeling, sensation and intuition (Jung, 1971). Jung arranged these as two pairs of opposites as follows:

thinking

intuition   sensation

feeling

Sensation tells us that a thing exists by means of conscious sensory perception; thinking tells us what it is by means of logical thought; feeling tells us whether it is right or wrong and helps to ascribe value to it; and intuition tells us what its deeper meaning is, as Jung says, 'by way of unconscious contents and connections'. Thinking interprets, feeling evaluates, whilst sensation and intuition are purely perceptive in that they make us aware of what is happening without interpretation or evaluation. Jung observed that each person has an innate conscious orientation towards one of the four functions, whilst the opposite function remains largely unconscious and undeveloped. The other two functions are only partially conscious and generally serve the dominant function as auxiliaries. Wholeness for Jung requires the conscious development of the undeveloped function, together with an awareness of the four functions in oneself so as to achieve a well-rounded personality. As Jung acknowledged, his classification corresponds to ancient systems, which also recognized the existence of four psychological predispositions, such as Ancient Greek medicine, which thought in terms of phlegmatic, sanguine, choleric and melancholic personalities. Correspondences with the systems of other cultures, such as that of the native North Americans or with the Mandala systems of India and the Himalayan Buddhists, strongly suggest that this four-fold system has universal, cross-cultural psychological validity.

Applying Jung's typology in the way I have suggested leads us to the insight that mainstream science has based itself principally on the deliberate cultivation of the thinking function, which is dominant not only in science and technology but in the culture as a whole. Feeling, the evaluative, ethical function, is left out of science and is its inferior, undeveloped function. In conventional science sensation and intuition serve thinking as auxiliary functions. Sensation, the raw perception of colours, sounds, tastes, touches and smells, is obviously essential for doing science since without it the world can make no impact on our minds. Unable to deny the importance of the senses, the conditioned response of conventional science is to convert raw sensory experience into numbers or abstractions as quickly as possible, often using sophisticated scientific instruments to gather numerical information about phenomena rather than use the naked senses directly. This mode of sensing marginalizes phenomena and is heavily influenced by the intellectual mind, which inhibits the possibility of the perception of depth and intrinsic value in the thing being studied.

Thinking, the function most valued by our culture, is used in conventional science to devise experiments and to construct mathematically coherent theories and accounts of how the world works for prediction and control. The predominant style of thinking used in conventional science is 'reductionism', by means of which, as Descartes taught, the scientist attempts to gain complete understanding and mastery of a phenomenon by breaking it down into its component parts. Once the behaviour of the parts is known, this style of thinking tells us that it should in principle be quite straightforward to predict the behaviour of the whole since it is assumed that it will consist of no more than the summed interactions of the individual components, which are unconditioned by the whole to which they belong. The reductionist approach is built on a broader set of important assumptions, such as that objects matter more than relationships, that the world is hierarchically ordered with physics as its foundation, that it is possible for us to gain objective knowledge, and that nature is inherently linear and predictable in that small disturbances give rise to small effects and vice versa.

Reductionism works very well if we want to design technologies like cars and computers, but it is not so successful in areas such as biology, ecology or in the realm of human social life where complex, non-linear interactions are the norm. In these areas a different style of thinking has developed within holistic science, which builds on and incorporates the undoubted intellectual advantages of reductionism whilst moving beyond it. Physicist Fritjof Capra points out that this new approach, now commonly referred to as complexity theory, involves shifting the emphasis from objects to processes and relationships, from hierarchies to networks, and from objective knowledge to contextual knowledge (Capra, 1997).

What does this mean? A key insight in complexity theory is that we can understand a great deal more about a system if we focus on the patterns of relationship between the parts rather than on the parts themselves as isolated entities. By so doing we come to realize that the properties of the parts depend on how they relate to each other and to the larger whole that they help to constitute. We also come to understand that there are no fundamental building blocks such as subatomic particles, at the base of a hierarchical ordering of nature, but that nature self-organizes into multi-levelled sets of networks within networks – such as cells within tissues within organs within organisms within ecosystems within the Earth system – and that no one level is fundamental. Finally we realize, as did Werner Heisenberg, one of the great physicists of the early part of the last century, that we cannot observe nature as she is in herself independent of human observers, for nature reveals her different aspect depending on how we frame our scientific experiments. Hence we cannot be objective observers, for our knowledge is utterly dependent on how we have chosen to interact with the world.

When we focus on relationships between parts rather than on parts in isolation we very quickly encounter surprising properties and 'qualities' at the level of the whole system, which cannot be understood through an understanding of the parts alone. This is the principle of 'emergence' (Goodwin, 1994; Reason and Goodwin, 1999). A common adage that expresses this insight is that the 'whole is greater than the sum of the parts'. Emergence is widespread in nature. Good examples abound in the realm of social insects where interactions between individual bees, ants or termites, each obeying simple rules for when to engage in activities such as searching for food, tending the brood or building, give rise to complex behaviours at the level of the colony as a whole that defy reductionist explanation. Individual ants of the genus *Leptothorax* are active or inactive according to a technically chaotic schedule, but when they interact together at the right density the colony throbs with collective rhythmic activity (Goodwin, 1994). The complex, baroque labyrinths that constitute the interior of termite mounds are built by individual termites each depositing little dollops of mud saturated with a pheromone or chemical signal, which gradually evaporates away into the surrounding air. Termites are attracted to the pheromone and deposit new dollops of mud wherever they encounter it at high concentrations. Computer models of this process show that an initially random pattern of mud dollops soon turns into an emergent, regular array of pillars and columns that look remarkably similar to the insides of real termite mounds. None of the ants or termites had the 'blueprint' for the right behaviour or mound structure – they emerged out of the interactions and relationships between the members of the insect social group (Sole and Goodwin, 2000).

Complexity theory moves us away from the notion that it is possible to predict and control nature in anything but very limited ways. This insight has come in part from the study of deterministic chaos that teaches us that it can be difficult to predict the outcome of interactions in even simple non-linear systems because such systems display sensitivity to initial conditions (Stewart, 1990). This means that slightly different starting points diverge exponentially fast, so that after just a short time they can be scattered anywhere within the overall domain, or 'strange attractor' of the system. A common image for this property is the 'butterfly effect'; a lepidopteran flapping its wings in Iowa could in principle lead to a typhoon in Indonesia.

Another important quality of complex systems is that they may well locate themselves on the 'edge of chaos', a state of organization in which inappropriate ordered patterns can be dissolved so that novel patterns of order better suited to new contexts and conditions can emerge (Reason and Goodwin, 1999). Furthermore, systems at the edge of chaos can suddenly change state due to fluctuations in their internal dynamics or because of external disturbances, a condition known as 'criticality'. If we cannot predict the exact nature of emergent properties, or the exact point at which a system at the edge of chaos will reorganize itself, and if small changes can have unforeseeable and potentially dramatic outcomes, we have to accept the possibly uncomfortable conclusion that nature is inherently unpredictable and uncontrollable. Hence complexity theory, which is the style of thinking adopted in holistic science, provides a cognitive justification for the shift from control to participation. In order to participate fully and properly we need to use quantitative methods appropriately but since these provide us with only a limited understanding we need to complement them by developing deep, intuitive sensitivity to the emergent qualities of whole systems. So what is intuition and how can it be developed in the practice of holistic science?

Intuition for philosopher-scientist Henri Bortoft is the direct, non-rational 'simultaneous perception of the whole' (Bortoft, 1996). It is concerned with meaning, with sensitivity to what a thing is in itself, to where it has come from and to where it is going, and often manifests as an 'oceanic feeling' of deep connectedness and belonging. As we noted earlier, for Jung intuition is informed by unconscious psychological processes to which consciousness has no access. Intuition has an important but largely unacknowledged role to play in conventional science, despite examples such as the dream of a snake biting its own tail, which led Kekule to solve the puzzle of the structure of the benzene molecule, and the fact that scientists often speak of getting 'hunches' about how to interpret a set of data.

In cultivating intuition, holistic scientists use a methodology that has been largely attributed to the German poet and scientist Johann Wolfgang von Goethe (1749–1832) but which in fact can be traced back several centuries before him to Pico della Mirandola, Ficino, Paracelsus and before them to the Hermetic tradition. In this method careful attention is paid to the phenomenon being studied through a process of *active looking* without attempting to reduce the experience to quantities or explanations. In active looking (Bortoft, 1996) the mind focuses on sense perception by cultivating the ability to suspend its pre-conceived notions and its habitual responses about the thing being perceived; one makes an effort to notice the specific details of the thing in all its particularity as they appear to the senses. This focus on the unique qualities of the thing before one's gaze enlivens and deepens perception, and allows the phenomenon, as Bortoft says 'to coin itself into thought', and 'to induce itself in the thinking mind as an idea'. One has the intuitive perception of the thing as a presence *within* oneself and not as an object *outside* one's own being; one has therefore developed a sense of deep relatedness to the object and has transformed consciousness into a means for holistic perception, in which we are able to apprehend the intrinsic qualities of things. This methodology develops what Bortoft (1996) calls 'non-informational perception', in contrast to the conventional approach, which stresses perception in the service of information gathering. Non-informational perception is the subtlest mode of perception and brings a sense of wholeness, whilst informational perception, the coarsest mode, brings only alienation and separation.

This kind of intuitive work is best practised by a group of practitioners, for consensus is as important in this approach as it is in quantitative science. Through the process of 'inter-subjective consensus', in which the intuitive findings of a group are carefully collated and examined, it becomes possible to discriminate between genuine perceptions of intrinsic qualities and what could be merely idiosyncratic fantasy and projection.

A similar approach has been used with great success by the animal welfare scientist Francoise Wemelsfelder who has found that subjective assessments of the mental and physical states of farm pigs are highly correlated amongst participants and are a very good indicator of the overall health of the animals (Wemelsfelder et al., 2000).

The intuitive, holistic perception of wholeness naturally connects us with Jung's feeling function; that is, with the domain of ethics. Ethics, simply put, is the ability to decide whether a thing is right or wrong, whether it is good or not. Conventional science ignores ethics, leaving it to society to decide how to use the fruits of scientific research in the world at large. However, in holistic science we realize that perceptions of wholeness are inseparable from a deep sensitivity to the intrinsic value in the being or entity we are interacting with, making it very difficult for us to do anything that will harm or disturb the 'inner necessity and truth' of that being. For holistic scientists, most (and possibly all) forms of genetic engineering are ethically unacceptable because intrinsic natures are violated when alien genes are transferred from one being to another. These perceptions of intrinsic value have practical consequences – because of them holistic scientists feel obliged to become actively involved in public debates that concern their areas of research.

Implications for environmental technology management

We have seen how holistic science is about reuniting fact and value in ways that enable our culture to explore new possibilities of living sustainably with the Earth. What would environmental technology management be like if it were to be practised in accordance with this key principle of holistic science? We can explore this question with reference to Jung's four psychological functions, beginning with thinking, our dominant function.

Environmental technology managers inspired by holistic thinking, as manifested in complexity theory, realize that it is not possible to fully predict the outcomes of a mechanistic intervention because natural systems are sensitive to initial conditions and often produce unexpected emergent behaviours. Thus issues of scale should be carefully considered by managers of environmental technologies. Even though as we have seen, small perturbations can have large effects in complex systems, large-scale projects are more likely to create unexpected emergent perturbations at a variety of spatial and temporal scales, which could lead to viscous positive feedbacks. Furthermore, subsequent attempts to implement remedial technologies could further exacerbate these unwanted emergent dynamics. Examples come from the field of Earth Systems Engineering (Allaby, 1999) whose supporters promote the use of large-scale environmental technologies for 'fixing' the crisis of climate change. Advocates of one such project suggest that we should increase the biological drawdown of atmospheric carbon dioxide by fertilizing the tropical oceans with iron in order to stimulate phytoplankton growth, but the effects of such massive inputs of this highly bioactive element on the complex system of the ocean are in principle impossible to predict (see discussion in Allaby, 1999). Other Earth Systems engineers are promoting the pumping of carbon dioxide into deep underground caverns or into the deep ocean (Herzog and Drake, 1996), an option that has recently become all the more viable due to the invention of an energy-efficient mineral membrane carbon dioxide filter, which can be attached to fossil fuel burning power stations (Cui et al., 2003). Such a large-scale intervention assumes that there will be virtually no leak of the sequestered carbon dioxide to the atmosphere, but once again, given that these so-called carbon stores are themselves part of the complex dynamical system of the Earth's crust, this cannot be predicted with complete certainty and indeed is highly unlikely. Furthermore, since carbon dioxide in the deep ocean has a residence time of about 1000 years, the implementation of this environmental technology will merely postpone the effects of the captured carbon dioxide on the atmosphere by ten centuries – a terrible legacy for future inhabitants of the Earth. Yet another proposal is to control the outflow from the Mediterranean to the Atlantic in order to modify oceanic and atmospheric circulation in the high northern latitudes, but once again, the effects of such a large-scale intervention on the ocean and the Earth as complex systems are very difficult to foresee (Johnson, 1997).

There are many more such gargantuan environmental technologies now under serious consideration. A particularly disturbing example is the effort to 'replumb the planet' by sequestering and redirecting water through canals and tunnels from some of the world's largest rivers to distant drought-stricken areas (Pearce, 2003). Proposals are now afoot to redirect China's Yangtze river northwards to Beijing and to channel waters from rivers in the foothills of the Indian Himalayas towards the arid south of the country. Australia is planning to move water from rivers in the water-rich north to the drier south, and in the United States plans are being mooted for diverting water from rivers in the Canadian Arctic to provide water for Los Angeles and Las Vegas. Similar projects are now being seriously considered for Spain and Africa. Environmentalists and holistic scientists are issuing warnings that such large-scale environmental technologies will create social and ecological havoc. In India the new canals and reservoirs will displace two million people and waters in the south will be contaminated with polluted water from northern industrial rivers. In Central Asia the Aral sea was converted into a salty wasteland by the diversion of two of its major inflowing rivers and in Australia ecologists warn that additional irrigation will serve only to further increase the salinization of agricultural land. Water transfers will shunt species from one river system to another, triggering ecosystem disruption and loss of biodiversity. About two-thirds of the transported water will evaporate en route and any water eventually arriving at its destination will be far too expensive for farmers.

Complexity theory and holistic science suggest that the alternative to such large-scale environmental technologies is to encourage the involvement of local people in the implementation of small-scale environmental technologies, as suggested by the economist E.F. Schumacher (1974). This approach is less likely to destabilize local ecosystems and local communities because contextual knowledge that has evolved as a result of extended interaction between people and their local environments is more likely to be sensitive to subtleties that are opaque to outside 'experts' schooled in the conventional interventionist approach. As a case in point, it is well known that simple, local solutions are readily available for solving the water crises outlined above. Better irrigation, particularly drip irrigation from perforated hoses and porous clay pots, traditional water harvesting ponds, plugging leaks in city water mains, using waste water to irrigate crops and water capture from roofs can all be implemented locally and would solve the problem without requiring a single huge water project (Pearce, 2003). Such solutions underline a key application of holistic science and complexity theory to the field of environmental technology management, namely that 'people's technologies' can in many cases provide simple solutions to complex problems. If foreign 'experts' are needed at all, their role should be to encourage the emergence of such technologies from the grassroots through listening to and implementing local traditional wisdom, through empowering local communities and by providing simple materials if they are needed.

Guided by the notion of participation with nature, managers of environmental technologies would, wherever possible, implement solutions that mimic nature's own ways of problem-solving. This approach, known as 'biomimicry', is defined by Benyus (1997) as 'the conscious emulation of life's genius', or 'innovation inspired by nature'. An excellent example of this approach is the 'living machine' concept developed by John and Nancy Jack Todd (1993). A living machine is an assemblage of species from various trophic levels placed together in a series of interconnected tanks with water flowing between them. Each tank has a different set of species, often collected from local ecosystems, which interact and together provide the service of water purification in much the same way as a natural wetland does. The Todds have found that living machines clean wastewater more effectively than conventional chemically-based, energy-hungry systems. Living machines are useful when the task at hand needs to be carried out indoors but when water needs to be purified outside it is often preferable to use a 'constructed wetland'. In this approach tanks and containers are not needed; rather, water-purifying organisms such as rushes and sedges are planted directly in the ground and the whole assemblage, coupled as it is to local ecosystems, provides effective water purification services as well as enhancing native biodiversity by providing habitats for wildlife species.

A holistic analysis of how environmental technology management can contribute to solving the crisis of ecology is incomplete without a careful consideration of the crucial but often ignored question of economic growth. A commonly held view is that we will be able to achieve ecological sustainability if we devise environmental technologies that make industry more efficient in its use of raw materials and in its ability to reduce the output of pollutants (Weizsacker et al., 1997). But several economists (Daly, 1978; Meadows et al., 1992) have pointed out that in a growing economy even the impressive efficiency gains that are undoubtedly made possible by ever more ingenious environmental technologies will eventually be overwhelmed and counteracted by the economy's demands for fresh raw materials from undisturbed nature. The answer is to develop what Herman Daly (1978) has called a 'steady state economy' in which scientifically agreed fixed stocks of raw materials are recycled through the economy. The absolute quantity of goods must then reach a limit determined by the extent to which these fixed amounts of materials can be shared out amongst a set of increasingly 'dematerialized' artefacts (Weizsacker et al., 1997). This means that we will have to accept limits to our possession of material goods. Efficiency gains are what I call the 'soft problem of sustainability', for our immense ingenuity as originators and managers of environmental technologies will see to it that efficiencies are constantly improved upon. The 'hard problem' is how to create a viable economy in which the throughput of materials is at an ecologically sustainable steady state, with no new materials extracted from nature. The problem lies in designing financial incentives consistent with steady state, given that the starting point must necessarily be our current, hugely destructive growth-centred economy. How is it possible to wean ourselves off our addiction to growth? Daly suggests that three major measures are necessary: humane coercion-free population control, a free market system that allocates resources once steady state levels have been set by scientific consensus and measures to ensure a far more even wealth distribution both locally and globally than is currently the case (Daly, 1978). That these conclusions are clearly consistent with the four principles of The Natural Step (Waage, 2003, and Chapter 13, this volume), perhaps one of the best set of operating rules for an ecologically viable industrial economy drafted to date, is heartening, although the connection between The Natural Step and steady state economics needs to be much more extensively contemplated in mainstream circles.

What about sensing and intuition in the practice of environmental technology management? Their use implies that managers of environmental technologies can no longer be detached 'experts', applying standardized knowledge to whatever problem they are trying to solve. Groups of practitioners, working consensually, would need to spend a great deal of time interacting closely with the situation they have been called to participate in before any remedial actions are taken. There would be time to allow the situation to live itself into the imagination through the medium of the senses, so that the practitioners become so closely identified with the problem at hand that it is given a chance to co-create a solution with their unconscious minds. This is no different from how good doctors work – they watch their patients carefully, picking up on subtle cues such as skin colour, body posture and tone of voice, and as a result often have intuitive insights into the underlying causes of the physical symptoms. Much time needs to be spent developing a 'conversation' with the situation by means of careful note-taking, talking to local people, drawing, walking, visualizing and, if appropriate, sleeping out in the field, but without at this stage being overly influenced by the promptings of the thinking mind. The aim is to clarify perception to such an extent that a genuine connection is forged between situation and practitioners so that pre-conceived notions involving standard 'textbook' technologies are not allowed to obscure any culturally and ecologically sustainable solutions that might otherwise emerge from the local context. In this way intuition is given the chance to suggest how to best deploy whatever technologies might be appropriate.

The holistic approach we have explored here suggests that it is no longer enough to focus merely on the delivery of technical excellence in solving the problems that confront us as managers of environmental technologies. We urgently need to experience the world as a whole, not only cognitively as a fully interactive system consisting of life, its abiotic environment and the human domain, but also as the source of deep feelings of connectedness that inspire us to engage in right action (Harding, 1997, 2006). We must widen and deepen our understanding of both the terrible crisis we have unleashed and of our deepest capacities for overcoming it, for the urgency of the task calls us to transform ourselves inwardly even as we transform the outer situation. Holistic science, which is the marriage of thinking, feeling, sensing and intuition in the service of healing our split with nature and within ourselves, is vital if we are to engage with the massive problems we now face with the full spectrum of our capabilities. Anything less will surely lead us further into the multiple catastrophes that our culture, in its unconsciousness, is so intent on creating.

References

Allaby, B. (1999), 'Earth System Engineering: The Role of Industrial Ecology in an Engineered World', *Journal of Industrial Ecology*, **2**(3), 73–93.
Benyus, J.M. (1997), *Biomimicry: Innovation Inspired by Nature*, New York: Morrow.
Berry, T. (1999), *The Great Work*, New York: Bell Tower.
Bortoft, H. (1996), *The Wholeness of Nature*, Edinburgh: Floris Books.
Capra, F. (1997), *The Web of Life*, London: Harper Collins.
Cui, Y., Kita, H. and Okamoto, K. (2003), 'Preparation and Gas Separation Properties of Zeolite T Membrane', *Chemical Communications*, 2154–5.
Daly, H.E. (1978), *Steady-state Economics: Economics of Biophysical Equilibrium and Moral Growth*, San Francisco: W.H. Freeman.
Goodwin, B. (1994), *How the Leopard Changed its Spots*, London: Weidenfield and Nicholson.
Harding, S.P. (1997) 'What is Deep Ecology?', *Resurgence*, **185**, 14–17.

Harding, S.P. (2006), *Animate Earth*, Totnes: Green Books.

Herzog, H.J. and Drake, E.M. (1996), 'Carbon Dioxide Recovery and Disposal from Large Energy Systems', *Annual Review of Energy and the Environment*, **21**, 145–66.

Johnson, R.G. (1997), 'Climate Control Requires a Dam at the Strait of Gibraltar', *Eos*, **78**(277), 280–81.

Jung, C.G. (1971), *The Collected Works, Volume 6*, London: Routledge and Kegan Paul, pp. 330–408.

Kuhn, T.S. (1962), *The Structure of Scientific Revolutions*, Chicago: University of Chicago Press.

Leopold, A. (1949), *A Sand County Almanac*, Oxford: Oxford University Press.

Meadows, D.H., Meadows, D.L. and Randers, J. (1992), *Beyond the Limits, Confronting Global Collapse, Envisioning a Sustainable Future*, Post Mills, VT: Chelsea Green.

Merchant, C. (1989), *The Death of Nature. Women, Ecology and the Scientific Revolution*, San Francisco: Harper.

Morris, W.M. (1995), 'Earth's Peeling Veneer of Life', *Nature*, **373**, 25.

Naess, A. (1986), 'The Deep Ecology Movement, Some Philsophical Aspects', *Philosophical Inquiry*, **8**, 1–2.

Pearce, F. (2003), 'Replumbing the Planet', *New Scientist*, **7**, June, 30–34.

Reason, P. and Goodwin, B. (1999), 'Towards a Science of Qualities in Organizations', *Concepts and Transformations*, **4**(3), 281–317.

Schumacher, E.F. (1974), *Small is Beautiful*, London: Sphere Books.

Sessions, G. (ed.) (1995), *Deep Ecology for the 21st Century*, Boston: Shambhala.

Sole, R. and Goodwin, B. (2000), *Signs of Life, How Complexity Pervades Biology*, New York: Basic Books.

Stewart, I. (1990), *Does God Play Dice?*, London: Penguin Books.

Todd, N.J. and Todd, J. (1993), *From Eco-cities to Living Machines*, Berkeley, California: North Atlantic Books.

Toulmin, S. (1990), *Cosmopolis: The Hidden Agenda of Modernity*, Chicago: University of Chicago Press.

Waage, S. (2003), *Ants, Galileo and Gandhi: Designing the Future of Business through Nature, Genius and Compassion*, Sheffield: Greenleaf Publishing.

Weizsacker, E., Lovins, A.B. and Lovins, L.H. (1997), *Factor Four. Doubling Wealth, Halving Resource Use*, London: Earthscan.

Wemelsfelder, F., Hunter, E.A., Mendl, M.T. and Lawrence, A.B. (2000), 'The Spontaneous Qualitative Assessment of Behavioural Expressions in Pigs: First Explorations of a Novel Methodology for Integrative Animal Welfare Measurement', *Applied Animal Behaviour Science*, **67**, 193–215.

Wilson, E.O. (2002), *The Future of Life*, New York: Little Brown.

Worldwatch Institute (2002), *Vital Signs*, New York and London: W.W. Norton & Company.

35 The value loop – a new framework for business thinking
Hardin Tibbs

Introduction

The survival of our natural environment in its present form, and the existing type of human society within it, is not assured. One of the main direct causes of this environmental and cultural risk, or unsustainability, is the pattern and scale of the worldwide industrial economy. As the industrial economy approaches the same physical scale as the biosphere, a new organizing pattern for industry is urgently becoming needed, one in which a new form of physical organization will be matched and supported by a new organizing template for business activity. This chapter expands on the concept of the 'value loop' as a potential solution.

In the present mode of industrial production the pattern of physical flows is putting ever-increasing pressure on nature. Raw materials are extracted from the environment (the biosphere and the Earth's crust), processed to create products and economic value and the waste is eventually dumped back into the natural environment. The assumption is that industry can act like a suction tube, pulling materials from nature, using them briefly to create economic value and then pumping the residue back into nature as waste and pollution. This is workable if industry is small and nature is vast, but as industry continues to grow rapidly and approaches the same scale as nature itself, it quickly becomes a major problem. For one thing, if the flow of materials being sucked from nature begins to be as large as the stocks and flows in nature itself, then the supply of materials will not last for very long. For another, if the flow of pollution is too large in comparison to the waste-processing capacity of nature, nature is likely to be overwhelmed.

The driving force behind this flow of materials is business activity around the world. Business operates by creating products and services it can sell and most of these economic transactions involve an exchange or flow of material. In the case of products, the material forms the product itself, and services usually need energy – which itself involves a flow of materials – and often depend on some exchange or consumption of materials.

Business activity therefore forms a flow of transactions that parallel the flow of materials through the industrial economy. Primary industry is responsible for extracting resources from the environment, manufacturing industry turns it into products, which the retail industry then sells to consumers. This flow of transactions is known in business as the 'value chain' – a concept first articulated by Michael Porter in 1985 in the *Harvard Business Review* (Porter, 1985).[1]

Each business is concerned about the viability of its place in the chain, and uses whatever materials and energy it needs to create value. Each one is concerned that it can earn marginally more money from the products and services it sells than the amount it must spend to buy the resources it needs. This is how businesses survive. Few if any businesses are concerned about the survival of the chain as a whole. Each one assumes that there will be another organization or activity upstream and downstream of its own operations that will provide the resources it needs and deal with the ultimate fate of the materials it sells or discards. However, upstream and downstream of the whole chain there is nature and nature's survival in its current form cannot be assured if the flow through the chain grows too big.

The whole industrial economy is already very large compared with nature, and is growing very rapidly in terms of physical flows – it is doubling in size every 20 years or so. Very soon now we will need to redesign the whole thing if we want our present way of life to survive. The good news is that there is a way of doing this. The essence of the redesign will be to shift from operating industry as a straight-through chain of activities, to operating it as a circular flow. The ends of the chain can be linked together to form a loop, which could then be known as a 'value loop', in place of the existing 'value chain'.

In the value loop, businesses connect to each other and to their customers in circles of transactions with no breaks, and the materials they need flow between them continuously in a sustained cycle. Instead of being put under increasing pressure at each end of the chain, nature can now be kept largely independent of the loop.

Linear flow is unsustainable

As described, the existing industrial economy is based on a straight-through, or source to sink, flow of materials. For convenience, the straight-through flow will be referred to as 'linear' in this chapter. This linear flow cannot be sustained for two reasons: scale and rate of growth. If the scale of industry was small compared with nature – meaning the biosphere – then its activities could be sustained without undue difficulty. Particularly if it was not only small but also growing slowly, then the linear flow of materials could be sustained for a long period.

Industry is not small however. Even compared with the vast size of the global biosphere, the industrial economy is now very large. Several measures can be used to show this. One example is the production rate of toxic heavy metals. Another is the annual rate of consumption of biomass.

The toxic heavy metals are elements that occur naturally in the Earth's crust and include lead, copper and arsenic. Because of natural processes such as the weathering of rock, a certain quantity of these elements does enter the biosphere naturally each year and is safely absorbed. Human activity, principally mining, has greatly increased the rate at which these elements are released into the biosphere. Lead, for example, a nerve poison, is still used in a number of countries as an anti-knock additive in petrol (gasoline), which accounts for its high rate of release into the atmosphere. The global industrial flow of most of the toxic heavy metals is now significantly greater than the natural background flows. This means that for these elements, which are particularly important because they are poisonous, industry is now, quite literally, larger than nature.

A similar but better-known example is the release of carbon dioxide into the atmosphere from burning fossil fuels. There is a natural flow of carbon dioxide into the atmosphere from the respiration of living things, including plants, animals and humans. This natural flow is very large since respiration is a basic activity of most living things, and carbon is one of the major nutrient cycles in the biosphere – it is one of the so-called 'grand cycle' elements that include nitrogen and sulphur. Every year roughly 200 billion tonnes of carbon is cycled through the atmosphere naturally – released to it and absorbed from it by the biosphere.

By the 1990s the burning of fossil fuel and deforestation were releasing about 30 billion tonnes of carbon dioxide, containing eight billion tonnes of carbon, into the atmosphere every year. The total anthropogenic or human-caused flow of carbon is equal to about a fifth of the natural background flow, and is increasing the total atmospheric 'reservoir' by some 4 billion tonnes a year (only half of the excess carbon is being successfully absorbed by the biosphere and ocean every year). The effect of this accumulation is directly measurable. The concentration of carbon dioxide in the atmosphere has increased from an estimated pre-industrial level of 280 parts per million (ppm) in 1750, to 315 ppm in 1958 and 381 ppm in 2005. So in the case of carbon dioxide, industry is still significantly smaller than nature, yet this flow has already reached a scale that is thought to be triggering changes in the world's climate.

Another 'greenhouse gas', methane, is released in far smaller amounts naturally than carbon dioxide but has a more potent influence on global temperature. The human-caused flow is about 375 megatonnes per year, whereas the natural flow is only about 160 megatonnes per year. So in the case of this gas, the human-caused flow is already 2.3 times larger than the natural flow (Houghton, 1997, p. 35).

The annual consumption rate of biomass provides another way of understanding the scale of the industrial economy. In 1986 a study was conducted at Stanford University to estimate how much of the total annual biological growth, or increase in biomass of the biosphere as a whole, was being consumed by the human economy. The research team calculated that in 1986, when the world's population was five billion, the human consumption of food and biomass resources had reached 40 per cent of the entire annual land-based product of photosynthesis (Vitousek et al., 1986). The world's population has now grown to over six billion people, and since the human consumption of biomass reflects this growth plus increasing wealth, the percentage would now be significantly more. So in terms of the basic productivity of the whole biosphere, the industrial economy is now in the order of half the size of nature and growing rapidly. This growing percentage requires very large-scale clearance of wild habitat for agriculture and forestry and it is unlikely that the natural ecosystem can tolerate usage rates much beyond 60 or 70 per cent.

A similar measure of relative size is provided by the 'world ecological footprint' that is calculated annually by WWF International. The ecological footprint compares human consumption of renewable natural resources with the ability of the biosphere to regenerate them. The WWF analysis compares the estimated area needed to provide for human consumption renewably with the size of the most biologically productive part of the Earth's surface. On this basis, human consumption of renewable natural resources grew by 80 per cent between 1961 and 1999, to a level 20 per cent above the Earth's biological capacity. About 11.4 billion hectares, just under a quarter of the Earth's surface, is the most biologically productive. This area corresponded to an average of 1.9 hectares per person in 1999, but the level of human consumption in 1999 would have required the renewable productivity of 13.7 billion hectares, or 2.3 hectares per person on average.

Human consumption therefore overshot the biosphere's renewable productive capacity by 20 per cent. This over-consumption is being achieved by depleting the natural capital base – an example would be the one-time clearing of forest instead of renewably harvesting from an ongoing forest, which would require a larger area to produce the same amount of timber but the renewable yield could be maintained over time instead of being taken as a one-off harvest. In contrast, consumption or draw-down of the natural capital base removes the source of further renewable production, making it dangerously unsustainable. So on the basis of WWF calculations, the human economy as a whole is already 20 per cent larger than nature can support on an ongoing basis. The WWF expresses this in numbers of planets, pointing out that we would now need more than 1.2 planets to sustain the current rate of human consumption (WWF, 2002).

These measures all tell the same basic story: that the size of the human economy is now close to or already is the same size as nature. This would be problematic enough if the size of the human economy was static, but it is not: it is growing rapidly.

The human economy is growing in size because the human population is growing in numbers and simultaneously becoming more affluent. The growth to today's population of over six billion was sudden, recent and historically unprecedented.

For thousands of years of pre-history, the human population remained at a few million people worldwide, growing on average at not more than 0.05 per cent a year and not rising above 100 million until after 1000 BC. The increase was not steady during this period – there were alternating surges and contractions – and there was no consistent exponential growth towards present levels as is often assumed. From AD 1 to 1750 the average growth rate was about 0.07 per cent a year. The historically recent and unprecedented change is that from 1750 to 1950 the average growth rate jumped to about 0.7 per cent a year, and between 1950 and 1960 it jumped again to an average of 1.88 per cent, reaching a peak rate of about 2.06 per cent around 1965. By the late 1990s it had fallen back to less than 1.5 per cent – but this is still 20 to 30 times higher than the pre-industrial baseline (Cohen, 1995).

In terms of absolute numbers the world population is still growing extremely fast, even though the underlying growth rate is now declining. More people were *added* to the world population in the 1990s than existed in the entire world in 1750. The world population has doubled during the lifetime of anyone now over 40 – a unique occurrence in human history – and if present rates of growth were maintained, the total population would double again within the next 40 to 50 years. If the growth rate continues to slow as it has since 1965, the population will still increase by about another three billion people before it stabilizes.

As rapid as this population increase is, the flow of materials through the industrial economy is growing twice as quickly. At its current rate of growth, materials consumption will double in 20 years, compared with the current doubling time of about 50 years for population. In broad terms this is because material affluence is rising twice as fast as the number of people. The consumption of materials in the United States has ballooned from 140 million metric tonnes a year in 1900 to 2.8 billion metric tonnes a year in 1990, up from about 1.6 tonnes a person to 10.6 tonnes a person. Environmental pollution has been rising in step with this, because in a straight-line-flow economy the flow of waste and pollution is proportional to industrial throughput. For example, the amount of carbon dioxide released into the atmosphere every year has doubled twice over since 1950, in line with the 20-year doubling time.

The Stanford University calculation described earlier, estimating the human consumption of annual biological growth, provides a way of thinking about what this rate of increase means for the biosphere. If we assume the take of biomass is increasing in line with the growth in the throughput of all materials, then the 40 per cent number calculated by the Stanford team in 1986 will have doubled by 2006. This means it would reach 80 per cent by 2006 and 100 per cent by 2013, which would mean no wild biosphere remaining.

These numbers are only speculative approximations but they do show that the problem might reach crisis point in the very near future, not in many decades time.

Why does this problem seem to be rushing at us so suddenly? The growth of materials consumption is exponential, which means it doubles with every tick of the time interval that corresponds to its growth rate, in this case every 20 years. Like a rocket taking off, at first this appears very slow and then suddenly accelerates into an extremely fast increase. Exponential growth in a finite space – like the surface of a planet – is highly deceptive psychologically as it appears to be reassuringly slow at the outset but later accelerates suddenly to use up remaining space or resources with startling speed. This is particularly dangerous for political and other decision-making as it is notoriously difficult to take exponential change seriously until it is almost too late.

There is a French riddle for children, quoted by Donella Meadows in *Beyond the Limits*, that illustrates the problem exponential growth poses for decision-makers:

> Suppose you own a pond on which a water lily is growing. The lily doubles in size each day. If the plant were allowed to grow unchecked, it would completely cover the pond in 30 days, choking off all other forms of life in the water. For a long time the lily plant seems small, so you decide not to worry about it until it covers half the pond. On what day will that be? (Meadows, 1992, p. 18)

The answer is surprising on first encounter: on the 29th day. On the 30th day the lily doubles for the last time, taking the pond from half-full to full. What is also worth noting is just how small the lily is for most of the month – as late as the 25th day it still covers only 1/32 of the pond, yet five days later the pond is full.

What might happen when human economic consumption reaches very high levels in terms of worldwide biological capacity? Are there likely to be any obvious warning signs if we are approaching danger point? The answer lies outside our experience because we have never pushed a whole planet to the limit before. As Professor Edward O. Wilson at Harvard University succinctly puts it: 'One planet, one experiment' (Wilson, 1992, p. 182). However, we do know what happens in the case of certain individual ecosystems.

Ecological systems are capable of absorbing very considerable amounts of stress over extended periods without showing apparent harm but while this is happening their capacity for resilience is gradually being depleted. The moment comes when a critical threshold is suddenly crossed, followed by very rapid and 'unexpected' breakdown in some aspect of the ecological system – a condition of acute unsustainability.

The recent history of Big Moose Lake in the Adirondack Mountains illustrates this kind of abrupt ecological breakdown (Stigliani and Salomons, 1993). From about 1880 onwards, huge amounts of sulphur were deposited as acid rain in the watershed of the lake, from coal burnt several hundred miles upwind in the Ohio River valley. The burden of sulphur climbed steeply until 1920, when it stabilized at around 3.5 million tonnes a year. Yet it was not for another 30 years after this that the acidity of the lake water showed any change. The acidity of the water had held steady for 200 years – as long as it had been measured – but between 1950 and 1980 it suddenly increased tenfold, from a pH of 5.5 to 4.5, killing all the fish in the lake.

The explanation for the delayed but sudden onset of this environmental breakdown turned out to be that the soils in the watershed of the lake had provided an enormous buffering capacity that was able to neutralize the acid rain for six decades. Finally when the buffering capacity was exhausted, the acidity of the lake abruptly registered a change that had in fact been initiated 60 years earlier.

If ecosystems responded to stress by showing continuous gradual degradation we might be justified in hoping for a similar pattern of response at the global level that could at least provide some degree of warning. Unfortunately many ecosystems don't respond like that – they absorb significant environmental impacts with no sign of degradation and then suddenly collapse without warning. The uncomfortable question that confronts us is whether the global ecosystem will show similar behaviour and the implication is that there could be abrupt ecosystem breakdown at the planetary level.

This possibility is explored in the book *Beyond the Limits* (Meadows, 1992), the 1992 update and revision of the controversial *The Limits to Growth* (Meadows et al., 1972). It describes a series of runs on a computer model originally developed at the Massachusetts Institute of Technology (MIT) called 'World3', which show in broad terms how the global system might react in the years ahead, based on a variety of different assumptions about resources and responses. When the World3 computer model was run to reflect continuing 'business as usual' economic activity, the result was a general collapse early in the twenty-first century. This is acute unsustainability with a vengeance – the equivalent of a fatal heart attack. As the book says (Meadows, 1992, p. 130):

> On a local scale, overshoot and collapse can be seen in the processes of desertification, mineral or groundwater depletion, poisoning of soils or forests by long-lived toxic wastes. Legions of failed civilizations, abandoned farms, busted boomtowns, and abandoned, toxic industrial lands testify to the 'reality' of this system behavior. On a global scale, overshoot and collapse could mean the breakdown of the great supporting cycles of nature that regulate climate, purify air and water, regenerate biomass, preserve biodiversity, and turn wastes into nutrients. Twenty years ago few people would have thought ecological collapse on that scale possible. Now it is the topic of scientific meetings and international negotiations.

Our worldwide situation today bears an uncanny resemblance to the situation in late medieval England, when the population had soared from roughly two million to around six million over a 200-year period, reaching the limits of environmental and social organization of the time, triggering livestock disease and famines. Then the climate changed, with the onset of the 'Little Ice Age', and the Black Death struck soon after. The 'Great Dying' of 1348/49 alone killed 40 per cent of the English population in a two-year period, and the plague returned again in the 1350s and the 1370s. By the end of the fourteenth century the English population had collapsed to only 2.5 million from its peak of six million in 1300. It took 450 years for the population to recover from the calamitous experience of the fourteenth century. Today's graph of rapid demographic growth over a 200-year period looks eerily similar to the graph of English population up to 1300, except that it reaches a level exactly three orders of magnitude higher, six billion instead of six million, and encompasses the whole globe (Heinzen, unpublished).[2]

Continued exponential growth of the global industrial economy based on a source-to-sink flow of materials is genuinely unsustainable and can only be maintained for another few years at best. All the evidence about the relative scale and rate of growth of the economy in fact points towards the very real risk of an abrupt environmental discontinuity in the near future. For this reason a new pattern of organization for the industrial system is now urgently needed.

Cyclic flow is sustainable

The alternative to a linear flow of materials is a 'cyclic' flow, in which materials continuously flow in circular loops and are reprocessed for reuse each time they reach the end of a period of use. A circular flow of materials has two main advantages over a linear flow, particularly when the linear flow reaches large scale.

One is that it avoids exhausting a finite environmental stock of materials, since it draws on the stock only once, to fill the loop, and then keeps circulating and reusing the material. In this way it ensures an indefinite supply of materials for the industrial economy.

The other advantage is that a cyclic flow avoids the pressures a linear flow exerts on the environment. At the start of the chain, extraction of materials often causes damage, for example, habitat and species loss, collateral pollution and changes to geological structures and hydrology. Similarly, the release of waste at the end of the linear flow puts a heavy and growing burden on the environment's limited capacity to act as a sink and processor for waste and pollution. The circular flow, or loop, creates a supply of materials that is independent of nature once it has been created and so does not exert pressure on nature.

A cyclic flow thus assures sustainability in two senses: the sustainability of nature and the sustainability of a society able to use a high level of technology. The creation of a physical 'cyclic loop' of materials is therefore one of the key aspects of securing sustainability at the material or physical level.

The 'loop' is the overall form of the pattern seen at the scale of the whole system but at a finer level of detail the cyclic flow would form a complex network or web of interlocking flows. This is similar to the way the overall circulation of the blood is evident at the scale of the whole body but seen through a microscope at the level of the capillaries it looks like a meshwork of small flows in a variety of directions.

At the most basic level the materials in the loop are the full set of elements we wish to use for practical purposes, such as copper, iron and silicon. In the case of minerals, it is easy to see that once they have been mined and refined, the resulting purified elements can then continue to be reprocessed and reused without requiring more mining. For example, enough gold has probably already been mined to meet all human uses indefinitely if it were to be put into circulation instead of being stockpiled in banks. In the case of organic compounds and structural materials from biological sources, the picture is more complicated, since the reprocessing and remanufacturing part of the loop may actively depend on nature for the tasks such as biodegradation and agricultural production. The scale of this production and the degree to which it may compete with food production then becomes a determining issue.

In contemplating an evolutionary transition of our industrial economy from a linear to a cyclic flow of materials we are following in nature's footsteps. The biosphere itself operates as a cyclic system, endlessly circulating and transforming materials and managing to run almost entirely on solar energy. The evolution of this global cyclic pattern from its early non-cyclic origins – about 2.5 thousand million years ago when oxygen first began to accumulate in the atmosphere – mirrors the transition that our global industrial materials processing system now needs to make.

The British scientist James Lovelock was one of the first to recognize that the web of life on Earth creates a set of flows and feedback loops that resemble the physiology of a regular organism – an inter-relationship he termed Gaia (Lovelock, 1979), the name of the Ancient Greek goddess of the Earth. Lovelock pioneered the study of planetary physiology, or geophysiology as he calls it. Geophysiology can be understood as including geological processes, climatic and hydrological cycles, and ecology, all of which are responsible for materials flow in the natural environment. The overall aim of creating a circular flow of materials for industry is to accommodate the global industrial system – or more generally the human use of advanced technology – within the physiology of the whole planet on an ongoing basis. Put another way, this means creating a global technological infrastructure able to harmonize with the unique biogeochemical processes and cycles of this planet.

As we learn more about the intricately interlocked workings of geophysiology, the detailed implications for industry are becoming clearer. Our own bodies provide an analogy for what needs to be achieved. We have within us biochemical processes that not only serve our own life, but also enable the biogeochemical processes of Gaia. In a very literal sense, we are a functional part of the planet. Industry needs to be structured the same way – to serve human needs as well as planetary needs. Industry must become a cooperative (or at the very least not a harmful) part of the planet, of the life of Gaia.

The cyclic loop: the physical design
From a practical point of view, the cyclic loop has two distinct aspects that need to be considered. One is the physical design of the loop itself. The other is the idea of a circular flow considered as the basis of creating value. It is important first to understand the main aspects of the physical design as this sets key parameters for value creation.

The first question of physical design relates to the choice of materials that the loop will contain. There are two reasons why a limited set of materials might be chosen – recycling volume and stability.

Recycling materials is more economic if there are large amounts of material available for reprocessing and one way to achieve this is to focus on a relatively small number of materials. At present this is simply not an objective for product engineers. For example, one study in Australia showed that the components of a typical domestic washing machine contained in the order of 200 different materials. Yet it would almost certainly be possible to design all the components of a fully functional washing machine using only a dozen or so different materials. This has not been done up to now because there has been no motivation for engineers to focus on reducing the materials count of manufactured products.

The choice of materials could also affect the stability or resilience of the industrial facility that makes use of them. At the scale of an industrial production zone, the cyclic flow would be likely to take the form of an 'industrial ecosystem'. Industrial ecosystems are complex food webs between companies and industries that optimize the use of materials and make use of byproducts that would otherwise become wastes. A complex of industrial producers applying these principles has been referred to as an 'eco-industrial park'.

If an eco-industrial park is optimized to reduce materials use and waste, what happens if or when one of its component processes becomes obsolete, perhaps because the market for that firm's product has declined? Does the whole industrial ecosystem then become unviable? Or can structures of interlock be designed that allow for change, such as cleaner future technology?

One way of addressing these recycling and stability concerns would be to select a suite of elements and materials that offers an adequate range of physical properties for design engineers to use, allows production processes to interlock and provides ease of reprocessing. This suite of materials would ideally also be bio-compatible and eco-compatible (in other words, in the chemical form used, not toxic to organisms or ecosystems) and abundant in nature. The objective would be a fairly small set of acceptable materials, with long-term geo-physiological compatibility that could be used to provide say 80 per cent or more of all production needs. A first step towards identifying these materials might be to start by considering the 16 or so elements that are the most abundant in the biosphere, as these constitute both the large-scale cycles in nature and comprise the living molecules such as enzymes that both process the flows and form the physical basis of life itself.

The next step might be to devise clusters of production processes that use some or all these materials and which can be interlocked ecosystem-style. Focusing production activity on a small number of 'basic materials' would enhance resilience, because if one firm in an eco-industrial cluster failed, it would be more likely that another that used the same materials could be found to take its place. The resulting industrial clusters or industrial ecosystems would then stand a reasonable chance of being stable over time. The Zero Emissions Research and Initiatives (ZERI) at the United Nations University in Japan has shown that focused industrial clusters of this sort that aim for zero waste can make very good business, social and environmental sense.

The second physical design question relates to the amount of material the loop would contain. Once the flow of materials in the cyclic loop is established, does the volume of materials flowing in the loop need to be maintained at a given level, or can it be allowed to vary or grow?

At present the demand for industrial goods is rising exponentially. If all materials flowed in a cyclic loop and demand continued to grow exponentially, additional new materials would need to be added to the loop continuously to meet the growing demand. Under these circumstances the demand for virgin materials would be equal to the margin by which current demand for new materials exceeded the return flow of recycled material from previous demand. Even if demand was not growing when a new loop was being created, the demand for virgin materials would continue during the period when the loop was being primed or filled with materials, before the materials in use began to return for reuse.

Being able to avoid the use of virgin non-renewable resources would mean either that demand has ceased to grow, or if not that the materials intensity of economic output – the physical mass consumed per dollar of economic value created – is decreasing fast enough to offset the growth. In principle either of these would allow the amount of material in a cyclic loop to remain stable over time.

The phenomenon of declining physical mass per unit of value created is referred to as dematerialization, sometimes dubbed 'doing more with less'. As it happens, this kind of decline in the materials and energy intensity of industrial production is an existing 'mega-trend' in industrially developed economies, and in fact constitutes one of the long-run drivers of fundamental economic growth. Both materials and energy use (measured as quantity per constant dollar of Gross National Product or GNP) have been falling since the 1970s (Larson et al., 1986). This is because the market for basic products has saturated in developed economies, while the weight and size of many other products has fallen. For instance, the progressive miniaturization of information technology allows more product features to be added as product bulk is reduced. However, in spite of the dematerialization trend, the absolute growth in demand is still outstripping the rate of dematerialization, which is why the total amount of materials used in the global industrial economy is still growing rapidly.

Assuming that the global population increases by up to 50 per cent and also becomes more affluent during the time a global industrial cyclic loop is established (say during the next 20 to 40 years), then reducing or just holding steady the amount of material in the loop will require accelerated dematerialization. The rate of dematerialization would have to equal or exceed the absolute rate of growth in demand, which is the product of population times per capita consumption. If the rate of dematerialization exceeded the rate of demand growth, this would represent a complete decoupling of materials flow rates from economic growth rates, allowing goods to be provided equitably for a growing population from a fixed or even diminishing flow of materials in the cyclic loop.

To achieve this level of dematerialization will require not only deliberate efforts by design engineers to reduce the mass of materials in use but also significant technological advances. Fortunately emerging materials technologies such as nanotechnology – which is beginning to enable assembly of materials atom by atom – promise to accelerate the dematerialization phenomenon by vastly improving the performance-to-weight ratio of materials.

This means that continued exponential growth in demand could potentially be offset by a dramatic dematerialization of the useful products created, to the point where the volume of materials in the loop could at least be kept stable. Assuming that this could be done, would this stabilization then allow a cyclic loop flow of materials to provide all the physical materials the economy would require without harm to the environment?

The answer depends on several further factors. One is the level of leaks from the loop. The cyclic flow of materials, like anything flowing in a large-scale engineered system, would inevitably suffer from unintentional losses – the failure to fully recapture all the material for recycling – which would then require replacement virgin material. The level of these losses could be reduced either by progressively improving the completeness of recycling or by reducing the volume of material flowing in the loop.

Another class of 'leaks' arise from design, not accident. At present the way many materials are used results in them being 'dissipated' or dispersed into the environment as they are used, with no hope of recovery for recycling. This problem can be overcome by designing differently. For instance, as car brake pads wear down they leave behind a finely dispersed toxic powder. In electric or hybrid electric cars, such as the Toyota Prius, this can be avoided by generating electricity as a means of braking. By having the wheels turn the electric motors during braking, the energy of forward motion of the car is converted back into electricity that can be returned to the battery or captured by a rapid storage device such as an ultracapacitor. This process, known as regenerative braking, causes the car to slow down without involving heat-producing friction or the dissipation of brake pad material in the form of toxic dust.

In general, if a particular usage of a material results in dissipation, then as far as possible that use should be minimized by a different (least damaging) choice of materials or by a design reconfiguration. This however, implies a broad assumption that all dissipative uses are negative. There are some counter-instances where dissipation is positive. For example, if zinc is added to mineral-deficient soil this involves the dispersion of zinc into the biosphere, but in this case there is a beneficial outcome. However, even this may run the risk of overdosing the soil, in which case monitoring is required, and the dissipative use needs to be curtailed at the point when it starts to have negative environmental or health impacts. Similarly, there may not necessarily be a problem releasing carbon dioxide to the atmosphere if emissions are in balance with absorption elsewhere. This is after all what nature does: dispersion of carbon dioxide into the atmosphere is the biosphere's means of moving carbon from sources to sinks. So the general principle is not as simple as 'zero emissions' but instead that the cyclic flow of materials should be maximized and any non-cyclic dispersion of materials into the biosphere should be consistent with zero biospheric harm (which includes human health). An extreme example of violating this principle would be the use of depleted uranium in armour-penetrating munitions, as on impact the uranium burns to a fine radioactive powder that disperses into the environment.

Energy supply is another factor determining the level of materials that can flow in the loop. The energy requirement for keeping materials moving around the loop can be significantly less than for producing new materials. To recover copper from scrap takes as little as 15 per cent of the energy requirement for smelting copper from ore, and atmospheric emissions of carbon dioxide and sulphur dioxide are also substantially reduced. Even in a loop, materials do need to be transported and reprocessed to keep them useful as they move around the loop, and this requires energy. For a given process or activity, the amount of energy needed will depend on the volume of materials in the loop that need processing.

If the type of energy in use produces a dissipative mass flow (such as carbon dioxide from fossil fuels), then the materials flowing in the loop would need to be reduced over time to reduce the dissipative flow from the energy consumption. For example, if the energy is provided by fossil fuels that cause an unbalanced flow of carbon into the atmosphere, this flow could be reduced if the demand in the loop for materials processing was reduced. So the type of energy used in the future for processing materials as they flow around the loop will determine whether the volume of the cyclic flow needs to be progressively reduced over time. If future energy technology moves towards low-carbon, low-waste sources, and if leaks from the loop can be minimized, then in principle simply stabilizing the volume of materials flowing in the loop would protect the environment.

As it happens, energy technology has been evolving in this direction over time. Energy sources have been 'decarbonizing' – their carbon content has been falling – for more than 150 years. As countries successively industrialized they moved from high- to low-carbon fuel sources, starting with firewood, then going to coal, to oil and recently to the lowest carbon so far, natural gas (methane) (Ausubel, 1989, p. 83). A completely carbon-free energy supply could in principle be provided using pure hydrogen gas produced from renewable energy sources, if the technical and economic challenges can be met. In one possible scenario this could lead to the creation of a future 'hydrogen-electric economy'. This would involve hydrogen gas (the lightest element), used as a clean, low-mass carbon-free carrier and store of renewable energy. Fuel cells could convert the hydrogen into electricity at the point of use, meshing with the increasing use of electricity-based technologies for efficiency, precision and convenience in final energy use.

In fact, a hydrogen-electric economy would itself form part of the overall cyclic loop since the hydrogen would flow in a continuous loop through the technosphere and back via the biosphere. The hydrogen would be generated from water using clean electricity from a wind farm or solar plant, transported by pipeline or tanker to the point of energy use, and used to generate electricity by fuel cells that release almost pure water vapour into the atmosphere as exhaust. From there the water (hydrogen plus oxygen) would move through the natural water cycle, returning to the wind farm or solar plant as rain or surface water.

A further point related to energy for materials processing needs to be noted. Energy is required for materials processing when it involves the application of heat or pressure for a physical transformation, as in the melting of metal scrap for recasting or the movement of materials from place to place. However, when the processing of materials involves a chemical change or reaction, as in the case of processing mineral ore into metal, the situation is usually more complicated. In this case the energy carrier is often not simply supplying energy but also some mass that takes part in the reaction. For example, if the ore is iron oxide and the energy carrier is carbon, the chemical energy potential of the carbon (known as exergy) drives a chemical reaction that produces pure iron with carbon dioxide as a waste product. In cases of primary processing of minerals like this where the fuel supplied takes part in a process reaction, the laws of thermodynamics indicate that some level of waste may be inevitable (Baumgärtner and de Swaan Arons, 2003). However, in dealing with a cyclic flow of say, iron, after the primary processing of ore has been completed, it should be possible to continue cycling the refined material by incorporating it into an energy and materials 'industrial ecosystem' that has no net waste.

A further factor influencing both the level of leaks and the amount of energy required to run the cyclic loop is the velocity of the materials flowing in the loop. If the velocity increases, this will increase the energy requirements for reprocessing as there will be more material moving through the loop in a given time. High velocity would mean rapid capital stock turnover and new generations of products rapidly replacing earlier products, and would be needed for accelerated product dematerialization and improvements in the energy efficiency of products and production equipment. At a time of rapid technological change and high rates of innovation, a high velocity of materials flow through the loop may therefore be desirable as a means of reducing overall energy consumption and materials intensity. This follows the principle that the velocity should only increase where technological development is rapid enough to offset the additional energy consumption for increased materials cycling and reprocessing.

At present, while population growth is still rapid, the marginal energy efficiency of new technology is improving, but absolute energy consumption is increasing. This is partly because total market growth is outstripping the marginal gains in efficiency, and partly because the large potential for efficiency improvement is not being fully deployed in new production and infrastructure. If the full potential for improvement that now exists were to be applied in all new product design the growth in absolute energy consumption could be slowed. Actually reducing absolute energy demand will become more feasible as the world population continues to decelerate. Similarly, absolute materials consumption continues to grow rapidly, but is likely to be reduced in the future by a combination of advancing materials technology, design-driven dematerialization and population deceleration.

This overview of the physical characteristics of the cyclic loop sets the stage for thinking about the way business will need to change. The emergence of the cyclic loop will require a new way of thinking about business models. The existing way of thinking, which involves creating a product concept that can be scaled up to achieve high levels of physical growth, will no longer be the key to success when materials flow in a loop and the volume of cyclic flow is continually falling.

The value loop: creating business models based on the cyclic loop

Business activity around the world is the force driving the existing linear flow of materials through the industrial economy. Each business creates economic value by transforming materials to create goods and services people will buy. The result is a linear progression of materials through the industrial economy. Primary industry extracts materials from the Earth's crust or from the biosphere, then sells them to manufacturing industry, which converts them into products, which are then sold by retailers to consumers, who later discard them. This post-consumer waste, along with the waste generated at each previous stage, generally flows back to the environment.

Businesses are therefore linked by a linear sequence of transactions that Michael Porter described as the 'value chain'. Each business in the chain must have a 'business model' that enables it to earn marginally more money from its offerings than the amount it spends to buy inputs. The inputs embody materials on which work has been done by the preceding business in the chain.

Each business uses whatever materials and energy it needs to create value. The only limit placed on this consumption is its ability to purchase these inputs in the market. If the ultimate sourcing and disposal of materials, at the start and end of the chain causes problems, these are not considered to be the responsibility of the businesses using the materials in the middle of the chain. These environmental or social impacts may result in costs that economists call 'externality costs' (external to market costs) that must often be paid by society as a whole, but in general they are not linked back to the way materials are priced throughout the value chain.

The amount of material that a business in the middle of the chain must buy or sell to create a specific type of economic value is in general continually being reduced by technological advances. In effect these technology-driven gains in productivity represent the whole history of industrial development since the 1750s. However, as has been described, they are not enough on their own to eliminate the pressure on nature at each end of the linear flow, since technological capability also gives rise to these flows in the first place.

If the flow of materials through the economy were to be reorganized into a loop, the corresponding flow of transactions in business would also form a loop. The important question for business is what difference this would make to business activity. The most obvious adaptive challenges would be for businesses at the start and end of the value chain.

Mining companies, for example, would no longer be able to view themselves as a one-way source point for materials in the industrial economy. Their role in the value loop would still be to supply concentrated, purified materials in specified physical forms to manufacturers who will convert these materials into finished goods. Instead of sourcing the materials by digging holes in the ground, they would be 'mining' the scrap flows from urban areas. Future business viability of these 'urban miners' would depend on having well-located plants in or near urban areas, much as today's miners rely on having concessions in highly prospective mineral provinces. This logic already determines the location of steel mini-mills that process scrap steel from old automobiles, as with the 40 hectare Rooty Hill steel mini-mill in the western suburbs of Sydney designed by BHP to achieve very stringent requirements for environmental performance and noise reduction.

The linear flow of material along the value chain progresses from 'raw materials' to 'end products' such as consumer goods. Businesses generally add value by putting materials in progressively more organized states, so increasing their usefulness. In physics, the idea of less or more ordered states of matter is referred to as an increase or decrease in entropy. If no work is done on materials, their condition will gradually deteriorate towards thermodynamic disorder, and this is said to be an increase in entropy (the term is used counterintuitively). At each step in the value chain, energy is expended and work is done on materials, reducing their entropy as they move along the chain. At any point when it is not economically useful to reduce the entropy of materials any further, they become waste, either immediately or after a period of use.

One strategy for reducing the negative environmental impact of materials is to retain materials in a useful low entropy state as long as possible. Examples include extending the useful lifetime of products or finding ways to directly reuse components. This strategy has its limits, and if advancing technology is making new more efficient generations of products possible, extending the useful lifetime may simply prolong, say, high energy use. It can therefore be necessary for materials to fall back to less-ordered, higher entropy states before they can once more be reorganized into new low entropy states. This is the 'return' path of the value loop. The 'forward' half of the loop corresponds to the value chain, in that it involves a reduction in entropy, while the return half must generate economic value from a transient increase in the entropy of materials.

A key challenge of the value loop is to create business models that can generate value as the entropy of materials increases, or even better, to design loops in which the entropy is kept low as the material moves through a sequence of transformations, each one having some intrinsic value. The generic activity involved on the return path is to collect waste material and return it to a point where it can be reprocessed into new 'raw' material. This can be thought of as a sequence of steps, from product take-back, to product demanufacture and materials deprocessing. Some materials may move directly from product demanufacture to reprocessing, or a material deprocessing step may be needed first, depending on the formulation of the material. The demanufactured or deprocessed material will have value to the reprocessor as an input, although it is likely to compete unfavourably with mined raw material unless externality costs are taken into account.

At present, rising externality costs are leading indirectly to increasing waste disposal fees and return path business models are possible based on charging a fee for product take-back that is competitive with waste collection fees. This in effect subsidizes the demanufacturing and deprocessing steps, allowing the new 'raw' material to be competitive in price with mined material. This type of business model depends on legislation that has a narrow focus on regulating waste disposal. If in the future there is legislation that more broadly enables the value loop as a whole, the opportunities can be expected to shift.

It is useful to consider what form this legislation might take. New and proposed legislation in the EU for product take-back indicates that policy-makers are beginning to think in terms of mandating end-of-life product take-back for consumer products, while simultaneously requiring a minimum percentage of recycled content in new products. The EU End of Life Vehicles Directive of 2000 requires that carmakers pay for the cost of taking back and recycling old cars, starting with cars sold in 2001. One of the aims of this directive is to raise the reuse or recycling of the metal content of vehicles to 85 per cent by 2015, while also banning the use of hazardous heavy metals. This is leading car manufacturers such as BMW and Volkswagen to redesign their products to make them easier to demanufacture, and to use materials that require less deprocessing. The carmakers are adopting a cyclic loop business model that focuses on a target for recycled content, and uses redesign of the product to keep down the costs of the return loop, making it economically viable for third-party reprocessors. The benchmark for this viability will not, however, be the comparison with waste disposal costs, but the lowest competitive bid for fulfilling the manufacturer's take-back responsibilities. The manufacturer has an incentive for keeping this cost low, hence the product redesign, and it can be expected that product manufacturers and third-party reprocessors will jointly develop innovative systems for doing this at steadily lower cost.

Another approach is to avoid having the product ever become waste. One way of doing this is to lease the product to its user rather than sell it. Possibly the best-known proponent of this cyclic loop business model is Interface Inc., based in Atlanta, Georgia. Interface manufactures carpets but sells the services carpets provide, for example, replacing carpet tiles when necessary as part of the service contract. In the company's words:

> Client carpet needs can be met initially and rejuvenated periodically, prolonging the useful life of the carpet. Large initial capital expenditures may be reduced and replaced by predetermined monthly billings. Monthly billings are actually a 'lease' on the flooring systems and the services associated with them . . . Interface continues to own the means of delivery (i.e. carpet) for its useful life in the building, ensuring that as product is replaced, it is either used again or reclaimed and recycled, and that it never ends up in a landfill.

However, the company does note that full cost pricing is necessary if the used carpet is to be financially worth salvaging to replace virgin petrochemicals. They comment: 'When the price of oil reflects its true cost, we intend to be ready.'

Yet another possible business model would be to brand and manage each material itself throughout the cyclic loop. At present it is not possible to supply certain global metal requirements from recycling alone, so mining is still a valid starting point for a transition to cyclic flows. For instance, suppose a major new copper mine were being established in say, central Kalimantan (Indonesian Borneo) – a plausible location given that it is one of the few remaining largely unexplored mineral provinces in the world – with the intention of developing a world-class example of 'sustainable mining'. The metal from the new mine would be branded as 'cyclic copper' and linked to the mine and the story of its location, its exemplary environmental performance, and the inclusion of the local people in the social ownership strategy of the business. The cyclic copper would be used by manufacturers willing to include the cyclic copper as a branded product feature, and able to retrieve and return it at product take-back. This would work initially in markets with take-back legislation where, say, luxury carmakers could highlight their environmental performance by using branded cyclic copper in their cars. Given that the copper would be embodied in products for about 10–15 years, after this time substantial amounts of copper would begin to return to the cyclic copper company. Supposing the mine had an expected working life of 30 years, the rising flow of recycled copper would smoothly replace the declining output of the mine during the second half of its life. From this point forward, the cyclic copper company would be the ongoing steward of a sustained cyclic loop of copper with a total mass equal to the entire output of the mine over its lifetime, providing an indefinitely sustained livelihood to shareholders and local people at the minesite. This would contrast with conventional mining operations that often leave a legacy of environmental, social and economic problems when a mine is exhausted.

Already, producers of toxic metals are being forced to think in this way, though without the benefit of ownership or control of the material throughout its lifecycle. Lead is toxic, and the lead industry could face significant constraints on its future markets if it cannot convince regulators and communities that its products will be handled in the safest possible way. The industry has launched a 'Green Lead' initiative to address the difficulty facing one or a few companies acting alone in meeting all of their environmental objectives, or in producing a genuinely 'green' product, without the involvement and cooperation of all the other companies they interact with. This collective industry response is paving the way for the development of a set of interlocking cyclic loop business models.

The need for new cyclic loop business models also applies to energy producers. The biggest linear mass flow from industry into the biosphere today is carbon from the use of fossil fuels. Alternative business models that can create economic value from selling less energy are therefore important while fuel sources contain carbon and other pollutants.

One approach is for energy companies to carry the capital cost of investments that improve energy efficiency on the customer's side and then to share the value of resulting energy savings with the customer. This approach to offering energy efficiency services can be effective but ultimately has limitations, as it is in tension with the company's basic role as a seller of energy. The sharing of savings will only work as long as it can be related to a pre-existing high energy-use context and will only come at the cost of lower long-term energy sales. If energy efficiency technology advances rapidly, this approach would be counter-productive as it would shrink the remaining market opportunity for both energy sales and the money value of available savings. Another way of putting this is that 'negawatts' (saved watts) can only be delivered in a system that delivers watts, and that every negawatt sold reduces the remaining market for watts, even if demand is growing.

Overcoming this dilemma would require a different business model, one that moved from selling energy plus savings of energy, to one that sold energy end-use benefits rather than energy. An electric utility could sell lighting services rather than electricity, measuring the service provided, light, in its own intrinsic units, lumens. The utility would own the customer's lighting system and would have a clear incentive to install the most efficient lighting technology available at any given time, as this would require the least amount of energy, reducing the utility's operating costs.

This approach would create value over the long term by riding the accelerating trend towards more efficient energy end-use technology. The economic gains from this long-run technology trajectory will flow to the owners and operators of the technology assets. Charging for energy service performance supplied (lumens, air temperature and so on), and creating a margin by consistently investing in and operating the latest and most efficient technology for providing the service (light, heat and so on), would yield a sustained revenue flow while reducing emissions of carbon dioxide and pollutants. Although this model would contribute to cyclic loop objectives by reducing linear mass flows, a more radical approach would be needed to achieve a true cyclic loop business model for fossil fuel use.

A further type of business model is possible, based on merging technological and business innovation, and co-designing the material flow and the business logic. In the energy field, an American technology-based social enterprise called Eprida is offering a sustainable energy technology that creates hydrogen rich bio-fuels and a restorative high-carbon fertilizer from biomass alone, or a combination of coal and biomass, while removing net carbon dioxide from the atmosphere. The process involves making charcoal from biomass in a sealed vessel, producing heat, steam and hydrogen. The hydrogen can be used directly as a non-fossil energy source, or it can be used to make fuels such as GTL (gas to liquid) biodiesel. Charcoal on its own can be used as a soil enhancer, increasing soil fertility and returning the carbon from the biomass to the soil and increasing fertility. The charcoal provides a substrate for microorganisms and fungi, which bind organic carbon to minerals to enrich soil. In the Eprida process, the charcoal is also brought together with ammonia, carbon dioxide and water, forming a nitrogen-based fertilizer that binds inside the pores of the charcoal. In comparison, conventional nitrogen fertilizer releases one molecule of carbon dioxide for each molecule of ammonia made, and is not retained in the soil, typically washing away and causing algal blooms in waterways. The carbon dioxide for this fertilizer step can come from burning coal, reducing the carbon footprint of this still extensively-used fossil fuel. The charcoal, plus ammonia made from about 30 per cent of the hydrogen from burning the biomass, will together remove 60 per cent of the carbon dioxide from burning coal, plus all of the sulphur and nitrogen oxides that would otherwise cause acid rain. The process makes it possible to burn coal and biomass and produce a slow-release nitrogen fertilizer bound to charcoal. This returns almost all of the carbon from the biomass to the soil in a stable form, enriching the soil and sequestering the carbon for up to several hundred years, thanks to the cooperative action of soil fungi. In addition, when used alongside a coal-fired power plant, it can remove much of the carbon dioxide from the flue gas.

This innovative energy system forms a cyclic loop for carbon when used with biomass alone, and also simultaneously forms the basis for a viable energy business and fertilizer operation. It may well be that this kind of combined technology system and business model design will prove to be the most powerful kind of cyclic loop business model in the future, simultaneously meeting environmental and economic objectives without the need for supporting regulations. It is also a loop in which the entropy remains low as the material moves through a sequence of transformations, each one having some intrinsic value in the overall process.

Conclusion

The aim of this chapter is to offer a new template for business thinking to supersede the familiar concept of the value chain. The concept of the value loop corresponds to the transition that needs to take place in the flow of materials through the industrial economy if sustainability is to be achieved. Although the objective of 'closing the loop' is well established in the field of environmental technology, it is not well known in the business community, possibly because no correspondingly simple value-creation principle has been articulated. This chapter explains the environmental pressure resulting from the existing linear flow of materials and presents the rationale for a shift to a cyclic flow of materials. The examples of value-loop business models offered here are only early indications of the way the value loop can be applied but hopefully they do provide a sense of what can be achieved.

Ideally, the concepts of the cyclic loop and the value loop will provide a framework that can enable business innovators and entrepreneurs to create economic value at the same time as addressing significant global environmental issues. This kind of business activity is vital if we are to create a sustainable future economy.

Notes

1. In his original article, Michael Porter refers to the value flow within businesses as the 'value chain' and the flow between businesses as the 'value system' but this distinction was subsequently overlooked.
2. The land area of Great Britain is approximately 24 million hectares and the land area of the Earth is approximately 13 billion hectares.

References

Ausubel, J.H. (1989), 'Regularities in Technological Development: An Environmental View', in Ausubel, J.H. and Sladovich, H.E. (eds), *Technology and Environment*, Washington, DC: National Academy Press, pp. 70–91.

Baumgärtner, S. and de Swaan Arons, J. (2003), *Journal of Industrial Ecology*, 7(2), 113–23.

Cohen, J.E. (1995), *How Many People Can the Earth Support?*, New York: W.W. Norton.

Heinzen, B., *Feeling for Stones*, unpublished.

Houghton, J. (1997), *Global Warming: the Complete Briefing*, Cambridge, UK: Cambridge University Press.

Larson, E.D., Ross, M.H. and Williams, R.H. (1986), 'Beyond the Era of Materials', *Scientific American*, June, 24–31.

Lovelock, J.E. (1979), *Gaia: A New Look at Life on Earth*, Oxford: Oxford University Press.

Meadows, D.H. (1992), *Beyond the Limits*, Post Mills, Vermont: Chelsea Green Publishing.

Meadows, D.H., Meadows, D.L., Randers, J. and Behrens III, W.W. (1972), *The Limits to Growth*, New York: Universe Books.

Porter, M. (1985), *Competitive Advantage: Creating and Sustaining Superior Performance*, New York: Free Press, pp. 33–61.

Stigliani, W. and Salomons, W. (1993), 'Our Fathers' Toxic Sins', *New Scientist*, 11 December, 38–42.

Vitousek, P.M., Ehrich, P.R., Ehrich, A.H. and Matson, P.A. (1986), 'Human Appropriation of the Products of Photosynthesis', *BioScience*, 36(6), 368–73.

Wilson, E.O. (1992), *The Diversity of Life*, Boston: Belknap, Harvard University Press.

World Wide Fund for Nature (WWF) (2002), *Living Planet Report*, Cambridge: Banson.

PART VII

CONCLUSION

36 Environmental technology management – lessons from today for a more sustainable future
Dora Marinova

As the work on this *Handbook* progressed, major changes occurred in the global political, social, economic and natural environment. The threat and reality of terrorism, including suicide bombings in London's public transport and Bali's restaurants, and continuous anti-terrorist interventions and measures, such as the extended western presence in Iraq and Afghanistan as well as tougher national legislations, became a high political and economic priority. The outer arrondissements of Paris experienced racially fired riots in late 2005, as did the southern suburb of Cronulla in Sydney. The environmental, social and infrastructure damage caused to New Orleans and other American sites by hurricane Katrina, which devastated the Gulf Coast, was seen as a demonstration of the severe consequences from climate change. The shrinking of the polar ice cover observed and measured by scientists is another warning as to how the planet is changing.

The gap between the rich and the poor has increased even further between and within national economies. For example, in Australia the changes in industrial relations putting additional pressure on individual employees and giving more power to corporations drove 500 000 people to the streets of the country's capital cities in protest, but the government paid no attention to these large demonstrations. The world seems to have become a bleaker place that offers less optimism and hope for the future.

Against this background the power of the sustainability agenda appears to be the only positive way to go against the fear and destruction caused by the prevailing socially, economically and environmentally unsustainable practices of global capitalism. Sustainability is not a magic wand (although sometimes it does feel that it has the capacity to make impossible things happen) and requires conceptual changes and a different outlook at the world. This is particularly felt in places where business, government and the community have got together and have been able to develop sustainability strategies that provide an agenda for change. An example of this is the Western Australian State Sustainability Strategy discussed in several sections in this publication.

This is exactly the message that the *Handbook* brings – take the positive examples and lessons from today to build a better more sustainable future. Its purpose was multifold and included among others the following: (1) to present the new developments that have occurred in the three related areas of environmental management, environmental technologies and technology managements; (2) to present new approaches in understanding sustainability, ecomodernization and technology transformation; (3) to give some examples of what has successfully worked; and (4) to indicate directions of inquiry that should be followed in the future.

The evidence presented here covers theoretical approaches and models as well as numerous practical examples and the 36 chapters with 58 contributors from ten countries and four continents have captured the imagination of the readers. It is hoped that the practical approach taken will lead theoreticians as well as practitioners in contributing in their own individual way for changes that will help arrest those practices leading to detrimental treatment of the physical environment, other people and economies. For example, the problem of waste seems to be a major focus of environmental concern in the management of technology but it also has enormous social and economic implications in terms of employment, design and development practices, doing business within localities and educating the future generations. Another prominent issue is that of sustainable production and consumption that again reflects in an integrated way care for the way resources are used, including the role of environmental technologies.

The point about education is not explicitly highlighted by the *Handbook* but has been an implied purpose of its existence and nature. It is impossible to expect any changes in the future unless education in environmental technology management is taken seriously as part of the United Nations' 2005–14 Decade of Education for Sustainable Development. What we actually mean is education for all ages, at all levels and in all environments, including at home, work, teaching institutions or recreational activities. A major aspect of this is for sustainability to impregnate the value systems that define the actions of people as well as corporations and governments creating a new culture that endorses care and respect for the human and natural world.

There are four main lessons that we want the reader to take into the future:

1. Every small step in the right direction matters and the examples of the leading companies and individuals will set the scene for others to follow.
2. The approach to transforming our current unsustainable practices should be holistic, involving inter- and transdisciplinary perspectives, complete lifecycle responsibilities as well as people's passion, senses, intuition and emotions.
3. Measuring progress in ecomodernization and environmental technology management is important in order to raise awareness of the aims and give recognition to what has been achieved.
4. The sustainability culture should be developed at all levels (individual, organizational, local, regional, national and global) and should be seen as a unity of the environmental, economic and social considerations.

Maintaining a long-term vision is at the core of the sustainability concept and the technologies that we employ and how we manage them should be aimed at facilitating and keeping this vision alive. They should empower people and encourage active engagement and participation, should foster collaboration and building of networks and communities who work together. The transition to a more sustainable future is a multidimensional, multifaceted, personally and organizationally demanding process but it is also extremely rewarding as evidenced by the examples in this book.

There certainly will be more challenges presented by the increasing complexity and turbulence of the world in which we exist. The theoretical stances advocated by the contributors to this *Handbook* will facilitate the way of handling this but it will also require creativity, critical thinking and innovation. The frontiers of knowledge will constantly be moved, and the flow of new ideas and new ways of understanding and doing things will continue. It is important, however, that we take sustainability as the main value from today into the future.

Index